Physics of Ionic Conduction in Narrow Biological and Artificial Channels

Physics of Ionic Conduction in Narrow Biological and Artificial Channels

Editors

Peter V E McClintock

Dmitry G. Luchinsky

MDPI • Basel • Beijing • Wuhan • Barcelona • Belgrade • Manchester • Tokyo • Cluj • Tianjin

Editors
Peter V E McClintock
Department of Physics
Lancaster University
Lancaster
United Kingdom

Dmitry G. Luchinsky
Department of Physics
Lancaster University
Lancaster
United Kingdom

Editorial Office
MDPI
St. Alban-Anlage 66
4052 Basel, Switzerland

This is a reprint of articles from the Special Issue published online in the open access journal *Entropy* (ISSN 1099-4300) (available at: https://www.mdpi.com/journal/entropy/special_issues/physics_Ionchannels).

For citation purposes, cite each article independently as indicated on the article page online and as indicated below:

LastName, A.A.; LastName, B.B.; LastName, C.C. Article Title. *Entropy* **Year**, *Volume Number*, Page Range.

ISBN 978-3-0365-1646-2 (Hbk)
ISBN 978-3-0365-1645-5 (PDF)

The cover illustrates a partially hydrated potassium ion translocating through a sub-nanoscale pore in a monolayer graphene membrane. We thank Subin Sahu and Michael Zwolak for permission to use their image.

Contents

About the Editors

Peter V E McClintock is Emeritus Professor of Physics at Lancaster University. He was educated at Queen's University, Belfast (B.Sc.1962) and the University of Oxford (D.Phil. 1966). Following a postdoctoral appointment at Duke University, he came to Lancaster in 1968 where he has served in many roles including as Head of Department for six years. His contributions to the physics of ion channels, with collaborators, include the proposal of ionic Coulomb blockade as a mechanism of selectivity and its verification in models and experiments. His research interests also encompass the superfluidity of liquid helium, isotopic purification of He-4, stochastic dynamics of nonlinear oscillators, large fluctuations, rogue waves, wave turbulence, quantum turbulence, neutron dynamics in fluid helium, and the nonlinear dynamics of ageing, anaesthesia, autism, hypertension and other cardiovascular diseases. He is Editor-in-Chief of Fluctuation and Noise Letters.

Dmitry G. Luchinsky is Senior Research Fellow at Lancaster University and Senior Research Scientist with KBR Inc. He was educated at Moscow State University (M.Sc. 1983) and the Russian National Research Institute for Metrological Service (Ph.D. 1990). Following several appointments in the Institute, he came to Lancaster in 1995 and was transferred to the permanent staff in 2006. He is currently a contractor for KBR Inc at the NASA Ames Research Centre. He has been investigating the physics of ion channels for almost twenty years with contributions that include the statistical and linear response theory of ion channels and the energetics of ionic transitions through solid state nanopores. His research interests also include additive manufacturing in space and machine learning methods for the discovery of partial differential equations.

Preface to "Physics of Ionic Conduction in Narrow Biological and Artificial Channels"

This is a book about ion channels. It has been written mostly by physical scientists and mathematicians, even though the most widespread and important manifestation of ion channels is in biology, where they are essential to life in all its forms. How do non-biologists get involved in such investigations? Everyone will have their own particular story but, for ourselves, it was the heady combination of scientific curiosity, a wish to contribute to the fundamental understanding of natural phenomena that clearly have crucially important applications, and the realisation that some of our physics knowledge and expertise might be relevant.

The key properties underlying ion channel function are those of conductivity and selectivity –the ability to select between different kinds of ions, allowing the favoured species to pass easily while blocking others. It is now appreciated that an understanding of this selective conduction requires physics, and that the physics of biological ion channels has a great deal in common with that of artificial nanopores. We came to realise that, in each case, there are intriguing analogies with the physics of electrons in quantum dots. Although the ability to predict the function of a channel from its structure remains elusive, recent advances have brought us tantalisingly close to a fundamental theory of ionic permeation, based on the statistical physics of ions within the channel.

The book provides a collection of cutting-edge papers on ionic permeation through narrow water-filled channels, both biological and artificial, reprinted from a recent Special Issue of Entropy. The invited authors were selected as being leading scientists in the field with whose work we were already familiar. In some cases they are our past or present collaborators, or collaborators of collaborators, in what is quite a specialised scientific area. The book describes the statistical physics of the permeation process, mathematical aspects, modelling by molecular dynamics, and experiments. The time is ripe for bringing together these mutually complementary approaches. We hope and believe that they will facilitate major breakthroughs in understanding, enabling the design of nanopores to meet particular technological requirements as well as improvements in drug design.

August 2021

Peter V E McClintock, Dmitry G. Luchinsky
Editors

entropy

Editorial

Introduction to the Physics of Ionic Conduction in Narrow Biological and Artificial Channels

Dmitry G. Luchinsky [1,2] and Peter V. E. McClintock [1,*]

[1] Department of Physics, Lancaster University, Lancaster LA1 4YB, UK; d.luchinsky@lancaster.ac.uk
[2] KBR Inc., Ames Research Center, Moffett Field, CA 94035, USA
* Correspondence: p.v.e.mcclintock@lancaster.ac.uk; Tel.: +44-1524-593073

"There is plenty of room at the bottom"

Richard Feynman

The permeation of ions through narrow water-filled channels is essential to life and of rapidly growing importance in technology. Reaching an understanding of the mechanisms underlying the permeation process requires an interdisciplinary approach, where ideas drawn from physics are of particular importance and have brought encouraging progress in recent years. This introduction sets into context the several ground-breaking papers presented in the *Entropy* Special Issue on "The Physics of Ionic Conduction in Narrow Biological and Artificial Channels".

Understanding, predicting and optimising the ionic selective transport properties of nanopores remains a critical challenge, both to nanotechnology and to biophysics. The last few decades have witnessed substantial progress in the analysis of such transport based on the use of a variety of experimental, numerical, and theoretical methods. Indeed, it would require several books to do full justice to the current state of the art in the field.

In some cases, the crystal structures (e.g., those of potassium, sodium, and calcium voltage-gated channels) have been discovered. This has provided invaluable insight, but has also thrown into sharp relief the structure–function problem: how to predict the conduction/selectivity properties of a known structure; or, conversely, how to design a structure with the required properties. A reliable solution to the problem promises to open new horizons in terms of pharmaceutical applications and the improved fabrication of solid-state nanopores for the sensing of molecules, desalination, DNA sequencing, and the other developments that together are marking a new era in nanotechnology.

Novel numerical methods and computer hardware now enable microsecond-long simulations of systems with hundreds thousands of atoms and the exploration of polarisable and quantum mechanical force fields. They provide unprecedented capabilities for reaching an understanding of experimental data and for the development of novel devices and techniques. Theoretical advances not only underlie many developments in molecular dynamics, including enhanced sampling and advanced force fields, but are also opening up new research frontiers and shedding fresh light on a number of longstanding problems such as binding probabilities, knock-on mechanisms of conduction, gating, electric double-layers, and local dielectric permittivity, just to mention a few.

It is now appreciated that selective conduction in biological ion channels has a great deal in common with that in artificial nanopores. In each case, there are intriguing analogies with the physics of quantum dots leading to the development of the theory of ionic Coulomb blockade. We dedicate this Issue to the memory of our late colleague, Dr Igor Kh. Kaufman, who developed an elegant theory of ionic Coulomb blockade in biological ion channels and suggested a simple classification of voltage-gated channels based on the charge of the selectivity filter.

At the same time, it is known that specific features of ionic conduction (e.g., dehydration, ion-specific binding affinities, protonation, the multicomponent and competitive nature

Citation: Luchinsky, D.G.; McClintock, P.V.E. Introduction to the Physics of Ionic Conduction in Narrow Biological and Artificial Channels. *Entropy* **2021**, *23*, 644. https://doi.org/10.3390/e23060644

Received: 14 May 2021
Accepted: 19 May 2021
Published: 21 May 2021

of ion dynamics, the complex and adaptive structure of the ionic pathway, long-range interaction, local variation of the effective dielectric constant, highly correlated motion of more than one ion within a narrow channel, electric double layers, and water layering at the channel entrances) add many layers of complexity to the fundamental physics analogies.

This Special Issue brings together original high-quality papers on ionic permeation through narrow water-filled channels, both biological and artificial, from some of the best researchers in the field. It includes papers on the statistical physics of the process, on molecular dynamics and Brownian dynamics simulations, and on relevant experiments. Although any selection of papers can only be a narrow slice of the field, our aim is to emphasise the complexity and mutual interdependence of recent multifaceted progress in understanding the physics of ion channels and nanopores. The time is ripe for bringing together these complementary approaches, and we anticipate that they will facilitate major breakthroughs, enabling the design of nanopores to meet particular technological requirements as well as improvements in drug design and perhaps in personalised medicine.

Importantly, the Poisson–Nernst–Planck (PNP) and kinetic models remain among the principal tools for predicting current through nanopores, both in biology and nanotechnology. An example of the classical application of the PNP model to the analysis of reversal potentials and zero-current fluxes, in a system with a fixed profile of permanent charges and two mobile ion species, is provided by the paper by Mofidi et al. [1]. Rigorous analytic and numerical results establish the dependence of the electric and chemical potential profiles on voltage and permanent charge.

At the same time, it is well known that classical Poisson–Boltzmann (PB) and PNP theories do not take account of short-range ion–ion, ion–wall, or ion–water interactions in ion channels. Efforts to eliminate or ameliorate the effects of this deficiency of the continuum models have a long history. This stream of research is represented by the interesting paper of J.-L. Liu and R.S. Eisenberg [2], featuring the development of a molecular mean-field theory—a fourth-order Poisson–Nernst–Planck–Bikerman theory for modelling ionic and water flows in biological ion channels. The theory treats ions and water molecules, in channels of any volume or shape, with interstitial voids, polarisation of water, and ion–ion and ion–water correlations. It can be applied to electrolyte solutions in the nanopores of batteries and fuel cells.

The modelling of ionic currents with reduced models is extensively analysed by Boda et al. [3]. They show that channels are especially amenable to reduced modelling because their functions and the relationships between input parameters (e.g., applied voltage, bath concentrations) and output parameters (e.g., current, rectification, selectivity) are well-defined, allowing one to focus on the physics of input–output relationships rather than on the atomic-scale physics inside the pore. Based on decades of research, the authors propose four general rules for constructing good reduced models of ion channels and nanopores, focusing on the physics of input–output relationships rather than on atomic structure. The proposed rules relate to the importance of (1) the axial concentration profiles, (2) the pore charges, (3) choosing the right explicit degrees of freedom, and (4) creating the proper response functions. Examples demonstrating the application of these rules are provided. Further improvements in predicting the capabilities of reduced models can be achieved by incorporating into the solution of the one-dimensional electro-diffusion model the potential of the mean force obtained from MD simulations. The performance of two such methods is examined by A. Pohorille and M. A. Wilson1 [4] using stochastic simulations. These methods require neither knowledge of the diffusivity nor simulations at multiple voltages, which greatly reduces the computational effort needed to probe the electrophysiology of ion channels. They can be used to determine the free energy profiles from either forward or backward one-sided properties of ions in the channel, such as ion fluxes, density profiles, committor probabilities, or from their two-sided combination. In this work, large sets of stochastic trajectories were generated, individually designed to mimic the molecular dynamics crossing statistics of models for channels of trichotoxin, p7 from hepatitis C and a bacterial homolog of the pentameric ligand-gated ion channel

(LGIC). The authors found that the free energy profiles and the current–voltage curves obtained from the generated trajectories reproduce with good accuracy results obtained in molecular dynamics simulations.

The charged particles of which matter is composed move when an external electric field is applied, and their changed distribution is traditionally described in terms of a polarisation field. For insulators, it is usually possible to define a relative permittivity (dielectric constant) to quantify the material's responsiveness to the electric field. In ion channels, for example, the protein walls and the water are usually treated as dielectric continua with relative permittivities of around 2 and 80, respectively. This approach can be very helpful and revealing, but it involves greater approximation than spatial averaging because, as R.S. Eisenberg points out [5], the material's response to the electric field may be both nonlinear and time-dependent. In order to accommodate such phenomena, while simultaneously challenging physicists to review their knowledge of electromagnetism in biological dielectrics, he proposes and discusses an apparently minor change in Maxwell's first equation. It produces a major consequence when joined with Maxwell's second equation in that conservation of total current (including the displacement current) then emerges as a general principle. In one-dimensional systems like ion channels or electronic circuit components, the consequences are profound: there, total currents are equal at all locations at any given time, so the space variable does not appear in the description of total current.

There follow two papers reporting MD simulations of ion currents in biological and artificial channels. First, S.M. Cosseddu et al. [6] present an extended MD-based analysis of ion motion within the KcsA channel. They reveal complicated patterns of potassium currents that are governed by the structural variability of the selectivity filter. They show that ion motion involves the complex dynamics of a strongly correlated network of residues and water molecules. Intriguing features of self-organisation and readjustment of the network are analysed statistically and discussed in detail.

Secondly, we note that ionic transport in nano- to subnano-scale pores is highly dependent on translocation barriers and potential wells. These features in the free-energy landscape are primarily the result of ion dehydration and electrostatic interactions. For pores in atomically thin membranes, the ionic dynamics both inside and outside the geometrical volume of the pore can be critical in determining its transport properties. S. Sahu and M. Zwolak [7] examine regimes of transport that are highly sensitive to pore size due to the interplay of dehydration and interaction with pore charge, where picometer changes in the size (e.g., due to a minute strain), can lead to a large change in conductance.

We have already remarked upon the crucial importance of water, the electric double-layer, water-layering, polarisation, and the resultant changes of local dielectric permittivity at the entrances of nanopores. Another approach to this problem is illustrated in the paper by T.-L. Horng [8]. Starting from the classical Helmholtz free energy functional for an electrolyte, including the solvation energies for anions and cations, the author follows the Bikerman modification by adding an entropy term to the functional, and he then extends the Bikerman approach by introducing ion-size-specific corrections to the theory.

The approach based on density functional theory (DFT), which works well near charged walls and in bulk electrolytes, can be extended to the analysis of the orientational ordering of water dipoles in membrane nanotubes. M. Drab et al. [9] analyse water ordering in nanotubes by minimising the corresponding Helmholtz free energy functional, also including the orientational entropy contribution of water dipoles, and deriving the modified Langevin–Poisson–Boltzmann (MLPB) model of the electric double-layer. The MLPB equation is solved in cylindrical coordinates to determine the spatial dependences of the electric potential, relative permittivity, and average orientations of water dipoles within charged tubes of different radii. Results show that for tubes of large radius, the macroscopic (net) volume charge density of co-ions and counterions is zero on the geometrical axis. This is attributed to effective electrolyte charge screening in the vicinity of the charged inner surface of the tube. For tubes of small radius, the screening region extends into the whole inner space of the tube, leading to non-zero net volume charge density and non-zero orientational ordering of water dipoles near the axis.

The DFT results mentioned above are examples of statistical physics yielding insight into the function of ion channels and nanopores. This theme is continued and extended, first by Gibby et al. [10], who apply their recent derivation of an effective grand canonical ensemble and linear response theory of ion channels to analyse the conduction of the bacterial NaChBac selectivity filter. The authors compare their theory to experimental current–voltage and current–concentration dependences for a single channel and for a whole cell. They find that the statistical theory in the linear response regime correctly predicts many important properties of the NaChBac filter, including the concentration dependence of the reversal potential and the current–voltage relations. They also show that the theoretical results are consistent with MD simulations of the filter population at each binding site.

Secondly, the analysis of quantum mechanical effects in ion channels is another important direction, supported by the extended capabilities of modern quantum mechanics/molecular mechanics simulations. In this respect, interesting perspectives are opened by mapping the statistical mechanics of ion channels onto an effective quantum mechanics. Such investigations are reviewed by T. Gulden and A. Kamenev [11], who study the dynamics and thermodynamics of ion channels, considered as effective 1D Coulomb systems whose statistical mechanics is dominated by entropic effects that may be taken accurately into account by mapping onto an effective quantum mechanics. The corresponding semiclassical calculations for non-Hermitian Hamiltonians are conducted by applying tools from algebraic topology. The relationship of the solutions to the thermodynamics and correlation functions of multivalent solutions within long water-filled channels is discussed.

The actual properties of real nanopores are, of course, discovered by experiment, which has been leading the research in this area, especially since the discovery of the structure of the KcsA channel. In our Special Issue, experimental insight is provided by two of the leading research groups in the field.

O. Fedorenko et al. [12] discuss the properties of voltage-gated sodium channels (Navs). These channels play fundamental roles in eukaryotes but lack structural resolution, which renders understanding their structure–function relationships a challenging problem. Bacterial Navs, representing simplified homologues of their eukaryotic counterparts, have enabled both structural resolution and electrophysiological characterisation. However, their homotetrameric structure leads to an EEEE locus in the SF that is at odds with the DEKA locus of eukaryotic Navs. Indeed, prokaryotic Navs have long been considered more similar to eukaryotic calcium channels (Cavs) than to Navs, leading to the formulation of the "EEEE paradox". This was arguably solved by Kaufman et al. by the realisation that there is a critical D residue close to the EEEE ring of eukaryotic Cavs generating an effective EEEED locus of charge $-5e$. Fedorenko et al. present a follow-up of a previous study, aimed at mimicking the SF of eukaryotic Navs by engineering radial asymmetry into the SFs of bacterial channels. This goal was pursued with two approaches: co-expression of different monomers of the NaChBac bacterial channel in mammalian cells to induce the random assembly of heterotetramers, and the concatenation of four bacterial monomers to form a concatemer that can be targeted by mutagenesis on specific strands of the SF, thereby introducing asymmetry. Patch-clamp measurements and MD simulations showed that an additional gating charge in the SF leads to a significant increase of Na^+ and a modest increase in Ca^{2+} conductance in the NavMs concatemer, in agreement with the behaviour of the population of random heterotetramers with the highest proportion of channels with charge $-5e$. This study confirms that, although the charge at the SF is important, it is not the only factor affecting conduction and selectivity. It also offers new tools extending the use of bacterial channels as models of eukaryotic ones.

The work by A. Chernev et al. [13] reviews the most promising approaches to the fabrication of artificial nanofluidic devices capable of reproducing the properties of single ion channels. It is shown that modern technologies have great potential in allowing one to test various theoretical models of ion channels. The review aims to highlight ionic Coulomb blockade—the phenomenon which (see above) can often be a key player in ion channel selectivity. The authors discuss the most critical obstacles associated with these studies, and suggest possible solutions to further advance the field.

The rapid interdisciplinary advances in nanotechnology can be characterised as the beginning of a new industrial revolution, where novel devices and materials are fabricated and controlled on the atomic level. Ion- and water-selective nanopores represent an important frontier in these advances.

The selected papers in this Special Issue provide both a snapshot of the present as well as strong indications of how the subject is likely to evolve over the coming years. We may, for example, anticipate major developments in the theory at a fundamental level, based on statistical mechanics and quantum mechanics; substantial improvements in "intermediate-level" theories like PNP, modified CKE, and DFT which promise quantitative predictions of the properties of real channels; as well as much faster and more capacious MD modelling of larger ensembles of atoms on longer timescales, more accurate due to use of polarisable force fields and QM/MM, encompassing gating and permeation events at a statistically useful level. This progress is expected to lead to the first-principles design and fabrication of structures optimised for many important applications including ion pumps, energy harvesting, and field-effect ionic transistors as well as those mentioned at the beginning. Many of these will require theory and experiment on small scales where disciplinary distinctions have mostly faded away, but where physics predominates.

An additional impulse propelling these developments forward is expected due to the fusion of physics-based approaches with artificial intelligence. The latter has already been proven to be very useful for the accelerated learning of the force fields in MD, as well as for the reconstruction of the potentials of the mean force and neural-network-based discovery of partial differential equations. Remarkably, it also underlies a recent breakthrough in the solution of the protein-folding problem.

Author Contributions: The authors contributed equally to this paper. All authors have read and agreed to the published version of the manuscript.

Funding: This research was funded by the Leverhulme Trust (UK) grant number RPG-2017-134.

Institutional Review Board Statement: Not applicable.

Informed Consent Statement: Not applicable.

Data Availability Statement: Not applicable.

Acknowledgments: We are very grateful for much help and advice from Miroslav Barabash, Bob Eisenberg, William Gibby and other authors of the papers in this Special Issue and we acknowledge enumerable valuable discussions with Aneta Stefanovska.

Conflicts of Interest: The authors declare no conflict of interest.

Reference

1. Mofidi, H.; Eisenberg, R.S.; Liu, W. Effects of diffusion coefficients and permanent charge on reversal potentials in ionic channels. *Entropy* **2020**, *22*, 325.
2. Liu, J.L.; Eisenberg, B. Molecular mean-field theory of ionic solutions: A Poisson-Nernst-Planck-Bikerman model. *Entropy* **2020**, *22*, 550.
3. Boda, D.; Valiskó, M.; Gillespie, D. Modeling the device behavior of biological and synthetic nanopores with reduced models. *Entropy* **2020**, *22*, 1259.
4. Wilson, M.A.; Pohorille, A. Electrophysiological properties from computations at a single Voltage: Testing theory with stochastic simulations. *Entropy* **2021**, *23*, 571.
5. Eisenberg, R.S. Maxwell equations without a polarization field, Using a paradigm from biophysics. *Entropy* **2021**, *23*, 172.
6. Cosseddu, S.M.; Choe, E.J.; Khovanov, I.A. Unravelling of a strongly-correlated network of residues controlling the permeation of potassium in the KcsA ion channel. *Entropy* **2021**, *23*, 72.
7. Sahu, S.; Zwolak, M. Diffusion limitations and translocation barriers in atomically thin biomimetic pores. *Entropy* **2020**, *22*, 1326.

8. Horng, T.L. Review and modification of entropy modeling for steric effects in the Poisson-Boltzmann equation. *Entropy* **2020**, *22*, 632
9. Drab, M.; Gongadze, E.; Kralj-Iglič, V.; Iglič, A. Electric double layer and orientational ordering of water dipoles in narrow channels within a modified Langevin Poisson-Boltzmann model. *Entropy* **2020**, *22*, 1054.
10. Gibby, W.A.T.; Barabash, M.L.; Guardiani, C.; Luchinsky, D.G.; McClintock, P.V.E. Application of a statistical and linear response theory to multi-ion Na^{+} conduction in NaChBac. *Entropy* **2021**, *23*, 249.
11. Gulden, T.; Kamenev, A. Dynamics of Ion Channels via Non-Hermitian Quantum Mechanics. *Entropy* **2021**, *23*, 125.
12. Fedorenko, O.A.; Khovanov, I.A.; Roberts, S.K.; Guardiani, C. Changes in ion selectivity following the asymmetrical addition of charge to the selectivity filter of bacterial sodium channels. *Entropy* **2020**, *22*, 1390.
13. Chernev, A.; Marion, S.; Radenovic, A. Prospects of observing ionic Coulomb blockade in artificial ion confinements. *Entropy* **2020**, *22*, 1430.

entropy

Article

Effects of Diffusion Coefficients and Permanent Charge on Reversal Potentials in Ionic Channels

Hamid Mofidi [1], Bob Eisenberg [2] and Weishi Liu [1,*]

[1] Department of Mathematics, University of Kansas, Lawrence, KS 66045, USA; h.mofidi@ku.edu
[2] Department of Physiology and Biophysics, Rush University, Chicago, IL 60612, USA; beisenbe@rush.edu
* Correspondence: wsliu@ku.edu

Received: 3 January 2020; Accepted: 10 March 2020; Published: 12 March 2020

Abstract: In this work, the dependence of reversal potentials and zero-current fluxes on diffusion coefficients are examined for ionic flows through membrane channels. The study is conducted for the setup of a simple structure defined by the profile of permanent charges with two mobile ion species, one positively charged (cation) and one negatively charged (anion). Numerical observations are obtained from analytical results established using geometric singular perturbation analysis of classical Poisson–Nernst–Planck models. For 1:1 ionic mixtures with arbitrary diffusion constants, Mofidi and Liu (arXiv:1909.01192) conducted a rigorous mathematical analysis and derived an equation for reversal potentials. We summarize and extend these results with numerical observations for biological relevant situations. The numerical investigations on profiles of the electrochemical potentials, ion concentrations, and electrical potential across ion channels are also presented for the zero-current case. Moreover, the dependence of current and fluxes on voltages and permanent charges is investigated. In the opinion of the authors, many results in the paper are not intuitive, and it is difficult, if not impossible, to reveal all cases without investigations of this type.

Keywords: reversal potential; effects of diffusion coefficients; permanent charge

---

1. Introduction

Ion channels are proteins found in cell membranes that create openings in the membrane to allow cells to communicate with each other and with the outside to transform signals and to conduct tasks together [1,2]. They have an aqueous pore that becomes accessible to ions after a change in the protein structure that makes ion channels open. Ion channels permit the selective passage of charged ions formed from dissolved salts, including sodium, potassium, calcium, and chloride ions that carry electrical current in and out of the cell.

To unravel how ion channels operate, one needs to understand the physical structure of ion channels, which is defined by the channel shape and the spatial distribution of permanent and polarization charge. The shape of a typical ion channel is often approximated as a cylindrical-like domain with a non-uniform cross-sectional area. Within a large class of ion channels, amino acid side chains are distributed mainly over a "short" and "narrow" portion of the channel, with acidic side chains contributing permanent negative charges and basic side chains contributing permanent positive charges, analogous to the doping of semiconductor devices, e.g., bipolar PNP and NPN transistors.

The spatial distribution of side chains in a specific channel defines the permanent charge of the channel. The spatial distribution of permanent charge forms (most of) the electrical structure of the channel protein. The spatial distribution of mass forms the structure studied so successfully by molecular and structural biologists. Ions that move through channels are often only an Angstrom or so away from the permanent charges residing on acid and base side chains. In addition, electrical forces are in general much stronger than entropic forces. Thus, in most cases, the electrical structure is more important in determining how ions go through a channel than the mass structure. Sometimes, the dielectric properties ("polarization") of the protein contribute a charge that is significant. Then, the spatial distribution of dielectric properties becomes an important part of the electrical structure.

The most basic function of ion channels is to regulate the permeability of membranes for a given species of ions and to select the types of ions and to facilitate and modulate the diffusion of ions across cell membranes. At present, these permeation and selectivity properties of ion channels are usually determined from the current–voltage (I–V) relations measured experimentally [2,3]. Individual fluxes carry more information than the current, but it is expensive and challenging to measure them [4,5]. Indeed, the measurement of unidirectional fluxes by isotopic tracers allowed the early definition of channels and transporters and is a central subject in the history of membrane transport, as described in textbooks—for example, [6–9]. The precise definition and use of unidirectional fluxes are dealt with at length in the paper [5]. The I–V relation defines the function of the channel structure, namely the ionic transport through the channel. That transport depends on driving forces expressed mathematically as boundary conditions. The multi-scale feature of the problem with multiple physical parameters allows the system to have great flexibility and to exhibit vibrant phenomena/behaviors—a great advantage of "natural devices" [10]. On the other hand, the same multi-scale feature with multiple physical parameters presents an extremely challenging task for anyone to extract meaningful information from experimental data, also given the fact that the internal dynamics cannot be measured with present techniques. The general inverse problem is challenging, although specific inverse problems have been successfully solved with surprisingly little difficulty using standard methods and software packages [11].

To understand the importance of the relation of current and permanent charges, that is, the I–Q relation, we point out that the role of permanent charges in ionic channels is similar to the role of doping profiles in semiconductor devices. Semiconductor devices are similar to ionic channels in the way that they both use atomic-scale structures to control macroscopic flows from one reservoir to another. Ions move a lot like quasi-particles move in semiconductors. In a crude sense, holes and electrons are the cations and anions of semiconductors. Semiconductor technology depends on the control of migration and diffusion of quasi-particles of charge in transistors and integrated circuits. Doping is the process of adding impurities into intrinsic semiconductors to modulate its electrical, optical, and structural properties [12,13]. In a crude sense, doping provides the charges that acid and base side chains provide in a protein channel.

Ion channels are almost always passive and do not require a source of chemical energy (e.g., ATP hydrolysis) in order to operate. Instead, they allow ions to flow passively driven by a combination of the transmembrane electrical potential and the ion concentration gradient across the membrane. For other fixed physical quantities, the total current $\mathcal{I} = \mathcal{I}(\mathcal{V}, \mathcal{Q})$ depends on the transmembrane potential $\mathcal{V}$ and the permanent charge $\mathcal{Q}$. For fixed $\mathcal{Q}$, *a reversal potential* $\mathcal{V} = \mathcal{V}_{rev}(\mathcal{Q})$ is a transmembrane potential that produces zero current $\mathcal{I}(\mathcal{V}_{rev}(\mathcal{Q}), \mathcal{Q}) = 0$. Similarly, for fixed transmembrane potential $\mathcal{V}$, *a reversal permanent charge* $\mathcal{Q} = \mathcal{Q}_{rev}(\mathcal{V})$ is a permanent charge that produces zero current $\mathcal{I}(\mathcal{V}, \mathcal{Q}_{rev}(\mathcal{V})) = 0$.

The *Goldman–Hodgkin–Katz (GHK) equation* for reversal potentials involving multiple ion species [14,15] is used to determine the reversal potential across ion channels. The GHK equation is an extension of the Nernst equation—the latter is for one ion species. The classical derivations were based on the incorrect assumption that the electric potential $\Phi(X)$ is linear in X—the coordinate along the length of the channel. This assumption is particularly unfortunate because it is the change in the shape of the electrical potential $\Phi(X)$ that is responsible for so many of the fascinating behaviors of transistors or ionic systems [16–21]. There was no substitute for GHK equations until authors of [22,23] recently offered equations derived from self-consistent Poisson–Nernst–Planck (PNP) systems, to the best of our knowledge.

In this work, focusing on basic understanding of possible effects of unequal diffusion coefficients and, as a starting point, we will use the classical PNP model with a piecewise constant permanent charge and a cylinder-like channel with variable cross-sectional area. The classical PNP model treats ions as point charges. Among many limitations, gating and selectivity cannot be captured by the simple classical PNP model. However, the basic finding on reversal potentials and their dependence on permanent charges and on ratios of diffusion constants seems important and some are non-intuitive and deserving of further investigation. In the future, more structural detail and more correlations between ions should be taken into considerations in PNP models such as those including various potentials for ion-to-ion interaction accounting for ion sizes effects and voids [24–32].

There have been great achievements in analyzing the PNP models for ionic flows through ion channels [5,28,33–36], etc. Although mathematical analysis plays a powerful and unique role to explain mechanisms of observed biological phenomena and to discover new phenomena, numerical simulations are needed to fit actual experimental data and study cases where analytical solutions do not exist. Furthermore, numerical observations may give clues for more theoretical investigations. Indeed, numerical and analytical studies are linked; any progress in one catalyzes work in the other.

This paper is a mathematical study on some aspects of ionic flows via the PNP models. It uses established mathematical methods and analytical results [23,33] that are derived without further assumption from their underlying physical models. The numerical results, throughout the paper, are gained from the algebraic systems (15), (22), (23) and (27), obtained from reduced matching systems of analytical results in [23,33]. The nonlinear algebraic systems are then solved by the MATLAB® (Version 9.5) function *fsolve* that uses the trust–region dogleg algorithm. The trust–region algorithm is a subspace trust–region method and is based on the interior-reflective Newton method described in [37]. Our numerical results indicate that current–voltage and current-permanent charge and even zero-current relations depend on a rich interplay of boundary conditions and the channel geometry arising from the mathematical properties analyzed in [23,33,34,38]. Although the work here is presented in the context of biological ion channels, it is clear that the results apply to the artificial channels that are now being studied for their engineering applications.

The highlights of our studies in this paper as well as in [23,33,34,38] applied to the setup of this paper include:

(i) a mathematically derived system for the zero-current condition (see System (15)) that can be used to determine the reversal potential in terms of other parameters (see Display (22));
(ii) an examination on how the reversal potential depends on permanent charge: its sign and its monotonicity in permanent charge (see Section 2.2); and a comparison between this reversal potential and that from GHK in the special setting (see Section 2.3);
(iii) a characterization of monotonic dependencies of the reversal potential on the ratio of diffusion coefficients in terms of different conditions on the boundary concentrations (see Section 2.2), as well as effects of un-equal diffusion coefficients on signs of zero-current flux and its dependence on permanent charge (see Section 2.1);
(iv) numerical spatial profiles under the zero current condition of the concentrations and electric potential, and hence the profiles of the electrochemical potentials for several choices of permanent charges that reveal special features of permanent charge effects (see Section 2.4, particularly, Remark 3);
(v) numerical and analytical studies of I–V and I–Q relations, and zero-voltage current and its rich dependence on permanent charge (see Section 3.3).

Furthermore, there are several qualitatively important but non-intuitive results discussed in this work. These qualitative results may be helpful in guiding experimentation and some might not be apparent in intuitive thinking about ion channel behavior. Here are some examples:

a. The zero-current flux J has the same sign as that of $l - r$ (see Section 2.1).
b. The magnitude of the ratio between of the two diffusion coefficients affects the monotonicity of the zero-current flux J in Q (see Section 2.1).
c. I–Q curves are not monotonic in general (see Section 3.2).

d. Rich phenomena of interplay between boundary conditions and diffusion coefficients in terms of monotonicity of zero-voltage current on permanent charge (Section 3.3).

To this end, we would like to emphasize that applying the geometric analysis allows us to identify and formulate quantities and properties that are crucial to biology, while also providing quantitative and qualitative understanding and predictions.

This paper is organized as follows. The classical PNP model for ionic flows is recalled in Section 1.1 to prepare the stage for investigations in later sections. In Section 2, we study zero current problems to investigate the corresponding fluxes and reversal potentials $\mathcal{V}_{rev}$. In particular, we compare a special case of the reversal potential with the GHK equation. Some other numerical observations are also provided to study profiles of relevant physical quantities in Section 2.4. In Section 3, we first recall the analytical results in [33] when diffusion constants are also involved. Then, numerical observations are provided to examine behaviors of current, voltage, and permanent charge with respect to each other in some general cases. Some concluding remarks are provided in Section 4.

1.1. Poisson–Nernst–Planck Models for Ionic Flows

The PNP system of equations has been analyzed mathematically to some extent, but the equations have been simulated and computed to a much larger extent [39–43]. One can see from these simulations that macroscopic reservoirs must be included in the mathematical formulation to describe the actual behavior of channels [24,44]. For an ionic mixture of n ion species, the PNP type model is, for $k = 1, 2, ..., n$,

$$\begin{aligned} \text{Poisson:} \quad & \nabla \cdot \Big(\varepsilon_r(\overrightarrow{X})\varepsilon_0 \nabla\Phi\Big) = -e_0\Big(\sum_{s=1}^{n} z_s C_s + \mathcal{Q}(\overrightarrow{X})\Big), \\ \text{Nernst-Planck:} \quad & \partial_t C_k + \nabla \cdot \overrightarrow{\mathcal{J}}_k = 0, \quad -\overrightarrow{\mathcal{J}}_k = \frac{1}{k_B T}\mathcal{D}_k(\overrightarrow{X}) C_k \nabla \mu_k, \end{aligned} \tag{1}$$

where $\overrightarrow{X} \in \Omega$ with Ω being a three-dimensional cylindrical-like domain representing the channel of length $\hat{L}$ (nm $= \hat{L} \times 10^{-9}$m), $\mathcal{Q}(\overrightarrow{X})$ is the permanent charge density of the channel (with unit 1M $=$ 1Molar $=$ 1mol/L $= 10^3$mol/m^3), $\varepsilon_r(\overrightarrow{X})$ is the relative dielectric coefficient (with unit 1), $\varepsilon_0 \approx 8.854 \times 10^{-12}$ F m^{-1} is the vacuum permittivity, $e_0 \approx 1.602 \times 10^{-19}$C (coulomb) is the elementary charge, $k_B \approx 1.381 \times 10^{-23}$JK^{-1} is the Boltzmann constant, T is the absolute temperature ($T \approx$ 273.16 K =kelvin, for water); Φ is the electric potential (with the unit V = Volt = JC^{-1}), and, for the k-th ion species, C_k is the concentration (with unit M), z_k is the valence (the number of charges per particle with unit 1), and μ_k is the electrochemical potential (with unit J = CV) depending on electrical potential Φ and concentrations C_k. The flux density $\overrightarrow{\mathcal{J}}_k(\overrightarrow{X})$ (with unit mol m^{-2}s^{-1}) is the number of particles across each cross-section in per unit time, $\mathcal{D}_k(\overrightarrow{X})$ is the diffusion coefficient (with unit m^2/s), and n is the number of distinct types of ion species (with unit 1).

Ion channels have narrow cross-sections relative to their lengths. Therefore, three-dimensional PNP type models can be reduced to quasi-one-dimensional models. The authors of [45] first offered a reduced form, and, for a particular case, the reduction is precisely verified by the mathematical analysis of [46]. The quasi-one-dimensional steady-state PNP type is, for $k = 1, 2, ..., n$,

$$\begin{aligned} & \frac{1}{\mathcal{A}(X)}\frac{d}{dX}\left(\varepsilon_r(X)\varepsilon_0\mathcal{A}(X)\frac{d\Phi}{dX}\right) = -\, e_0\left(\sum_{s=1}^{n} z_s C_s + \mathcal{Q}(X)\right), \\ & \frac{d\mathcal{J}_k}{dX} = 0, \quad -\mathcal{J}_k = \frac{1}{k_B T}\mathcal{D}_k(X)\mathcal{A}(X) C_k \frac{d\mu_k}{dX}, \end{aligned} \tag{2}$$

where X is the coordinate along the channel, $\mathcal{A}(X)$ is the area of cross-section of the channel over location X, and $\mathcal{J}_k$ (with unit mol s^{-1}) is the total flux through the cross-section. Equipped with System (2), we impose the following boundary conditions, for $k = 1, 2, \cdots, n$,

$$\Phi(0) = \mathcal{V}, \quad C_k(0) = L_k > 0; \quad \Phi(\hat{L}) = 0, \quad C_k(\hat{L}) = R_k > 0. \tag{3}$$

One often uses the electroneutrality conditions on the boundary concentrations because the solutions are made from electroneutral solid salts,

$$\sum_{s=1}^{n} z_s L_s = \sum_{s=1}^{n} z_s R_s = 0. \tag{4}$$

The electrochemical potential $\mu_k(X)$ for the k-th ion species consists of the ideal component $\mu_k^{id}(X)$ and the excess component $\mu_k^{ex}(X)$, i.e., $\mu_k(X) = \mu_k^{id}(X) + \mu_k^{ex}(X)$. The excess electrochemical potential $\mu_k^{ex}(X)$ accounts for the finite size effect of ions. It is needed whenever concentrations exceed, say 50 mM, as they almost always do in technological and biological situations and often reach concentrations 1M or more. The classical PNP model only deals with the ideal component $\mu_k^{id}(X)$, which reflects the collision between ions and water molecules and ignores the size of ions; that is,

$$\mu_k(X) = \mu_k^{id}(X) = z_k e_0 \Phi(X) + k_B T \ln \frac{C_k(X)}{C_0}, \tag{5}$$

where C_0 is a characteristic concentration of the problems, and one may consider

$$C_0 = \max_{1 \le k \le n} \{L_k, R_k, \sup_{X \in [0,\hat{L}]} |\mathcal{Q}(X)|\}. \tag{6}$$

For given $\mathcal{V}$, $\mathcal{Q}(X)$, L_k's and R_k's, if $(\Phi(X), C_k(X), \mathcal{J}_k)$ is a solution of the boundary value problem (BVP) (2) and (3), then the electric current $\mathcal{I}$ is

$$\mathcal{I} = e_0 \sum_{s=1}^{n} z_s \mathcal{J}_s. \tag{7}$$

For an analysis of the BVP (2) and (3), we work on a dimensionless form. Set

$$\mathcal{D}_0 = \max_{1 \le k \le n} \{ \sup_{X \in [0,\hat{L}]} \mathcal{D}_k(X)\} \text{ and } \bar{\varepsilon}_r = \sup_{X \in [0,\hat{L}]} \varepsilon_r(X).$$

Let

$$\begin{aligned} &\varepsilon^2 = \frac{\bar{\varepsilon}_r \varepsilon_0 k_B T}{e_0^2 \hat{L}^2 C_0}, \quad \hat{\varepsilon}_r(x) = \frac{\varepsilon_r(X)}{\bar{\varepsilon}_r}, \quad x = \frac{X}{\hat{L}}, \quad h(x) = \frac{\mathcal{A}(X)}{\hat{L}^2}, \quad D_k(x) = \frac{\mathcal{D}_k(X)}{\mathcal{D}_0}, \\ &Q(x) = \frac{\mathcal{Q}(X)}{C_0}, \quad \phi(x) = \frac{e_0}{k_B T}\Phi(X), \quad c_k(x) = \frac{C_k(X)}{C_0}, \quad \hat{\mu}_k = \frac{1}{k_B T}\mu_k, \quad J_k = \frac{\mathcal{J}_k}{\hat{L} C_0 \mathcal{D}_0}. \end{aligned} \tag{8}$$

In terms of the new variables, the BVP (2) and (3) becomes, for $k = 1, 2, \cdots, n$,

$$\begin{aligned} \frac{\varepsilon^2}{h(x)} \frac{d}{dx}\left(\hat{\varepsilon}_r(x) h(x) \frac{d}{dx}\phi\right) &= -\sum_{s=1}^{n} z_s c_s - Q(x), \\ \frac{dJ_k}{dx} = 0, \quad -J_k &= h(x) D_k(x) c_k \frac{d}{dx}\hat{\mu}_k, \end{aligned} \tag{9}$$

with the boundary conditions

$$\phi(0) = V = \frac{e_0}{k_B T}\mathcal{V}, \quad c_k(0) = l_k = \frac{L_k}{C_0}; \quad \phi(1) = 0, \quad c_k(1) = r_k = \frac{R_k}{C_0}. \tag{10}$$

Remark 1. *The actual dimensional forms of quantities have been used for all figures throughout the paper, that is,*

$$\begin{aligned} C_k =& C_0 c_k\ (M), \qquad \mathcal{Q} = C_0 Q\ (M), \qquad \mu_k = \left(e_0\Phi + k_B T \ln(C_k/C_0)\right)\ (J), \\ \Phi =& \frac{k_B T}{e_0}\phi\ (V), \qquad \mathcal{J}_k = \hat{L} C_0 \mathcal{D}_0 J_k\ (mol/s), \qquad \mathcal{I} = \hat{L} C_0 \mathcal{D}_0 e_0 I\ (A), \end{aligned} \tag{11}$$

and we take $C_0 = 10\,M$, $\hat{L} = 2.5\,nm$ and $\mathcal{D}_0 = 2.032 \times 10^{-9}\ m^2/s$, and, for diffusion constants [31],

$$\begin{aligned} \mathcal{D}_k =& 1.334 \times 10^{-9}\ m^2/s \text{ for } Na^+, \text{ or} \\ \mathcal{D}_k =& 2.032 \times 10^{-9}\ m^2/s \text{ for } Cl^-, \text{ or} \\ \mathcal{D}_k =& 0.792 \times 10^{-9}\ m^2/s \text{ for } Ca^{2+}. \end{aligned} \tag{12}$$

1.2. Setup of the Problem

We now designate the case we will study in this paper. We will investigate a simple setup, the classical PNP model (9) with ideal electrochemical potential (5), and the boundary conditions (10). More precisely, we assume

(A0) The ionic mixture consists of two ion species with valences $z_1 = -z_2 = 1$;
(A1) $D_k(x) = D_k$ for $k = 1, 2$ is a constant and $\hat{\varepsilon}(x) = 1$;
(A2) Electroneutrality boundary conditions (4) hold;
(A3) The permanent charge Q is piecewise constant with one nonzero region; that is, for a partition $0 < a < b < 1$ of $[0, 1]$,

$$Q(x) = \begin{cases} Q_1 = Q_3 = 0, & x \in (0, a) \cup (b, 1), \\ Q_2 = 2Q_0, & x \in (a, b), \end{cases} \tag{13}$$

where Q_0 is a constant.

We assume that $\varepsilon > 0$ in System (14) is small. The assumption is reasonable since, if $\hat{L} = 2.5\,\text{nm} = 2.5 \times 10^{-9}\,\text{m}$ and $C_0 = 10\,\text{M}$, then $\varepsilon \approx 10^{-3}$ [47]. The assumption that ε is small enables one to treat System (14) of the dimensionless problem as a singularly perturbed problem that can be analyzed by the theory of geometrical singular perturbations (GSP). GSP uses the modern invariant manifold theory from nonlinear dynamical system theory to study the entire structure, i.e., the phase space portrait of the dynamical system, and is not to be confused with the classical singular perturbation theory that uses, for example, matched asymptotic expansions.

We rewrite the classical PNP system (9) into a standard form of singularly perturbed systems and turn the boundary value problem to a connecting problem. We refer the readers to the papers [33] and [36] (with insignificantly altered notations) for details. Denote the derivative with respect to x by overdot and introduce $u = \varepsilon\dot{\phi}$. System (9) becomes, for $k = 1, 2$,

$$\begin{aligned} \varepsilon\dot{\phi} =& u, \quad \varepsilon\dot{u} = -\sum_{s=1}^{2} z_s c_s - Q(x) - \varepsilon\frac{h_x(x)}{h(x)}u, \\ \varepsilon\dot{c}_k =& -z_k c_k u - \varepsilon\frac{J_k}{D_k h(x)}, \quad \dot{J}_k = 0. \end{aligned} \tag{14}$$

System (14) will be treated as a dynamical system with the phase space $\mathbb{R}^7$ and the independent variable x is viewed as time for the dynamical system.

A GSP framework for analyzing BVP of the classical PNP systems was developed first in [33,35] for ionic mixtures with two types of ion species. The model of ion channel properties involves coupled nonlinear differential equations. Until accomplished, it was not apparent that any analytical results could be found, let alone the powerful ones provided by geometrical singular perturbation. This GSP framework was extended to an arbitrary number of types of ion species successfully only when *two special mathematical structures* of the PNP system were revealed [36]. One special structure is *a complete set of integrals (or conserved quantities)* for the $\varepsilon = 0$ limit fast (or inner) system that allows a detailed analysis of a singular layer component of the full problem. It should pointed out that most of the integrals are NOT conserved for the physical problem since, no matter how small it is, ε is NOT zero. The GSP allows one to make conclusion about the BVP for $\varepsilon > 0$ small from information of $\varepsilon = 0$ limit systems. The other special structure is that *a state-dependent scaling* of the independent variable turns the nonlinear limit slow (or outer) system to a linear system with constant coefficients. The coefficients do depend on unknown fluxes to be determined as a part of the whole problem, and this is the mathematical reason for the rich dynamics of the problem. As a consequence of the framework, the existence, multiplicity, and spatial profiles of *the singular orbits*—zeroth order in ε approximations of the BVP—are reduced to a system of nonlinear algebraic equations that involves all relevant quantities altogether. This system of nonlinear **algebraic** equations defines the physical framework of the problem precisely. The system shows explicitly what has been guessed implicitly "everything interacts with everything else" and, in the cases analyzed in this paper, the system shows quantitatively how those interactions occur. This geometric framework with its extensions to include some of the effects of ion size [28,29,32] has produced a number of results that are central to ion channel properties [5,23,30,34,38,48]; for example, it was shown in [34] that *a positive permanent charge may enhance anion flux as well as cation flux*; and, *in order to optimize effects of the permanent charge, the channel should have a short and narrow neck within which the permanent charge is confined*; and, it was shown in [38] that *large permanent charge is responsible for the declining phenomenon—decreasing flux with increasing transmembrane electrochemical potential*. We refer the readers to the aforementioned papers for more details on geometric singular perturbation framework for PNP as well as concrete applications to ion channel problems.

In this paper, we will apply some results and follow the notations in [23,33] for analytical results where the quantities are all in their dimensionless forms. In addition, for simplicity, we use the letters l, r and Q_0 where $l_1 = l_2 = l$, $r_1 = r_2 = r$, $Q_2 = 2Q_0$.

Remark 2. *We remind the readers that the quantities $V, l, r, c_k, Q, \phi, \hat{\mu}_k, J_k, D_k$, and I are dimensionless quantities corresponding to the dimensional quantities $\mathcal{V}, L, R, C_k, \mathcal{Q}, \Phi, \mu_k, \mathcal{J}_k, \mathcal{D}_k$, and $\mathcal{I}$, respectively, obtained from Display (8). We switch from dimensional form to the dimensionless form and vice versa several times throughout the paper. Dimensionless variables are convenient for illustrating and analyzing mathematical and general physical relations. Dimensional variables are necessary for showing how evolution has exploited those general relations.*

2. Zero Current Problems with General Diffusion Constants

In this section, we study how boundary concentrations, electric potential, permanent charges, and diffusion constants work together to produce current reversal. Throughout this section, in order to express the effects of diffusion constants on zero-current flux and reversal potential, we study and compare the results for different cases of diffusion constants where $\mathcal{D}_1 = \mathcal{D}_2$ and where $\mathcal{D}_1 \neq \mathcal{D}_2$, to indicate and emphasize the differences.

Diffusion is the phenomenon through which the spatial distribution of solute particles varies as a result of their potential energy. It is a spontaneous process that acts to eliminate differences in concentration and eventually leads a given mixture to a state of uniform composition. Fick's first law [49] describes diffusion of uncharged particles by $\partial_t c = \mathcal{D}\partial^2_{xx} c$, where c is the concentration, $\mathcal{D}$ is the diffusion constant, and t is time. Frequently, the determination of diffusion constants involves measuring sets of simultaneous values of t, c, and x. These measured values are then applied to a solution of Fick's law to get the diffusion constants. Many techniques are available for the determination of diffusion constants of ions (charge particles) in aqueous solutions [31,50–53], etc. When diffusion constants are equal, classical electrochemistry tells that many electrical phenomena "disappear" altogether, e.g., the "liquid junction" is zero. If the diffusion constants of potassium and chloride are equal, classical electrochemistry says that KCl acts nearly as an uncharged species. Indeed, this is the basis for the saturated KCl salt bridge used in a broad range of electrochemical experiments for many years. Therefore, the equal diffusion constants case is quite degenerate. Experimental measurements are exclusively performed under isothermal conditions to avoid deviation of $\mathcal{D}$ values. Nevertheless, even diffusion constants of certain ionic species may differ from one method to another, even when all other parameters are held constant. Everything becomes much more complicated mathematically when the diffusion constants are not equal, however. This complexity is what makes many biological and technological devices interesting, useful, and valuable. Some kinds of selectivity depend on the non-equality of diffusion constants as well.

Applying GSP theory to the classical PNP system (2) for two ion species with diffusion constants $\mathcal{D}_k$, $k = 1,2$, the authors of [23] obtained an algebraic matching system with eleven equations and eleven unknowns for zero current problems and singular orbits on $[0,1]$. They further reduced the matching system for the case where two ion valences satisfy $z_1 = -z_2$. It follows that the reduced matching system for zero current $I = J_1 - J_2 = 0$ when $z_1 = -z_2 = 1$ is

$$G_1(A, Q_0, \theta) = V \text{ and } G_2(A, Q_0, \theta) = 0, \tag{15}$$

where

$$\begin{aligned} G_1(A, Q_0, \theta) =& \theta\Big(\ln\frac{S_a + \theta Q_0}{S_b + \theta Q_0} + \ln\frac{l}{r}\Big) - (1+\theta)\ln\frac{A}{B} + \ln\frac{S_a - Q_0}{S_b - Q_0}, \\ G_2(A, Q_0, \theta) =& \theta Q_0 \ln\frac{S_a + \theta Q_0}{S_b + \theta Q_0} - N, \end{aligned} \tag{16}$$

and, A is the geometric mean of concentrations at $x = a$, that is,

$$A = \sqrt{c_1(a)c_2(a)}, \tag{17}$$

$$B = \frac{1-\beta}{\alpha}(l - A) + r, \quad S_a = \sqrt{Q_0^2 + A^2}, \quad S_b = \sqrt{Q_0^2 + B^2}, \quad N = A - l + S_a - S_b, \tag{18}$$

and

$$\theta = \frac{D_2 - D_1}{D_2 + D_1}, \quad \alpha = \frac{H(a)}{H(1)}, \quad \beta = \frac{H(b)}{H(1)} \text{ where } H(x) = \int_0^x \frac{1}{h(s)} ds. \tag{19}$$

Note that, if $h(x)$ is uniform, then $H(x)$ is the ratio of the length with the cross-section area of the potion of the channel over $[0, x]$. The original of this quantity $H(x)$ has its root in Ohm law for resistance of a uniform resistor. It turns out that the quantities α and β together with the value Q_0 are key characteristics for the shape and the permanent charge of the channel structure (see Section 4 in [34] for more detailed and concrete results about the roles of α and β on the fluxes).

To this end, we recall three relevant results from [23] on which most of our analytical and numerical studies are based.

For fixed Q_0 and θ, A can actually be solved from $G_2(A, Q_0, \theta) = 0$, where G_2 is defined in Display (16) with the properties stated in the next theorem.

Theorem 1 (Theorem 3.4 in [23]). *The solution $A = A(Q_0, \theta)$ of $G_2(A, Q_0, \theta) = 0$ satisfies*

(a) $A(0, \theta) = (1-\alpha)l + \alpha r$ *and* $\lim_{Q_0 \to \pm\infty} A(Q_0, \theta) = l$,
(b) *if* $l > r$, *then* $l > A(Q_0, \theta) > A^* > B(Q_0, \theta) > r$,
(c) *if* $l < r$, *then* $l < A(Q_0, \theta) < A^* < B(Q_0, \theta) < r$,

(d) *if* $\theta Q_0 \ge 0$, *then* $\partial_{Q_0} A(Q_0, \theta)$ *has the same sign as that of* $(l - r)Q_0$,

where $A^* = \dfrac{(1-\beta)l + \alpha r}{1 - \beta + \alpha}$.

For fixed Q_0 and θ, the reversal potential $V_{rev} = V_{rev}(Q_0, \theta)$ can also be determined and enjoy properties stated in the next two theorems. Recall that we denote $J_1 = J_2$ by J.

Theorem 2 (Theorem 4.2 in [23]). *For the reversal potential* $V_{rev} = V_{rev}(Q_0, \theta)$, *one has*

(i) *if* $l > r$, *then* $J > 0$, *and, hence,* $-\dfrac{1}{z_1}\ln\dfrac{l}{r} < V_{rev}(Q_0,\theta) < \dfrac{1}{z_1}\ln\dfrac{l}{r}$;

(ii) *if* $l < r$, *then* $J < 0$, *and, hence,* $\dfrac{1}{z_1}\ln\dfrac{l}{r} < V_{rev}(Q_0,\theta) < -\dfrac{1}{z_1}\ln\dfrac{l}{r}$;

(iii) $V_{rev}(0,\theta) = \dfrac{\theta}{z_1}\ln\dfrac{l}{r}$ *and* $\lim_{Q_0\to\pm\infty} V_{rev}(Q_0,\theta) = \pm\dfrac{1}{z_1}\ln\dfrac{l}{r}$.

Theorem 3 (Theorem 4.4 in [23]). *For any given* $\theta \in (-1, 1)$, *one has*

(i) *if* $\theta = 0$, *then* $V_{rev}(Q_0, \theta)$ *is increasing in* Q_0 *for* $l > r$ *and decreasing in* Q_0 *for* $l < r$;

(ii) *if* $\theta > 0$, *then, for* $Q_0 \ge 0$, $V_{rev}(Q_0, \theta)$ *is increasing in* Q_0 *for* $l > r$ *and decreasing in* Q_0 *for* $l < r$;

(iii) *if* $\theta < 0$, *then, for* $Q_0 \le 0$, $V_{rev}(Q_0, \theta)$ *is increasing in* Q_0 *for* $l > r$ *and decreasing in* Q_0 *for* $l < r$.

In what follows, numerical simulations are conducted with the help of analysis on System (15). The combination of numerics and analysis gives a better understanding of the zero-current problems and compliments some analytical results obtained in [23]. For our numerical simulations, we choose $a = 1/3$, $b = 2/3$ in Display (13) and $h(x) = 1$ for simplicity and for definiteness.

2.1. Zero-Current Flux $J = J_1 = J_2$.

We aim to clarify the relationships of ion fluxes with permanent charge and diffusion constants when current is zero.

Recall that fluxes J_1 and J_2 are equal for this case and let J denote it. For any permanent charge $Q = 2Q_0$, once a solution (A, V) of System (15) is obtained, it follows from matching equations (see Appendix in [23]) that J is given by

$$J = -\frac{6D_1D_2(A-l)}{(D_1+D_2)} = -\frac{6D_1D_2(r-B)}{(D_1+D_2)}. \tag{20}$$

2.1.1. Sign of Zero-Current Flux $\mathcal{J}$

It was observed in [22] that the Nernst–Planck equation in Display (9) (with dimensionless quantities) gives, for $k = 1, 2$,

$$\frac{J_k}{D_k}\int_0^1 \frac{1}{h(x)c_k(x)}dx = z_k V + \ln\frac{l}{r}. \tag{21}$$

Therefore, the sign of flux J_k depends only on the boundary conditions l, r and V. Note that Equation (21) holds for any condition, not just zero-current condition.

For zero-current problem, $V = V_{rev}$ depends on l, r, D_1, D_2, and Q as well, in general. Thus, the sign of zero-current flux J seems to depend on all quantities and to be difficult to figure out. It is not the case. A consequence of Display (20) together with Theorem 1 is that:

The zero-current flux J has the same sign as that of $l - r$

The latter follows directly from Theorem 1 that, for zero-current, $l - A$ has the same sign as that of $l - r$. This is consistent with observations in Figure 1 where $\mathcal{D}_1 = 1.334 \times 10^{-9}$ m^2/s is fixed, and $\mathcal{D}_2$ varies from the same value to $\mathcal{D}_2 = 2.032 \times 10^{-9}$ m^2/s, and to a random large value.

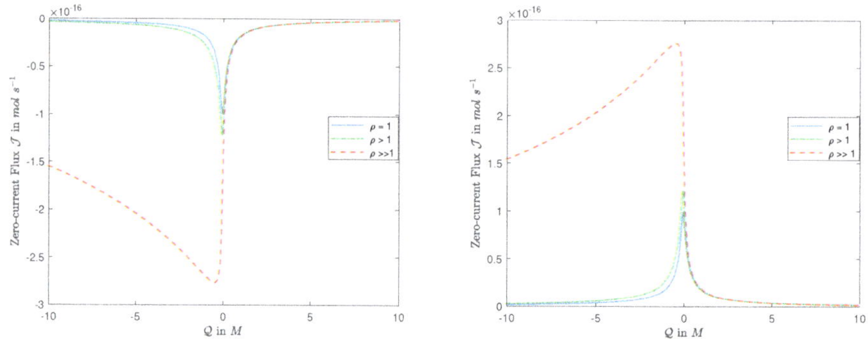

Figure 1. The function $\mathcal{J} = \mathcal{J}(\mathcal{Q})$ for various values of $\rho = D_2/D_1$: The left panel for $L = 2$ mM and $R = 5$ mM; the right panel for $L = 5$ mM and $R = 2$ mM.

2.1.2. Dependence of Zero-Current Flux J on Q_0 and D_k's

Concerning the dependence of the zero-current flux J on Q_0, we have the following:

(i) *If $D_1 = D_2$, then the zero-current flux J is an even function in Q_0, and it is monotonic for $Q_0 > 0$.*

In this case, $\theta = 0$ and, hence, it follows from Theorem 1 that A is an even function in Q_0 and is monotonic in Q_0 for $Q_0 > 0$, and thus is the zero-current flux J from Display (20).

(ii) *If $D_1 \neq D_2$, then the zero-current flux J is not an even function in Q_0 and the monotonicity of the zero-current flux J in Q_0 seems to be more complicated.*

In this case, it can be seen that G_2 in Display (16) is not an even function in Q_0, and, hence, the zero-current flux J is not. We would like to point out that, it follows from [38], for fixed $\rho = D_2/D_1$, no matter how large, one always has $J \to 0$ as $Q_0 \to \pm\infty$ that is consistent with the observations in Figure 1.

(iii) Another fascinating result is that *the magnitude of $\rho = D_2/D_1$ affects the monotonicity of the zero-current flux J in Q_0.*

In this case, if one fixes D_1, and let D_2 increases from small values to $D_2 \to \infty$, (i.e., $\rho \to \infty$), then it follows from Display (20) that there is a meaningful change in the monotonicity of the zero-current flux J, for small values of Q_0 that is not intuitive.

Let us consider the case where $L < R$ and $Q_0 < 0$ is small. Recall that A is the geometric mean of concentrations at $x = a$. It follows from System (15) and (16) that, as Q_0 increases,

(a) A increases if $\rho \approx 1$ (that is $\theta \approx 0$), and consequently the zero-current flux J decreases;
(b) A decreases if $\rho \gg 1$ (that is $\theta \gg 1$), and, hence, the zero-current flux J increases.

Thus, depending on the size of ρ, the zero-current flux J may increase or decrease in $Q_0 < 0$, which is also consistent with the observations in Figure 1. The analysis for the case with $L > R$ is similar.

It seems likely that the engineering, like evolution, will use these mathematical properties to control the qualitative properties of channels, technological, and biological.

2.2. Reversal Potential V_{rev}.

Experimentalists have long identified reversal potential as an essential characteristic of ion channels [54,55]. Reversal potential is the potential at which the current reverses direction, i.e., $\mathcal{V} = \Phi(0) - \Phi(\hat{L})$ that produces zero current $\mathcal{I}$. Using dimensionless form of quantities (see Remark 2), it follows from System (15) and (16) (where there are two ion species with valences $z_1 = -z_2 = 1$) that for general permanent charge $Q = 2Q_0 \neq 0$ with arbitrary diffusion constants [23], the variable A (the geometric mean of concentrations at $x = a$) can be solved uniquely from $G_2 = 0$ in System (15), and the reversal potential is then

$$V_{rev} = \theta\Big(\ln\frac{S_a + \theta Q_0}{S_b + \theta Q_0} + \ln\frac{l}{r}\Big) - (1+\theta)\ln\frac{A(Q_0,\theta)}{B(Q_0,\theta)} + \ln\frac{S_a - Q_0}{S_b - Q_0}, \tag{22}$$

where B, S_a, S_b, and θ are defined in Displays (18) and (19).

2.2.1. Range of Reversal potential V_{rev}

For fixed l, r, and for any given Q_0, it follows from Theorem 2 that there exists a unique reversal potential V_{rev} such that $V_{rev} \leq |\ln\frac{l}{r}|$. As $Q_0 \to \pm\infty$, then V_{rev} gets close to the boundary values, i.e., $V_{rev} \to \pm\ln\frac{l}{r}$.

2.2.2. Zero Reversal Potential

One particular case is when the reversal potential is zero. To examine under what conditions one obtains $V_{rev} = V_{rev}(Q_0) = 0$, it follows Theorem 2 that,

(i) if $D_1 = D_2$, then $V_{rev}(Q_0) = 0$ for $Q_0 = 0$,
(ii) if $D_1 < D_2$, then there is a $Q_0 < 0$, such that $V_{rev}(Q_0) = 0$,
(iii) if $D_1 > D_2$, then there is a $Q_0 > 0$, such that $V_{rev}(Q_0) = 0$.

Considering the second case above, the observations in Figure 2 show that, as $\rho = D_2/D_1$ increases, magnitude of the corresponding Q_0 becomes larger. In fact, as $\rho \to \infty$, then $Q_0 \to -\infty$.

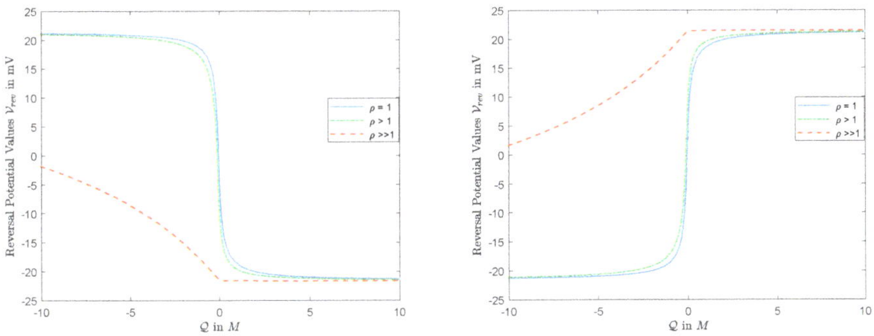

Figure 2. The function $\mathcal{V} = \mathcal{V}_{rev}(\mathcal{Q})$: The left panel for $L = 2$ mM and $R = 5$ mM; the right panel for $L = 5$ mM and $R = 2$ mM.

2.2.3. Reversal Potential $V_{rev}(Q_0)$ for $Q_0 = 0$

For $Q_0 = 0$, one has $V_{rev}(0) = \theta\ln\frac{l}{r}$ from Theorem 2. Therefore,

(i) if $D_1 = D_2$, then $V_{rev}(0) = 0$,
(ii) if $D_1 \neq D_2$, then $V_{rev}(0)$ has the same sign as that of $\theta(l - r)$.

Let us consider the case where $D_1 < D_2$. In that case, $V_{rev}(0)$ has the same sign as that of $l - r$. This is reasonable, since, for $V = 0$, we have $|J_1| < |J_2|$ (since all but J_k/D_k are independent of D_k in Equation (21)), and to help $|J_1|$ more than $|J_2|$ to get $J_1 = J_2$ for zero current conditions, one needs to increase V when $l > r$ (and decrease V when $l < r$), and that is why $V_{rev}(0) > 0$ for $l > r$ (and $V_{rev}(0) < 0$ for $l < r$). This is consistent with observations in Figure 2 as well. The analysis for the other case with $D_1 > D_2$ is similar.

2.2.4. Monotonicity of $\mathcal{V}_{rev}$ with Respect to $\mathcal{Q}$

It follows from Theorem 3 that

$$\textit{For } \theta Q > 0,\ \partial_Q \mathcal{V}_{rev} \textit{ has the same sign as that of } l - r.$$

This analytical result does not allow conclusions about the case for $\theta Q < 0$, however. The observations in Figures 2 and 3 show that the result holds for any θ and Q. Thus, we have

$$\textit{Conjecture}: V_{rev} \textit{ is increasing in } Q \textit{ for } l > r \textit{ and decreasing in } Q \textit{ for } l < r.$$

We remark that, in Figure 3, we take $L = 20\,\text{mM}$, $R = 50\,\text{mM}$, and $\mathcal{D}_1 = 1.334 \times 10^{-9}\ \text{m}^2/\text{s}$ and $\mathcal{D}_2 = 2.032 \times 10^{-9}\ \text{m}^2/\text{s}$ which are diffusion constants of, say, Na^+ and Cl^-, respectively (see the solid line), and $\mathcal{D}_1 = 1.334 \times 10^{-9}\text{m}^2/\text{s}$ and $\mathcal{D}_2 = 0.792 \times 10^{-9}\text{m}^2/\text{s}$, where $\mathcal{D}_2$ is the diffusion constants of Ca^{2+} (see the dashed line).

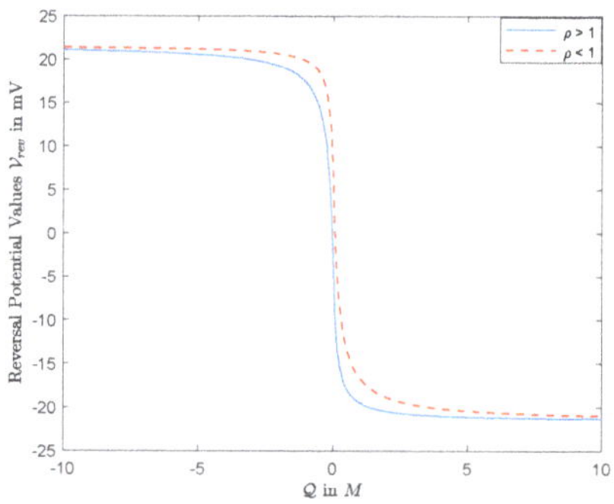

Figure 3. $\mathcal{V} = \mathcal{V}_{rev}(\mathcal{Q})$ decreases when $L < R$, independent of values of diffusion constants.

2.2.5. Dependence of $\mathcal{V}_{rev}$ on $\rho = D_2/D_1$

Let us discuss the dependence of V_{rev} on $\rho = D_2/D_1$ for effects of D_1 and D_2. It follows from Proposition 4.6 in [23] that

$$\textit{The reversal potential } V_{rev} \textit{ is increasing in } \rho \textit{ if } l > r \textit{ and is decreasing in } \rho \textit{ if } l < r.$$

This feature reveals a fantastic aspect that is not intuitive immediately. Recall Equation (21). Given the boundary values and diffusion constants, the values one obtains for all terms in Equation (21) except J_k are independent of D_k [36]. The relation surely holds for the zero-current condition, i.e., $J_1 = J_2$ with $V = V_{rev}$. Now, let us fix D_1 and increase D_2 (so ρ is increasing). Then, $|J_2|$ increases since all but J_2/D_2 in Equation (21) are independent of D_2. Consequently, to meet the zero-current condition, we need to increase $|J_1|$. Intuitively increasing V_{rev} seems to lead to an increase in $|J_1|$. This intuition agrees with the result for $l > r$. However, for the case with $l < r$, the result is the exact opposite of the intuitive result. That is, for $l < r$, it says, as ρ increases, V_{rev} decreases. This counterintuitive behavior could be explained by the fact that $c_1(x)$ depends on V_{rev}, and reducing V_{rev} could still increase $|J_1|$. In fact, $l < r$ will result in reducing V_{rev}, but $c_1(x)$ changes in a way that consequently increases $|J_1|$.

To illustrate the counterintuitive behavior, we provide a numerical result in Figure 4. We choose C_0, $\hat{L}$ and $\mathcal{D}_1$ for Na^+ as mentioned in Remark 1. Now, suppose that $\mathcal{D}_2^1 = 0.792 \times 10^{-9} m^2/s$, and consider the boundary concentrations $L = 20\,mM$, $R = 50\,mM$ and $\mathcal{Q} = 1\,M$. In this case, $\mathcal{V}_{rev} = -16.7657\,mV$ and $\mathcal{J} = -1.7632 \times 10^{-17} mol\,s^{-1}$. Now, if we increase $\mathcal{D}_2^1$ to $\mathcal{D}_2^2 = 2.032 \times 10^{-9} m^2/s$, which is Cl^- diffusion constant, then $\mathcal{V}_{rev} = -19.5527\,mV$ and $\mathcal{J} = -1.8788 \times 10^{-17} mol\,s^{-1}$. These values make sense now, based on the above discussion. Note that we just pictured the middle part of the channel in Figure 4 since the sides are almost identical. One should notice that it is hard to realize, from Figure 4, how $L < R$ will result in reducing $\mathcal{V}_{rev}$. The complicated behavior discussed above convinces us that detailed analytical studies, even for special cases, could be critical for the design and interpretation of numerical results.

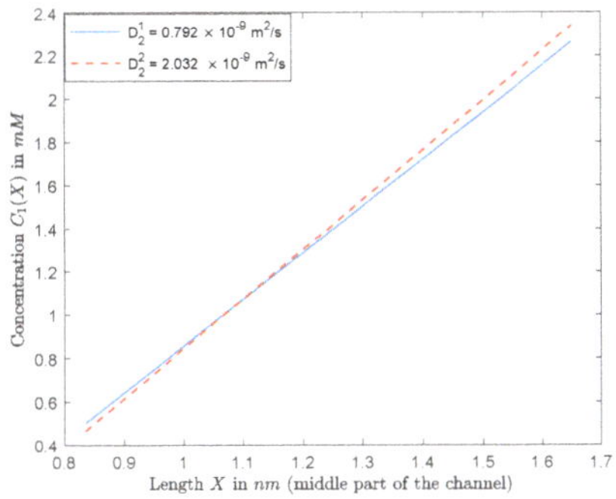

Figure 4. Graphs of $C_1(X)$ when $\mathcal{D}_1$ is fixed, but we increase $\mathcal{D}_2$.

2.3. A Comparison with Goldman–Hodgkin–Katz Equation for $\mathcal{V}_{rev}$.

In this section, we first recall the GHK equation [14,15], which relates the reversal potential with the boundary concentrations and the permeabilities of the membrane to the ions. If the membrane is permeable to only one ion, then that ion's Nernst potential is the reversal potential at which the electrical and chemical driving forces balance. The GHK equation is a generalization of the Nernst equation in which the membrane is permeable to more than just one ion. The derivation of GHK equation assumes that the electric field across the lipid membrane is constant (or, equivalently, the electric potential $\phi(x)$ is linear in x in the PNP model). Under the assumption, the I–V (current–voltage) relation is given by

$$I = V \sum_{k=1}^{n} z_k^2 D_k \frac{r_k - l_k e^{z_k V}}{1 - e^{z_k V}}.$$

For the case where $n = 2$ and $z_1 = -z_2 = 1$, the GHK equation for the reversal potential is

$$V_{rev}^{GHK}(\rho) = \ln \frac{r + \rho l}{l + \rho r}. \tag{23}$$

The assumption that the electric potential $\phi(x)$ is linear is not correct when applied to channels in proteins. This is because proteins have specialized structure and spatial distributions of permanent charge (acid and base side chains) and polarization (polar and nonpolar side chains). *Experimental manipulations of the structure of channel proteins show that these properties control the biological function of the channel.* The GHK equation does not contain variables to describe any of these properties and so cannot account for the biological functions they control. A linear $\phi(x)$ is widely believed to make sense without channel structure presumably, in particular, where $Q_0 = 0$. However, this is not correct either. It follows from Formula (22) for $Q_0 = 0$ that the zeroth order in ε approximation of the reversal potential in this case is

$$V_{rev}(0,\rho) = \frac{\rho - 1}{\rho + 1} \ln \frac{l}{r}. \tag{24}$$

Figure 5 compares $V_{rev}(0,\rho)$ in Formula (24) with V_{rev}^{GHK} from the GHK-equation in Display (23). It can be seen from the left panel that, when l and r are not far away from each other (for example $L = C_0 l = 20\,\text{mM}$, $R = C_0 r = 50\,\text{mM}$), then the two curves have almost the same behavior. However, when we reduce L from 20 mM to 1 mM, then the right panel shows a significant difference between the two graphs.

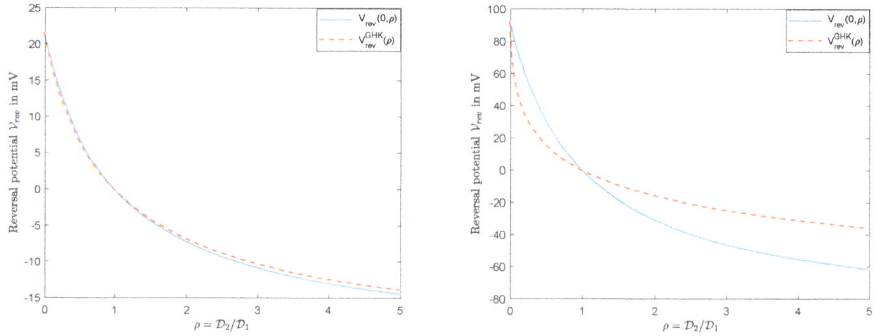

Figure 5. $\mathcal{V}_{rev}(\mathcal{Q} = 0,\rho)$ vs. $\mathcal{V}_{rev}^{GHK}(\rho)$: The left panel for $L = 20\,\text{mM}$ and $R = 50\,\text{mM}$; the right panel for $L = 1\,\text{mM}$ and $R = 50\,\text{mM}$.

In Figure 6, we arrange a simple numerical result for the case where $\mathcal{Q} \neq 0$ to compare the graphs of $\mathcal{V}_{rev}(\mathcal{Q},\rho)$, obtained from Formula (22), for various values of permanent charge $\mathcal{Q}$. We consider $L = 20\,\text{mM}$, $R = 50\,\text{mM}$, and $0 < \rho < 5$ for some values of $\mathcal{Q}$, i.e., $\mathcal{Q} = 0\,\text{M}, 1\,\text{M}, 10\,\text{M}$.

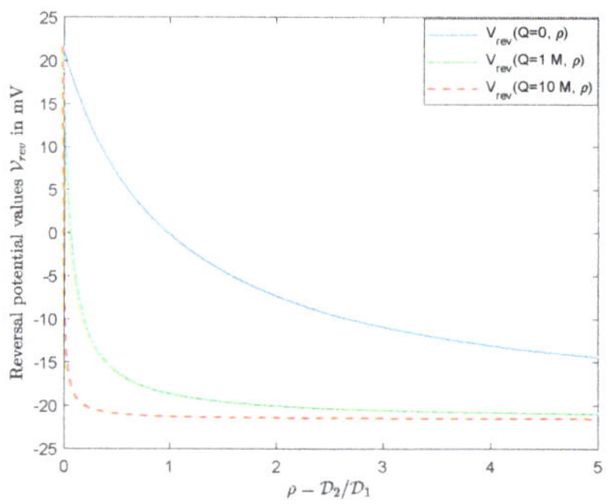

Figure 6. $\mathcal{V}_{rev}(\mathcal{Q},\rho)$ with various values of permanent charges.

2.4. Profiles of Relevant Physical Quantities

It follows that, for any given Q, once a solution (A, V) of Equations (15) and (16) is determined, all the other unknowns can be settled, and, hence, the approximation of the solution of boundary value problem can be obtained. We consider the dimensional form of quantities, and fix $(\mathcal{Q}, L, R, \mathcal{D}_1, \mathcal{D}_2)$ to numerically investigate the behavior of $C_k(X)$ and $\Phi(X)$ throughout the channel. Figures 7 and 8 graph the cases with small permanent charge $\mathcal{Q} = 0.1\,\text{mM}$ when $L = 20\,\text{mM}$, $R = 50\,\text{mM}$, $\mathcal{D}_1 =$

$1.334 \times 10^{-9}\,\text{m}^2/\text{s}$, and $\mathcal{D}_2 = 2.032 \times 10^{-9}\,\text{m}^2/\text{s}$. In this case, we obtain $\mathcal{J} = -1.2079 \times 10^{-16}\,\text{mol s}^{-1}$ and $\mathcal{V}_{rev} = -4.4820\,\text{mV}$.

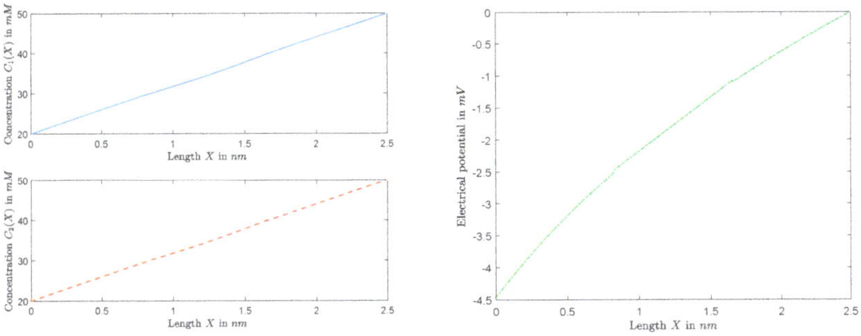

Figure 7. The functions $C_k(X)$ (**left**) and $\Phi(X)$ (**right**) with $\mathcal{Q} = 0.1\,\text{mM}$.

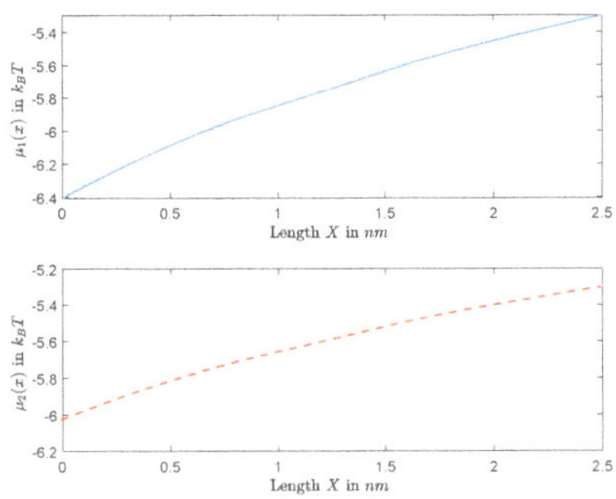

Figure 8. The functions $\mu_1(X)$ and $\mu_2(X)$ are increasing for $\mathcal{Q} = 0.1\,\text{mM}$.

Furthermore, Figures 9 and 10 show graphs of concentrations, electrical potential, and electrochemical potentials versus X, where $L = 20\,\text{mM}$, $R = 50\,\text{mM}$, $Q = 2\,\text{M}$, and diffusion constants are the same as the previous one. In this case, we obtain $\mathcal{J} = -1.8789 \times 10^{-17}\,\text{mol s}^{-1}$ and $\mathcal{V}_{rev} = -19.5527\,\text{mV}$.

Remark 3. *We end this section with a few of the remarks on some important features captured in Figures 7–10. It follows from the Nernst–Planck equation that $\mu_k'(x)$ has the same sign as that of $\mu_k(1) - \mu_k(0)$ or the opposite sign as that of J_k; in particular, $\mu_k(x)$ is always monotonically increasing or decreasing. For the zero-current situation, the reversal potential depends on ALL other parameters; and so it seems that it would be hard to make general conclusions about $\mu_k(x)$, for example, about its monotonicity. This is not true. In fact, in Section 2.1, we have concluded that the sign of zero-current flux J is the same as that of $L - R$, and, hence, $\mu_k'(x)$ has the opposite sign as that of $L - R$. For the case considered in this part, $L = 20\,mM < R = 50\,mM$, one has $J < 0$, independent of the value of Q. Therefore, $\mu_k'(x) > 0$ for $k = 1, 2$, and, hence, $\mu_k(x)$'s are increasing for zero-current situation when $L < R$, independent of Q, as shown in Figure 8 for $Q = 0.1\,mM$ and in Figure 10 for $Q = 1\,mM$. On the other hand, as one changes the value of Q, the profiles of concentrations $c_k(x)$'s and electrical potential $\phi(x)$ may vary from monotone to non-monotone, as shown between Figure 7 for $Q = 0.1\,mM$ and Figure 9 for $Q = 1\,M$.*

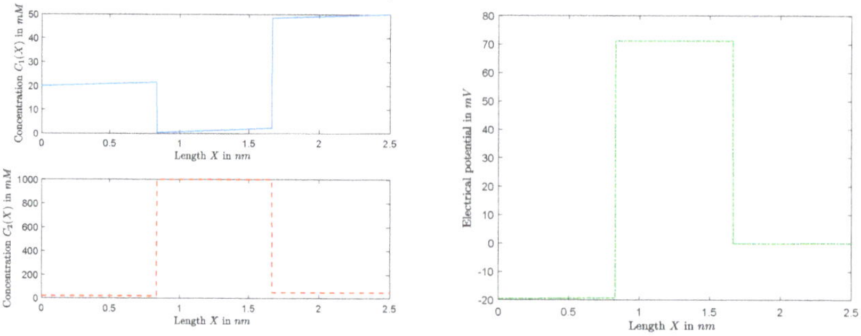

Figure 9. The functions $C_1(X)$ and $C_2(X)$ (**left**) and the function $\Phi(X)$ (**right**) for $\mathcal{Q} = 1$ M.

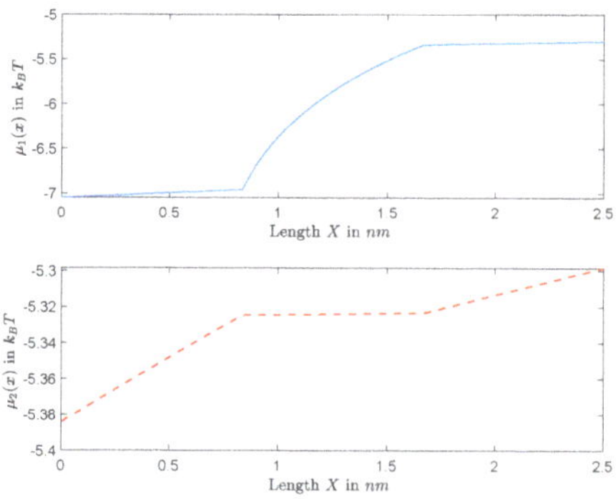

Figure 10. The functions $\mu_1(X)$ and $\mu_2(X)$ are increasing for $\mathcal{Q} = 1$ M.

3. Current–Voltage and Current-Permanent Charge Behaviors

Ionic movements across membranes lead to the generation of electrical currents. The current carried by ions can be examined through *current–voltage* relation or I–V curve. In such a case, voltage refers to the voltage across a membrane potential, and current is the flow of ions through channels in the membrane. Another important piece of data are *current-permanent charge* (I–Q) relation. Dependence of current on membrane potentials and permanent charge is investigated in this section for arbitrary values of diffusion constants.

To derive the I–V and I–Q relations, we rely on [33] where the authors showed that the set of nonlinear algebraic equations is equivalent to one nonlinear equation for A, the geometric mean of concentrations at $x = a$ defined in Equation (17). All other quantities or variables such as fluxes, profiles of electric potential $\phi(x)$ and concentrations $c_k(x)$ can then be obtained in terms of A. It is crucial to realize that this is a specific result, not available for general cases. One can only imagine that the resulting simplification produces controllable and robust behavior that proved useful as evolution designed and refined protein channels. The reduction allowed by this composite variable can be postulated to be a "design principle" of channel construction, in technological (engineering) language, or an evolutionary adaptation, in biological language. In particular, the current I can be explicitly expressed in terms of boundary conditions, permanent charge, diffusion constants, and transmembrane potential in the special case that allows the determination of A. In what follows, we derive flux and current equations—when diffusion constants are involved as well—in terms of boundary concentrations, membrane potential, and permanent charge. The I–V, I–Q, J–V, and J–Q relations are investigated afterward in Section 3.2.

3.1. Reduced Flux and Current Equations

In this section, for simplicity, in addition to the assumptions at the beginning of the setup section (Section 1.2), we will also assume that $h(x) = 1$, $a = 1/3$ and $b = 2/3$, in particular, $\alpha = 1/3$ and $\beta = 2/3$ (see Display (19)). It was shown in [33] that the BVP (9) and (10) can be reduced to the algebraic equation

$$\eta \ln \frac{S_b - \eta}{S_a - \eta} - N = 0, \tag{25}$$

where $B = l - A + r$, S_a, S_b and N are defined in Display (18), and

$$\eta = Q_0 - \frac{Q_0}{\ln \frac{Bl}{Ar}} \Big(V + \ln \frac{l(S_b - Q_0)}{r(S_a - Q_0)} \Big) + \frac{N}{\ln \frac{Bl}{Ar}}. \tag{26}$$

Once A is solved from Equation (25), we can obtain the flux densities and current equations as follows:

$$\begin{aligned} J_k := J_k(V, l, r, D_1, D_2) =& 3D_k(l - A)\Big(1 + (-1)^k \frac{\eta}{Q_0}\Big), \quad \text{for } k = 1, 2, \\ I := I(V, l, r, D_1, D_2) =& J_1 - J_2 = 3(l - A)\Big(D_1 - D_2 - \frac{\eta}{Q_0}(D_1 + D_2)\Big). \end{aligned} \tag{27}$$

For any given (l, r, D_1, D_2, Q_0, V), there exists a solution for the flux J and current I. The numerical results in the next section give us more information on "current–voltage" and "current-permanent charge" relations.

3.2. Current–Voltage and Current-Permanent Charge Relations

3.2.1. Dependence of Current on Diffusion Constants

Now, we reveal a feature of the theoretical results that is not intuitive. Suppose that (l, r, Q_0) is given (V is still free and is allowed to take any value!). It follows from Display (18) for the definition of N that there exists an A so that $N = 0$. It consequently follows from Equation (26) that, if $V = \ln \frac{B(S_a - Q_0)}{A(S_b - Q_0)}$, then $\eta = 0$. Therefore, from Display (27), $I = 3(l - A)(D_1 - D_2)$, which implies

For special values of parameters (l, r, V, Q_0), *the sign of* I *is determined by the sign of* $D_1 - D_2$.

3.2.2. I–V Curves and I–Q Curves

Figure 11 is a numerical simulation from Equations (25) and (26) of the I–V curves for several values of $\mathcal{Q}$ with $\mathcal{D}_1 = 1.334 \times 10^{-9}\,\text{m}^2/\text{s}$ and $\mathcal{D}_2 = 2.032 \times 10^{-9}\,\text{m}^2/\text{s}$. One may suspect, based on the numerical observations, that the value of current $\mathcal{I}$, obtained from Display (27), is unique for any $\mathcal{V}$, and is monotonically increasing in $\mathcal{V}$. However, this may not be correct, in general. This is important since the opening and closing properties of channels might be thought to arise from non-unique solutions [16,17].

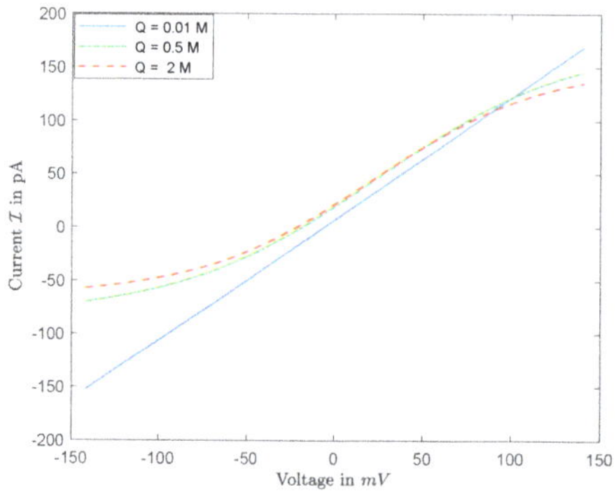

Figure 11. The function $\mathcal{I} = \mathcal{I}(\mathcal{V})$ for $L = 20\,\text{mM}$ and $R = 50\,\text{mM}$.

Now, for I–Q relations, our numerical experiments shows that:

I–Q curves are not monotonic in general.

Recall that Equation (21), in dimensional form, gives

$$\mathcal{J}_k \int_0^{\hat{L}} \frac{k_B T}{\mathcal{D}_k \mathcal{A}(X) C_k(X)} dX = \mu_k(0) - \mu_k(\hat{L}), \quad k = 1, 2.$$

The sign of $\mathcal{J}_k$ is determined by the boundary conditions, independent of the permanent charge. Nevertheless, as expected and seen in Figure 12, the magnitudes of $\mathcal{J}_k$'s, and, consequently, the sign and the size of the current $\mathcal{I}$ do depend on $\mathcal{Q} = 2Q_0C_0$ in general. (Here, $\mathcal{Q}$ would be the nonzero value of the permanent charge in dimensional form.) Treating $\mathcal{V}$ as a parameter, the current $\mathcal{I}$ is a function of $\mathcal{Q}$. The numerical observations in Figure 12 indicate that,

(i) there exists some $\mathcal{V}^*$ such that, for $\mathcal{V} > \mathcal{V}^*$, $\mathcal{I}(\mathcal{Q})$ has a unique maximum;
(ii) there exists some $\mathcal{V}_*$ such that, for $\mathcal{V} < \mathcal{V}_*$, $\mathcal{I}(\mathcal{Q})$ has a unique minimum.

In particular, I–Q curves are not monotonic in general.

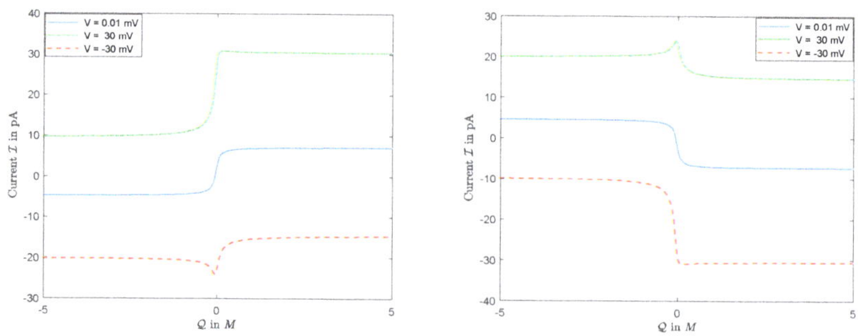

Figure 12. The function $\mathcal{I} = \mathcal{I}(\mathcal{Q})$ with $\mathcal{D}_1 < \mathcal{D}_2$: The left panel for $L = 20\,\text{mM}$ and $R = 30\,\text{mM}$; the right panel for $L = 30\,\text{mM}$ and $R = 20\,\text{mM}$.

In addition, we claim based on numerical observations (not proven though) that there exists $\hat{\mathcal{V}}(\mathcal{D}_1, \mathcal{D}_2) = \min\{\mathcal{V}^*, |\mathcal{V}_*|\}$, such that

(i) for any given $\mathcal{V}$ where $|\mathcal{V}| > \hat{\mathcal{V}}$, the corresponding current $\mathcal{I}$ is non-monotonic in $\mathcal{Q}$, but
(ii) for any $\mathcal{V}$ where $|\mathcal{V}| < \hat{\mathcal{V}}$, the corresponding current $\mathcal{I}$ is monotonic in $\mathcal{Q}$.

In particular, it can be seen in Section 3.3 that current is monotonic in $\mathcal{Q}$ for $\mathcal{V}=0$. In the end, we would like to mention that the diffusion constants affect the values V^* and V_* above.

3.3. Zero-Voltage Current

The different permeability of the membrane determines the zero membrane potential (voltage) to different types of ions, as well as the concentrations of the ions, the permanent charge, and the shape of the channel. Denote current $I(V;Q)$, and the fluxes $J_k(V;Q)$, for $k=1,2$, to include the dependence on Q too. We call $I(0;Q)$ the *zero-potential current* and $J_k(0;Q)$ the *zero-potential fluxes*, respectively, when $V=0$. For any given value of membrane potential V, approximation formulas for the current $I(V;Q)$, for small and large values of permanent charge Q, are provided in [34,38], respectively.

It follows from [34] that, for *small* values of Q, applying $V=0$, zero-potential current $I^s(0;Q)$, and zero-potential fluxes $J_k^s(0;Q)$ (in dimensionless forms as mentioned in Remark 2) are

$$\begin{aligned} I^s(0;Q) =&(l-r)(D_1-D_2)-\frac{3}{2}\frac{(D_1+D_2)(l-r)^2}{(2l+r)(l+2r)\ln\frac{l}{r}}Q+O(Q^2),\\ J_k^s(0;Q) =&(l-r)D_k+(-1)^k\frac{3(l-r)^2D_k}{2(2l+r)(l+2r)\ln\frac{l}{r}}Q+O(Q^2),\quad k=1,2. \end{aligned} \tag{28}$$

For *large* positive values of $Q=2Q_0$, with $\nu=1/Q_0$ (where ν is small), it follows from [38] that zero-potential current $I^l(\nu)=I^l(0;Q)$ and zero-potential fluxes $J_k^l(\nu)=J_k^l(0;Q)$ are

$$\begin{aligned} I^l(\nu) =&-6D_2\sqrt{lr}\frac{(\sqrt{l}-\sqrt{r})}{\sqrt{l}+\sqrt{r}}+\frac{3}{2}D_1\Big(\frac{l+r}{\sqrt{l}+\sqrt{r}}\Big)^2(l-r)\nu\\ &+\frac{3}{2}D_2\frac{l+r}{(\sqrt{l}+\sqrt{r})^2}f(l,r)(l-r)\nu+O(\nu^2),\\ J_1^l(\nu) =&\frac{3}{2}D_1\Big(\frac{l+r}{\sqrt{l}+\sqrt{r}}\Big)^2(l-r)\nu,\\ J_2^l(\nu) =&6D_2\sqrt{lr}\frac{\sqrt{l}-\sqrt{r}}{\sqrt{l}+\sqrt{r}}-\frac{3}{2}D_2\frac{l+r}{(\sqrt{l}+\sqrt{r})^2}f(l,r)(l-r)\nu+O(\nu^2), \end{aligned} \tag{29}$$

where

$$f(l,r)=\frac{2lr}{(\sqrt{l}+\sqrt{r})^2}+l+r-\frac{1}{2}\frac{\ln l-\ln r}{l-r}(l+r)^2. \tag{30}$$

It can be readily seen from Equations (28) that, for small values of Q, the zero-potential current $I^s(0;Q)$ is increasing in Q when $l<r$ and is decreasing in Q if $l>r$.

However, for large values of the permanent charge Q, the zero-potential current $I^l(0;Q)$ depends on Q in a much richer way. To state the results, we need the following lemma.

Lemma 1. *There are t_1 and t_2 with $0<t_1<1<t_2$ so that $f(l,r)>0$ for $l/r\in(t_1,t_2)$ and $f(l,r)<0$ for $l/r\notin[t_1,t_2]$.*

Note that
$$\frac{d}{d\nu}I^l(0)=\frac{3D_2}{2}\Big(\frac{l+r}{\sqrt{l}+\sqrt{r}}\Big)^2\Big(\frac{D_1}{D_2}+\frac{f(l,r)}{l+r}\Big)(l-r).$$

It follows from Equations (29) and Lemma 1 that, for large values of Q (small values of ν),

(i) if $l/r\in[t_1,t_2]$ (so that $f(l,r)\geq 0$), then, for arbitrary D_1 and D_2, the zero-potential current $I^l(\nu)$ is decreasing in ν (increasing in Q) when $l/r\in[t_1,1)$, and is increasing in ν (decreasing in Q) when $l/r\in(1,t_2]$;

(ii) if $l/r\notin[t_1,t_2]$ (so that $f(l,r)<0$), then,

(a) for $\frac{D_1}{D_2} + \frac{f(l,r)}{l+r} > 0$, the zero-potential current $I^l(\nu)$ is decreasing in ν (increasing in Q) for $l < r$, and is increasing in ν (decreasing in Q) for $l > r$;

(b) for $\frac{D_1}{D_2} + \frac{f(l,r)}{l+r} < 0$, the zero-potential current $I^l(\nu)$ is increasing in ν (decreasing in Q) for $l < r$, and is decreasing in ν (increasing in Q) for $l > r$.

Figure 13 illustrates some of the above conclusions. In addition, it suggests that the monotonicity of $I(0)$ holds for all values of permanent charge, not only for small or large values. We emphasize that the monotonicity of current I with respect to permanent charge Q is just true for zero membrane potential, i.e., $V = 0$. Indeed, one should recall from Section 3.2 that, when $V \neq 0$, then the current I is not monotonic in Q.

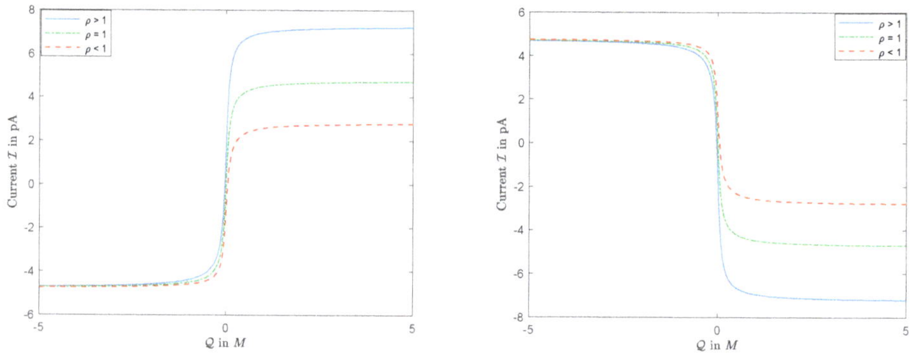

Figure 13. The function $\mathcal{I} = \mathcal{I}(\mathcal{Q})$ for $\mathcal{V} = 0$: The left panel for $L = 20$ mM and $R = 30$ mM; the right panel for $L = 30$ mM and $R = 20$ mM.

4. Conclusions

In this paper, we first recall the analytical results in [23] for arbitrary diffusion constants. To investigate the reversal potential problem for which the current is zero, we do numerical investigations based on the analytical results in [23], where many cases are studied analytically. We derive several remarkable properties of biological significance, from the analysis of these governing equations that hardly seem intuitive.

Biophysicists are also interested in the relation of current–voltage (I–V), and current-permanent charge (I–Q), as well as reversal potential problems. To do that, we first recall the analytical results in [33], for arbitrary diffusion constants, to drive the flux densities and current equations explicitly. One way to characterize channels is the current at zero electric potential, that is, when $V = 0$, which has practical advantages. Since it is usually easier to measure a large current than a vanishing one, we analyzed this case as well. Furthermore, we briefly study the special cases of small and large permanent charge for zero voltage case, based on the analytical results of [34,38], respectively. To bridge between small and large values of permanent charges, we numerically study I–V and I–Q relations for this case as well.

Author Contributions: All authors of this paper contributed equally. All authors have read and agreed to the published version of the manuscript.

Acknowledgments: We thank the anonymous referees for suggestions that significantly improved the paper. The research is partially supported by Simons Foundation Mathematics and Physical Sciences-Collaboration Grants for Mathematicians #581822.

Conflicts of Interest: The authors declare no conflict of interest.

References

1. Boda, D.; Nonner, W.; Valisko, M.; Henderson, D.; Eisenberg, B.; Gillespie, D. Steric selectivity in Na channels arising from protein polarization and mobile side chains. *Biophys. J.* **2007**, *93*, 1960–1980.
2. Eisenberg, B. Crowded charges in ion channels. In *Advances in Chemical Physics*; Rice, S.A., Ed.; John Wiley & Sons: Hoboken, NJ, USA, 2011; pp. 77–223.
3. Gillespie, D. Energetics of divalent selectivity in a calcium channel: the Ryanodine receptor case study. *Biophys. J.* **2008**, *94*, 1169–1184.
4. Hodgkin, A.L.; Keynes, R.D. The potassium permeability of a giant nerve fibre. *J. Physiol.* **1955**, *128*, 61–88.
5. Ji, S.; Eisenberg, B.; Liu, W. Flux ratios and channel structures. *J. Dynam. Diff. Equ.* **2019**, *31*, 1141–1183.
6. Boron, W.; Boulpaep, E. *Medical Physiology*; Saunders: New York, NY, USA, 2008.
7. Ruch, T.C.; Patton, H.D. The Brain and Neural Function. In *Physiology and Biophysics*; W.B. Saunders Company: Philadelphia, PA, USA, 1973; Volume 1.
8. Ruch, T.C.; Patton, H.D. Circulation, Respiration and Balance. In *Physiology and Biophysics*; W.B. Saunders Company: Philadelphia, PA, USA, 1973; Volume 2.
9. Ruch, T.C.; Patton, H.D. Digestion, Metabolism, Endocrine Function and Reproduction. In *Physiology and Biophysics*; W.B. Saunders Company: Philadelphia, PA, USA, 1973; Volume 3.
10. Eisenberg, B. Ion Channels as Devices. *J. Comp. Electron.* **2003**, *2*, 245–249.
11. Burger, M.; Eisenberg, R.S.; Engl, H. Inverse Problems Related to Ion Channel Selectivity. *SIAM J. Appl. Math.* **2007**, *67*, 960–989.
12. Rouston, D.J. *Bipolar Semiconductor Devices*; McGraw-Hill: New York, NY, USA, 1990.
13. Warner, R.M., Jr. Microelectronics: Its unusual origin and personality. *IEEE Trans. Electron. Devices* **2001**, *48*, 2457–2467.
14. Goldman, D.E. Potential, impedance, and rectification in membranes. *J. Gen. Physiol.* **1943**, *27*, 37–60.
15. Hodgkin, A.L.; Katz, B. The effect of sodium ions on the electrical activity of the giant axon of the squid. *J. Physiol.* **1949**, *108*, 37–77.
16. Eisenberg, R.S. Atomic biology, electrostatics and ionic channels. New developments and theoretical studies of proteins. *R. Elber. Philadel. World Sci.* **1996**, *7*, 269–357.
17. Eisenberg, R.S. Computing the field in proteins and channels. *J. Membr. Biol.* **1996**, *150*, 1–25.
18. Markowich, P.A.; Ringhofer, C.A.; Schmeiser, C. *Semiconductor Equations*; Springer: New York, NY, USA, 1990.
19. Selberherr, S. *Analysis and Simulation of Semiconductor Devices*; Springer: New York, NY, USA, 1984.
20. Shockley, W. *Electrons and Holes in Semiconductors to Applications in Transistor Electronics*; van Nostrand: New York, NY, USA, 1950.
21. Vasileska, D.; Goodnick, S.M.; Klimeck, G. *Computational Electronics: Semiclassical and Quantum Device Modeling and Simulation*; CRC Press: New York, NY, USA, 2010.
22. Eisenberg, B.; Liu, W.; Xu, H. Reversal permanent charge and reversal potential: Case studies via classical Poisson–Nernst–Planck models. *Nonlinearity* **2015**, *28*, 103–128.
23. Mofidi, H.; Liu, W. Reversal potential and reversal permanent charge with unequal diffusion coefficients via classical Poisson–Nernst–Planck models. *arXiv* **2019**, arXiv:1909.01192.
24. Gillespie, D.; Nonner, W.; Eisenberg, R.S. Coupling Poisson–Nernst–Planck and density functional theory to calculate ion flux. *J. Phys. Condens. Matter* **2002**, *14*, 12129.
25. Kilic, M.S.; Bazant, M.Z.; Ajdari, A. Steric effects in the dynamics of electrolytes at large applied voltages. II. Modified Poisson–Nernst–Planck equations. *Phys. Rev. E* **2007**, *75*, 021503.
26. Hyon, Y.; Eisenberg, R.; Liu, C. A mathematical model for the hard sphere repulsion in ionic solutions. *Commun. Math. Sci.* **2011**, *9*, 459–475.
27. Maffeo, C.; Bhattacharya, S.; Yoo, J.; Wells, D.; Aksimentiev, A. Modeling and simulation of ion channels. *Chem. Rev.* **2012**, *112*, 6250–6284.
28. Ji, S.; Liu, W. Poisson–Nernst–Planck systems for ion flow with density functional theory for hard-sphere potential: I–V relations and critical potentials. Part I: Analysis. *J. Dyn. Differ. Equ.* **2012**, *24*, 955–983.
29. Liu, W.; Tu, X.; Zhang, M. Poisson–Nernst–Planck systems for ion flow with density functional theory for hard-sphere potential: I–V relations and critical potentials. Part II: Numerics. *J. Dyn. Differ. Equ.* **2012**, *24*, 985–1004.

30. Lin, G.; Liu, W.; Yi, Y.; Zhang, M. Poisson–Nernst–Planck systems for ion flow with a local hard-sphere potential for ion size effects. *SIAM J. Appl. Dyn. Syst.* **2013**, *12*, 1613–1648.
31. Liu, J.-L.; Eisenberg, B. Poisson–Nernst–Planck-Fermi theory for modeling biological ion channels. *J. Chem. Phys.* **2014**, *141*, 12B640.
32. Sun, L.; Liu, W. Non-localness of Excess Potentials and Boundary Value Problems of Poisson–Nernst–Planck Systems for Ionic Flow: A Case Study. *J. Dyn. Differ. Equ.* **2018**, *30*, 779–797.
33. Eisenberg, B.; Liu, W. Poisson–Nernst–Planck systems for ion channels with permanent charges. *SIAM J. Math. Anal.* **2007**, *38*, 1932–1966.
34. Ji, S.; Liu, W.; Zhang, M. Effects of (small) permanent charge and channel geometry on ionic flows via classical Poisson–Nernst–Planck models. *SIAM J. Appl. Math.* **2015**, *75*, 114–135.
35. Liu, W. Geometric singular perturbation approach to steady-state Poisson–Nernst–Planck systems. *SIAM J. Appl. Math.* **2005**, *65*, 754–766.
36. Liu, W. One-dimensional steady-state Poisson–Nernst–Planck systems for ion channels with multiple ion species. *J. Differ. Equ.* **2009**, *246*, 428–451.
37. Coleman, T.F.; Li, Y. An interior, trust region approach for nonlinear minimization subject to bounds. *SIAM J. Optim.* **1996**, *6*, 418–445.
38. Zhang, L.; Eisenberg, B.; Liu, W. An effect of large permanent charge: Decreasing flux with increasing transmembrane potential. *Eur. Phys. J. Spec. Top.* **2019**, *227*, 2575–2601.
39. Chen, D.P.; Eisenberg, R.S. Charges, currents and potentials in ionic channels of one conformation. *Biophys. J.* **1993**, *64*, 1405–1421.
40. Chung, S.; Kuyucak, S. Predicting channel function from channel structure using Brownian dynamics simulations. *Clin. Exp. Pharmacol. Physiol.* **2001**, *28*, 89–94.
41. Barcilon, V.; Chen, D.-P.; Eisenberg, R.S. Ion flow through narrow membrane channels: Part II. *SIAM J. Appl. Math.* **1992**, *52*, 1405–1425.
42. Hollerbach, U.; Chen, D.P.; Eisenberg, R.S. Two- and three-dimensional Poisson–Nernst–Planck simulations of current flow through gramicidin-A. *J. Comput. Sci.* **2001**, *16*, 373–409.
43. Im, W.; Roux, B. Ion permeation and selectivity of OmpF porin: A theoretical study based on molecular dynamics, Brownian dynamics, and continuum electrodiffusion theory. *J. Mol. Biol.* **2002**, *322*, 851–869.
44. Nonner, W.; Catacuzzeno, L.; Eisenberg, B. Binding and selectivity in L-type Calcium channels: A mean spherical approximation. *Biophys. J.* **2000**, *79*, 1976–1992.
45. Nonner, W.; Eisenberg, R.S. Ion permeation and glutamate residues linked by Poisson–Nernst–Planck theory in L-type Calcium channels. *Biophys. J.* **1998**, *75*, 1287–1305.
46. Liu, W.; Wang, B. Poisson–Nernst–Planck systems for narrow tubular-like membrane channels. *J. Dyn. Differ. Equ.* **2010**, *22*, 413–437.
47. Eisenberg, B.; Liu, W. Relative dielectric constants and selectivity ratios in open ionic channels. *Mol. Based Math. Biol.* **2017**, *5*, 125–137.
48. Liu, W. A flux ratio and a universal property of permanent charges effects on fluxes. *Comput. Math. Biophys.* **2018**, *6*, 28–40.
49. Fick, A. On liquid diffusion. *Philos. Mag. J. Sci.* **1855**, *10*, 31–39.
50. Bard, A.J.; Faulkner, L.R. *Electrochemical Methods, Fundamentals and Applications*; Wiley: New York, NY, USA, 1980.
51. Brooks, R.E.; Heflinger, L.O.; Wuerker, R.F. Interferometry with a holographically reconstructed comparison beam. *J. Appl. Phys. Lett.* **1965**, *7*, 248–249.
52. Gerhardt, G.; Adams, R.N. Determination of diffusion constants by flow injection analysis. *Anal. Chem. J.* **1982**, *54*, 2618–2620.
53. Smith, H.M. *Principles of Holography*; Wiley (Interscience): New York, NY, USA, 1969.
54. Hodgkin, A.L. The ionic basis of nervous conduction. *Science* **1964**, *145*, 1148.
55. Huxley, A.F. The quantitative analysis of excitation and conduction in nerve (reprint of Nobel lecture). *Science* **1966**, *145*, 1154–1159.

entropy

MDPI

Review

Molecular Mean-Field Theory of Ionic Solutions: A Poisson-Nernst-Planck-Bikerman Model

Jinn-Liang Liu [1,*] and Bob Eisenberg [2,3]

[1] Institute of Computational and Modeling Science, National Tsing Hua University, Hsinchu 300, Taiwan
[2] Department of Physiology and Biophysics, Rush University, Chicago, IL 60612, USA; beisenbe@rush.edu
[3] Department of Applied Mathematics, Illinois Institute of Technology, Chicago, IL 60616, USA
* Correspondence: jinnliu@mail.nd.nthu.edu.tw

Received: 16 April 2020; Accepted: 12 May 2020; Published: 14 May 2020

Abstract: We have developed a molecular mean-field theory—fourth-order Poisson–Nernst–Planck–Bikerman theory—for modeling ionic and water flows in biological ion channels by treating ions and water molecules of any volume and shape with interstitial voids, polarization of water, and ion-ion and ion-water correlations. The theory can also be used to study thermodynamic and electrokinetic properties of electrolyte solutions in batteries, fuel cells, nanopores, porous media including cement, geothermal brines, the oceanic system, etc. The theory can compute electric and steric energies from all atoms in a protein and all ions and water molecules in a channel pore while keeping electrolyte solutions in the extra- and intracellular baths as a continuum dielectric medium with complex properties that mimic experimental data. The theory has been verified with experiments and molecular dynamics data from the gramicidin A channel, L-type calcium channel, potassium channel, and sodium/calcium exchanger with real structures from the Protein Data Bank. It was also verified with the experimental or Monte Carlo data of electric double-layer differential capacitance and ion activities in aqueous electrolyte solutions. We give an in-depth review of the literature about the most novel properties of the theory, namely Fermi distributions of water and ions as classical particles with excluded volumes and dynamic correlations that depend on salt concentration, composition, temperature, pressure, far-field boundary conditions etc. in a complex and complicated way as reported in a wide range of experiments. The dynamic correlations are self-consistent output functions from a fourth-order differential operator that describes ion-ion and ion-water correlations, the dielectric response (permittivity) of ionic solutions, and the polarization of water molecules with a single correlation length parameter.

Keywords: bioelectricity; electrochemistry; thermodynamics; electrokinetics; molecular mean-field theory; Boltzmann and Fermi distributions; Poisson–Boltzmann; Poisson–Fermi; Poisson–Bikerman; Nernst–Planck; steric and correlation effects; ion channels; ion activity; double-layer capacitance; nanofluidics

1. Introduction

Water and ions give life. Their electrostatic and kinetic interactions play essential roles in biological and chemical systems such as DNA, proteins, ion channels, cell membranes, physiology, nanopores, supercapacitors, lithium dendrite growth, porous media, corrosion, geothermal brines, environmental applications, and the oceanic system [1–34]. Poisson, Boltzmann, Nernst, and Planck laid the foundations of classical electrostatic and kinetic theories of ions in 1813–1890 [35–39]. Gouy [40] and Chapman [41] formulated the Poisson–Boltzmann (PB) equation in 1910 and 1913, respectively [9]. Bikerman proposed a modified PB equation in 1942 for binary ionic liquids to account for **different-sized** ions with **voids** [42]. Eisenberg puns PNP for Poisson-Nernst-Planck and Positive-Negative-Positive semiconductor transistors to emphasize nonequilibrium flows of ions through ion channels as life's transistors [43]. Ions in classical PB and PNP theories are treated as volumeless point charges like the 'ions' of semiconductors, namely holes and electrons in semiconductor electronics [44–55]. Water molecules are treated as a dielectric medium (constant) without volumes either. However, advanced technologies in ion channel experiments [56,57] and material science [33,34,58] have raised many challenges for classical continuum theories to describe molecular mechanisms of ions and water (or solvents) with specific size effects in these systems at nano or atomic scale [9,12,14,16–19,30,33].

There is another important property that classical continuum theories fail to describe, namely short-range **ion-ion** or **ion-water correlations** in ion channels [8,9], charge-induced thickening and density oscillations near highly charged surfaces [14], correlation-induced charge inversion on macroions (DNA, actin, lipid membranes, colloidal particles) [59], the phase structure of plasma and polar fluids [60], colloidal charge renormalization [60], etc. Several other properties related to correlations such as the dielectric response of electrolytes solutions and the polarization of water in various conditions or external fields are usually modeled differently from the correlation perspective [61–63].

We have recently developed a molecular mean-field theory called—Poisson-Nernst-Planck-Bikerman (PNPB) theory—that can describe the size, correlation, **dielectric**, and **polarization** effects of ions and water in aqueous electrolytes at equilibrium or nonequilibrium all within a unified framework [64–76]. Water and ions in this theory can have *different shapes* and *volumes* necessarily with intermolecular voids. The theory generalizes and unifies the second-order Poisson–Bikerman equation [42] of binary ionic liquids for different-sized ions with identical steric energies [72] and the **fourth-order** differential permittivity operator in Santangelo's model of one component plasma [77] or in the Bazant, Storey, and Kornyshev theory of general nonlocal permittivity for equal-sized ions in ionic liquids [78].

Ion-ion and ion-water correlations are modeled by the permittivity operator with a correlation length that depends on the diameter of ions or water and the valence of ions of interest [78]. The fourth-order operator yields a permittivity as an *output function* of spatial variables, salt concentration, and hydration shell structure including water diameter from solving the PNPB model and thus describes the **dehydration** of ions from bath to channel pore or from bulk to charged wall, the polarization of water, and the change of permittivities of electrolyte solutions at different locations in response to different configurations and conditions. Water densities also change with configurations and conditions.

The fourth-order operator introduces correlations into the mean-field equations so they can deal more realistically with real systems in which the correlations are of the greatest importance. A remark should be made here that simulations containing only particles do not automatically deal with correlations better than mean-field theories with fourth-order operators like this. It is not at all clear that simulations widely done in biophysics actually compute correlations well. Indeed, it is difficult to see how simulations that use conventions to approximate the electric field, and periodic boundary conditions to approximate macroscopic systems could deal with correlations correctly. The dearth of direct checks of the role of periodic boundary conditions, and of the accuracy of the conventional treatment of electrostatics, does little to assuage these concerns. The detailed direct checks found necessary in computational electronics are not easily found in simulations of ions in electrolyte solutions (see Chap. 6, particularly Figures 6.34–35 of [55] for some details that are found to be necessary in the simulations of computational electronics).

It is important to reiterate the obvious. Our model includes water as a molecule and depends on the hydration structure around ions. Our model uses partial differential equations (PDEs) to describe these essentially discrete properties of ionic solutions, and uses the physical parameters of individual atoms and water molecules, NOT just their mean-field description. This use of PDEs to describe inherently discrete processes is hardly new: most of probability theory [79,80] and the entire theory of wave equations, including the wave equation of the electron called the Schrödinger equation [81], treat discrete processes the same way, using PDEs that measure (in probability theory) the underlying discrete system, or represent it exactly as the discrete solutions of a continuum PDE (e.g., the Schrödinger equation describing a hydrogen atom).

The most important contribution of our work is to include water as discrete molecules by using **Fermi distributions** [82] of classical particles having excluded volumes with interstitial voids. We show that the treatment of water as finite size molecules requires, as a matter of mathematics, not physics, the existence of voids. This is demonstrated by mathematics and simple ways to compute the voids and their role are presented. These Fermi-like distributions yield **saturation** of all particles (ions and water) even under mathematically infinite large external fields. These distributions also satisfy **mass conservation** in the region of interest such as channel pores, which classical theories fail to describe as well. This Fermi distribution of classical particles obeying volume exclusion is reminiscent of the Fermi distribution of identical particles obeying the Pauli exclusion principle [83] in quantum mechanics.

We also introduce a new concept of distance-dependent potential between non-bonded particles for different-sized particles similar to the electric potential for different-charged ions and name it the **steric potential**. The void distribution function describes the van der Waals potential [84] of paired particles [85,86] in the system in a mean-field sense. The steric potential can be written as a distribution function of voids, emphasizing the crucial role of voids in our theory. The specific sizes of particles and the distance-dependent steric potential allow us to calculate steric energies at the **atomic scale**. Using Coulomb's law allows calculation of electric energies at the atomic scale as well. Therefore, our theory applies to biological or chemical systems with explicit atomic structures, as well as classical mean-field representations of bulk solutions, for example. We have shown that solving the PNPB model in different continuum and molecular domains yields *self-consistent* electric and steric potentials in many examples of biological ion channels or chemical systems in [64–76]. The theory is also *consistent* with classical theories in the sense that its model converges to the corresponding classical one when the volume of all particles and the correlation length tend to zero, i.e., steric and correlation effects vanish asymptotically to classical cases.

In this review article, we explain the above bold-face terms in detail and compare them with those of earlier theories in a precise but limited way. The precision means that we display explicitly, to the best of our ability, the significant differences between analogous concepts in our theory and previous treatments. It is obviously impossible to do complete comparisons in this vast and formidable field. No doubt we are ignorant of significant relevant papers. We apologize to those inadvertently slighted and ask them to help us remedy our oversight. The remaining of this article consists as follows.

Section 2 describes the physical meaning of Fermi distributions and the steric potential of ions and water with excluded volumes. We also explain the differences between Fermi and Boltzmann distributions in the context of statistical thermodynamics.

Section 3 unifies Fermi distributions and correlations into the simple and concise 4th-order Poisson–Bikerman (4PBik) equation. The simplicity refers to the correlation length being the only empirical parameter in the equation. The conciseness means that the fourth-order differential operator can describe the complex and correlated properties of ion-ion and ion-water interactions, polarization of water, and dielectric response of electrolytes solutions all in a single model setting.

Section 4 presents a Gibbs free energy functional for the 4PBik equation. We show that minimization of the functional yields the equation and Fermi distributions that reduce to Boltzmann distributions when the volumes of particles vanish in limiting case. This functional is critical to explain a major shortcoming of earlier modified PB models that cannot yield Boltzmann distributions in the limit. These models are thus not consistent with classical theories and may poorly estimate steric energies and other physical properties due to their coarse approximation of size effects.

Section 5 generalizes the 4PBik equation to the PNPB model to describe flow dynamics of ions and water in the system subject to external fields. The most important feature in this section is the introduction of the steric potential to the classical Nernst–Planck equation. Electric and steric potentials describe the dynamic **charge/space competition** between ions and water. We also show that the PNPB model reduces to the 4PBik equation at equilibrium.

Section 6 presents a generalized Debye-Hückel theory from the 4PBik equation for thermodynamic modeling. The theory yields an equation of state that analytically models ion activities in all types of binary and multi-component electrolyte solutions over wide ranges of concentration, temperature, and pressure. It is also useful to study the size, correlation, dielectric, and polarization effects in a clear comparison with those ignoring these effects.

Section 7 discusses numerical methods for solving the PNPB model that is highly nonlinear and complex when coupled with the electrical field generated by protein charges in ion channels, for example. It is very challenging to numerically solve the model with tolerable accuracy in 3D protein structures that generate extremely large electric field, e.g., 0.1 V in 1 Angstrom, in parts of the molecule of great biological importance where crowded charges directly control biological function, in the same sense that a gas pedal controls the speed of a car.

Section 8 demonstrates the usefulness of the PNPB theory for a wide range of biological and chemical systems, where the steric and correlation effects are of importance. We choose a few examples of these systems, namely electric double layers, ion activities, and biological ion channels.

Section 9 summarizes this review with some concluding remarks.

2. Fermi Distributions and Steric Potential

The total volume of an aqueous electrolyte system with K species of ions in a solvent domain Ω_s is

$$V = \sum_{i=1}^{K+1} v_i N_i + V_{K+2}, \tag{1}$$

where $K+1$ and $K+2$ denote water and voids, respectively, v_i is the volume of each species i particle, N_i is the total number of species i particles, and V_{K+2} is the total volume of all the voids [68]. The volume of each particle v_i will play a central role in our analysis, as well that the limit v_i goes to zero. This limit defines the solution of point particles of classical PB and PNP theory. We must include the voids as a separate species if we treat ions and water having volumes in a model. This necessity can be proven by mathematics (see below). It is also apparent to any who try to compute a model of this type with molecular water, as it was to us [68].

Dividing the volume Equation (1) in bulk conditions by V, we get the bulk volume fraction of voids

$$\Gamma^B = 1 - \sum_{i=1}^{K+1} v_i C_i^B = \frac{V_{K+2}}{V}, \tag{2}$$

where $C_i^B = \frac{N_i}{V}$ are bulk concentrations. If the system is spatially inhomogeneous with variable electric or steric fields, as in realistic systems, the constants C_i^B then change to functions $C_i(\mathbf{r})$ and so does Γ^B to a void volume function

$$\Gamma(\mathbf{r}) = 1 - \sum_{i=1}^{K+1} v_i C_i(\mathbf{r}). \tag{3}$$

We define the concentrations of particles (i.e., the distribution functions of the number density) in Ω_s [72] as

$$C_i(\mathbf{r}) = C_i^B \exp\left(-\beta_i \phi(\mathbf{r}) + \frac{v_i}{v_0} S^{trc}(\mathbf{r})\right), \quad S^{trc}(\mathbf{r}) = \ln \frac{\Gamma(\mathbf{r})}{\Gamma^B}, \tag{4}$$

where $\phi(\mathbf{r})$ is an electric potential, $S^{trc}(\mathbf{r})$ is called a *steric potential*, $\beta_i = q_i/k_BT$ with q_i being the charge on species i particles and $q_{K+1} = 0$, k_B is the Boltzmann constant, T is an absolute temperature, and $v_0 = \left(\sum_{i=1}^{K+1} v_i\right)/(K+1)$ is an average volume. The following inequalities

$$\begin{aligned} C_i(\mathbf{r}) &= C_i^B \exp\left(-\beta_i\phi(\mathbf{r})\right) \left[\frac{\Gamma(\mathbf{r})}{\Gamma^B}\right]^{v_i/v_0} = \alpha_i \left[1 - \sum_{j=1}^{K+1} v_j C_j(\mathbf{r})\right]^{v_i/v_0} \\ &= \alpha_i \left[1 - v_i C_i(\mathbf{r}) - \sum_{j=1, j\neq i}^{K+1} v_j C_j(\mathbf{r})\right]^{v_i/v_0} < \alpha_i \left[1 - v_i C_i(\mathbf{r})\right]^{v_i/v_0} \\ &\leq \alpha_i \left[1 - \frac{v_i^2}{v_0} C_i(\mathbf{r})\right] \text{ if } v_i/v_0 \leq 1, \text{ by Bernoulli's inequality,} \end{aligned} \tag{5}$$

$$\begin{aligned} C_i(\mathbf{r}) &< \alpha_i \left[1 - v_i C_i(\mathbf{r})\right]^{v_i/v_0} = \alpha_i \left[1 - v_i C_i(\mathbf{r})\right]^{\gamma} \left[1 - v_i C_i(\mathbf{r})\right]^{v_i/v_0 - \gamma} \\ &< \alpha_i \left[1 - v_i C_i(\mathbf{r})\right] \left[1 - \left(v_i/v_0 - \gamma\right) v_i C_i(\mathbf{r})\right] \\ &< \alpha_i \left[1 - v_i C_i(\mathbf{r})\right] \text{ if } v_i/v_0 > 1, \end{aligned} \tag{6}$$

imply that the distributions are of Fermi-like type [87]

$$C_i(\mathbf{r}) < \lim_{\alpha_i \to \infty} \frac{\alpha_i}{1 + \alpha_i v_i^2/v_0} < \frac{v_0}{v_i^2} \text{ if } v_i/v_0 \leq 1, \tag{7}$$

$$C_i(\mathbf{r}) < \lim_{\alpha_i \to \infty} \frac{\alpha_i}{1 + \alpha_i v_i} < \frac{1}{v_i} \text{ if } v_i/v_0 > 1, \tag{8}$$

i.e., $C_i(\mathbf{r})$ cannot exceed the maximum value $1/v_i^2$ or $1/v_i$ for any arbitrary (or even infinite) potential $\phi(\mathbf{r})$ at any location $\mathbf{r}$ in the domain Ω_s, where $i = 1, \cdots, K+1$, $\alpha_i = C_i^B \exp\left(-\beta_i\phi(\mathbf{r})\right) / \left(\Gamma^B\right)^{v_i/v_0} > 0$, $0 < v_i/v_0 - \gamma < 1$, and $\gamma \geq 1$.

The classical Boltzmann distribution appears if all particles are treated as volumeless points, i.e., $v_i = 0$ and $\Gamma(\mathbf{r}) = \Gamma^B = 1$. The classical Boltzmann distribution may produce an infinite concentration $C_i(\mathbf{r}) \to \infty$ in crowded conditions when $-\beta_i\phi(\mathbf{r}) \to \infty$, close to charged surfaces for example, which is physically impossible [64–66]. This is a major, even crippling deficiency of PB theory for modeling a system with strong local electric fields or interactions. The difficulty in the application of classical Boltzmann distributions to saturating systems has been avoided in the physiological literature (apparently starting with Hodgkin, Huxley, and Katz [88]) by redefining the Boltzmann distribution to deal with systems that can only exist in two states. This redefinition has been vital to physiological research and is used in hundreds of papers [89,90], but confusion results when the physiologists' saturating two-state Boltzmann is not kept distinct from the unsaturating Boltzmann distribution of statistical mechanics [91].

It should be clearly understood that as beautiful as is Hodgkin's derivation it begs the question of what physics creates and maintains two states. Indeed, it is not clear how one can define the word state in a usefully unique way in a protein of enormous molecular weight with motions covering the scale from femtoseconds to seconds.

The steric potential $S^{trc}(\mathbf{r})$ in Equation (4) first introduced in [64] is an entropic measure of crowding or emptiness of particles at $\mathbf{r}$. If $\phi(\mathbf{r}) = 0$ and $C_i(\mathbf{r}) = C_i^B$ then $S^{trc}(\mathbf{r}) = 0$. The factor v_i/v_0 shows that the steric energy $\frac{-v_i}{v_0} S^{trc}(\mathbf{r}) k_BT$ of a type i particle at $\mathbf{r}$ depends not only on the steric potential $S^{trc}(\mathbf{r})$ but also on its volume v_i similar to the electric energy $\beta_i\phi(\mathbf{r})k_BT$ depending on both $\phi(\mathbf{r})$ and q_i [72]. The steric potential $S^{trc}(\mathbf{r})$ is especially relevant to determining selectivity of specific ions by certain biological ion channels [65,66,68,70,72].

In this mean-field Fermi distribution, it is impossible for a volume v_i to be completely filled with particles, i.e., it is impossible to have $v_iC_i(\mathbf{r}) = 1$ (and thus $\Gamma(\mathbf{r}) = 0$) since that would make

$S^{trc}(\mathbf{r}) = -\infty$ and hence $C_i(\mathbf{r}) = 0$, a contradiction. Therefore, *we must include the voids as a separate species if we treat ions and water having volumes in a model* for which $C_i(\mathbf{r}) < 1/v_i$ and $\Gamma(\mathbf{r}) \neq 0$ for all $i = 1, \cdots, K+1$ and $\mathbf{r} \in \Omega_s$. This is a critical property distinguishing our theory from others that do not consider water as a molecule with volume and so do not have to consider voids. We shall elaborate this property in Section 4.

Our theory is consistent with the classical theory of van der Waals in molecular physics, which describes nonbond interactions between any pair of atoms as a distance-dependent potential such as the Lennard-Jones (L-J) potential that cannot have zero distance between the pair [85,86]. Indeed, the steric potential $S^{trc}(\mathbf{r})$ can be written as a function of the volume of all molecular species (of course, including water as well as ions). Classical extensions of van der Waals theories often use this variable, but seem not to mention the existence or importance of voids.

The steric potential $S^{trc}(\mathbf{r})$ lumps all van der Waals potential energies of paired particles in a mean-field sense. It is an approximation of L-J potentials that describe local variations of L-J distances (and thus empty voids) between any pair of particles. L-J potentials are highly oscillatory and extremely expensive and unstable to compute numerically [66]. Calculations that involve L-J potentials [92–98], or even truncated versions of L-J potentials [99–101] must be extensively checked to be sure that results do not depend on irrelevant parameters. Any description that uses L-J potentials has a serious problem specifying the combining rule. The details of the combining rule *directly* change predictions of effects of different ions (selectivity) and so predictions depend on the reliability of data that determines the combining rule and its parameters.

The steric potential does not require combining rules. Since we consider specific sizes of ions and water with voids, the steric potential is valid on the *atomic scale* of L-J potentials. It is also *consistent* with that on the *macroscopic scale* of continuum models as shown in Sections 6 and 8.

To our surprise during the writing of this article, we found Equation (2) in Bikerman's 1942 paper [42] is *exactly* the *same* as Equation (4) for a special case of binary ionic liquids with the identical steric energies of different-sized ions, i.e., the factor $v_i/v_0 = 1$ in (4). The steric potential in Equation (4) is however not explicitly expressed in Bikerman's paper. Therefore, Bikerman's concentration function is a Fermi distribution, a generic term used in statistical mechanics. We do NOT use exactly the Fermi distribution as Fermi derived in 1926 for identical particles now called fermions in quantum mechanics. So it is both more precise and historically correct to use the name "Poisson–Bikerman" equation for finite-sized ions as a generalization of the Poisson–Boltzmann equation for volumeless ions in electrochemical and bioelectric systems.

As noted by Bazant et al. in their review paper [14], Bikerman's paper has been poorly cited in the literature until recently. In our intensive and extensive study of the literature since 2013 [64], we have never found any paper specifically using Bikerman's formula as Equation (4), although of course there may be an instance we have not found. We thus now change the term "Poisson-Fermi" used in our earlier papers to "Poisson-Bikerman" in honor of Bikerman's brilliant work. We present here mathematical as well as physical justifications of a very general treatment of different-sized ions and water molecules in the mean-field framework based on Bikerman's pioneer work.

3. Fourth-Order Poisson-Bikerman Equation and Correlations

Electrolytes have been treated mostly in the tradition of physical chemistry of isolated systems that proved so remarkably successful in understanding the properties of ideal gases in atomic detail, long before the theory of partial differential equations, let alone numerical computing was developed. Most applications of ionic solutions however involve systems that are not at all isolated. Rather, most practical systems include electrodes to deliver current and control potential, and reservoirs to manipulate the concentrations and types of ions in the solution. Indeed, all biology occurs in ionic solutions and nearly all of biology involves large flows. It is necessary then to extend classical approaches so they deal with external electric fields and other boundary conditions and allow flow so the theory can give useful results that are applicable to most actual systems.

When the electrolyte system in Ω_s is subject to external fields such as applied voltages, surface charges, and concentration gradients on the boundary $\partial\Omega_s$, the electric field $\mathbf{E}(\mathbf{r})$ of the system, the displacement field $\mathbf{D}(\mathbf{r})$ of free ions, and the polarization field $\mathbf{P}(\mathbf{r})$ of water are generated at all $\mathbf{r}$ in Ω_s. In Maxwell's theory [102,103], these fields form a constitutive relation

$$\mathbf{D}(\mathbf{r}) = \epsilon_0\mathbf{E}(\mathbf{r}) + \mathbf{P}(\mathbf{r}) \tag{9}$$

and the displacement field satisfies

$$\nabla \cdot \mathbf{D}(\mathbf{r}) = \rho_{ion}(\mathbf{r}) = \sum_{i=1}^{K} q_i C_i(\mathbf{r}), \tag{10}$$

where ϵ_0 is the vacuum permittivity, $\rho_{ion}(\mathbf{r})$ is the charge density of ions, and $C_i(\mathbf{r})$ are the concentrations defined in (4). See [104] for a modern formulation of Maxwell's theory applicable wherever the Bohm version of quantum mechanics applies [105,106].

The electric field $\mathbf{E}(\mathbf{r})$ is thus screened by water (Bjerrum screening) and ions (Debye screening) in a correlated manner that is usually characterized by a correlation length l_c [77,78,107]. The screened force between two charges in ionic solutions (at $\mathbf{r}$ and $\mathbf{r}'$ in Ω_s) has been studied extensively in classical field theory and is often described by the van der Waals potential kernel [71,72,84,107,108]

$$W(\mathbf{r}-\mathbf{r}') = \frac{e^{-|\mathbf{r}-\mathbf{r}'|/l_c}}{|\mathbf{r}-\mathbf{r}'|/l_c} \tag{11}$$

that satisfies the Laplace-Poisson equation [108]

$$-\Delta W(\mathbf{r}-\mathbf{r}') + \frac{1}{l_c^2}W(\mathbf{r}-\mathbf{r}') = \delta(\mathbf{r}-\mathbf{r}'),\ \mathbf{r}, \mathbf{r}' \in R^3 \tag{12}$$

in the whole space R^3, where $\Delta = \nabla \cdot \nabla = \nabla^2$ is the Laplace operator with respect to $\mathbf{r}$ and $\delta(\mathbf{r}-\mathbf{r}')$ is the Dirac delta function at $\mathbf{r}'$.

The potential $\widetilde{\phi}(\mathbf{r})$ defined in

$$\mathbf{D}(\mathbf{r}) = -\epsilon_s\nabla\widetilde{\phi}(\mathbf{r}) \tag{13}$$

describes an electric potential of free ions [72,107] that are correlated only by the mean electric field according to the Poisson equation

$$-\epsilon_s\Delta\widetilde{\phi}(\mathbf{r}) = \rho_{ion}(\mathbf{r}), \tag{14}$$

a second-order partial differential equation, where $\epsilon_s = \epsilon_w\epsilon_0$ and ϵ_w is the dielectric constant of water. This potential does not account for correlation energies between individual ions or between ion and polarized water in high field or crowded conditions under which the size and valence of ions and the polarization of water play significant roles [17,65–68,77,78,107].

The correlations implicit in Maxwell's equations are of the mean-field and can be summarized by the statement that current is conserved perfectly and universally on all scales that the Maxwell equations are valid, where current includes the term $\epsilon_0\frac{\partial\mathbf{E}(\mathbf{r},t)}{\partial t}$. This term allows the Maxwell equations to describe the propagation of light through a vacuum, and it allows charge to be relativistically invariant, i.e., independent of velocity unlike mass, length, and time all of which vary dramatically as velocities approach the speed of light [104,106].

We introduce the *correlated* electric potential

$$\phi(\mathbf{r}) = \int_{R^3} \frac{1}{l_c^2}W(\mathbf{r}-\mathbf{r}')\widetilde{\phi}(\mathbf{r}')d\mathbf{r}' \tag{15}$$

in [72] as a convolution of the displacement potential $\widetilde{\phi}(\mathbf{r}')$ with $W(\mathbf{r}-\mathbf{r}')$ to deal with the correlation and polarization effects in electrolyte solutions. However, it would be too expensive to calculate $\phi(\mathbf{r})$ using (15). Multiplying (12) by $\widetilde{\phi}(\mathbf{r}')$ and then integrating over R^3 with respect to $\mathbf{r}'$ [71], we obtain

$$-l_c^2\Delta\phi(\mathbf{r})+\phi(\mathbf{r})=\widetilde{\phi}(\mathbf{r}) \tag{16}$$

a Laplace-Poisson equation [107,108] that satisfies (15) in the whole unbounded space R^3 with the boundary conditions $\phi(\mathbf{r}) = \widetilde{\phi}(\mathbf{r}) = 0$ at infinity. From (14) and (16), we obtain the *4th-order* Poisson–Bikerman equation

$$\epsilon_s\left[l_c^2\Delta-1\right]\Delta\phi(\mathbf{r})=\rho_{ion}(\mathbf{r}),\ \mathbf{r}\in\Omega_s, \tag{17}$$

a PDE that is an approximation of (16) in a bounded domain $\Omega_s \subset R^3$ with suitable boundary conditions (see below) of $\phi(\mathbf{r})$ on $\partial\Omega_s$. We can thus use (9) to find the polarization field

$$\mathbf{P}(\mathbf{r})=\epsilon_s l_c^2\nabla(\Delta\phi(\mathbf{r}))-(\epsilon_w-1)\epsilon_0\nabla\phi(\mathbf{r}) \tag{18}$$

with $\mathbf{E}(\mathbf{r}) = -\nabla\phi(\mathbf{r})$. If $l_c = 0$, we recover the standard Poisson Equation (14) and the standard polarization $\mathbf{P} = \epsilon_0(\epsilon_w-1)\mathbf{E}$ with the electric susceptibility $\epsilon_w - 1$ (and thus the dielectric constant ϵ_w) if water is treated as a time independent, isotropic, and linear dielectric medium [103]. In this case, the field relation $\mathbf{D} = \epsilon_w\epsilon_0\mathbf{E}$ with the scalar constant permittivity $\epsilon_s\epsilon_0$ is an approximation of the exact relation (9) due to the simplification of the dielectric responses of the medium material to the electric field $\mathbf{E}$ [109–111].

The exponential van der Waals potential $W(\mathbf{r}-\mathbf{r}') = \frac{e^{-|\mathbf{r}-\mathbf{r}'|/l_c}}{|\mathbf{r}-\mathbf{r}'|/l_c}$ [84] is called the Yukawa [112] potential in [71,72] and usually in physics, which is an anachronism [108,113]. Van der Waals derived this potential in his theory of capillarity based on the proposition that the intermolecular potential of liquids and gases is shorter-ranged, but much stronger than Coulomb's electric potential [108]. Ornstein and Zernike (OZ) introduced short- (direct) and long-ranged (indirect) correlation functions in their critical point theory [114]. There are three important properties of the van der Waals potential: (i) it satisfies the Laplace-Poisson Equation (12), (ii) it generates the same functional form for short- and long-ranged correlations in the OZ theory, and (iii) it solves van der Waals's problem for the intermolecular potential [108].

Therefore, the potential $\phi(\mathbf{r})$ in (15) includes *correlation* energies of *ion-ion* and *ion-water* interactions in *short* as well as *long* ranges in our system. The *correlation length* l_c can be derived from the OZ equation, see Equation (13) in [108], but the derivation is not very useful. The correlation length becomes an unknown functional of $\rho_{ion}(\mathbf{r})$ in (10) and the OZ direct correlation function, and is hence usually chosen as an empirical parameter to fit experimental, molecular dynamics (MD), or Monte Carlo (MC) data [14,64–73,75,77,78,107]. It seems clear that it would be useful to have a theory that showed the dependence of correlation length on ion composition and concentration, and other parameters.

There are several approaches to fourth-order Poisson-Boltzmann equations for modeling ion-ion and ion-water correlations from different perspectives of physics [71,77,78,115,116]. In [77], a decomposed kernel acts on a charge density of counterions in a binary liquid without volumes and water (ion-ion correlations) in contrast to the potential $\widetilde{\phi}(\mathbf{r})$ in (15) that is generated by different-sized ions and water with voids in (14) (ion-ion and ion-water correlations in a multi-component aqueous electrolyte). The kernel consists of short-range (of van der Waals type) and long-range components from a decomposition of Coulomb's interactions. In [78], the kernel is a general nonlocal kernel that acts on a charge density of equal-sized ions in a binary liquid without water (ion-ion correlations). The kernel is a series expansion of the gradient operator ∇ and thus can yield not only a fourth-order PB but even higher-order PDEs. The fourth-order PB is the first-order approximation of the energy expansion that converges only with small wavenumbers k in Fourier frequency domain for the dielectric response of ionic liquids [78].

Derived from the framework of nonlocal electrostatics for modeling the dielectric properties of water in [107], the kernel acting on $\widetilde{\phi}(\mathbf{r})$ in [71] (ion-ion and ion-water correlations) consists of a van der Waals function and the Dirac delta function that correspond to the limiting cases $k = 0$ and $k = \infty$, respectively. In [115], a system of three PDEs derived from electrostatics and thermodynamic pressure has electric potential and concentration gradients of equal-sized cations and anions in a binary fluid as three unknown functions. Linearization and simplification of the nonlinear system can yield a linear fourth-order PB (ion-ion correlations). In [116], the fourth-order PB is derived from a free energy functional that models ion-ion correlations in a binary liquid using volume-fraction functions of equal-sized cations and anions with two additional parameters associated with the interaction energies of these two functions and their gradients.

The *dielectric operator* $\epsilon_s\left(l_c^2\Delta - 1\right)$ in (17) describes changes in dielectric response of water with salt concentrations (ion-water correlations), ion-ion correlations, and water polarizations all via the mean-field charge density function $\rho_{ion}(\mathbf{r})$ provided that we can solve (4) and (17) for a consistent potential function $\phi(\mathbf{r})$. Therefore, the operator (a mapping) depends not only on ion and water concentrations (C_i^B for all arbitrary species $i = 1, \cdots, K+1$ of particles with any arbitrary shapes and volumes) but also on the location $\mathbf{r}$ and the voids at $\mathbf{r}$. The operator thus produces a *dielectric function* $\widehat{\epsilon}(\mathbf{r},C_i^B)$ as an *output* from the solution $\phi(\mathbf{r})$ that satisfies the 4PBik (17) that saturates as a function of concentration (4), as we shall repeatedly emphasize. This dielectric function $\widehat{\epsilon}(\mathbf{r},C_i^B)$ is not an additional model for $\widetilde{\epsilon}(\mathbf{r})$, $\widetilde{\epsilon}(k)$, or $\widetilde{\epsilon}(C_i^B)$ as it often is in other models in the literature [62,63,117–127]. Here the dielectric function is an output, as we have stated.

The 4PBik Equation (17) with (4) is a very general model using only one extra parameter l_c in the fourth-order operator to include many physical properties ignored by the classical Poisson-Boltzmann equation. We shall illustrate these properties of our model in Section 8.

4. Generalized Gibbs Free Energy Functional

To generalize the Gibbs free energy functional for Boltzmann distributions that satisfy the classical Poisson–Boltzmann equation [3,128,129], we introduce a functional in [72] for saturating Fermi distributions (4) that satisfy the 4th-order Poisson–Bikerman Equation (17)

$$F(\mathbf{C}) = F_{el}(\mathbf{C}) + F_{en}(\mathbf{C}), \tag{19}$$

$$F_{el}(\mathbf{C}) = \frac{1}{2}\int_{\Omega_s} \rho_{ion}(\mathbf{r})L^{-1}\rho_{ion}(\mathbf{r})d\mathbf{r}, \tag{20}$$

$$F_{en}(\mathbf{C}) = k_BT\int_{\Omega_s}\left\{\sum_{i=1}^{K+1} C_i(\mathbf{r})\left(\ln\frac{C_i(\mathbf{r})}{C_i^B} - 1\right) + \frac{\Gamma(\mathbf{r})}{v_0}\left(\ln\frac{\Gamma(\mathbf{r})}{\Gamma^B} - 1\right)\right\}d\mathbf{r}, \tag{21}$$

where $F_{el}(\mathbf{C})$ is an electrostatic functional, $F_{en}(\mathbf{C})$ is an entropy functional, $\mathbf{C} = (C_1(\mathbf{r}), C_2(\mathbf{r}), \cdots, C_{K+1}(\mathbf{r}))$, and L^{-1} is the inverse of the self-adjoint positive linear operator $L = \epsilon_s\left(l_c^2\Delta - 1\right)\Delta$ [71] in (17), i.e., $L\phi(\mathbf{r}) = \rho_{ion}(\mathbf{r})$. $\mathbf{C}$ is a 'concentration vector' that specifies the number density, i.e., concentration of each species in the ionic solution, including water. $\mathbf{C}$ plays a central role in any theory of ionic solutions because it specifies the main property of a solution, namely its composition.

Taking the variations of $F(\mathbf{C})$ at $C_i(\mathbf{r})$, we have

$$\frac{\delta F(\mathbf{C})}{\delta C_i} = \int_{\Omega_s}\left\{k_BT\left[\ln\frac{C_i(\mathbf{r})}{C_i^B} - \frac{v_i}{v_0}\ln\frac{\Gamma(\mathbf{r})}{\Gamma^B}\right] + \frac{1}{2}\left(q_iL^{-1}\rho_{ion}(\mathbf{r}) + \rho_{ion}(\mathbf{r})L^{-1}q_i\right)\right\}d\mathbf{r},$$

$$\frac{1}{2}\left(q_iL^{-1}\rho_{ion}(\mathbf{r}) + \rho_{ion}(\mathbf{r})L^{-1}q_i\right) = q_i\phi(\mathbf{r}),$$

$$\frac{\delta F(\mathbf{C})}{\delta C_i} = 0 \Rightarrow k_BT\left[\ln\frac{C_i(\mathbf{r})}{C_i^B} - \frac{v_i}{v_0}\ln\frac{\Gamma(\mathbf{r})}{\Gamma^B}\right] + q_i\phi(\mathbf{r}) = 0 \tag{22}$$

that yields the saturating Fermi distributions in (4) for all $i = 1, \cdots, K+1$. Moreover, we have

$$\frac{\delta^2 F(\mathbf{C})}{\delta C_i^2} = \int_{\Omega_s} \left\{ k_B T \left[\frac{1}{C_i(\mathbf{r})} + \frac{v_i^2}{v_0} \frac{\Gamma^B}{\Gamma(\mathbf{r})} \right] + q_i^2 L^{-1} C_i \right\} d\mathbf{r} > 0 \tag{23}$$

implying that the saturating Fermi distribution vector $\mathbf{C}$ is a unique minimizer of the functional $F(\mathbf{C})$.

The Gibbs-Bikerman free energy functional $F(\mathbf{C})$ has two important properties. First, its electrostatic part $F_{el}(\mathbf{C})$ is defined in terms of the composition vector $\mathbf{C}$ only. It depends only on concentrations and nothing else. If an electrostatic functional $\widetilde{F}_{el}(\widetilde{\phi}(\mathbf{r}))$ is defined in terms of $|\nabla\widetilde{\phi}(\mathbf{r})|^2$ for the PB equation [64,78,124,130–137], the corresponding concentration vector $\widetilde{\mathbf{C}}$ and the potential $\widetilde{\phi}(\mathbf{r})$ do *not* minimize the corresponding functional $\widetilde{F}(\widetilde{\mathbf{C}}, \widetilde{\phi}(\mathbf{r}))$ [128,129], i.e., $\widetilde{F}$ is not a Gibbs free energy functional [3,128]. Second, the limit of its entropic part

$$\lim_{v_i \to 0} F_{en}(\mathbf{C}) = k_B T \int_{\Omega_s} \sum_{i=1}^{K+1} C_i^0(\mathbf{r}) \left(\ln \frac{C_i^0(\mathbf{r})}{C_i^B} - 1 \right) d\mathbf{r} \tag{24}$$

exists (F_{en} converges) when the volume v_i tends to zero for all $i = 1, \cdots, K+1$. This implies that all ionic species have Boltzmann distributions $C_i^0(\mathbf{r}) = C_i^B \exp(-\beta_i \phi(\mathbf{r}))$, $i = 1, \cdots, K$, the water concentration $C_{K+1}^0(\mathbf{r}) = C_{K+1}^B$ is a constant, and the void fraction $\Gamma(\mathbf{r}) = \Gamma^B = 1$ since all particles are volumeless in PB theory. Therefore, the 4PBik model (4) and (17) is physically and mathematically *consistent* with the classical PB model in the limiting case when we ignore the steric ($v_i = 0$) and correlation ($l_c = 0$) effects.

There are many shortcomings of the lattice approach [138] frequently used to account for steric effects in lattice-based PB models [14,61,64,78,124,129,133–136,139–141]. For example, (i) it assumes equal-sized ions and thus cannot distinguish non-uniform particles as in (1), (ii) its effective ion size needs to be unrealistically large to fit data [14], (iii) its correction over Boltzmann's point charge approach appears only at high surface charges [125], (iv) its pressure term diverges very weakly (is greatly underestimated) at close packing [142], and (v) its entropy functional may diverge as the volume of ions tends to zero, i.e., the corresponding *lattice-based PB* model is *not* physically and mathematically *consistent* with the classical PB model in the limiting case [66].

The importance of the restriction in Point (i) is hard to overstate. Almost all the interesting properties of ionic solutions arise because of their selectivity (as it is called in biology) or specificity between species, and those different properties arise in large measure because of the different diameters of the ions. The equal diameter case is dull and degenerate.

Point (v) is a critical problem that is closely related to Points (ii)–(iv). The divergence is obvious for an entropy term $\widetilde{F}_{en}$ in Equation (2) in [133] as

$$\lim_{v \to 0} \widetilde{F}_{en} = \lim_{v \to 0} \sum_{i=1}^{K} \widetilde{C}_i(\mathbf{r}) \ln \left(v \widetilde{C}_i(\mathbf{r}) \right) = -\infty, \tag{25}$$

which also appears in [61,64,78,124,129,133–136,139–141]. It is impossible to derive Boltzmann distributions $\widetilde{C}_i(\mathbf{r}) = C_i^B \exp(-\beta_i \widetilde{\phi}(\mathbf{r}))$ from $\widetilde{F}_{en}$ as $v \to 0$ without extra assumptions, see (2.6) in [129], for example. In fact, the assumption (2.6), i.e., $v\widetilde{C}_i(\mathbf{r}) > 0$, actually forbids us from taking v to the limit zero.

Our derivation of $F_{en}(\mathbf{C})$ does not employ any lattice models but simply uses the exact volume Equation (1). Our theory should not be classified then as a lattice model as sometimes is the case, at least in informal discussions. The void function $\Gamma(\mathbf{r})$ is an analytical generalization of the void fraction $1 - \Phi$ in (20) in [14] with all volume parameters v_i (including the bulk fraction Γ^B) being physical instead of empirical as Φ. The excess chemical potential in [14] is $-k_B T \ln(1 - \Phi)$ whereas ours is $F_{en}(\mathbf{C})$ in (21).

These expressions are different in important respects. Our model is not a lattice-based model because its differences are crucial both mathematically and physically. Indeed, the lattice-based model is in a certain sense internally inconsistent with classical statistical mechanics since a fundamental result of classical statistical mechanics $v\widetilde{C}_i(\mathbf{r}) > 0$ prevents the model from satisfying the classical imperative of the Boltzmann distribution in the limit of zero v.

The Langmuir-type distribution

$$C_i(x) = \frac{C_i^B \exp\left(-\beta_i \phi(x)\right)}{1 + \sum_{j=1}^{K} \frac{C_j^B}{C_j^{\max}} \left(\exp\left(-\beta_j \phi(x)\right) - 1\right)} \tag{26}$$

of different-sized ions (without water) proposed in [125] also reduces to a Boltzmann distribution as $v_j \to 0$, $\forall j$, where $C_j^{\max} = p/v_j$ and $p \leq 1$ is a packing parameter. This distribution saturates and thus is of Fermi type, i.e., $C_i(x) \leq C_i^{\max}$ and $v_i C_i(x) \leq 1$. The entropy term $-\ln\left(1 + \sum_{j=1}^{K} \frac{C_j^B}{C_j^{\max}} \left(\exp\left(-\beta_j \phi(x)\right) - 1\right)\right)$ does not involve voids so it is different from the $S^{trc}(\mathbf{r})$ in (4). Our distribution in (4) does not need any packing parameters and satisfies $v_i C_i(\mathbf{r}) < 1$.

5. Poisson-Nernst-Planck-Bikerman Model of Saturating Phenomena

For nonequilibrium systems, we can also generalize the classical Poisson-Nernst-Planck model [38,39,43,143,144] to the Poisson-Nernst-Planck–Bikerman model by coupling the flux density equation

$$\frac{\partial C_i(\mathbf{r}, t)}{\partial t} = -\nabla \cdot \mathbf{J}_i(\mathbf{r}, t), \ \mathbf{r} \in \Omega_s \tag{27}$$

of each particle species $i = 1, \cdots, K+1$ (including water) to the 4PBik Equation (17), where the flux density is defined as

$$\mathbf{J}_i(\mathbf{r}, t) = -D_i \left[\nabla C_i(\mathbf{r}, t) + \beta_i C_i(\mathbf{r}, t) \nabla \phi(\mathbf{r}, t) - \frac{v_i}{v_0} C_i(\mathbf{r}, t) \nabla S^{trc}(\mathbf{r}, t)\right], \tag{28}$$

D_i is the diffusion coefficient, and the time variable t is added to describe the dynamics of electric $\phi(\mathbf{r}, t)$ and steric $S^{trc}(\mathbf{r}, t)$ potentials.

The flux Equation (27) is called the Nernst-Planck-Bikerman equation because the steric potential $S^{trc}(\mathbf{r}, t)$ is introduced into the classical NP equation so it can deal with saturating phenomena including those that arise from the unequal volumes of ions and the finite volume of molecular water. The PNPB model can be extended to include hydrodynamic kinetic and potential energies in the variational treatment of energy processes (i.e., EnVarA) by Hamilton's least action and Rayleigh's dissipation principles [145,146]. We shall however consider this as a topic for future work.

At equilibrium, the net flow of each particle species is a zero vector, i.e., $\mathbf{J}_i(\mathbf{r}) = \mathbf{0}$ (in a steady state), which implies that

$$\begin{aligned} \nabla C_i(\mathbf{r}) + \beta_i C_i(\mathbf{r}) \nabla \phi(\mathbf{r}) - \frac{v_i}{v_0} C_i(\mathbf{r}) \nabla S^{trc}(\mathbf{r}) &= \mathbf{0}, \\ \nabla \left[C_i(\mathbf{r}) \exp(\beta_i \phi(\mathbf{r}) - \frac{v_i}{v_0} S^{trc}(\mathbf{r}))\right] &= \mathbf{0}, \\ C_i(\mathbf{r}) \exp(\beta_i \phi(\mathbf{r}) - \frac{v_i}{v_0} S^{trc}(\mathbf{r})) &= c_i, \end{aligned} \tag{29}$$

where the constant $c_i = C_i^B$ for $\phi(\mathbf{r}) = S^{trc}(\mathbf{r}) = 0$. Therefore, (29) = (4), i.e., the NPB Equation (27) reduces to the saturating Fermi distribution (4) as the classical NP equation reduces to the Boltzmann distribution at equilibrium.

The gradient of the steric potential $\nabla S^{trc}(\mathbf{r},t)$ in (28) represents an entropic force of vacancies exerted on particles. The negative sign in $-C_i(\mathbf{r},t)\nabla S^{trc}(\mathbf{r},t)$ means that the steric force $\nabla S^{trc}(\mathbf{r},t)$ is in the opposite direction to the diffusion force $\nabla C_i(\mathbf{r},t)$.

Larger $S^{trc}(\mathbf{r},t) = \ln \frac{\Gamma(\mathbf{r},t)}{\Gamma^B}$ implies lower pressure because the ions occupy more space (less crowded) as implied by the numerator $\Gamma(\mathbf{r},t)$. The larger the $S^{trc}(\mathbf{r},t)$ the lower pressure at the location $\mathbf{r}$, the more the entropic force (the higher pressure) pushes particles to $\mathbf{r}$ from neighboring locations. The steric force is the opposite of the diffusion force $\nabla C_i(\mathbf{r},t)$ that pushes particles away from $\mathbf{r}$ if the concentration at $\mathbf{r}$ is larger than that at neighboring locations.

Moreover, the Nernst-Einstein relationship between diffusion and mobility [9] implies that the steric flux $D_i \frac{v_i}{v_0} C_i(\mathbf{r},t)\nabla S^{trc}(\mathbf{r},t)$ is greater if the particle is more mobile. The Nernst-Einstein relationship is generalized to

$$\mu_i = v_i q_i D_i/(v_0 k_B T), \tag{30}$$

where the mobility coefficient μ_i of an ion depends on its size v_i in addition to its charge q_i. The mobility coefficient of water is $\mu_{K+1} = v_{K+1}D_{K+1}/(v_0 k_B T)$. The drift term in (28) is thus $-D_i\beta_i C_i(\mathbf{r},t)\nabla\phi(\mathbf{r},t) = -\mu_i(v_0/v_i)C_i(\mathbf{r},t)\nabla\phi(\mathbf{r},t)$.

Therefore, the gradients of electric and steric potentials ($\nabla\phi(\mathbf{r},t)$ and $\nabla S^{trc}(\mathbf{r},t)$) describe the *charge/space competition* mechanism of particles in a crowded region within a mean-field framework. Since $S^{trc}(\mathbf{r},t)$ describes the dynamics of void movements, the dynamic crowdedness (pressure) of the flow system can also be quantified. A large amount of experimental data exists concerning the dependence of diffusion coefficient on the concentration and size of solutes. Comparing our model with this data is an important topic of future work.

The motion of water molecules, i.e., the *osmosis* of water [147,148] is directly controlled by the steric potential in our model and their distributions are expressed by $C_{K+1}(\mathbf{r},t) = C^B_{K+1}\exp\left(v_{K+1}S^{trc}(\mathbf{r},t)/v_0\right)$. Nevertheless, this motion is still implicitly changed by the electric potential $\phi(\mathbf{r},t)$ via the correlated motion of ions described by other $C_j(\mathbf{r},t)$ in the void fraction function $\Gamma(\mathbf{r},t)$ and hence in the charge density $\rho_{ion}(\mathbf{r},t)$ in (17).

In summary, the PNPB model accounts for (i) the *steric* (*pressure*) effect of ions and water molecules, (ii) the *correlation* effect of crowded ions, (iii) the *screening* (*polarization*) effect of polar water, and (iv) the *charge/space competition* effect of ions and water molecules of different sizes and valences. These effects are all closely related to the interstitial voids between particles and described by two additional terms, namely the *correlation length* and the *steric potential*. The steric potential is most naturally written as a function of the volume of voids, but it can also be written as a function of the total volume of all molecules, including water and ions.

6. Generalized Debye-Hückel Theory

Thermodynamic modeling is of fundamental importance in the study of chemical and biological systems [1,6,9,11–13,16,32]. Since Debye and Hückel (DH) proposed their theory in 1923 [149] and Hückel extended it to include Born energy effects in 1925 [150], a great variety of extended DH models (equations of state) have been developed for modeling aqueous or mixed-solvent solutions over wide ranges of composition, temperature, and pressure [6,19,151–155]. Despite these intense efforts, robust thermodynamic modeling of electrolyte solutions still presents a difficult challenge for extended DH models due to an enormous number of parameters that need to be adjusted carefully and often subjectively [19,152–154,156].

It is indeed a frustrating despair (the word *frustration* on p. 11 in [16] and the word *despair* on p. 301 in [1]) that about *22,000* parameters [19] need to be extracted from the available experimental data for one temperature for combinatorial solutions of the most important 28 cations and 16 anions in salt chemistry by the Pitzer model [6], which is the most widely used DH model with unmatched precision for modeling electrolyte solutions [153]. The JESS (joint expert speciation system) is the world's largest system of thermodynamic information relating to electrolytes, reactions in aqueous media, and hydrocarbon phase equilibria [157]. The total number of Pitzer's fitting parameters in JESS is 95 [158].

By contrast, we propose in [75,76] a generalized Debye-Hückel theory from the 4PBik Equation (17) to include (i) steric effects, (ii) correlation effects, (iii) Born solvation energy, and (iv) ion hydration [159–166] that are missing in the original DH theory. The generalized theory can be used to calculate ion activities in all types of binary and multi-component solutions over wide ranges of concentration, temperature, and pressure with only 3 fitting parameters [69,73,75,76].

We briefly outline the derivation of a generalized DH equation of state and refer to [76] for more details. The activity coefficient γ_i of an ion of species i in an aqueous electrolyte solution with a total of K species of ions describes deviation of the chemical potential of the ion from ideality ($\gamma_i = 1$) [11]. The excess chemical potential $\mu_i^{ex} = k_B T \ln \gamma_i$ can be calculated by [69,167]

$$\mu_i^{ex} = \frac{1}{2} q_i \phi(\mathbf{0}) - \frac{1}{2} q_i \phi^0(\mathbf{0}), \tag{31}$$

where q_i is the charge of the hydrated ion (also denoted by i), $\phi(\mathbf{r})$ is a reaction potential [167] function of spatial variable $\mathbf{r}$ in the domain $\overline{\Omega} = \overline{\Omega}_i \cup \overline{\Omega}_{sh} \cup \overline{\Omega}_s$ shown in Figure 1, Ω_i is the spherical domain occupied by the ion i, Ω_{sh} is the hydration shell domain of the ion, Ω_s is the rest of solvent domain, $\mathbf{0}$ denotes the center (set to the origin) of the ion, and $\phi^0(\mathbf{r})$ is a potential function when the solvent domain Ω_s does not contain any ions at all with pure water only, i.e., when the solution is ideal. The radii of Ω_i and the outer boundary of Ω_{sh} are denoted by R_i^{Born} (ionic cavity radius [160]) and R_i^{sh}, respectively.

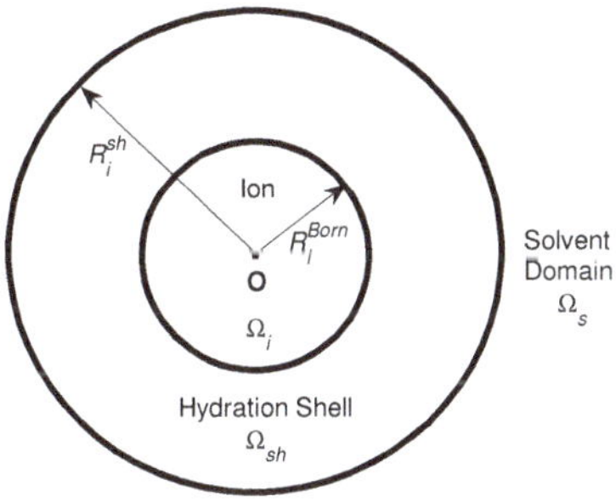

Figure 1. The model domain Ω is partitioned into the ion domain Ω_i (with radius R_i^{Born}), the hydration shell domain Ω_{sh} (with radius R_i^{sh}), and the remaining solvent domain Ω_s.

The potential function $\phi(\mathbf{r})$ can be found by solving the 4PBik Equation (17) and the Laplace equation [69,73]

$$\Delta\phi(\mathbf{r}) = 0 \text{ in } \Omega_i \cup \Omega_{sh}, \tag{32}$$

where ϵ_s is defined in $\overline{\Omega}_{sh} \cup \Omega_s$, the correlation length $l_c = \sqrt{l_B l_D / 48}$ is a density-density correlation length independent of specific ionic radius [168], l_B and l_D are the Bjerrum and Debye lengths, respectively, the concentration $C_k(\mathbf{r})$ function (4) is defined in $\overline{\Omega}$ for all $k = 1, \cdots, K+1$ in molarity (M), and $v_k = 4\pi a_k^3/3$ with radius a_k. Since the steric potential takes particle volumes and voids into account, the shell volume V_{sh} of the shell domain Ω_{sh} can be determined by the steric potential $S_{sh}^{trc} = \frac{v_0}{v_w} \ln \frac{O_i^w}{V_{sh} C_{K+1}^B} = \ln \frac{V_{sh} - v_w O_i^w}{V_{sh} \Gamma^B}$ [69,73], where the occupant (coordination) number O_i^w of water molecules is given by experimental data [166]. The shell radius R_i^{sh} is thus determined and depends not only on O_i^w but also on the bulk void fraction Γ^B, namely *on all salt and water bulk concentrations* (C_k^B) [69,73].

We reduce the complexity of higher-order approximations, and make them easier to implement by transforming the fourth-order PDE (17) to the following two second-order PDEs [64]

$$\left(l_c^2\Delta - 1\right)\psi(\mathbf{r}) = \rho_{ion}(\mathbf{r}) \text{ in } \Omega_s, \quad (33)$$

$$\epsilon_s\Delta\phi(\mathbf{r}) = \psi(\mathbf{r}) \text{ in } \Omega_s, \quad (34)$$

where the extra unknown function $\psi(\mathbf{r})$ is a density-like function as seen from (33) by setting $l_c = 0$. The boundary and interface conditions for $\phi(\mathbf{r})$ and $\psi(\mathbf{r})$ in (32)–(34) are [64]

$$\phi(\mathbf{r}) = \psi(\mathbf{r}) = 0 \text{ on } \partial\Omega_s \backslash \partial\Omega_{sh}, \quad (35)$$

$$\psi(\mathbf{r}) = -\rho_s(\mathbf{r}) \text{ on } \partial\Omega_{sh} \cap \partial\Omega_s, \quad (36)$$

$$[\phi(\mathbf{r})] = 0 \text{ on } \partial\Omega_i \cup (\partial\Omega_{sh} \cap \partial\Omega_s), \quad (37)$$

$$[\nabla\phi(\mathbf{r})\cdot\mathbf{n}] = 0 \text{ on } \partial\Omega_{sh} \cap \partial\Omega_s, \quad (38)$$

$$[\epsilon(\mathbf{r})\nabla\phi(\mathbf{r})\cdot\mathbf{n}] = \epsilon_i\nabla\phi^*(\mathbf{r})\cdot\mathbf{n} \text{ on } \partial\Omega_i, \quad (39)$$

where ∂ denotes the boundary of a domain, the jump function $[\phi(\mathbf{r})] = \lim_{\mathbf{r}_{sh}\to\mathbf{r}}\phi(\mathbf{r}_{sh}) - \lim_{\mathbf{r}_i\to\mathbf{r}}\phi(\mathbf{r}_i)$ at $\mathbf{r} \in \partial\Omega_i$ with $\mathbf{r}_{sh} \in \Omega_{sh}$ and $\mathbf{r}_i \in \Omega_i$, $\epsilon(\mathbf{r}) = \epsilon_s$ in Ω_{sh} and $\epsilon(\mathbf{r}) = \epsilon_{ion}\epsilon_0$ in Ω_i, ϵ_{ion} is a dielectric constant in Ω_i, $\mathbf{n}$ is an outward normal unit vector at $\mathbf{r} \in \partial\Omega_i$, and $\phi^*(\mathbf{r}) = q_i/(4\pi\epsilon_i|\mathbf{r}-\mathbf{0}|)$. Equation (32) avoids large errors in a direct approximation of the delta function $\delta(\mathbf{r}-\mathbf{0})$ in the singular charge $q_i\delta(\mathbf{r}-\mathbf{0})$ of the solvated ion at the origin $\mathbf{0}$ by transforming the singular charge to the Green's function $\phi^*(\mathbf{r})$ on $\partial\Omega_i$ in (39) as an approximation source of the electric field produced by the solvated ion [169,170].

For simplicity, we consider a general binary ($K = 2$) electrolyte $C_{z_2}A_{z_1}$ with the valences of the cation C^{z_1+} and anion A^{z_2-} being z_1 and z_2, respectively. The first-order Taylor approximation of the charge density functional $\rho_{ion}(\phi(\mathbf{r}))$ in (17) with respect to the electric potential $\phi(\mathbf{r})$ yields

$$\rho_{ion}(\phi(\mathbf{r})) \approx \frac{-C_1^B q_1}{k_B T}\left[(q_1 - q_2) - \Lambda q_1\right]\phi(\mathbf{r}), \quad (40)$$

where $\Lambda = C_1^B(v_1 - v_2)^2 / \left[\Gamma^B v_0 + \left(v_1^2 C_1^B + v_2^2 C_2^B + v_3^2 C_3^B\right)\right]$ which is a quantity corresponding to a linearization of the steric potential $S^{trc}(\mathbf{r})$ [76]. Consequently, we obtain a *generalized Debye length*

$$l_{D4PBik} = \left(\frac{\epsilon_s k_B T}{C_1^B((1-\Lambda)q_1^2 - q_1 q_2)}\right)^{1/2} \quad (41)$$

that reduces to the original Debye length l_D [11] if $v_1 = v_2 \neq 0$ (two ionic species with equal radius and thus $\Lambda = 0$) or $v_1 = v_2 = v_3 = 0$ (all particles treated as volumeless points in standard texts for PB [11]). The nonlinear value of $\Lambda \neq 0$ for $v_1 = v_2 \neq 0$ can be obtained by Newton's method [76].

Equation (33) is a second-order PDE that requires two boundary conditions like (35) and (36) for a unique solution $\psi(\mathbf{r})$. Since $\psi(\mathbf{r}) = \epsilon_s\nabla^2\phi(\mathbf{r}) = -\rho(\mathbf{r}) \approx \epsilon_s\kappa^2\phi(\mathbf{r})$ if $l_c = 0$, Equation (36) is a simplified (approximate) boundary condition for $\psi(\mathbf{r})$ on $\partial\Omega_{sh} \cap \partial\Omega_s$ without involving higher-order derivatives of $\psi(\mathbf{r})$ (or the third-order derivative of $\phi(\mathbf{r})$). The approximations in (36) and (40) do not significantly affect our generalized DH model's ability to fit activity data. However, these assumptions should be carefully scrutinized in other applications such as highly charged surfaces. Bazant et al. have recently developed more consistent and general boundary conditions for their fourth-order model by enforcing continuity of the Maxwell stress at a charged interface [171,172].

In [76], we analytically solve the linear 4PBik PDEs (32), (33), and (34) with (40) in a similar way as Debye and Hückel solved the linear PB equation for a spherically symmetric system. However, the spherical domain shown in Figure 1 and the boundary and interface conditions in (35)–(39) are different from those of the standard method for the linear PB equation in physical chemistry texts [11]. The analysis consists of the following steps: (i) The nonlinear term $\rho_{ion}(\mathbf{r})$ in (33) is linearized to the linear term $-\epsilon_s \phi / l^2_{D4PBik}$ in (40) as that of Debye and Hückel. (ii) The linear PDEs corresponding to (33) and (34) are then formulated into a system of eigenvalue problems with eigenfunctions $(\phi(\mathbf{r}), \psi(\mathbf{r}))$ and eigenvalues (λ_1, λ_2), where the general solution of $\phi(\mathbf{r})$ is equal to that of Debye and Hückel in the solvent domain Ω_s (not the entire domain) when $l_c = v_1 = v_2 = v_3 = 0$. (iii) A unique pair of eigenfunctions $\left(\phi^{4PBik}(\mathbf{r}), \psi^{4PBik}(\mathbf{r})\right)$ is found under conditions (35)–(39), where $\phi^{4PBik}(\mathbf{r})$ is equal to that of Debye and Hückel in Ω_s when $l_c = v_1 = v_2 = v_3 = 0$.

The analytical potential function that we found [76] is

$$\phi^{4PBik}(r) = \begin{cases} \frac{q_i}{4\pi\epsilon_s R_i^{Born}} + \frac{q_i}{4\pi\epsilon_s R_i^{sh}} (\Theta - 1) \text{ in } \Omega_i \\ \frac{q_i}{4\pi\epsilon_s r} + \frac{q_i}{4\pi\epsilon_s R_i^{sh}} (\Theta - 1) \text{ in } \Omega_{sh} \\ \frac{q_i}{4\pi\epsilon_s r} \left[\frac{\lambda_1^2 e^{-\sqrt{\lambda_2}\left(r - R_i^{sh}\right)} - \lambda_2^2 e^{-\sqrt{\lambda_1}\left(r - R_i^{sh}\right)}}{\lambda_1^2\left(\sqrt{\lambda_2} R_i^{sh} + 1\right) - \lambda_2^2\left(\sqrt{\lambda_1} R_i^{sh} + 1\right)} \right] \text{ in } \Omega_s, \end{cases} \tag{42}$$

where

$$\Theta = \frac{\lambda_1^2 - \lambda_2^2}{\lambda_1^2\left(\sqrt{\lambda_2} R_i^{sh} + 1\right) - \lambda_2^2\left(\sqrt{\lambda_1} R_i^{sh} + 1\right)}, \tag{43}$$

$r = |\mathbf{r}|$, $\lambda_1 = \left(1 - \sqrt{1 - 4l_c^2/l^2_{D4PBik}}\right) / \left(2l_c^2\right)$, and $\lambda_2 = \left(1 + \sqrt{1 - 4l_c^2/l^2_{D4PBik}}\right) / \left(2l_c^2\right)$. Please note that $\lim_{l_c \to 0} \lambda_1 = 1/l^2_{D4PBik}$, $\lim_{l_c \to 0} \lambda_2 = \infty$, and $\lim_{l_c \to 0} \Theta = \lim_{C_1^B \to 0} \Theta = \lim_{l_{D4PBik} \to \infty} \Theta = 1$ [76]. The linearized 4PBik potential $\phi^{4PBik}(r)$ reduces to the linearized PB potential $\phi^{PB}(r) = q_i e^{-r/l_D} / (4\pi\epsilon_s r)$ as in standard texts (e.g., Equation (7.46) in [11]) by taking $\lim_{l_c \to 0} \phi^{4PBik}(r)$ with $v_k = 0$ for all k, $R_i^{sh} = 0$, and $r > 0$ [76].

As discussed in [173], since the solvation free energy of an ion i varies with salt concentrations, the Born energy $q_i^2 \left(\frac{1}{\epsilon_w} - 1\right) / 8\pi\epsilon_0 R_i^0$ in pure water (i.e., $C_i^B = 0$) with a constant Born radius R_i^0 should change to depend on $C_i^B \geq 0$. Equivalently, the Born radius R_i^{Born} in (42) is variable and we can model it from R_i^0 by a simple formula [69,73]

$$R_i^{Born} = \theta R_i^0, \;\; \theta = 1 + \alpha_1^i \left(\overline{C}_i^B\right)^{1/2} + \alpha_2^i \overline{C}_i^B + \alpha_3^i \left(\overline{C}_i^B\right)^{3/2}, \tag{44}$$

where $\overline{C}_i^B = C_i^B/\text{M}$ is a dimensionless bulk concentration and α_1^i, α_2^i, and α_3^i are parameters for modifying the experimental Born radius R_i^0 to fit experimental activity coefficient γ_i that changes with the bulk concentration C_i^B of the ion. The Born radii R_i^0 given below are from [173] obtained from the experimental hydration Helmholtz free energies of those ions given in [12]. The three parameters in (44) have physical or mathematical meanings unlike many parameters in the Pitzer model [19,153,156]. The first parameter α_1^i adjusts R_i^0 and accounts for the real thickness of the ionic atmosphere (Debye length), which is proportional to the square root of the ionic strength in the DH theory [11]. The second α_2^i and third α_3^i parameters are adjustments in the next orders of approximation beyond the DH treatment of ionic atmosphere [73].

The potential value $\phi^0(\mathbf{0}) = \lim_{C_1^B \to 0} \phi^{4PBik}(\mathbf{0}) = q_i / \left(4\pi\epsilon_s R_i^0\right)$ by $\lim_{C_1^B \to 0} \Theta = 1$ and $\lim_{C_1^B \to 0} R_i^{Born} = R_i^0$. From (31) and (42), we thus have a generalized activity coefficient γ_i^{4PBik} in

$$\ln \gamma_i^{4PBik} = \frac{q_i^2}{8\pi\epsilon_s k_B T} \left(\frac{1}{R_i^{Born}} - \frac{1}{R_i^0} + \frac{\Theta - 1}{R_i^{sh}} \right), \tag{45}$$

which satisfies the DH limiting law, i.e., $\gamma_i^{4PBik} = \gamma_i^{DH} = 1$ for infinite dilute (ideal) solutions as $C_i^B \to 0$. The generalized activity coefficient γ_i^{4PBik} reduces to the classical γ_i^{DH} proposed by Debye and Hückel in 1923 [149], namely

$$\ln \gamma_i^{DH} = \frac{-q_i^2}{8\pi\epsilon_s k_B T l_D (1 + R_i^{sh}/l_D)} \quad (46)$$

if $R_i^{Born} = R_i^0$ (without considering Born energy effects), $R_i^{sh} = R_i$ (an effective ionic radius [149]), $l_{D4PBik} = l_D$ (no steric effect), and $l_c = 0$ (no correlation effect). The reduction shown in [76] is by taking the limit of the last term in (45) as $l_c \to 0$, i.e., $\lim_{l_c \to 0} \frac{\Theta - 1}{R_i^{sh}} = \frac{-1}{R_i + l_D}$.

Hückel soon realized that the DH formula (46) failed to fit experimental data at high ionic strengths and modified it in 1925 [150] by adding one more parameter η_1 to become (see Equation (7.115) in [11])

$$\ln \gamma_i^{DHB} = \frac{-q_i^2}{8\pi\epsilon_s k_B T l_D (1 + \eta_0 \sqrt{I})} + \eta_1 I, \quad (47)$$

where η_0 (an approximation of R_i^{sh}) and η_1 account for the distance of closest approach to the ion i and the salting-out effect (an approximation of the Born energy), respectively [11], where $I = \frac{1}{2}\sum_i C_i^B z_i^2$ is the ionic strength of the solution. Consequently, a variety of extended DH models γ_i^{DHBx} [153,174] in the form similar to

$$\ln \gamma_i^{DHBx} = \frac{-q_i^2}{8\pi\epsilon_s k_B T l_D (1 + \eta_0 \sqrt{I})} + \sum_{k \neq 0} \eta_k I^k \quad (48)$$

have been proposed in the literature to express other thermodynamic properties such as temperature and pressure by a power expansion of I with more and more parameters η_k that can increase combinatorially with various composition, temperature, and pressure to a frustrating amount [1,16,19]. Please note that η_k may also depend on ionic strength I in a complicated way, see e.g., Equation (2) in [153]. Many expressions of those parameters are rather long and tedious and do not have clear physical meaning [19,153,156]. The Davies equation [175] is a special form of (47) with a linear term in I.

The R_i^{Born} term in (45) *differs significantly* from the last term in (48) as they are the *inverse* of each other in terms of I and parameters, i.e., I, α_1, α_2, and α_3 are in the denominator in (45) whereas I and η_k are in the numerator in (48). This implies that γ_i^{4PBik} and γ_i^{DHBx} vary oppositely with I. Consequently, the values of α_1, α_2, and α_3 are totally different from those of η_k when we use γ_i^{4PBik} and γ_i^{DHBx} to fit experimental activity coefficients with I varying from low to high values [76]. This may explain why the empirical nature of extended DH models requires a great deal of effort to extract parameters (without physical hints) from existent thermodynamic databases by regression analysis [19,153].

7. Numerical Methods

Numerical simulations are indispensable to study chemical, physical, and mathematical properties of biological and chemical systems in realistic applications, especially with experimental details at atomic scale such as ion channels in the Protein Data Bank (PDB) [57]. Continuum PDE models have substantial advantages over Monte Carlo, Brownian dynamics (BD), or molecular dynamics in physical insights and computational efficiency that are of great importance in studying a range of conditions and concentrations especially for large nonequilibrium or inhomogeneous systems, as are present in experiments and in life itself [10,17,21,95,121,176–185].

The literature on numerical methods for solving PB and PNP models is vast [64,68,74]. We summarize here some important features of the methods proposed in [64,68,74] for Poisson-Bikerman and Poisson-Nernst-Planck-Bikerman models, which may be useful for workers in numerical analysis and coding practice. Since PNPB including 4PBik is highly nonlinear and the geometry of protein structures is very complex, we emphasize two different types of methods, namely nonlinear iterative methods and discretization methods for these two problems as follows.

7.1. Nonlinear Iterative Methods

For the PNPB system of $K+1$ NP equations in (27), Laplace Equation (32), and two 4PBik equations in (33) and (34), the total number of second-order PDEs that we need to solve is $K+4$. These PDEs are coupled together and highly nonlinear except (32). Numerically solving this kind of nonlinear systems is not straightforward [64,68,74]. We use the following algorithm to explain essential procedures for solving the steady-state PNPB system, where Ω_m denotes the biomolecular domain that contains a total of Q fixed atomic charges q_j located at $\mathbf{r}_j$ in a channel protein as shown in Figure 2L for the gramicidin A channel downloaded from PDB with $Q = 554$, for example, $\partial\Omega_m$ denotes the molecular surface of the protein and the membrane lipids through which the protein crosses as shown in Figure 2R, and Ω_s is the solvent domain consisting of the channel pore and the extracellular and intracellular baths for mobile ions and water molecules.

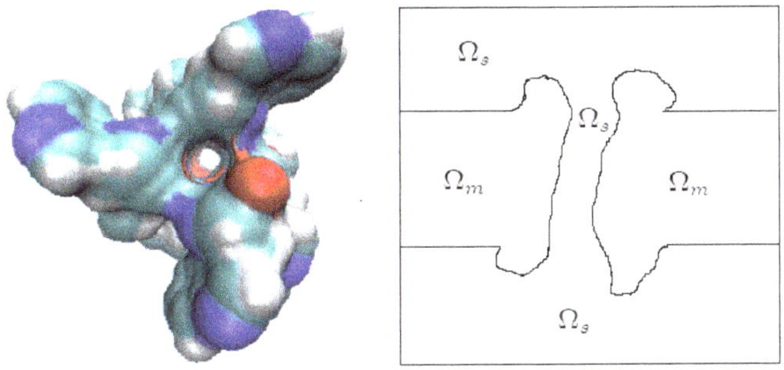

Figure 2. Left (L): Top view of the gramicidin A channel. **Right (R)**: A cross section of 3D simulation domain for the channel placed in a rectangular box, where Ω_m is the biomolecular domain consisting of the channel protein and the membrane and Ω_s is the solvent domain consisting of the channel pore and the baths.

Nonlinear Iterative Algorithm [68]:

1. Solve Laplace Equation $-\nabla^2\phi(\mathbf{r}) = 0$ for $\phi^0(\mathbf{r})$ in Ω_m once for all with $\phi^0(\mathbf{r}) = \phi^*(\mathbf{r}) = \sum_{j=1}^{Q} q_j/(4\pi\epsilon_m\epsilon_0\left|\mathbf{r}-\mathbf{r}_j\right|)$ on $\partial\Omega_m$.
2. Solve Poisson Equation $-\nabla\cdot(\epsilon\nabla\phi(\mathbf{r})) = \rho_{ion}(\mathbf{r})$ for $\phi^{Old}(\mathbf{r})$ in Ω_s with $\rho_{ion}(\mathbf{r}) = 0$, $\phi^{Old} = V = 0$ on $\partial\Omega$, and the jump condition $\left[\epsilon\nabla\phi^{Old}\cdot\mathbf{n}\right] = -\epsilon_m\epsilon_0\nabla(\phi^*+\phi^0)\cdot\mathbf{n}$ on $\partial\Omega_m$ as (39), where V denotes applied voltage.
3. $V = V_0 \neq 0$ an initial voltage.
4. Solve 4PBik1 Equation $\epsilon_s\left(\lambda_c l_c^2\nabla^2 - 1\right)\Psi(\mathbf{r}) = \sum_{i=1}^{K} q_i C_i^{Old}(\mathbf{r})$ for $\Psi^{New}(\mathbf{r})$ in Ω_s with $\nabla\Psi^{New}\cdot\mathbf{n} = 0$ on $\partial\Omega_m$, $\Psi^{New} = 0$ on $\partial\Omega$, $C_i^{Old}(\mathbf{r}) = C_i^B\exp\left(-\beta_i\phi^{Old}(\mathbf{r}) + \frac{v_i}{v_0}S^{trc}(\mathbf{r})\right)$, $S^{trc}(\mathbf{r}) = \ln\frac{\Gamma^{Old}(\mathbf{r})}{\Gamma^B}$, and $\Gamma^{Old}(\mathbf{r}) = 1 - \sum_{j=1}^{K+1}\lambda_s v_j C_j^{Old}(\mathbf{r})$.
5. Solve 4PBik2 Equation $-\nabla\cdot(\epsilon_s\nabla\phi(\mathbf{r})) + \rho'(\phi^{Old})\phi(\mathbf{r}) = -\epsilon\Psi^{New} + \rho'(\phi^{Old})\phi^{Old}$ for $\phi^{New}(\mathbf{r})$ in Ω_s with $\phi^{New} = V$ on $\partial\Omega$ and the same jump condition in Step 2, where $\rho'(\phi)$ is the derivative of $\rho(\phi)$ with respect to ϕ.
6. If the maximum error norm $\left\|\phi^{New} - \phi^{Old}\right\|_\infty > Tol$, a preset tolerance, then set $\phi^{Old} = \omega_{4PBik}\phi^{Old} + (1-\omega_{4PBik})\phi^{New}$ and go to Step 4, else go to Step 7.
7. Solve NP Equation $-\nabla\cdot\mathbf{J}_i(\mathbf{r}) = 0$ for $C_i^{New}(\mathbf{r})$ in Ω_s for all $i = 1,\cdots,K+1$ with $\mathbf{J}_i(\mathbf{r}) = -D_i\left[\nabla C_i(\mathbf{r}) + \beta_i C_i(\mathbf{r})\ \nabla\phi^{Old}(\mathbf{r}) - \lambda_s\frac{v_i}{v_0}C_i(\mathbf{r})\nabla S^{trc}(\mathbf{r})\right]$, $S^{trc}(\mathbf{r}) = \ln\frac{\Gamma^{Old}(\mathbf{r})}{\Gamma^B}$, $C_i^{New}(\mathbf{r}) = 0$ on $\partial\Omega$, and $\mathbf{J}_i(\mathbf{r})\cdot\mathbf{n} = 0$ on $\partial\Omega_m$.

8. Solve 4PBik1 Equation for Ψ^{New} as in Step 4 with C_i^{New} in place of C_i^{Old}.
9. Solve 4PBik2 Equation for ϕ^{New} as in Step 5.
10. If $\left\|\phi^{New} - \phi^{Old}\right\|_\infty > Tol$, then set $\phi^{Old} = \omega_{PNPB}\phi^{Old} + (1 - \omega_{PNPB})\phi^{New}$ and go to Step 7, else go to Step 11.
11. $V = V + \Delta V$ and go to Step 4 until the desired voltage is reached.

Linearizing the nonlinear 4PBik (17) yields two second-order linear 4PBik1 and 4PBik2 in Steps 4 and 5 that differ from the nonlinear (33) and (34). Newton's iterative Steps 4–6 for solving 4PBik1 and 4PBik2 dictates convergence that also depends on various mappings from an old solution ϕ^{Old} to a new solution ϕ^{New}. This algorithm uses two relaxation and three continuation mappings for which we need to carefully tune two relaxation parameters ω_{4PBik} and ω_{PNPB} and three continuation parameters λ_c (related to correlation effects), λ_s (steric effects), and ΔV (incremental voltage for applied voltage). For example, the parameter λ_s in $\Gamma^{Old}(\mathbf{r}) = 1 - \sum_{j=1}^{K+1} \lambda_s v_j C_j^{Old}(\mathbf{r})$ can be chosen as $\lambda_s = k\Delta\lambda$, $k = 0, 1, 2, \cdots, \frac{1}{\Delta\lambda}$, an incremental continuation from 0 (no steric effects) to 1 (fully steric effects) with a tuning stepping length $\Delta\lambda$. The algorithm can fail to converge if we choose $\Delta\lambda = 1$ (without continuation) for some simulation cases, since we may have $\Gamma^{Old}(\mathbf{r}) < 0$ resulting in numerically *undefined* $S^{trc}(\mathbf{r}) = \ln \frac{\Gamma^{Old}(\mathbf{r})}{\Gamma^B}$ at some $\mathbf{r}$ where the potential $\phi^{Old}(\mathbf{r})$ is large.

7.2. Discretization Methods

All PDEs in Steps 1, 2, 4, 5, 8, and 9 are of Poisson type $-\nabla^2\phi(\mathbf{r}) = f(\mathbf{r})$. We use the central finite difference (FD) method [64]

$$\frac{-\phi_{i-1,j,k} + 2\phi_{ijk} - \phi_{i+1,j,k}}{\Delta x^2} + \frac{-\phi_{i,j-1,k} + 2\phi_{ijk} - \phi_{i,j+1,k}}{\Delta y^2} + \frac{-\phi_{i,j,k-1} + 2\phi_{ijk} - \phi_{i,j,k+1}}{\Delta z^2} = f_{ijk}, \tag{49}$$

to discretize it at all grid points $\mathbf{r}_{ijk} = (x_i, y_j, z_k)$ in a domain, where $\phi_{ijk} \approx \phi(x_i, y_j, z_k)$, $f_{ijk} = f(x_i, y_j, z_k)$, and Δx, Δy, and Δz are mesh sizes on the three axes from a uniform partition $\Delta x = \Delta y = \Delta z = h$. The domains in Steps 1 and 2 are Ω_m and Ω_s, respectively. The discretization leads to a sparse matrix system $A\vec{\phi} = \vec{f}$ with the compressed bandwidth of the matrix A being 7, where the matrix size can be millions for sufficiently small h to obtain sufficiently accurate ϕ_{ijk}.

The matrix system consists of four subsystems, two by the FD method (49) in Ω_m and Ω_s, one by another method (see below) to discretize the jump condition in Step 2 on the interface $\partial\Omega_m$ between Ω_s and Ω_m, and one by imposing boundary conditions on $\partial\Omega$. We need to solve the matrix system in the entire domain $\overline{\Omega} = \overline{\Omega}_m \cup \overline{\Omega}_s$.

The convergence order of (49) is $O(h^2)$ (optimal) in maximum error norm for sufficiently smooth function $\phi(\mathbf{r})$. However, this optimal order can be easily degraded to $O(h^{0.37})$ [186], for example, by geometric singularities if the jump condition is not properly treated. In [64], we propose the interface method

$$\frac{-\epsilon_{i-\frac{3}{2}}\phi_{i-2} + \left(\epsilon_{i-\frac{3}{2}} + (1 - A_1)\,\epsilon^-_{i-\frac{1}{2}}\right)\phi_{i-1} - A_2\epsilon^-_{i-\frac{1}{2}}\phi_i}{\Delta x^2} = f_{i-1} + \frac{\epsilon^-_{i-\frac{1}{2}} A_0}{\Delta x^2} \tag{50}$$

$$\frac{-B_1\epsilon^+_{i-\frac{1}{2}}\phi_{i-1} + \left((1 - B_2)\,\epsilon^+_{i-\frac{1}{2}} + \epsilon_{i+\frac{1}{2}}\right)\phi_i - \epsilon_{i+\frac{1}{2}}\phi_{i+1}}{\Delta x^2} = f_i + \frac{\epsilon^+_{i-\frac{1}{2}} B_0}{\Delta x^2}, \tag{51}$$

where

$$A_1 = \frac{-(\epsilon_m - \epsilon_s)}{\epsilon_m + \epsilon_s},\ A_2 = \frac{2\epsilon_m}{\epsilon_m + \epsilon_s},\ A_0 = \frac{-2\epsilon_m[\phi] - \Delta x\,[\epsilon\phi']}{\epsilon_m + \epsilon_s},$$

$$B_1 = \frac{2\epsilon_s}{\epsilon_m + \epsilon_s},\ B_2 = \frac{\epsilon_m - \epsilon_s}{\epsilon_m + \epsilon_s},\ B_0 = \frac{2\epsilon_s[\phi] - \Delta x\,[\epsilon\phi']}{\epsilon_m + \epsilon_s},$$

to discretize the 1D Poisson equation $-\frac{d}{dx}\left(\epsilon(x)\frac{d\phi(x)}{dx}\right) = f(x)$ at every jump position $\gamma \in \partial\Omega_m$ that is at the middle of its two neighboring grid points, i.e., $x_{i-1} < \gamma = x_{i-\frac{1}{2}} < x_i$, where $x_{i-\frac{1}{2}} = (x_{i-1} + x_i)/2$ and x_{i-1} and x_i belong to different domains Ω_s and Ω_m. The corresponding cases in y- and z-axis follow obviously in a similar way. This method yields *optimal* convergence [64].

Since the matrix system is usually very large in 3D simulations and we need to repeatedly solve such systems updated by nonlinear iterations as shown in the above algorithm, linear iterative methods such as the bi-conjugate gradient stabilized (bi-CG) method are used to solve the matrix system [74]. We propose two parallel algorithms (one for bi-CG and the other for nonlinear iterations) in [74] and show that parallel algorithms on GPU (graphic processing unit) over sequential algorithms on CPU (central processing unit) can achieve 22.8× and 16.9× speedups for the linear solver time and total runtime, respectively.

Discretization of Nernst–Planck equations in Step 7 is different from (49) because the standard FD method

$$\frac{C_{i+1} - C_i}{\Delta x} = \frac{C_{i+1} + C_i}{2}\left(-\beta\frac{\Delta\phi_i}{\Delta x} + \frac{\Delta S_i^{trc}}{\Delta x}\right) \tag{52}$$

for the zero flux ($J(x) = -D(x)\left(\frac{dC(x)}{dx} + \beta C(x)\frac{d\phi(x)}{dx} - \frac{v}{v_0}C(x)\frac{dS^{trc}(x)}{dx}\right) = 0$) can easily yield

$$C_{i+1} - C_i > C_{i+1} + C_i \tag{53}$$

and thereby a negative (*unphysical*) concentration $C_i < 0$ at x_i if

$$\frac{1}{2}\left(-\beta\Delta\phi_i + \Delta S_i^{trc}\right) > 1, \tag{54}$$

where $\Delta\phi_{i-1} = \phi_i - \phi_{i-1}$, $\phi_i \approx \phi(x_i)$ etc. Therefore, it is crucial to check whether the *generalized Scharfetter–Gummel* (SG) condition [68]

$$-\beta\Delta\phi_i + \Delta S_i^{trc} \le 2 \tag{55}$$

is satisfied by any discretization method in implementation. This condition generalizes the well-known SG stability condition in semiconductor device simulations [187,188] to include the steric potential function $S^{trc}(\mathbf{r})$.

We extend the classical SG method [187] of the flux $J(x)$ in [68] to

$$J_{i+\frac{1}{2}} = \frac{-D}{\Delta x}\left[B(-t_i)C_{i+1} - B(t_i)C_i\right] \tag{56}$$

where $t_i = \beta\Delta\phi_i - \Delta S_i^{trc}$ and $B(t) = \frac{t}{e^t - 1}$ is the Bernoulli function [188]. Equation (56), an exponential fitting scheme, satisfies (55) and is derived from assuming that the flux J, the local electric field $\frac{-d\phi}{dx}$, and the local steric field $\frac{dS^{trc}}{dx}$ are all constant in the sufficiently small subinterval (x_i, x_{i+1}), i.e.,

$$\frac{J}{D} = \frac{-dC(x)}{dx} - kC(x), \text{ for all } x \in (x_i, x_{i+1}), \tag{57}$$

where $k = \beta\frac{d\phi}{dx} - \frac{dS^{trc}}{dx}$. Solving this ordinary differential equation (ODE) with a boundary condition C_i or C_{i+1} yields the well-known Goldman-Hodgkin-Katz flux equation in ion channels [9], which is exactly the same as that in (56) but with the subinterval (x_i, x_{i+1}) being replaced by the height of the entire box in Figure 2R.

The generalized Scharfetter-Gummel method for Nernst-Planck equations is thus

$$\begin{aligned}
\frac{dJ(x_i)}{dx} &\approx \frac{J_{i+\frac{1}{2}} - J_{i-\frac{1}{2}}}{\Delta x} = \frac{b_{i-1}C_{i-1} + b_iC_i + b_{i+1}C_{i+1}}{\Delta x^2} = 0 \\
J_{i-\frac{1}{2}} &= \frac{-D}{\Delta x}\left[B(-t_{i-1})C_i - B(t_{i-1})C_{i-1}\right] \\
J_{i+\frac{1}{2}} &= \frac{-D}{\Delta x}\left[B(-t_i)C_{i+1} - B(t_i)C_i\right] \\
t_i &= \beta\Delta\phi_i - \Delta S_i^{trc}, B(t) = \frac{t}{e^t - 1} \\
b_{i-1} &= -B(t_{i-1}), b_i = B(-t_{i-1}) + B(t_i), b_{i+1} = -B(-t_i).
\end{aligned} \tag{58}$$

The SG method is *optimal* in the sense that it integrates the ODE (57) *exactly* at *every* grid point with a suitable boundary condition [189]. Therefore, the SG method can resolve sharp layers very accurately [189] and hence needs few grid points to obtain tolerable approximations when compared with the primitive FD method. Moreover, the exact solution of (57) for the concentration function $C(x)$ yields an exact flux $J(x)$. Consequently, the SG method is *current preserving*, which is particularly important in nonequilibrium systems, where the current is possibly the most relevant physical property of interest [190].

It is difficult to overstate the importance of the current preserving feature and it must be emphasized for workers coming from fluid mechanics that preserving current has a significance quite beyond the preserving of flux in uncharged systems. Indeed, conservation of current (defined as Maxwell did to include the displacement current of the vacuum $\epsilon_0\frac{\partial\mathbf{E}(\mathbf{r},t)}{\partial t}$) is an unavoidable consequence, nearly a restatement of the Maxwell equations themselves [104,106]. The electric field is so strong that the tiniest error in preserving current, i.e., the tiniest deviation from Maxwell's equations, produces huge effects. The third paragraph of Feynman's lectures on electrodynamics makes this point unforgettable [191]. Thus, the consequences of a seemingly small error in preserving the flow of charge are dramatically larger than the consequences of the same error in preserving the flux of mass.

We have developed a C++ code for solving 4PBik and PNPB models on laptop and high-performance (with GPU) computers. For solving a 4PBik problem with a matrix system of size 4,096,000, for example, the code requires about 300 MB memory to store the compressed matrix system with double precision. It took about 2 min and 47 s on a laptop computer equipped with 1.3 GHz Intel CPU and 2 GB RAM to solve the linear system once by the successive overrelaxation method with an error tolerance of 10^{-6} [64].

8. Applications

We have used the saturating Poisson-Nernst-Planck-Bikerman theory to study ion activities, electric double layers, and biological ion channels in the past. The theory accounts for the steric effect of ions and water molecules, the effects of ion-ion and ion-water correlations, the screening and polarization effects of polar water, and the charge/space competition effect of ions and water molecules of different sizes and valences. These effects are all closely related to the dielectric operator in (17) and the steric potential in (4) that works on both macroscopic and atomic scales. We now illustrate these properties in the following three areas using mostly experimental data to verify the theory.

8.1. Ion Activities

The curves in Figure 3 obtained by the generalized Debye-Hückel Formula (45) [75] fit well to the experimental data by Wilczek–Vera et al. [192] for single-ion activities in 8 1:1 electrolytes. There are only three fitting parameters in the formula, namely α_1^i, α_2^i and α_3^i, which we reiterate have specific physical meaning as parameters of the water shell around ions. The values of the parameters are given in Table 1 from which we observe that R_i^{Born} deviates from R_i^0 slightly. For example, $R_{\text{Cl}^-}^{Born}/R_{\text{Cl}^-}^0 =$ 1.007∼1.044 (not shown) for Figure 3a with [LiCl] = 0∼2.5 M, since the cavity radius $R_{\text{Cl}^-}^{Born}$ is an atomic measure from the infinite singularity $\delta(\mathbf{r}-\mathbf{0})$ at the origin, i.e., $\phi^{4PBik}(r)$ and thus γ_i^{4PBik} are very sensitive to R_i^{Born}. On the other hand, γ_i^{4PBik} is not very sensitive to R_i^{sh} ($R_{Cl^-}^{sh}$ = 5.123∼5.083 Å), i.e., the fixed choice of $O_i^w = 18$ (an experimental value in [166]) for all curves is not critical but reasonable [69]. The error between the estimated O_i^w and its unknown true value can always be compensated by small adjustments of R_i^{Born}. The values of other symbols are $a_{Li^+} = 0.6$ Å, $a_{Na^+} = 0.95$ Å, $a_{K^+} = 1.33$ Å, $a_{F^-} = 1.36$ Å, $a_{Cl^-} = 1.81$ Å, $a_{Br^-} = 1.95$, $a_{H_2O} = 1.4$ Å, $R_{Li^+}^0 = 1.3$ Å, $R_{Na^+}^0 = 1.618$ Å, $R_{K^+}^0 = 1.95$ Å, $R_{F^-}^0 = 1.6$ Å, $R_{Cl^-}^0 = 2.266$, $R_{Br^-}^0 = 2.47$ Å [173], $\epsilon_w = 78.45$, $\epsilon_{ion} = 1$, $T = 298.15$ K, where a_i is the (Pauling) radius of type i particle (ion) [173].

Table 1 also shows the significant order of these parameters, i.e., $|\alpha_1^i| > |\alpha_2^i| > |\alpha_3^i|$ in general cases. Please note that the three sets of the values of $\alpha_1^{Na^+}$, $\alpha_2^{Na^+}$, and $\alpha_3^{Na^+}$ for the same Na^+ in three different salts NaCl, NaBr, and NaF are different because the diameters of the anions are different. Figure 4 shows single-ion activities in 6 2:1 electrolytes by experiments [192] and 4PBik, where the significant order (not shown) of three fitting parameters is similar to that in Table 1.

Table 1. Values of α_1^i, α_2^i, α_3^i in (44).

Fig.#	i	α_1^i	α_2^i	α_3^i	Fig.#	i	α_1^i	α_2^i	α_3^i
3a	Li^+	−0.006	−0.037	0.004	3e	Na^+	−0.049	0.042	−0.013
3a	Cl^-	0.052	−0.015	0	3e	Br^-	0.071	−0.048	0.006
3b	Li^+	−0.006	−0.011	−0.004	3f	K^+	0.005	0.051	−0.015
3b	Br^-	0.026	−0.057	0.010	3f	F^-	0.033	−0.028	0.003
3c	Na^+	0	0	0	3g	K^+	0.031	0.022	−0.005
3c	F^-	0.027	0	0	3g	Cl^-	0.020	−0.025	0.004
3d	Na^+	−0.045	0.009	−0.002	3h	K^+	0.025	−0.062	0.018
3d	Cl^-	0.063	−0.014	−0.002	3h	Br^-	0.001	0.082	0

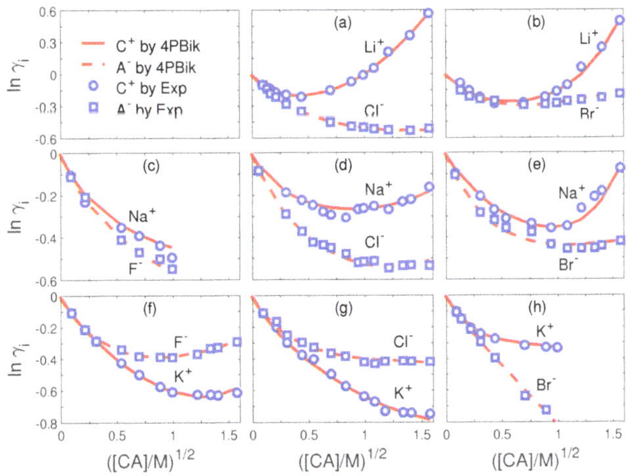

Figure 3. Single-ion activity coefficients of (**a**) LiCl (**b**) LiBr (**c**) NaF (**d**) NaCl (**e**) NaBr (**f**) KF (**g**) KCl (**h**) KBr electrolytes. Comparison of 4PBik results (curves) with experimental data (symbols) [192] on $i = C^+$ (cation) and A^- (anion) activity coefficients γ_i in various [CA] from 0 to 1.6 M.

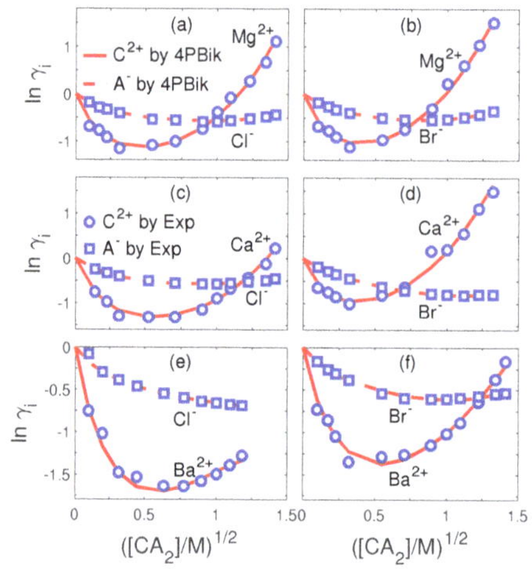

Figure 4. Single-ion activity coefficients of (**a**) $MgCl_2$ (**b**) $MgBr_2$ (**c**) $CaCl_2$ (**d**) $CaBr_2$ (**e**) $BaCl_2$ (**f**) $BaBr_2$ electrolytes. Comparison of 4PBik results (curves) with experimental data (symbols) [192] on $i = C^{2+}$ (cation) and A^- (anion) activity coefficients γ_i in various $[CA_2]$ from 0 to 1.5 M.

The electric potential and other physical properties of ionic activity can be studied in detail according to the partitioned domain in Figure 1 characterized by R_i^{Born} and R_i^{sh}. For example, we observe from Figure 5 that the electric potential ($\phi_{Br^-}^{4PBik}(0) = -2.4744\ k_BT/e$) and the Born radius ($R_{Br^-}^{Born}(2\text{ M}) = 2.0637$ Å) generated by Br^- at [LiBr] = 2 M are significantly different from that ($\phi_{Br^-}^{4PBik}(0) = -0.6860\ k_BT/e$, $R_{Br^-}^{Born}(2\text{ M}) = 4.2578$ Å) at [KBr] = 2 M. The only difference between these two solutions is the size of cations, i.e., the size of different positive ions significantly changes the activity of the same negative ion at high concentrations. The difference between $\phi_{Li^+}^{4PBik}(0)$ and $\phi_{K^+}^{4PBik}(0)$ is due to the sizes of Li^+ and K^+ not Br^- as it is the same for both solutions.

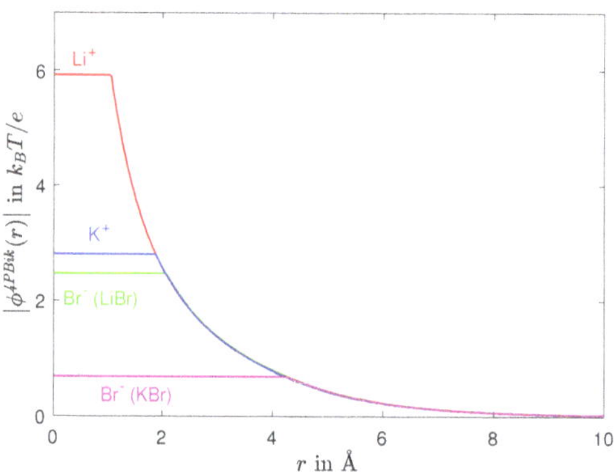

Figure 5. Electric potential $\phi^{4PBik}(r)$ profiles by (42) near the solvated ions Li^+ and Br^- at [LiBr] = 2 M, and K^+ and Br^- at [KBr] = 2 M, where r is the distance from the center of the respective ion.

This example clearly shows the atomic properties of 4PBik theory in the ion Ω_i and shell Ω_{sh} domains and the continuum properties in the solvent domain Ω_s. The Born radius R_i^{Born} in (42) determined by (44) changes with (i) *ion-water* interactions in $\Omega_i \cup \Omega_{sh}$ and (ii) *ion-ion* interactions in $\Omega_i \cup \Omega_s$ via $\phi^{4PBik}(r)$ in (42) that is self-consistently determined by the interface conditions in (35)–(39)

and by (iii) *multi-salt* [73,76] concentrations in Ω_s, (iv) the *screening* effects of water in Ω_{sh} and ions and water in Ω_s, (v) the *polarization* effect of water in Ω_s, (vi) the *correlation* effect between ions in Ω_s, (vii) the *steric* effects of all ions and water in the entire domain $\overline{\Omega} = \overline{\Omega}_i \cup \overline{\Omega}_{sh} \cup \overline{\Omega}_s$, (viii) *temperatures* [73,76], and (ix) *pressures* [73,76]. The generalized Debye-Hückel formula (45) includes all these 9 physical properties with only 3 fitting parameters. However, we look forward to the day when we can derive the three fitting parameters for particular types of ions, from independently determined experimental data.

8.2. Electric Double Layers

We consider a charged surface in contact with a 0.1 M 1:4 aqueous electrolyte, where the charge density is $\sigma = 1e/(50\ \text{Å}^2)$, the radius of both cations and anions is $a = 4.65$ Å (in contrast to an edge length of 7.5 Å of cubical ions in [133]), and $\epsilon_s = 80$ [72]. The multivalent ions represent large polyanions adsorbed onto a charged Langmuir monolayer in experiments [133]. We solve (33) and (34) using (49) in the rectangular box $\overline{\Omega} = \overline{\Omega}_s = \left\{(x,y,z) : 0 \leq x \leq 40, -5 \leq y \leq 5, -5 \leq z \leq 5\ \text{Å}\right\}$ such that $\phi(\mathbf{r}) \approx 0$ within the accuracy to 10^{-4} near and on the surface $x = 40$ Å. The boundary conditions on the surface and its adjacent four planes are $-\epsilon_s \nabla\phi \cdot \mathbf{n} = \sigma$ with $\mathbf{n} = \langle -1, 0, 0\rangle$ and $-\epsilon_s \nabla\phi \cdot \mathbf{n} = 0$ with $\mathbf{n}$ defined similarly, respectively.

The classical PB model (with $a = a_{H_2O} = l_c = 0$, i.e., no size, void, and correlation effects) produces unphysically high concentrations of anions (A^{4-}) near the surface as shown by the dashed curve in Figure 6L. The dotted curve in Figure 6L is similar to that of the modified PB in [133] and is obtained by the 4PBik model with $l_c = 0$ (no correlations), $V_{K+2} = 0$ (no voids), and $a_{H_2O} = 0$ (water is volumeless as in [133] and hence $\Gamma^B = 1 - \sum_{i=1}^{K} v_i C_i^B$ is the bulk water volume fraction). The voids ($V_{K+2} \neq 0$) and water molecules ($a_{H_2O} \neq 0$) have slight effects on anion concentration (because of saturation) and electric potential (because water and voids have no charges) profiles as shown by the thin solid curves in Figure 6L,R, respectively, when compared with the dotted curves. However, ion-ion correlations (with $l_c = 1.6a$ [78]) have significant effects on ion distributions as shown by the thick solid and dash-dotted curves in Figure 6L, where the saturation layer is on the order of ionic radius a and the *overscreening* layer [78] ($C_{A^{4-}}(x) \approx 0 < C_{A^{4-}}^B = 0.1$ M) of excess co-ions ($C_{C^+}(x) > C_{C^+}^B = 0.4$ M) is about 18 Å in thickness.

The *saturation layer* is an *output* (not an imposed condition) of our model unlike a Stern layer [193] imposed by most EDL models to account for size effects near charge surfaces [194–196]. The electric potentials $\phi(0) = 5.6\ k_BT/e$ at $x = 0$ and $\phi(11.5) = -1.97\ k_BT/e$ in Figure 6R obtained by 4PBik with voids and correlations deviate dramatically from those by previous models for nearly 100% at $x = 0$ (in the saturation layer) and 70% at $x = 11.5$ Å (in the screening layer) when compared with the maximum potential $\phi(0) = 2.82\ k_BT/e$ of previous models. The 4PBik potential depth $\phi(11.5) = -1.97\ k_BT/e$ of the overscreening layer is very sensitive the size a of ions and tends to zero as $a \to 0$.

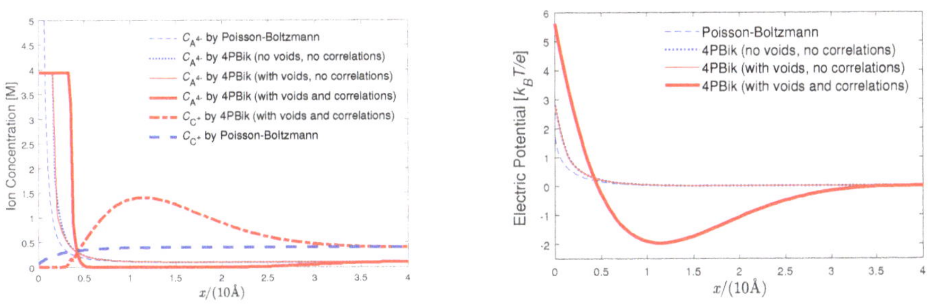

Figure 6. Left (L): Concentration profiles of anions $C_{A^{4-}}(x)$ and cations $C_{C^+}(x)$ obtained by various models in a C_4A electrolyte solution with the charge density $\sigma = 1e/(50\ \text{Å}^2)$ at $x = 0$. **Right (R)**: Electric potential profiles $\phi(x)$.

8.3. Biological Ion Channels

Biological ion channels are a particularly appropriate test of a model of concentrated ionic solutions.

The data available for tens to hundreds of different types of channels and transporters is breathtaking: it is often accurate to a few per cent (because signal to noise ratios are so large and biological variation hardly exists for channels of known amino acid sequence, which means nearly every channel presently). The data is always nonequilibrium, i.e., current voltage relations in a wide range of solutions of different composition and concentration, or (limiting zero voltage) conductance in those solutions. Indeed, many of the channels do not function if concentrations are equal on both sides and the electrical potential is zero. They are said to inactivate.

The data is often available for single channels recorded individually in patch clamp or bilayer configuration. Data is available for a range of divalent (usually calcium ion) concentrations because calcium concentration is often a controller of channel, transporter, and biological activity in the same sense that a gas pedal is the controller of the speed of a car. The structure of the ion channel or transporter is often known in breathtaking detail. The word 'breathtaking' is appropriate because similar structures are rarely if ever known of strictly physical systems. The structure and the structure of the permanent and polarization charge of the channel protein (that forms the pore through which ions move) can be modified by standard methods of site directed mutagenesis, for example that are available in 'kit' form usable by most molecular biology laboratories. Thus, models can be tested from atomic detail to single-channel function to ensemble function to cellular and physiological function, even to the ultimate biological function (like the rate of the heartbeat). Few other systems allow experimental measurement at each level of the hierarchy linking the atomic composition of genes (that encode the channel's amino acid composition), to the atomic structure of the channel, right to the function of the cell. The hierarchy here reaches from 10^{-11} to 10^{-5} m. When the channel controls the biological function of an organ like the heart, the hierarchy reaches to $2 \cdot 10^{-1}$ m, in humans for example.

The biological significance of ion channels is hard to exaggerate since they play a role in organisms analogous to the role of transistors in computers. They are the device that execute most of the physical controls of current and ion movement that are then combined in a hierarchy of structures to make biological cells, tissues, and organisms, if not populations of organisms.

From a physical point of view, ion channels provide a particularly crowded environment in which the effects of the steric potential (crowding in more traditional language) and electrical potential can combine to produce striking characteristics of selectivity and rectification. Theories that do not deal explicitly with ion channel data, i.e., that do not predict current voltage relations from known structures, seem to us to be begging central PHYSICAL questions that might falsify their approach. In fact, as a matter of history it is a fact that we learned how to construct our model of bulk solutions from our earlier work on ion channels.

8.3.1. Gramicidin A Channel

We use the gramicidin A (GA) channel in Figure 2L to illustrate the full Poisson–Nernst–Planck–Bikerman theory consisting of Equations (4), (27), (28), (32)–(34), and conditions (35)–(39) with—steric, correlation, polarization, dielectric, charge/space competition, and nonequilibrium effects—at steady state using the algorithm and methods in Section 7 to perform numerical simulations. The union domain $\overline{\Omega}_i \cup \Omega_{sh}$ in Figure 1 is replaced by the biomolecular domain Ω_m in Figure 2R.

Figure 7L shows I-V curves obtained by PNPB and compared with experimental data (symbols) by Cole et al. [197] with bath K^+ and Cl^- concentrations $C^B = 0.1, 0.2, 0.5, 1, 2$ M and membrane potentials $\Delta V = 0, 50, 100, 150, 200$ mV. The PNPB currents in pico ampere (pA) were obtained with $\theta = 1/4.7$ in the pore diffusion coefficients θD_i from (30) for all particle species. The reduction parameter θ has been used in all previous PNP papers and is necessary for continuum is comparable to MD, BD, or experimental data [198]. The values of other model parameters are listed in Table I in [68].

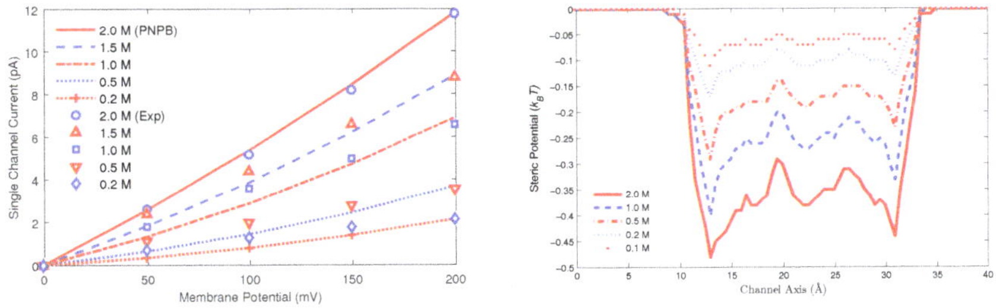

Figure 7. Left (L): A comparison of PNPB (curves) and experimental [197] (symbols) I-V results with bath K^+ and Cl^- concentrations $C^B = 0.1, 0.2, 0.5, 1, 2$ M and membrane potentials $\Delta V = 0, 50, 100, 150, 200$ mV. **Right (R)**: Averaged steric potential $S^{trc}(\mathbf{r})$ profiles at each cross section along the pore axis with $C^B = 0.1, 0.2, 0.5, 1, 2$ M and $\Delta V = 200$ mV. The same averaging method applies to the following profiles.

We summarize the novel results of PNPB in [68] when compared with those of earlier continuum models for ion channels: (i) The pore *diffusion* parameter $\theta = 1/4.7$ agrees with the range 1/3 to 1/10 obtained by many MD simulations of various channel models [199–201] indicating that the steric (Figure 7R), correlation, dehydration (Figure 8L), and dielectric (Figure 8R) properties have made PNPB simulations closer (realistic) to MD than previous PNP for which θ differs from MD values by an order to several orders of magnitude [200]. (ii) Figures 7R and 8L,R, which are all absent in earlier work, show that these properties *correlate* to each other and *vary* with salt concentration and protein charges in a *self-consistent* way by PNPB. (iii) The steric potential profiles in Figure 7R clearly illustrate the *charge/space competition* between ions and water under dynamic and variable conditions. For example, the global minimum value in Figure 7R at $\widehat{r} = 13.1$ on the channel axis, where the channel protein is most negatively charged, is $S^{trc}(\widehat{r}) = \ln \frac{\Gamma(\widehat{r})}{\Gamma^B} = -0.485$ yielding $\Gamma(\widehat{r})/\Gamma^B = 0.616$. Namely it is 38.4% more crowded at $\widehat{r}$ than in the bath and mainly occupied by K^+ as shown in Figures 8L and 9L. It is important to *quantify voids* ($\Gamma(\mathbf{r}) = 1 - \sum_{i=1}^{K+1} v_i C_i(\mathbf{r})$) at highly charged locations in channel proteins and many more biological, chemical, and nano systems. The charge space competition has been a central topic in the study of ion channels since at least [202–206]. The literature is too large to describe in detail here. Recent reviews can help [207–210]. (iv) PNPB preserves *mass conservation* due to void and size effects in contrast to PNP as shown in Figure 9R, where the total number of H_2O and K^+ in the channel pore is 8 [211].

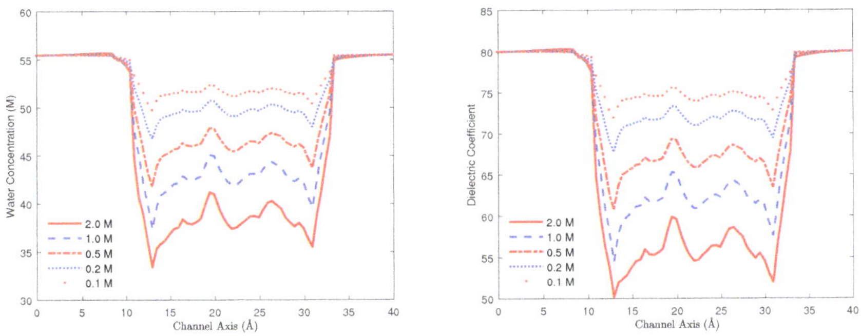

Figure 8. Left (L): Water concentration $C_{H_2O}(\mathbf{r})$ profiles. **Right (R)**: Dielectric function $\widetilde{\varepsilon}(\mathbf{r})$ profiles.

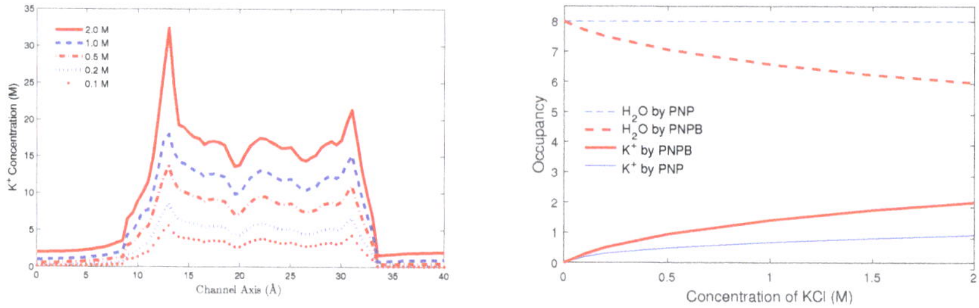

Figure 9. Left (L): K^+ concentration $C_{K^+}(\mathbf{r})$ profiles. **Right (R)**: Occupancy of H_2O and K^+ in the GA channel pore by PNPB and PNP as [KCl] increases from 0 to 2 M. The total number of H_2O and K^+ in the pore is 8 [211], which is conserved by PNPB but not by PNP (without steric and correlation effects).

8.3.2. L-Type Calcium Channel

L-type calcium channels operate very delicately in physiological and experimental conditions. They exquisitely tune their conductance from Na^+-flow, to Na^+-blockage, and to Ca^{2+}-flow when bath Ca^{2+} varies from trace to high concentrations as shown by the single-channel currents in femto ampere in Figure 10L (circle symbols) recorded by Almers and McCleskey [212], where the range of extracellular concentrations $[Ca^{2+}]_o$ is 10^8-fold from 10^{-10} to 10^{-2} M.

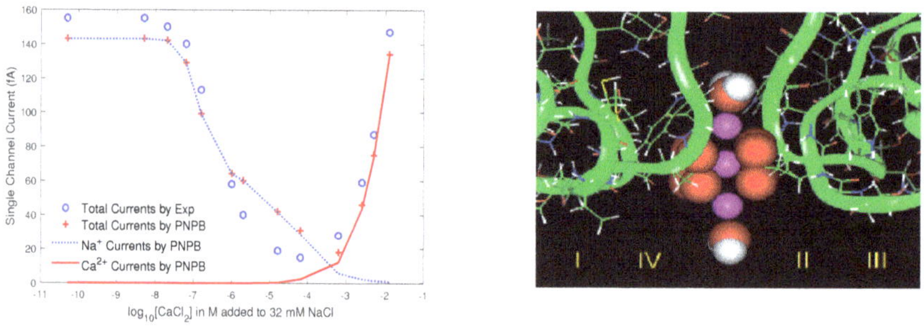

Figure 10. Left (L): Single-channel currents in femto ampere (fA) plotted as a function of $\log_{10}[Ca^{2+}]_o$. Experimental data of [212] are marked by small circles and PNPB data are denoted by the plus sign and lines. **Right (R)**: The Lipkind–Fozzard pore model of L-type calcium channel, where 3 Ca^{2+} are shown in violet, 8 $O^{1/2-}$ in red, 2 H_2O in white and red. Reprinted with permission from (G. M. Lipkind and H. A. Fozzard, Biochem. 40, 6786 (2001)). Copyright (2001) American Chemical Society.

We used the Lipkind-Fozzard molecular model [213] shown in Figure 10R to perform PNPB simulations with both atomic and continuum methods (Algorithm 2 in [68]) for this model channel, where the EEEE locus (four glutamate side chains modeled by 8 $O^{1/2-}$ ions shown by red spheres) forms a high-affinity Ca^{2+} binding site (center violet sphere) that is essential to Ca^{2+} selectivity, blockage, and permeation. Water molecules are shown in white and red. More realistic structures would be appropriate if the work were done now, but the analysis here shows the ability of PNPB to deal with experimental data using even a quite primitive model of the structure.

PNPB results (plus symbols) in Figure 10L agree with the experimental data at $[Na^+]_i = [Na^+]_o =$ 32 mM, $[Ca^{2+}]_i = 0$, $V_o = 0$, and $V_i = -20$ mV (the intracellular membrane potential), where the partial Ca^{2+} and Na^+ currents are denoted by the solid and dotted line, respectively. These two ionic currents show the anomalous mole fraction effect of the channel at nonequilibrium, i.e., trace concentrations of Ca^{2+} ions effectively block the flow of abundant monovalent cations [212].

8.3.3. Potassium Channel

Potassium channels conduct K^+ ions very rapidly (nearly at the diffusion rate limit (10^8 per second) in bulk water) and selectively (excluding, most notably, Na^+ despite their difference in radius is only $a_{K^+} - a_{Na^+} = 1.33 - 0.95 = 0.38$ Å in sub-Angstrom range) [9]. Figure 11 shows the structure of KcsA (PDB ID 3F5W) crystallized by Cuello et al. [214], where the spheres denote five specific cation binding sites (S0 to S4) [215] in the solvent domain Ω_s and the channel protein in Ω_m consists of N = 31,268 charged atoms. The exquisite selectivity of K^+ over Na^+ by K channels can be quantified by the free energy (G) differences of K^+ and Na^+ in the pore and in the bulk solution, i.e., by $\Delta G(K^+) = [G_{\text{pore}}(K^+) - G_{\text{bulk}}(K^+)]$ and $\Delta G(Na^+) = G_{\text{pore}}(Na^+) - G_{\text{bulk}}(Na^+)$ [215]. Experimental measurements [216–218] showed that the relative free energy

$$\Delta\Delta G(K^+ \to Na^+) = \Delta G(Na^+) - \Delta G(K^+) = 5{\sim}6 \text{ kcal/mol} \tag{59}$$

unfavorable for Na^+.

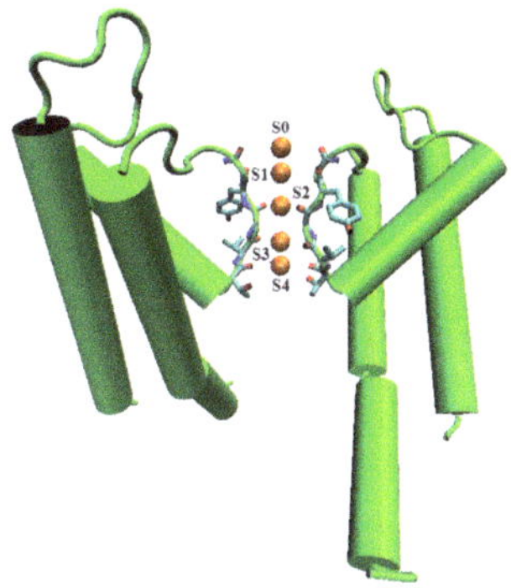

Figure 11. The crystal structure of the K channel KcsA (PDB ID 3F5W) [214] with five cation binding sites S0, S1, S2, S3, and S4 [215] marked by spheres.

Free energies can be calculated by the electric and steric potentials [72]

$$\phi_{S2} = \frac{1}{4\pi\epsilon_0}\left(\frac{1}{6}\sum_{k=1}^{6}\sum_{j=1}^{N}\frac{q_j}{\epsilon_p(r)|c_j - A_k|} + \frac{q_{S2}}{\epsilon_b a_{S2}}\right),\ S_{S2}^{trc} = \ln\frac{1-\frac{v_{S2}}{V_{S2}}}{\Gamma^B}, \tag{60}$$

at the binding site S2 [215] on the atomic scale, where S2 also denotes Na^+ or K^+ (the site is occupied by a Na^+ or K^+), q_j is the charge on the atom j in the protein given by PDB2PQR [219], $\epsilon_p(r) = 1 + 77r/(27.7 + r)$ [119], $r = |c_j - c_{S2}|$, c_j is the center of atom j, A_k is one of six symmetric surface points on the spherical S2, $\epsilon_b = 3.6$, and $V_{S2} = 1.5v_{K^+}$ is a volume containing the ion at S2. We obtained $\Delta\Delta G = 5.26$ kcal/mol [72] in accord with the MD result 5.3 kcal/mol [215], where $G_{\text{pore}}(Na^+) = 4.4$, $G_{\text{bulk}}(Na^+) = -0.26$, $G_{\text{pore}}(K^+) = -0.87$, $G_{\text{bulk}}(K^+) = -0.27$ kcal/mol, $\phi_{Na^+} = 7.5\ k_BT/e$, $\frac{v_{Na^+}}{v_0}S_{Na^+}^{trc} = 0.23$, $\phi_{K^+} = -1.93\ k_BT/e$, $\frac{v_{K^+}}{v_0}S_{K^+}^{trc} = -0.59$, and $C_{Na^+}^B = C_{K^+}^B = 0.4$ M.

The crucial parameter in (60) is the ionic radius $a_{S2} = 0.95$ or 1.33 Å (also in $|c_j - A_k|$) that affects ϕ_{S2} very strongly but S_{S2}^{trc} weakly. Another important parameter in (60) is the bulk void fraction Γ^B that depends on the bulk concentrations of all ions and water and links the total energy of the ion at S2 to these bulk conditions measured very far away ($\sim 10^6$ Å) in the baths on the atomic scale.

8.3.4. Sodium Calcium Exchanger

The Na^+/Ca^{2+} exchanger (NCX) is the major cardiac mechanism that extrudes intracellular Ca^{2+} across the cell membrane against its chemical gradient by using the downhill gradient of Na^+ [28].

The molecular basis of Na^+/Ca^{2+} interactions in NCX so striking to Lüttgau and Niedegerke [220] have been revealed by the cloning of NCX gene [221] and the structure of the ancient archaebacterial version NCX_Mj determined by Liao et al. [222]. Figure 12L illustrates NCX_Mj that consists of 10 transmembrane (TM) helices in which eight helices (TMs 2 to 5 and 7 to 10 labeled numerically in the figure) form a binding pocket of three putative Na^+ (green spheres) and one Ca^{2+} (blue sphere) binding sites [222].

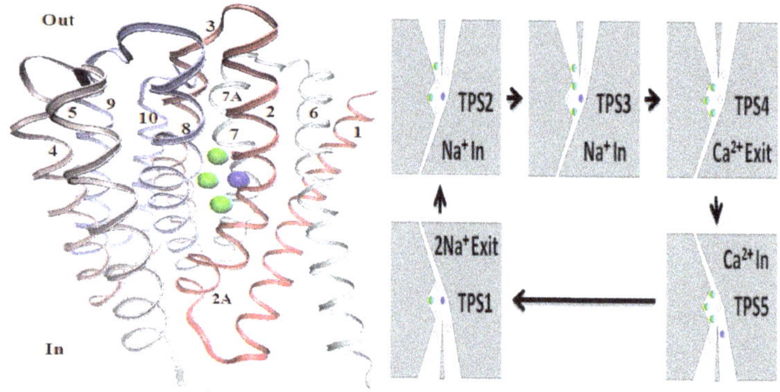

Figure 12. Left (L): Structure of NCX_Mj consisting of ten transmembrane helices that form a binding pocket of three Na^+ (green spheres) and one Ca^{2+} (blue sphere) binding sites [222]. **Right (R)**: Schematic diagram of a cycle of Na^+/Ca^{2+} exchange in NCX consisting of five total potential states (TPS). Two Na^+ and one Ca^{2+} ions enter the binding pocket in the outward- (TPS2 $\rightarrow$ TPS3 $\rightarrow$ TPS4) and inward-facing (TPS5 $\rightarrow$ TPS1) conformations, respectively. They exit in opposite conformations [70].

We developed a cyclic model of Na^+/Ca^{2+} exchange mechanism in NCX [70] using (60) to calculate five total (electric and steric) potential states (TPS) of various Na^+ and Ca^{2+} ions shown in Figure 12R, where TPS1 and TPS4 are stable (with negative values) and TPS2, TPS3, and TPS5 are unstable (positive). Four extra sites in Figure 12R are determined empirically and close to entrance or exit locations of the binding pocket. The green and blue dots in the diagram represent Na^+ and Ca^{2+} ions occupying the respective sites. Two Na^+ and one Ca^{2+} ions enter the binding pocket in the outward- (TPS2 $\rightarrow$ TPS3 $\rightarrow$ TPS4) and inward-facing (TPS5 $\rightarrow$ TPS1) conformations, respectively. They exit in opposite conformations. The cycle consists of five steps.

Step 1: A conformational change is hypothetically activated [70] by a binding Ca^{2+} at the blue site (S1) in TPS1 from inward-facing to outward-facing in TPS2.

Step 2: One Na^+ enters the binding pocket from the access site in TPS2 to the top Na^+ binding site (S2) in TPS3 followed by another Na^+ to the access site. These two coming Na^+ ions move the existing Na^+ ion from the middle Na^+ site (S3) to the bottom site (S4) by their Coulomb forces. TPS2 and TPS3 are unstable meaning that the two coming Na^+ ions have positive energies and are thus mobile. The selectivity ratio of Na^+ to Ca^{2+} by NCX from the extracellular bath to the binding site S2 is $C_{Na^+}(S2)/C_{Ca^{2+}}(S2) = 55.4$ under the experimental conditions of the extracellular bath $[Na^+]_o = 120$ mM and $[Ca^{2+}]_o = 1$ µM [70].

Step 3: The vacant site S3 in TPS3 is a deep potential well with TP = $-8.89\ k_BT/e$ that pulls the two unstable Na^+ ions to their sites in TPS4. Meanwhile, these two moving Na^+ and the stable Na^+ at S4 extrude the Ca^{2+} (with unstable TP = 1.65) at S1 out of the pocket to become TPS4.

Step 4: Now, all three Na^+ ions in TPS4 are stable with negative TP and the vacant site S1 has an even deeper TP = -16.02. We conjecture that this TP value may trigger a conformational change from outward-facing in TPS4 to inward-facing in TPS5. The mechanism of conformational changes in NCX is yet to be studied.

Step 5: Furthermore, this large negative TP in TPS5 yields a remarkably large *selectivity* ratio of Ca^{2+} to Na^{+} by NCX from the intracellular bath to S1, i.e., $C_{Ca^{2+}}(S1)/C_{Na^{+}}(S1) = 4986.1$ at $\left[Ca^{2+}\right]_i = 33$ μM and $\left[Na^{+}\right]_i = 60$ mM. A coming Ca^{2+} in TPS5 then extrudes two Na^{+} ions out of the packet when it settles at S1 in stable TPS1.

Assuming that the total time T of an exchange cycle is equally shared by the 5 TPS, this model also infers that the stoichiometry of NCX is $\frac{3}{5}T \cdot 2\ Na^{+} : \frac{2}{5}T \cdot 1\ Ca^{2+} = 3\ Na^{+} : 1\ Ca^{2+}$ in transporting Na^{+} and Ca^{2+} ions [70], which is the generally accepted stoichiometry (see reviews of Blaustein and Lederer [223] and Dipolo and Beaugé [224]) since the pivotal work of Reeves and Hale [225] and other subsequent experimental results. Please note that our model does not consider the electrogenic property of NCX [223], i.e., the driving force of the electric potential gradient.

9. Conclusions

We have covered a range of aspects of the fourth-order Poisson-Nernst-Planck-Bikerman theory from physical modeling, mathematical analysis, numerical implementation, to applications and verifications for aqueous electrolyte systems in chemistry and biology. The theory can describe many properties of ions and water in the system that classical theories fail to describe such as steric, correlation, polarization, variable permittivity, dehydration, mass conservation, charge/space competition, overscreening, selectivity, saturation, and more. All these properties are accounted for in a single framework with only two fundamental parameters, namely the dielectric constant of pure water and the correlation length of empirical choice. Ions and water have their physical volumes as those in molecular dynamic simulations. The theory applies to a system at both continuum and atomic scales due to the exact definition of the total volume of all ions, water molecules, and interstitial voids.

The most important features of PNPB are that (i) ions and water have unequal volumes with interstitial voids, (ii) their distributions are saturating of the Fermi type, (iii) these Fermi distributions approach Boltzmann distributions as the volumes tend to zero, and (iv) all the above physical properties appear self-consistently in a single model not separately by various models. Most existing modified Poisson-Boltzmann models consider ions of equal size and fail to yield Boltzmann distributions in limiting cases, i.e., the limit is divergent indicating that steric energies are poorly estimated. Numerous models for different properties such as steric, correlation, polarization, permittivity are proposed separately in the past.

We have shown how to solve 4PBik analytically and PNPB numerically. The generalized Debye-Hückel theory derived from the 4PBik model gives valuable insights into physical properties and leads to an electrolyte (analytical) equation of state that is useful to study thermodynamic activities of ion and water under wide ranges of composition, concentration, temperature, and pressure.

Numerically solving the fourth-order PNPB model in 3D for realistic problems is a challenging task. There are many pitfalls that one must carefully avoid in coding. For that reason, we have particularly mentioned some methods for handling the convergence issues of the highly nonlinear PNPB system of partial differential equations and the discretization problems concerning the complicate interface between molecular and solvent domains and the Scharfetter-Gummel stability condition to ensure positivity of numerical concentrations and current preservation.

Finally, we have shown novel results obtained by PNPB for chemical and biological systems on ion activities, electric double layers, gramicidin A channel, L-type calcium channel, potassium channel, and sodium calcium exchanger. These results agree with experiments or molecular dynamics data and show not only continuum but also atomic properties of the system under far-field conditions. The fourth-order PNPB model is consistent and applicable to a great variety of systems on a vast scale from *meter* to *Angstrom*.

Author Contributions: J.-L.L. developed the theory, performed the computations, and wrote the manuscript. B.E. conceived the topics, verified the results, and helped shape the research, analysis, and manuscript. All authors have read and agreed to the published version of the manuscript.

Funding: This work was funded by the Ministry of Science and Technology, Taiwan under Grant MOST107-2115-M-007-017-MY2 to J.-L.L.

Conflicts of Interest: The authors declare no conflict of interest.

Abbreviations

The following abbreviations are used in this manuscript:

4PBik	Fourth-Order Poisson-Bikerman
BD	Brownian Dynamics
bi-CG	bi-Conjugate Gradient Stabilized
DH	Debye-Hückel
EDL	Electric Double-Layer
FD	Finite Difference
GA	Gramicidin A
JESS	Joint Expert Speciation System
L-J	Lennard–Jones
MC	Monte Carlo
MD	Molecular Dynamics
NCX	Sodium Calcium Exchanger
NP	Nernst-Planck
ODE	Ordinary Differential Equation
OZ	Ornstein-Zernike
PB	Poisson–Boltzmann
PDE	Partial Differential Equation
PDB	Protein Data Bank
PNP	Poisson-Nernst-Planck
PNPB	Poisson-Nernst-Planck-Bikerman
SG	Scharfetter-Gummel
TPS	Total Potential State

References

1. Robinson, R.A.; Stokes, R.H. *Electrolyte Solutions*; Courier Corporation: North Chelmsford, MA, USA, 2002.
2. Zemaitis, J.F.; Clark, D.M.; Rafal, M.; Scrivner, N.C. *Handbook of Aqueous Electrolyte Thermodynamics*; Design Institute for Physical Property Data, American Institute of Chemical Engineers: New York, NY, USA, 1986.
3. Sharp, K.A.; Honig, B. Electrostatic interactions in macromolecules: Theory and applications. *Annu. Rev. Biophys. Biophys. Chem.* **1990**, *19*, 301–332.
4. Newman, J. *Electrochemical Systems*; Prentice-Hall: Upper Saddle River, NJ, USA, 1991.
5. Honig, B.; Nicholls, A. Classical electrostatics in biology and chemistry. *Science* **1995**, *268*, 1144–1149.
6. Pitzer, K.S. *Thermodynamics*; McGraw Hill: New York, NY, USA, 1995.
7. Attard, P. Electrolytes and the electric double layer. *Adv. Chem. Phys.* **1996**, *92*, 1–159.
8. Eisenberg, R.S. Computing the field in protein and channels. *J. Membr. Biol.* **1996**, *150*, 1–25.
9. Hille, B. *Ionic Channels of Excitable Membranes*; Sinauer Associates Inc.: Sunderland, MA, USA, 2001.
10. Fogolari, F.; Brigo, A.; Molinari, H. The Poisson-Boltzmann equation for biomolecular electrostatics: A tool for structural biology. *J. Mol. Recognit.* **2002**, *15*, 379–385.
11. Laidler, K.J.; Meiser, J.H.; Sanctuary, B.C. *Physical Chemistry*; Brooks Cole: Belmont, CA, USA, 2003.
12. Fawcett, W.R. *Liquids, Solutions, and Interfaces: From Classical Macroscopic Descriptions to Modern Microscopic Details*; Oxford University Press: New York, NY, USA, 2004.
13. Lebon, G.; Jou, D.; Casas-Vazquez, J. *Understanding Non-Equilibrium Thermodynamics: Foundations, Applications, Frontiers*; Springer: Berlin, Germany, 2008.
14. Bazant, M.Z.; Kilic, M.S.; Storey, B.D.; Ajdari, A. Towards an understanding of induced-charge electrokinetics at large applied voltages in concentrated solutions. *Adv. Coll. Interf. Sci.* **2009**, *152*, 48–88.
15. Kontogeorgis, G.M.; Folas, G.K. *Thermodynamic Models for Industrial Applications: From Classical and Advanced Mixing Rules to Association Theories*; John Wiley & Sons: Hoboken, NJ, USA, 2009.

16. Kunz, W. *Specific Ion Effects*; World Scientific: Singapore, 2010.
17. Eisenberg, B. *Crowded Charges in Ion Channels, Advances in Chemical Physics*; John Wiley & Sons, Inc.: Hoboken, NJ, USA, 2011; pp. 77–223.
18. Hunenberger, P.; Reif, M. *Single-Ion Solvation. Experimental and Theoretical Approaches to Elusive Thermodynamic Quantities*; Royal Society of Chemistry: London, UK, 2011.
19. Voigt, W. Chemistry of salts in aqueous solutions: Applications, experiments, and theory. *Pure Appl. Chem.* **2011**, *83*, 1015–1030.
20. Merlet, C.; Rotenberg, B.; Madden, P.A.; Taberna, P.L.; Simon, P.; Gogotsi, Y.; Salanne, M. On the molecular origin of supercapacitance in nanoporous carbon electrodes. *Nat. Mater.* **2012**, *11*, 306.
21. Wei, G.-W.; Zheng, Q.; Chen, Z.; Xia, K. Variational multiscale models for charge transport. *SIAM Rev.* **2012**, *54*, 699–754.
22. Fedorov, M.V.; Kornyshev, A.A. Ionic liquids at electrified interfaces. *Chem. Rev.* **2014**, *114*, 2978–3036.
23. Köpfer, D.A.; Song, C.; Gruene, T.; Sheldrick, G.M.; Zachariae, U.; de Groot, B.L. Ion permeation in K^+ channels occurs by direct Coulomb knock-on. *Science* **2014**, *346*, 352–355.
24. MacFarlane, D.R.; Tachikawa, N.; Forsyth, M.; Pringle, J.M.; Howlett, P.C.; Elliott, G.D.; Davis, J.H., Jr.; Watanabe, M.; Simon, P.; Angell, A. Energy applications of ionic liquids. *Energy Environ. Sci.* **2014**, *7*, 232–250.
25. Tang, L.; El-Din, T.M.G.; Payandeh, J.; Martinez, G.Q.; Heard, T.M.; Scheuer, T.; Zheng, N.; Catterall, W.A. Structural basis for Ca^{2+} selectivity of a voltage-gated calcium channel. *Nature* **2014**, *505*, 56–62.
26. Perreault, F.; De Faria, A.F.; Elimelech, M. Environmental applications of graphene-based nanomaterials. *Chem. Soci. Rev.* **2015**, *44*, 5861–5896.
27. Pilon, L.; Wang, H.; d'Entremont, A. Recent advances in continuum modeling of interfacial and transport phenomena in electric double layer capacitors. *J. Electrochem. Soci.* **2015**, *162*, A5158–A5178.
28. Shattock, M.J.; Ottolia, M.; Bers, D.M.; Blaustein, M.P.; Boguslavskyi, A.; Bossuyt, J.; Chen-Izu, Y.; Clancy, C.E.; Edwards, A.; Goldhaber, J.; et al. Na^+/Ca^{2+} exchange and Na^+/K^+-ATPase in the heart. *J. Physiol.* **2015**, *593*, 1361–1382.
29. Zamponi, G.W.; Striessnig, J.; Koschak, A.; Dolphin, A.C. The physiology, pathology, and pharmacology of voltage-gated calcium channels and their future therapeutic potential. *Pharmacol. Rev.* **2015**, *67*, 821–870.
30. Zheng, J.; Trudeau, M.C. *Handbook of Ion Channels*; CRC Press: Boca Raton, FL, USA, 2015.
31. Smith, A.M.; Lee, A.A.; Perkin, S. The electrostatic screening length in concentrated electrolytes increases with concentration. *J. Phys. Chem. Lett.* **2016**, *7*, 2157–2163.
32. Vera, J.H., Wilczek-Vera, G. *Classical Thermodynamics of Fluid Systems: Principles and Applications*; CRC Press: Boca Raton, FL, USA, 2016.
33. Faucher, S.; Aluru, N.; Bazant, M.Z.; Blankschtein, D.; Brozena, A.H.; Cumings, J.; de Souza, J.P.; Elimelech, M.; Epsztein, R.; Fourkas, J.T.; et al. Critical knowledge gaps in mass transport through single-digit nanopores: A review and perspective. *J. Phys.Chem. C* **2019**, *123*, 21309–21326.
34. Lian, T.; Koper, M.T.; Reuter, K.; Subotnik, J.E. Special topic on interfacial electrochemistry and photo(electro)catalysis. *J. Chem. Phys.* **2019**, *150*, 041401.
35. Chisholm, H.; Poisson, S.D. *Encyclopadia Britannica*; Cambridge University Press: Cambridge, UK, 1911; p. 896.
36. Bjerrum, N.J.; Bohr. N. *Niels Bjerrum: Selected Papers*; Edited by Friends and Coworkers on the Occasion of His 70th Birthday the 11th of March, 1949; Munksgaard: Copenhagen, Denmark, 1949.
37. Cercignani, C. *The Boltzmann Equation and Its Applications*; Springer: New York, NY, USA, 1988.
38. Nernst, W. Die elektromotorische wirksamkeit der jonen. *Z. Phys. Chem.* **1889**, *4*, 129–181.
39. Planck, M. Ueber die erregung von electricität und warme in electrolyten. *Ann. Der Phys.* **1890**, *275*, 161–186.
40. Gouy, M. Sur la constitution de la charge electrique a la surface d'un electrolyte (Constitution of the electric charge at the surface of an electrolyte). *J. Phys.* **1910**, *9*, 457–468.
41. Chapman, D.L. A contribution to the theory of electrocapillarity. *Phil. Mag.* **1913**, *25*, 475–481.
42. Bikerman, J.J. Structure and capacity of electrical double layer. *Philos. Mag.* **1942**, *33*, 384–397.
43. Eisenberg, R.; Chen, D. Poisson-Nernst-Planck (PNP) theory of an open ionic channel. *Biophys. J.* **1993**, *64*, A22.
44. Shockley, W. *Electrons and Holes in Semiconductors to Applications in Transistor Electronics*; van Nostrand: New York, NY, USA, 1950.

45. Van Roosbroeck, W. Theory of flow of electrons and holes in germanium and other semiconductors. *Bell Syst. Tech. J.* **1950**, *29*, 560–607.
46. Blotekjaer, K. Transport equations for electrons in two-valley semiconductors. *IEEE Trans. Elec. Dev.* **1970**, *17*, 38–47.
47. Kahng, D. A historical perspective on the development of MOS transistors and related devices. *IEEE Trans. Elec. Dev.* **1976**, *23*, 655–657.
48. Shockley, W. The path to the conception of the junction transistor. *IEEE Trans. Elec. Dev.* **1976**, *23*, 597–620.
49. Teal, G.K. Single crystals of germanium and silicon-Basic to the transistor and integrated circuit. *IEEE Trans. Elec. Dev.* **1976**, *23*, 621–639.
50. Selberherr, S. *Analysis and Simulation of Semiconductor Devices*; Springer: New York, NY, USA, 1984.
51. Jacoboni, C.; Lugli, P. *The Monte Carlo Method for Semiconductor Device Simulation*; Springer: New York, NY, USA, 1989.
52. Markowich, P.A.; Ringhofer, C.A.; Schmeiser, C. *Semiconductor Equations*; Springer: New York, NY, USA, 1990.
53. Jerome, J.W. *Analysis of Charge Transport. Mathematical Theory and Approximation of Semiconductor Models*; Springer: New York, NY, USA, 1995.
54. Ferry, D.K.; Goodnick, S.M.; Bird, J. *Transport in Nanostructures*; Cambridge University Press: New York, NY, USA, 2009.
55. Vasileska, D.; Goodnick, S.M.; Klimeck, G. *Computational Electronics: Semiclassical and Quantum Device Modeling and Simulation*; CRC Press: New York, NY, USA, 2010.
56. Sakmann, B.; Neher, E. *Single Channel Recording*; Plenum: New York, NY, USA, 1995.
57. Berman, H.M.; Westbrook, J.; Feng, Z.; Gilliland, G.; Bhat, T.N.; Weissig, H.; Shindyalov, I.N.; Bourne, P.E. The protein data bank. *Nucleic Acids Res.* **2000**, *28*, 235-242.
58. Gross, L.; Mohn, F.; Moll, N.; Liljeroth, P.; Meyer, G. The chemical structure of a molecule resolved by atomic force microscopy. *Science* **2009**, *325*, 1110–1114.
59. Shklovskii, B.I. Screening of a macroion by multivalent ions: Correlation-induced inversion of charge. *Phys. Rev. E* **1999**, *60*, 5802–5811.
60. Levin, Y. Electrostatic correlations: From plasma to biology. *Rep. Prog. Phys.* **2002**, *65*, 1577–1632.
61. Abrashkin, A.; Andelman, D.; Orland, H. Dipolar Poisson-Boltzmann equation: Ions and dipoles close to charge interfaces. *Phys. Rev. Lett.* **2007**, *99*, 077801.
62. Levy, A.; Andelman, D.; Orland, H. Dielectric constant of ionic solutions: A field-theory approach. *Phys. Rev. Lett.* **2012**, *108*, 227801.
63. Gavish, N.; Promislow, K. Dependence of the dielectric constant of electrolyte solutions on ionic concentration: A microfield approach. *Phys. Rev. E* **2016**, *94*, 012611.
64. Liu, J.-L. Numerical methods for the Poisson-Fermi equation in electrolytes. *J. Comput. Phys.* **2013**, *247*, 88–99.
65. Liu, J.-L.; Eisenberg, B. Correlated ions in a calcium channel model: A Poisson-Fermi theory. *J. Phys. Chem. B* **2013**, *117*, 12051–12058.
66. Liu, J.-L.; Eisenberg, B. Poisson-Nernst-Planck-Fermi theory for modeling biological ion channels. *J. Chem. Phys.* **2014**, *141*, 22D532.
67. Liu, J.-L.; Eisenberg, B. Analytical models of calcium binding in a calcium channel. *J. Chem. Phys.* **2014**, *141*, 075102.
68. Liu, J.-L.; Eisenberg, B. Numerical methods for a Poisson-Nernst-Planck-Fermi model of biological ion channels. *Phys. Rev. E* **2015**, *92*, 012711.
69. Liu, J.-L.; Eisenberg, B. Poisson-Fermi model of single ion activities in aqueous solutions. *Chem. Phys. Lett.* **2015**, *637*, 1–6.
70. Liu, J.-L.; Hsieh, H.-j.; Eisenberg, B. Poisson-Fermi modeling of the ion exchange mechanism of the sodium/calcium exchanger. *J. Phys. Chem. B* **2016**, *120*, 2658–2669.
71. Xie, D.; Liu, J.-L.; Eisenberg, B. Nonlocal Poisson-Fermi model for ionic solvent. *Phys. Rev. E* **2016**, *94*, 012114.
72. Liu, J.-L.; Xie, D.; Eisenberg, B. Poisson-Fermi formulation of nonlocal electrostatics in electrolyte solutions. *Mol. Based Math. Biol.* **2017**, *5*, 116–124.
73. Liu, J.-L.; Eisenberg, B. Poisson-Fermi modeling of ion activities in aqueous single and mixed electrolyte solutions at variable temperature. *J. Chem. Phys.* **2018**, *148*, 054501.

74. Chen, J.-H.; Chen, R.-C.; Liu, J.-L. A GPU Poisson-Fermi solver for ion channel simulations. *Comput. Phys. Commun.* **2018**, *229*, 99–105.
75. Liu, J.-L.; Li, C.-L. A generalized Debye-Huckel theory of electrolyte solutions. *AIP Adv.* **2019**, *9*, 015214.
76. Li, C.-L.; Liu, J.-L. Analysis of generalized Debye-Hückel equation from Poisson-Fermi theory. *arXiv* **2018**, arXiv:1801.03470.
77. Santangelo, C.D. Computing counterion densities at intermediate coupling. *Phys. Rev. E* **2006**, *73*, 041512.
78. Bazant, M.Z.; Storey, B.D.; Kornyshev, A.A. Double layer in ionic liquids: Overscreening versus crowding. *Phys. Rev. Lett.* **2011**, *106*, 046102.
79. Feller, W. *An Introduction to Probability Theory and Its Applications*; John Wiley & Sons: Hoboken, NJ, USA, 2008.
80. Karlin, S.; Taylor, H.E. *A Second Course in Stochastic Processes*; Elsevier: Amsterdam, The Netherlands, 1981.
81. Schrödinger, E. An undulatory theory of the mechanics of atoms and molecules. *Phys. Rev.* **1926**, *28*, 1049.
82. Fermi, E. Sulla quantizzazione del gas perfetto monoatomico. *Rend. Lincei* **1926**, *3*, 145–149. Translated as Zannoni, A. On the quantization of the monoatomic ideal gas. *arXiv* **1999**, arXiv:9912229.
83. Pauli, W. Über den Zusammenhang des Abschlusses der Elektronengruppen im Atom mit der Komplexstruktur der Spektren. *Z. Phys.* **1925**, *31*, 765–783.
84. Van der Waals, J.D. Thermodynamische Theorie der Capillariteit in de Onderstelling van Continue Dichtheidsverandering; Verhand. Kon. Akad. V Wetensch. Amst. Sect. 1, 1893. (English Translation, The thermodynamik theory of capillarity under the hypothesis of a continuous variation of density. *J. Stat. Phys.* **1979**, *20*, 197.
85. Hill, T.L. Steric effects. I. Van der Waals potential energy curves. *J. Chem. Phys.* **1948**, *16*, 399–404.
86. Regan, C.K.; Craig, S.L.; Brauman, J.I. Steric effects and solvent effects in ionic reactions. *Science* **2002**, *295*, 2245–2247.
87. Kornyshev, A.A. Double-layer in ionic liquids: Paradigm change? *J. Phys. Chem. B* **2007**, *111*, 5545–5557.
88. Hodgkin, A.L.; Huxley, A.F.; Katz, B. Ionic currents underlying activity in the giant axon of the squid. *Arch. Sci. Physiol.* **1949**, *3*, 129–150.
89. Bezanilla, F. The voltage sensor in voltage-dependent ion channels. *Physiol. Rev.* **2000**, *80*, 555–592.
90. Bezanilla, F.; Villalba-Galea, C.A. The gating charge should not be estimated by fitting a two-state model to a Q-V curve. *J. Gen. Physiol.* **2013**, *142*, 575–578.
91. McQuarrie, D.A. *Statistical Mechanics*; Harper and Row: New York, NY, USA, 1976.
92. Verlet, L. Computer "experiments" on classical fluids. I. Thermodynamical properties of Lennard-Jones molecules. *Phys. Rev.* **1967**, *159*, 98–103.
93. Heinz, H.; Vaia, R.A.; Farmer, B.L.; Naik, R.R. Accurate simulation of surfaces and interfaces of face-centered cubic metals using 12- 6 and 9- 6 Lennard-Jones potentials. *J. Phys. Chem. C* **2008**, *112*, 17281–17290.
94. Lu, B.; McCammon, J.A. Molecular surface-free continuum model for electrodiffusion processes. *Chem. Phys. Lett.* **2008**, *451*, 282–286.
95. Simakov, N.A.; Kurnikova, M.G. Soft wall ion channel in continuum representation with application to modeling ion currents in α-Hemolysin. *J. Phys. Chem. B* **2010**, *114*, 15180–15190.
96. Hyon, Y.K.; Fonseca, J.E.; Eisenberg, B.; Liu, C. Energy variational approach to study charge inversion (layering) near charged walls. *Discret. Cont. Dyn. Sys. Ser. A* **2012**, *17*, 2725–2743.
97. Horng, T.-L.; Lin, T.-C.; Liu, C.; Eisenberg, B. PNP equations with steric effects: A model of ion flow through channels. *J. Phys. Chem. B* **2012**, *116*, 11422–11441.
98. Maffeo, C.; Bhattacharya, S.; Yoo, J.; Wells, D.; Aksimentiev, A. Modeling and simulation of ion channels. *Chem. Rev.* **2012**, *112*, 6250–6284.
99. Lin, T.-C.; Eisenberg, B. A new approach to the Lennard-Jones potential and a new model: PNP-steric equations. *Commun. Math. Sci.* **2014**, *12*, 149–173.
100. Gavish, N. Poisson–Nernst–Planck equations with steric effects—non-convexity and multiple stationary solutions. *Physica D* **2018**, *368*, 50–65.
101. Gavish, N.; Elad, D.; Yochelis, A. From solvent-free to dilute electrolytes: Essential components for a continuum theory. *Phys. Chem. Lett.* **2018**, *9*, 36–42.
102. Jackson, J.D. *Classical Electrodynamics*; Wiley: Hoboken, NJ, USA, 1999.
103. Zangwill, A. *Modern Electrodynamics*; Cambridge University Press: New York, NY, USA, 2013.
104. Eisenberg, B. Updating Maxwell with electrons, charge, and more realistic polarization. *arXiv* **2019**, arXiv:1904.09695.

105. Liu, J.-L. A quantum corrected Poisson-Nernst-Planck model for biological ion channels. *Mol. Based Math. Biol.* **2015**, *3*, 70–77.
106. Eisenberg, B.; Oriols, X.; Ferry, D. Dynamics of current, charge, and mass. *Mol. Based Math. Biol.* **2017**, *5*, 78–115.
107. Hildebrandt, A.; Blossey, R.; Rjasanow, S.; Kohlbacher, O.; Lenhof, H.-P. Novel formulation of nonlocal electrostatics. *Phys. Rev. Lett.* **2004**, *93*, 108104.
108. Rowlinson, J.S. The Yukawa potential. *Physica A* **1989**, *156*, 15–34.
109. Eisenberg, B. Dielectric dilemma. *arXiv* **2019**, arXiv:1901.10805.
110. Barthel, J.; Buchner, R.; Münsterer, M. *Electrolyte Data Collection Vol. 12, Part 2: Dielectric Properties of Water and Aqueous Electrolyte Solutions*; Frankfurt am Main: Dechema, Czech Republic, 1995.
111. Buchner, R.; Barthel, J. Dielectric relaxation in solutions. *Annu. Rep. Prog. Chem. Sect. C Phys. Chem.* **2001**, *97*, 349–382.
112. Yukawa, H. On the interaction of elementary particles. I. *Proc. Phys.-Math. Soc. Jpn. Ser.* **1935**, *17*, 48–57.
113. Rowlinson, J.S. *Cohesion: A Scientific History of Intermolecular Forces*; Cambridge University Press: Cambridge, UK, 2005.
114. Ornstein, L.S.; Zernike, F. Accidental deviations of density and opalescence at the critical point of a single substance. *R. Netherlands Acad. Arts Sci. Proc.* **1914**, *17*, 793–806.
115. Blossey, R.; Maggs, A.C.; Podgornik, R. Structural interactions in ionic liquids linked to higher-order Poisson-Boltzmann equations. *Phys. Rev. E* **2017**, *95*, 060602.
116. Downing, R.; Berntson, B.K.; Bossa, G.V.; May, S. Differential capacitance of ionic liquids according to lattice-gas mean-field model with nearest-neighbor interactions. *J. Chem. Phys.* **2018**, *149*, 204703.
117. Kornyshev, A.A.; Sutmann, G. The shape of the nonlocal dielectric function of polar liquids and the implications for thermodynamic properties of electrolytes: A comparative study. *J. Chem. Phys.* **1996**, *104*, 1524.
118. Schutz, C.N.; Warshel, A. What are the dielectric "constants" of proteins and how to validate electrostatic models? *Proteins Struct. Funct. Bioinf.* **2001**, *44*, 400–417.
119. Mallik, B.; Masunov, A.; Lazaridis, T. Distance and exposure dependent effective dielectric function. *J. Comput. Chem.* **2002**, *23*, 1090–1099.
120. Corry, B.; Kuyucak, S.; Chung, S.-H. Dielectric self-energy in Poisson-Boltzmann and Poisson-Nernst-Planck models of ion channels. *Biophys. J.* **2003**, *84*, 3594–3606.
121. Graf, P.; Kurnikova, M.G.; Coalson, R.D.; Nitzan, A. Comparison of dynamic lattice Monte Carlo simulations and the dielectric self-energy Poisson-Nernst-Planck continuum theory for model ion channels. *J. Phys. Chem. B* **2004**, *108*, 2006–2015.
122. Cheng, M.H.; Coalson, R.D. An accurate and efficient empirical approach for calculating the dielectric self-energy and ion-ion pair potential in continuum models of biological ion channels. *J. Phys. Chem. B* **2005**, *4*, 81–93.
123. Ng, J.A.; Vora, T.; Krishnamurthy, V.; Chung, S.-H. Estimating the dielectric constant of the channel protein and pore. *Eur. Biophys. J.* **2008**, *37*, 213–222.
124. Silalahi, A.R.J.; Boschitsch, A.H.; Harris, R.C.; Fenley, M.O. Comparing the predictions of the nonlinear Poisson-Boltzmann equation and the ion size-modified Poisson-Boltzmann equation for a low-dielectric charged spherical cavity in an aqueous salt solution. *J. Chem. Theory Comput.* **2010**, *6*, 3631.
125. Lopéz-García, J.J.; Horno, J.; Grosse, C. Poisson-Boltzmann description of the electrical double layer including ion size effects. *Langmuir* **2011**, *27*, 13970–13974.
126. Nakamura, I. Effects of dielectric inhomogeneity and electrostatic correlation on the solvation energy of ions in liquids. *J. Phys. Chem. B* **2018**, *122*, 6064–6071.
127. Kjellander, R. Focus Article: Oscillatory and longrange monotonic exponential decays of electrostatic interactions in ionic liquids and other electrolytes: The significance of dielectric permittivity and renormalized charges. *J. Chem. Phys.* **2018**, *148*, 193701.
128. Fogolari, F.; Briggs, J.M. On the variational approach to Poisson–Boltzmann free energies. *Chem. Phys. Lett.* **1997**, *281*, 135–139.
129. Li, B. Minimization of electrostatic free energy and the Poisson-Boltzmann equation for molecular solvation with implicit solvent. *SIAM J. Math. Anal.* **2009**, *40*, 2536–2566.

130. Sharp, K.A.; Honig, B. Calculating total electrostatic energies with the nonlinear Poisson-Boltzmann equation. *J. Phys. Chem.* **1990**, *94*, 7684–7692.
131. Reiner, E.S.; Radke, C.J. Variational approach to the electrostatic free energy in charged colloidal suspensions: General theory for open systems. *J. Chem. Soci. Faraday Trans.* **1990**, *86*, 3901–3912.
132. Gilson, M.K.; Davis, M.E.; Luty, B.A.; McCammon, J.A. Computation of electrostatic forces on solvated molecules using the Poisson-Boltzmann equation. *J. Phys. Chem.* **1993**, *97*, 3591–3600.
133. Borukhov, I.; Andelman, D.; Orland, H. Steric effects in electrolytes: A modified Poisson-Boltzmann equation. *Phys. Rev. Lett.* **1997**, *79*, 435–438.
134. Ben-Yaakov, D.; Andelman, D.; Podgornik, R. Ion-specific hydration effects: Extending the Poisson-Boltzmann theory. *Curr. Opin. Colloid Interface Sci.* **2011**, *16*, 542–550.
135. Lu, B.; Zhou, Y.C. Poisson-Nernst-Planck equations for simulating biomolecular diffusion-reaction processes II: Size effects on ionic distributions and diffusion-reaction rates. *Biophys. J.* **2011**, *100*, 2475–2485.
136. Zhou, S.; Wang, Z.; Li, B. Mean-field description of ionic size effects with non-uniform ionic sizes: A numerical approach. *Phys. Rev. E* **2011**, *84*, 021901.
137. Qiao, Y.; Tu, B.; Lu, B. Ionic size effects to molecular solvation energy and to ion current across a channel resulted from the nonuniform size-modified PNP equations. *J. Chem. Phys.* **2014**, *140*, 174102.
138. Grimley, T.B.; Mott, N.F. The contact between a solid and a liquid electrolyte. *Discuss. Faraday Soc.* **1947**, *1*, 3–11.
139. Tresset, G. Generalized Poisson-Fermi formalism for investigating size correlation effects with multiple ions. *Phys. Rev. E* **2008**, *78*, 061506.
140. Bohinc, K.; Kralj-Iglič, V.; Iglič, A. Thickness of electrical double layer. Effect of ion size. *Electrochim. Acta* **2001**, *46*, 3033–3040.
141. McEldrew, M.; Goodwin, Z.A.; Kornyshev, A.A.; Bazant, M.Z. Theory of the double layer in water-in-salt electrolytes. *Phys. Chem. Lett.* **2018**, *9*, 5840–5846.
142. Maggs, A.C.; Podgornik, R. General theory of asymmetric steric interactions in electrostatic double layers. *Soft Matter* **2016**, *12*, 1219–1229.
143. Chen, D.P.; Barcilon, V.; Eisenberg, R.S. Constant fields and constant gradients in open ionic channels. *Biophys. J.* **1992**, *61*, 1372–1393.
144. Eisenberg, R.S.; Kłosek, M.M.; Schuss, Z. Diffusion as a chemical reaction: Stochastic trajectories between fixed concentrations. *J. Chem. Phys.* **1995**, *102*, 1767–1780.
145. Eisenberg, B.; Hyon, Y.K.; Liu, C. Energy variational analysis EnVarA of ions in water and channels: Field theory for primitive models of complex ionic fluids. *J. Chem. Phys.* **2010**, *133*, 104104.
146. Liu, C.; Wu, H. An energetic variational approach for the Cahn–Hilliard equation with dynamic boundary condition: Model derivation and mathematical analysis. *Arch. Ration. Mech. Anal.* **2019**, *233*, 167–247.
147. Xu, S.; Eisenberg, B.; Song, Z.; Huang, H. Osmosis through a semi-permeable membrane: A consistent approach to interactions. *arXiv* **2018**, arXiv:1806.00646.
148. Zhu, Y.; Xu, S.; Eisenberg, B.; Huang, H. A bidomain model for lens microcirculation. *Biophys. J.* **2019**, *116*, 1171–1184.
149. Debye, P.; Hückel, E. Zur Theorie der Elektrolyte. I. Gefrierpunktserniedrigung und verwandte Erscheinunge (The theory of electrolytes. I. Lowering of freezing point and related phenomena). *Phys. Z.* **1923**, *24*, 185–206.
150. Hückel, E. Zur Theorie konzentrierterer wässeriger L ösungen starker Elektrolyte. *Phys. Z.* **1925**, *26*, 93–147.
151. Myers, J.A.; Sandler, S.I.; Wood, R.H. An equation of state for electrolyte solutions covering wide ranges of temperature, pressure, and composition. *Ind. Eng. Chem. Res.* **2002**, *41*, 3282–3297.
152. Voigt, W.; Brendler, V.; Marsh, K.; Rarey, R.; Wanner, H.; Gaune-Escard, M.; Cloke, P.; Vercouter, T.; Bastrakov, E.; Hagemann, S. Quality assurance in thermodynamic databases for performance assessment studies in waste disposal. *Pure Appl. Chem.* **2007**, *79*, 883–894.
153. Rowland, D.; Königsberger, E.; Hefter, G.; May, P.M. Aqueous electrolyte solution modelling: Some limitations of the Pitzer equations. *Appl. Geochem.* **2015**, *55*, 170.
154. Kontogeorgis, G.M.; Maribo-Mogensen, B.; Thomsen, K. The Debye-Hückel theory and its importance in modeling electrolyte solutions. *Fluid Phase Equil.* **2018**, *462*, 130–152.
155. Bell, I.H.; Mickoleit, E.; Hsieh, C.M.; Lin, S.T.; Vrabec, J.; Breitkopf, C.; Jager, A. A Benchmark Open-Source Implementation of COSMO-SAC. *J. Chem. Theory Comput.* **2020**.

156. Fraenkel, D. Simplified electrostatic model for the thermodynamic excess potentials of binary strong electrolyte solutions with size-dissimilar ions. *Mol. Phys.* **2010**, *108*, 1435.
157. May, P.M.; Rowland, D.; Murray, K.; May, E.F. JESS: Joint Expert Speciation System. Available online: http://jess.murdoch.edu.au/jess_home.htm (accessed on 13 May 2020).
158. May, P.M.; Rowland, D.; Königsberger, E.; Hefter, G. JESS, a Joint Expert Speciation System—IV: A large database of aqueous solution physicochemical properties with an automatic means of achieving thermodynamic consistency. *Talanta* **2010**, *81*, 142–148.
159. Stokes, R. H.; Robinson, R.A. Ionic hydration and activity in electrolyte solutions. *J. Am. Chem. Soc.* **1948**, *70*, 1870–1878.
160. Rashin, A.A.; Honig, B. Reevaluation of the Born model of ion hydration. *J. Phys. Chem.* **1985**, *89*, 5588–5593.
161. Marcus, Y. Thermodynamics of solvation of ions. Part 5.— Gibbs free energy of hydration at 298.15 K. *J. Chem. Soc. Faraday Trans.* **1991**, *87*, 2995–2999.
162. Ohtaki, H.; Radnai, T. Structure and dynamics of hydrated ions. *Chem. Rev.* **1993**, *93*, 1157–1204.
163. Babu, C.S.; Lim, C. Theory of ionic hydration: Insights from molecular dynamics simulations and experiment. *J. Phys. Chem. B* **1999**, *103*, 7958–7968.
164. Varma, S.; Rempe, S.B. Coordination numbers of alkali metal ions in aqueous solutions. *Biophys. Chem.* **2006**, *124*, 192–199.
165. Mähler, J.; Persson, I. A study of the hydration of the alkali metal ions in aqueous solution. *Inorg. Chem.* **2011**, *51*, 425–438.
166. Rudolph, W.W.; Irmer, G. Hydration of the calcium(II) ion in an aqueous solution of common anions (ClO_4^-, Cl^-, Br^-, and NO_3^-). *Dalton Trans.* **2013**, *42*, 3919.
167. Bashford, D.; Case, D.A. Generalized Born models of macromolecular solvation effects. *Annu. Rev. Phys. Chem.* **2000** , *51*, 129–152.
168. Lee, B.P.; Fisher, M.E. Density fluctuations in an electrolyte from generalized Debye-Hueckel theory. *Phys. Rev. Lett.* **1996**, *76*, 2906.
169. Chern, I.-L.; Liu, J.-G.; Wang, W.-C. Accurate evaluation of electrostatics for macromolecules in solution. *Methods Appl. Anal.* **2003**, *10*, 309–328.
170. Geng, W.; Yu, S.; Wei, G. Treatment of charge singularities in implicit solvent models. *J. Chem. Phys.* **2007**, *127*, 114106.
171. de Souza, J.P.; Bazant, M.Z. Continuum theory of electrostatic correlations at charged surfaces. *arXiv* **2019**, arXiv:1902.05493.
172. Misra, R.P.; de Souza, J.P.; Blankschtein, D.; Bazant, M.Z. Theory of surface forces in multivalent electrolytes. *Langmuir* **2019**, *35*, 11550–11565.
173. Valiskó, M.; Boda, D. Unraveling the behavior of the individual ionic activity coefficients on the basis of the balance of ion-ion and ion-water interactions. *J. Phys. Chem. B* **2015**, *119*, 1546.
174. Grenthe I.; Puigdomenech I. *Modelling in Aquatic Chemistry*; OECD Nuclear Energy Agency: Paris, France, 1997.
175. Davies, C.W. The extent of dissociation of salts in water. Part VIII. An equation for the mean ionic activity coefficient of an electrolyte in water, and a revision of the dissociation constants of some sulphates. *J. Chem. Soc. (Resumed)* **1938**, 2093–2098.
176. Im, W.; Roux, B. Ion permeation and selectivity of ompf porin: A theoretical study based on molecular dynamics, Brownian dynamics, and continuum electrodiffusion theory. *J. Mol. Biol.* **2002**, *322*, 851–869.
177. Lu, B.; Holst, M.J.; McCammon, J.A.; Zhou, Y.C. Poisson-Nernst-Planck Equations for simulating biomolecular diffusion-reaction processes I: Finite element solutions. *J. Comput. Phys.* **2010**, *229*, 6979–6994.
178. Zheng, Q.; Chen, D.; Wei, G.-W. Second-order Poisson Nernst-Planck solver for ion channel transport. *J. Comput. Phys.* **2011**, *230*, 5239–5262.
179. Eisenberg, B. Multiple scales in the simulation of ion channels and proteins. *J. Phys. Chem. C* **2010**, *114*, 20719–20733.
180. Eisenberg, B. A leading role for mathematics in the study of ionic solutions. *SIAM News* **2012**, *45*, 11–12.
181. Eisenberg, B. Ionic interactions are everywhere. *Physiology* **2013**, *28*, 28–38.
182. Berti, C.; Furini, S.; Gillespie, D.; Boda, D.; Eisenberg, R.S.; Sangiorgi, E.; Fiegna, C. Three-dimensional Brownian dynamics simulator for the study of ion permeation through membrane pores. *J. Chem. Theory Comput.* **2014**, *10*, 2911–2926.

183. Kaufman, I.K.; McClintock, P.V.E.; Eisenberg, R.S. Coulomb blockade model of permeation and selectivity in biological ion channels. *New J. Phys.* **2015**, 17, 083021.
184. Luchinsky, D.G.; Gibby, W.A.T.; Kaufman, I.K.; McClintock, P.V.E.; Timucin, D.A. Relation between selectivity and conductivity in narrow ion channels. In Proceedings of the International Conference on Noise and Fluctuations (ICNF), Vilnius, Lithuania, 20–23 June 2017; pp. 1–4.
185. Catacuzzeno, L.; Franciolini, F. Simulation of gating currents of the Shaker K channel using a Brownian model of the voltage sensor. *Biophys. J.* **2019**, *117*, 2005–2019.
186. Hou, S.M.; Liu, X.-D. A numerical method for solving variable coefficient elliptic equation with interfaces. *J. Comput. Phys.* **2005**, *202*, 411–445.
187. Scharfetter, D.L.; Gummel, H.K. Large-signal analysis of a silicon Read diode oscillator. *IEEE Trans. Elec. Dev.* **1969**, *16*, 64–77.
188. Snowden, C.M. *Semiconductor Device Modelling*; Peter Peregrinus Ltd.: London, UK, 1988.
189. Markowich, P.A.; Ringhofer, C.A.; Selberherr, S.; Lentini, M. A singular perturbation approach for the analysis of the fundamental semiconductor equations. *IEEE Trans. Elec. Dev.* **1983**, *30*, 1165–1180.
190. Brezzi, F.; Marini, L.D.; Pietra, P. Two-dimensional exponential fitting and applications to drift-diffusion models. *SIAMJ Numer. Anal.* **1989**, *26*, 1342–1355.
191. Feynman, R.P.; Leighton, R. B.; Sands, M. *The Feynman Lectures on Physics, Volume II, Mainly Electromagnetism and Matter*; Addison-Wesley Publishing Co.: New York, NY, USA, 1963.
192. Wilczek-Vera, G.; Rodil, E.; Vera, J.H. On the activity of ions and the junction potential: Revised values for all data. *AIChE J.* **2004**, *50*, 445.
193. Stern, O. Zur theorie der electrolytischen doppelschicht. *Z. Elektrochem.* **1924**, *30*, 508–516.
194. Oldham, K.B. A Gouy–Chapman–Stern model of the double layer at a (metal)/(ionic liquid) interface. *J. Electr. Chem.* **2008**, *613*, 131–138.
195. Gongadze, E.; Van Rienen, U.; Iglič, A. Generalized stern models of the electric double layer considering the spatial variation of permittvity and finite size of ions in saturation regime. *Cell. Mol. Biol. Lett.* **2011**, *16*, 576.
196. Brown, M.A.; Goel, A.; Abbas, Z. Effect of electrolyte concentration on the stern layer thickness at a charged interface. *Angewandte Chem. Int. Ed.* **2016**, *55*, 3790–3794.
197. Cole, C.D.; Frost, A.S.; Thompson, N.; Cotten, M.; Cross, T.A.; Busath, D.D. Noncontact dipole effects on channel permeation. VI. 5f- and 6F-Trp gramicidin channel currents. *Biophys. J.* **2002**, *83*, 1974–1986.
198. Gillespie, D. Energetics of divalent selectivity in a calcium channel: The ryanodine receptor case study. *Biophys. J.* **2008** , *94*, 1169–1184.
199. Smith, G.R.; Sansom, M.S.P. Dynamic properties of Na^+ ions in models of ion channels: A molecular dynamics study. *Biophys. J.* **1998**, *75*, 2767–2782.
200. Allen, T.W.; Kuyucak, S.; Chung, S.H. Molecular dynamics estimates of ion diffusion in model hydrophobic and KcsA potassium channels. *Biophys. Chem.* **2000**, *86*, 1–14.
201. Mamonov, A.; Coalson, R.D.; Nitzan, A.; Kurnikova, M.G. The role of the dielectric barrier in narrow biological channels: A novel composite approach to modeling single channel currents. *Biophys. J.* **2003**, *84*, 3646–3661.
202. Chen, D.P.; Nonner, W.; Eisenberg, R.S. PNP theory fits current-voltage (IV) relations of a neuronal anion channel in 13 solutions. *Biophys. J.* **1995**, *68*, A370.
203. Nonner, W.; Eisenberg, B. Ion permeation and glutamate residues linked by Poisson-Nernst-Planck theory in L-type calcium channels. *Biophys. J.* **1998**, *75*, 1287–1305.
204. Nonner, W.; Catacuzzeno, L.; Eisenberg, B. Binding and selectivity in L-type calcium channels: A mean spherical approximation. *Biophys. J.* **2000**, *79*, 1976–1992.
205. Boda, D.; Busath, D.; Eisenberg, B.; Henderson, D.; Nonner, W. Monte Carlo simulations of ion selectivity in a biological Na+ channel: Charge-space competition. *Phys. Chem. Chem. Phys.* **2002**, *4*, 5154–5160.
206. Eisenberg, B. Proteins, channels, and crowded ions. *Biophys. Chem.* **2003**, *100*, 507–517.
207. Boda, D.; Henderson, D.; Gillespie, D. The role of solvation in the binding selectivity of the L-type calcium channel. *J. Chem. Phys.* **2013**, *139*, 055103–055110.
208. Boda, D. Monte Carlo simulation of electrolyte solutions in biology: In and out of equilibrium. *Annu. Rev. Comput. Chem.* **2014**, *10*, 127–164.
209. Gillespie, D. A review of steric interactions of ions: Why some theories succeed and others fail to account for ion size. *Microfluid. Nanofluid.* **2015**, *18*, 717–738.

210. Matejczyk, B.; Valisko, M.; Wolfram, M.T.; Pietschmann, J.F.; Boda, D. Multiscale modeling of a rectifying bipolar nanopore: Comparing Poisson-Nernst-Planck to Monte Carlo. *J. Chem. Phys.* **2017**, *146*, 124125.
211. Roux, B.; Prod'hom, B.; Karplus, M. Ion transport in the gramicidin channel: Molecular dynamics study of single and double occupancy. *Biophys. J.* **1995**, *68*, 876–892.
212. Almers, W.; McCleskey, E.W. Non-selective conductance in calcium channels of frog muscle: Calcium selectivity in a single-file pore. *J. Physiol.* **1984**, *353*, 585–608.
213. Lipkind, G.M.; Fozzard, H.A. Modeling of the outer vestibule and selectivity filter of the L-type Ca^{2+} channel. *Biochemistry* **2001**, *40*, 6786–6794.
214. Cuello, L.G.; Jogini, V.; Cortes, D.M.; Perozo, E. Structural mechanism of C-type inactivation in K+ channels. *Nature* **2010**, *466*, 203–208.
215. Noskov, S.Y.; Berneche, S.; Roux, B. Control of ion selectivity in potassium channels by electrostatic and dynamic properties of carbonyl ligands. *Nature* **2004**, *431*, 830–834.
216. Neyton, J.; Miller, C. Discrete Ba^{2+} block as a probe of ion occupancy and pore structure in the high-conductance Ca^{2+} activated K^+ channel. *J. Gen. Physiol.* **1988**, *92*, 569–596.
217. LeMasurier, M.; Heginbotham, L.; Miller, C. KcsA: It's a potassium channel. *J. Gen. Physiol.* **2001**, *118*, 303–314.
218. Nimigean, C.M.; Miller, C. Na^+ block and permeation in K^+ channel of known structure. *J. Gen. Physiol.* **2002**, *120*, 323–325.
219. Dolinsky, T.J.; Czodrowski, P.; Li, H.; Nielsen, J.E.; Jensen, J.H.; Klebe, G.; Baker, N.A. PDB2PQR: Expanding and upgrading automated preparation of biomolecular structures for molecular simulations. *Nucleic Acids Res.* **2007**, *35*, W522–W525.
220. Lüttgau, H.-C.; Niedergerke, R. The antagonism between Ca and Na ions on the frog's heart. *J. Physiol.* **1958**, *143* , 486–505.
221. Nicoll, D.A.; Longoni, S.; Philipson, K.D. Molecular cloning and functional expression of the cardiac sarcolemmal Na(+)-Ca2+ exchanger. *Science* **1990**, *250*, 562–565.
222. Liao, J.; Li, H.; Zeng, W.; Sauer, D.B.; Belmares, R.; Jiang, Y. Structural insight into the ion-exchange mechanism of the sodium/calcium exchanger. *Science* **2012**, *335*, 686–690.
223. Blaustein, M.P.; Lederer, W.J. Sodium/calcium exchange: Its physiological implications. *Physiol. Rev.* **1999**, *79*, 763–854.
224. Dipolo, R.; Beaugé, L. Sodium/calcium exchanger: Influence of metabolic regulation on ion carrier interactions. *Physiol. Rev.* **2006**, *86*, 155–203.
225. Reeves, J.P.; Hale, C.C. The stoichiometry of the cardiac sodium-calcium exchange system. *J. Biol. Chem.* **1984**, *259*, 7733–7739.

entropy MDPI

Article

Modeling the Device Behavior of Biological and Synthetic Nanopores with Reduced Models

Dezső Boda [1,*,†], Mónika Valiskó [1,†] and Dirk Gillespie [2,†]

1 Department of Physical Chemistry, University of Pannonia, P.O. Box 158, H-8201 Veszprém, Hungary; valisko@almos.uni-pannon.hu
2 Department of Physiology and Biophysics, Rush University Medical Center, Chicago, IL 60612, USA; Dirk_Gillespie@rush.edu
* Correspondence: boda@almos.uni-pannon.hu; Tel.: +36-88-624-000 (ext. 6041)
† These authors contributed equally to this work.

Received: 12 October 2020; Accepted: 2 November 2020; Published: 5 November 2020

Abstract: Biological ion channels and synthetic nanopores are responsible for passive transport of ions through a membrane between two compartments. Modeling these ionic currents is especially amenable to reduced models because the device functions of these pores, the relation of input parameters (e.g., applied voltage, bath concentrations) and output parameters (e.g., current, rectification, selectivity), are well defined. Reduced models focus on the physics that produces the device functions (i.e., the physics of how inputs become outputs) rather than the atomic/molecular-scale physics inside the pore. Here, we propose four rules of thumb for constructing good reduced models of ion channels and nanopores. They are about (1) the importance of the axial concentration profiles, (2) the importance of the pore charges, (3) choosing the right explicit degrees of freedom, and (4) creating the proper response functions. We provide examples for how each rule of thumb helps in creating a reduced model of device behavior.

Keywords: nanopores; ion channels; reduced models; Monte Carlo; classical density functional theory; Poisson-Nernst-Planck

---

We dedicate this paper to our distinguished colleague and dear friend, Douglas Henderson (1934–2020).

1. Introduction

When modeling anything, some approximations must be made, usually to make the calculations feasible. For example, molecular dynamics (MD) simulations use Lennard-Jones (LJ) interactions between atoms in lieu of quantum mechanical interactions. This still keeps the all-atomic nature of the simulations, but can now include more than a small number of atoms. Other models coarse-grain the system much more, reducing the physics to simply calculated properties. Electrical circuits are an example; the electrons are never explicitly modeled, nor are the quantum mechanical interactions that produce electrical resistance. Instead, the concept of resistance is reduced to a proportionality factor between the current and voltage, a kind of response function that (phenomenologically) encapsulates complicated physics in a number. A reduced model can thus be very powerful.

In many nanoscale systems, however, it is not a priori clear how (or even if) one can reduce the physics and still get sensible results. In this paper, we would like to better understand and define when and why reduced models work for certain problems, but not for others? (Ion channels can be considered as natural nanopores, a nomenclature often used in the literature. In this work, when we use the term "nanopore", we mean synthetic ones.) Why do reduced models work well for some biological ion channels and synthetic nanopores What are reduced models and what defines a "good" reduced model? Our attempt to answer these questions is based on the knowledge accumulated over 20 years [1–52] of modeling and computing permeation and selectivity in ion channels and nanopores.

1.1. The Device Approach

Reduced models are especially useful if we focus on a system as a simple device [53,54]. A device is a black box that responds to some incoming stimuli (input signals) by producing output signals. Our goal is to open the black box a little bit and peak into it to understand the inner mechanisms of the device that make the output. In the case of nanodevices, understanding necessarily means understanding molecular mechanisms due to the microscopic dimensions of the underlying processes. This is generally accomplished by modeling. In our model, we do not want to include everything; we focus on those components that are absolutely necessary to reproduce device behavior. By device behavior, we mean the relation of the input signal and output signal, also called device function.

By focusing on device function we reduce the problem at hand. We look at a complex system from an engineering point of view. While complex systems are called complex because the engineering approach tends to fail, there are systems where focusing on the important degrees of freedom allows us to reproduce and explain device behavior, which is an experimentally measurable quantity. The system gives the same response to a given signal in a reproducible manner no matter how complicated the underlying molecular processes are.

Let us take the example of a toy model of an airplane. If we want to reproduce the primary device function—the plane flies as a result of lift produced by a horizontal driving force—we do not need to model unimportant degrees of freedom like seats inside the plane and screens on the backs of the seats. We just need to model the proper shape of the plane, the wings especially. Those are the important degrees of freedom.

Similarly, in modeling ion channels, the knowledge of which amino acid residues are charged (and thus interact with the ions) is an important degree of freedom. The residues that are uncharged and are far from the pore are unimportant. For example, in our model of the 2.2 megadalton ryanodine receptor (RyR) channel (one of the largest ion channels known), we only include five charged amino acids. Moreover, as we describe later in Section 3.1.2, not having the surface charge pattern correct (because all the charged amino acids had not been identified yet) produces qualitatively incorrect results.

1.2. Ion Channels and Nanopores as Devices

A basic function of nanopores and open ion channels is to connect the bulk aqueous phases separated by a membrane and let ions through in a controlled manner [55–57]. The basic input signals of the baths+pore system are the concentrations and electrical potentials on the two sides of the membrane. A difference in any of these properties (concentration and/or electrical potential difference, for example, voltage) acts as a driving force for the diffusing ions and results in ionic currents that are the output signals of the system.

We can also consider the structural features of the nanopore as an input signal as soon as they can be changed easily. The most important feature is the surface charge pattern on the wall of the nanopore [58]. This can be modified very easily with pH [59–61] or an electrical potential [62,63] applied on the wall of a nanopore (a gate potential, to borrow a term from semiconductors) when it is made of a conducting material (typically, gold). Surface charge pattern can also be changed with chemical methods in the case of nanopores [64,65] and with point mutation techniques in the case of ion channels. Here, we restrict ourselves to bath concentrations and voltage (the boundary conditions of the problem of steady state transport) as the main input parameters also controlled by experiments.

The pore's structural features are important because they determine the current response of the device given to the driving force. That relation determines the various useful device functions that are commonly attributed to ion channels and nanopores. An especially important feature of ion channels is selectivity. Various ion channels with well-defined functions in the cell are often distinguished by the specific ion that they favor over other kinds of ions. Regardless of their strict selectivity properties, ions channels are often named on the basis of their physiological roles in the cell. This way, for example, we distinguish calcium channels, potassium channels, sodium channels, and so on.

Nanopores can be manipulated more easily, so they can exhibit a wider variety of functions [56,65–72]. They can also be selective if they distinguish cations from anions. They can behave as diodes if they let ions through at one sign of the voltage, but not at the opposite sign of voltage, a phenomenon called rectification. If we can modify the pore's properties by a third signal (gate voltage or pH, for example), we can use the pore as a transistor [45,48,73]. We can also decorate the nanopore's wall with molecules that bind certain ion species selectively. In this case, if that ion is present in the electrolyte, it can change the pore wall's properties by binding to these molecules and thus changing the current of the background electrolyte. In this way, the nanopore can be used as a sensor [43,47,49,51]. The range of applications of nanopores for specific tasks is much wider, well beyond the short list above, for example, DNA sequencing.

2. Reduced Models

The main idea of reduced models is in their name: the number of degrees of freedom that we treat in detail in the model is reduced. We build only those components into the reduced model that are necessary to reproduce and understand the device function. We call these degrees of freedom the *important* ones. The unimportant or implicit degrees of freedom are treated with less precision and are averaged into "response functions".

> *A good reduced model is defined by choosing the important degrees of freedom carefully and constructing sufficiently accurate response functions for the others.*

Our aim with this paper is to illustrate how to accomplish this, with ion channels and nanopores as worked examples.

The first question is how reduced our model should be? How much detail can we ignore? In this respect, the models shown in this paper belong to a "no man's land" between the really detailed all-atom models studied by MD simulations popular in ion channel studies and mean-field continuum models (the Poisson-Nernst-Planck (PNP) theory, for example) popular in nanopore studies. We believe that our position between these two limiting cases is especially suitable to shed light on the nature of good reduced models that are appropriate for a well-specified purpose, namely, studying a device behavior.

First, we explain in a few words, why reduced models can be more suitable for ion channel devices than all-atom models, at least, in certain cases. All-atom, in this context, means that we model all water molecules and every single atom of the protein and the membrane explicitly. There are various problems with these all-atom models. They cannot always cover the physiological parameter range, small voltages or small concentrations, for example. They have sampling issues, specifically regarding the simulation of ionic currents, because this means collecting samples of rare events, for example, ions passing through the pore. The applied force fields might be problematic; they tend to overestimate interactions between multiply charged entities due to missing electronic polarization in the models [74]. Also, the models are based on X-ray structures of the protein that are not always available, and even if they are, the crystal structures often do not represent native functional states. For example, the fact that they have been obtained for a frozen structure calls into question their applicability at room temperature. Such uncertainties might be overcome with reduced models with properly adjusted parameters if the goal is to reproduce the conductance properties of the pore.

Reduced models, as soon as they contain the necessary physics, do not suffer from these shortcomings as much (they have other kinds of shortcomings, naturally). They can be simulated faster, sampled better, and the model contains only the basic physics necessary to reproduce the device behavior. One can spare oneself from computing the unimportant details. What is the important physics and degrees of freedom is always determined by the problem at hand, the intention of the investigator (to what deepness are you interested in the details, for example), and the computational resources. Computation, namely, the simulation method with which we investigate the model is a crucial point of the research, but, from the point of view of the train of thoughts of this discussion, they have secondary importance.

2.1. Ionic Distribution in the Pore as a Determining Factor

One aspect of our methodologies, however, is important and should be discussed here. In our work, we usually use the Nernst-Planck (NP) transport equation [75,76] to compute the ionic flux:

$$\mathbf{j}_i(\mathbf{r}) = -\frac{1}{kT} D_i(\mathbf{r}) c_i(\mathbf{r}) \nabla \mu_i(\mathbf{r}), \tag{1}$$

where $\mathbf{j}_i(\mathbf{r})$, $D_i(\mathbf{r})$, $c_i(\mathbf{r})$, and $\mu_i(\mathbf{r})$ are the flux density, the diffusion coefficent profile, the concentration profile, and the electrochemical potential profile of ionic species i, respectively. One important principle (rule of thumb) of this paper follows from this equation:

> *1. The current carried by an ionic species as a result of a given driving force (conductance) is mainly determined by the axial concentration profile of that species inside the pore.*

One interpretation of this statement is the obvious one that if there are more ions in the pore, they will carry more current. The mechanism can, however, be more subtle than that. Pores working on the basis of excluding certain ions from the pore (sodium channels exclude K^+, while nanopores with overlapping double layers exclude the coion) are controlled by depletion zones of these excluded ionic species inside the pore somewhere. These depletion zones of low concentration act as high resistance elements in a equivalent circuit if we imagine the consecutive zones of the pore as resistors connected in series. These ideas will be fleshed out below, in our worked examples.

2.2. What Determines Local Concentration Inside the Pore?

The probability that a particle is found at a given position $\mathbf{r}$ in the system depends on the potential energy, $U(\mathbf{r})$, and the electrochemical potential, $\mu_i(\mathbf{r})$, of ionic species i at $\mathbf{r}$ (see the acceptance probability of the particle insertion/deletion step in a Grand Canonical Monte Carlo (GCMC) simulation [29]). The distribution of ions inside the pore, therefore, is influenced by (1) local interactions of the ions with pore charges, other ions, solvent molecules, and confining surfaces, and (2) external parameters such as concentration and electrical potential in the baths (the boundary conditions).

If local interactions dominate ($U(\mathbf{r})$ dominates over $\mu_i(\mathbf{r})$), such as in the crowded selectivity filters of calcium channels (see Section 3.1), the concentration profiles are not so sensitive to boundary conditions. In wide nanopores (Sections 3.2 and 3.3), on the other hand, changes in voltage or bath concentration can significantly influence the concentration profile. In bipolar nanopores, for example, changing the sign of the voltage reduces the depletion zones of ions even further, reducing current and resulting in a diode behavior.

Of these two factors, however, it is the local interactions that are more important for our discussion. These local effects determine the shape of the concentration profile, where it has peaks and where it has depletion zones. They determine the basic device characteristics of the pore and they determine how the pore responds to changes in the external conditions.

We can narrow what is important more specifically. Because free particles (ions and water) just respond to changes in $U(\mathbf{r})$ and $\mu_i(\mathbf{r})$, it is the features (structure) of the pore that determines device function. Moreover, because the ions are charged, their Coulomb interactions with pore charges are dominant; dipolar and higher-order terms in the multipole expansion are secondary both in strength and range. Concentration profiles, therefore, depend sensitively on the distribution of the pore charges. From all our work on channels and pores [5–31,33–52] up to this day, we can conclude the following principle:

> *2. We need to build the pore charges into the model properly if we want to reproduce local concentration, and, consequently, device function.*

In summary, pore charges are important degrees of freedom, as is the geometry of the pore (length, radius, shape). But what can we say about important and unimportant degrees of freedom?

2.3. *Important vs. Unimportant Degrees of Freedom*

Charges (monopoles) are the first, and strongest term of the multipole expansion. The second, and weaker term is the dipolar one that appears in the interaction of an ion with water molecules. The big question arises whether we need to take the water molecules into account explicitly (as in all-atom MD models), or can we replace them with response functions such as a dielectric constant or a diffusion coefficient?

The answer to this question also depends on the system at hand. In the case of ion channels, it is obvious that explicit water molecules are crucial in potassium channels; the selectivity of that channel is the result of a subtle balance between the interactions of the permeating ions with the atoms of the selectivity filter and with water molecules [77]. Calcium and sodium channels, however, as our model calculations imply, work on the basis of interactions with charged side chains inside the selectivity filter and volume exclusion (discussed below).

Using implicit water is not even a question in the nanopore world, where they abundantly use transport equations and the PB theory. In this world, there is no argument about the necessity of the implicit water model. Instead, we need to argue about the necessity of sophisticated statistical mechanical methods such as classical density functional theory (DFT) or MC.

Why can water be smeared into an implicit background in one case, but not in the other case? In other words, what decides whether explicit water is an important degree of freedom or not? Or, in general, what decides whether any degree of freedom is important or not? We give an explicit answer to this question that, we hope, will be a general recipe for building reduced models:

> *3. Those degrees of freedom are the important ones that depend on the input parameters of the device (voltage and concentration), while those that do not can be replaced by response functions.*

If a component of the system does not change considerably upon, for example, changing the voltage, then this component does not influence the mechanisms by which the model generates an output signal as a response to the input signal.

Let us use implicit water as an example to explain this, as this choice is sometimes controversial. Ions are screened by the surrounding water molecules no matter whether external conditions change or not. Certainly, an applied field or the presence of other ions distort the hydration shell around the ions, so screening is changed by changing voltage or concentration.

The effect of external conditions is small if they are small *relative* to primary effects, for example, to interactions with pore charges. If two degrees of freedom have a large relative difference in how they change with external conditions, then we can make the one with the small response implicit. This is a decision for the modeller, and, eventually, a matter of comparison of the model results with reference data. Reference data are primarily experimental data, but they can also be MD results for all-atom models (results will be shown for both cases).

Implicit water, although the most characteristic, is not the only way of reducing the number of explicit degrees of freedom. We can, for example, model the membrane with a slab between two hard walls. We can model the pore with a cylinder of hard wall. We can model the ion channel only by taking its selectivity filter into account, because that is the region that discriminates between ions. We can model protein side chains in a simplified way by taking only the oxygens of the carboxyl groups into account. There are a plenty of ways to simplify the model, but we need to ask ourselves at every step whether the details we just ignored are important or not.

As in the case of the mean-field PNP theory, it can happen that we ignore too much detail. It is well known that PNP cannot reproduce the selectivity behavior of calcium channels, because ionic correlations and volume exclusion that are so important in the highly charged and crowded selectivity filter of Ca channels are absent in PNP. We cannot use the approximations of PNP even in the case of the relatively wide nanopores if multivalent ions are present. Charge inversion, a feature that is common in charged confined systems with multivalent ions cannot be reproduced with PNP [78].

The bottom line is that we need to balance between too many and too few details when we create a model for a specific purpose. If one is curious about the detailed physics of the coordination of ions at binding sites, the reduced model is too crude. If one is studying a wide nanopore with a 1:1 electrolyte in it, PNP theory is probably all right. There is, however, a wide area in between, where ionic correlations (including finite size) matter, but explicit water does not matter.

2.4. What Are Good Response Functions?

If we managed to distinguish between important and less important degrees of freedom, the next step is to decide how to smear the less important ones into response functions. There are various possibilities and it is not always obvious which one we should choose. In this respect, we suggest the following principle.

> *4. When we create a response function, we should choose one whose parameters do not depend on external conditions, or, at least, we should minimize that dependence. In other words, those parameters should be transferable as much as possible.*

This rule might sound obvious because it seems quite ridiculous to refit the parameters for every state point (different values of input device parameters). A model is a model together with its parameters. If those parameters are not stable, meaning transferable between various state points, the model is probably missing some basic physics.

That is exactly the deeper meaning of the above rule. If the physics of the model is right, then it should describe the properties of the nanopore's wall or the ion channel's selectivity filter in a robust way. The model should be the same at another voltage or concentration. If the parameters depend on external conditions, they should do that in a physically well-based and explainable way. Otherwise, it is just an unsystematic fitting on the basis of a useless model. The model is useless in this case because it is unusable for prediction. Transferable parameters are the basis of predictions.

In the following, we present our results for three different case studies that illustrate the rules introduced above.

3. Case Studies

In the case studies presented in the following sections the system consists of two baths separated by a membrane that contains a pore connecting the two baths. Two electrodes in the two baths produce electrical potential difference (voltage) that is a part of the driving force of the transport of ions. Also, ionic concentrations can be different on the two sides of the membrane. Concentration difference and voltage add up to create an electrochemical potential difference that is the full driving force in the NP equation (Equation (1)).

In the model of this system we include the two baths, the membrane and the pore. The simulation cell is finite surrounded by a boundary at which different boundary conditions are prescribed for the ionic concentrations and the electrical potential on the two sides of the membrane. The electrolyte is modeled in the implicit water framework with the "Primitive Model" that, given the success of our models, is not so primitive after all.

The ions are modeled as charged hard spheres immersed in a dielectric continuum represented by the dielectric constant ϵ, one of the response functions. The interaction potential is

$$u_{ij}(r) = \begin{cases} \infty & \text{if} \quad r < R_i + R_j \\ \dfrac{1}{4\pi\epsilon_0\epsilon}\dfrac{z_i z_j e^2}{r} & \text{if} \quad r \geq R_i + R_j, \end{cases} \tag{2}$$

where R_i and R_j are the radii of ionic species i and j, respectively, z_i and z_j are the valences of ionic species i and j, respectively, ϵ_0 is the permittivity of vacuum, e is the elementary charge, and r is the distance between the two ions. The solvent also exerts its effect on the ions by hindering their diffusion via friction. This is taken into account by another response function, the diffusion coefficient $D_i(\mathbf{r})$ (see Equation (1)), which may include effects beyond interactions with waters, such as interactions with other ions and the confining geometry.

The membrane and the pore are defined by hard walls for simplicity. The most important difference between the test cases is that the pore is modeled differently in different cases. Basically, the shape of the pore and the representation of pore charges are different. By shape of the pore we mean an $R(z)$ function that defines the hard wall obtained by rotating this function around the z axis. The models of pore charges will be described in the different cases.

The models are studied with a hybrid simulation method in which the NP equation is coupled to the Local Equilibrium Monte Carlo (LEMC) method (NP+LEMC). The LEMC method is basically a generalization of the GCMC method [79,80] for the case of non-equilibrium systems where the chemical potential is not necessarily constant, so the system is not in global equilibrium. Instead, the input of the LEMC method is the $\mu_i(\mathbf{r})$ profile, while the output is the $c_i(\mathbf{r})$ profile. In practice, the system is divided into small subvolumes, V^α, in which the μ_i^α is constant (local equilibrium is assumed). The result of the simulation is the concentration in each subvolume, c_i^α. The resulting μ_i^α and c_i^α profiles are substituted into the NP equation providing a flux, $\mathbf{j}_i^\alpha$. An iteration process results in a self consistent μ_i^α and c_i^α pair that produces a flux density satisfying the continuity equation, $\nabla \cdot \mathbf{j}_i(\mathbf{r}) = 0$. It is an expression for local conservation of mass, while in our calculations we use the integrated form that states that the sum of inward and outward currents in and out of a volume element is zero. Details are found in previous papers [29,37,39,41].

The results of other models and computation methods will also be presented. Specifically, we will show results of DFT coupled to the NP equation and MD simulations for explicit water models. These models and methods will be described at the specific system, where they are used.

3.1. The Ryanodine Receptor Calcium Channel

The RyR is a biological ion channel that, in muscle, releases Ca^{2+} ions from the sarcoplasmic reticulum in response to an influx of Ca^{2+} through L-type calcium channels. In both cardiac and skeletal muscle cells, that RyR-mediated Ca^{2+} initiates muscle contraction. While its physiological importance is obvious, RyR is also interesting from a single-channel biophysics point of view. Experimentally, its large current allows for relatively easy single-channel current/voltage (IV) recordings. Theoretically, it is a Ca^{2+}-selective channel, but whose preference for Ca^{2+} is much lower than the L-type calcium channel, even though they share the same selectivity filter in amino acids.

What makes an ion channel a calcium channel is the abundance of negative carboxyl groups (COO^-) in and around the selectivity filter. Generally, four glutamate (E) and/or aspartate (D) amino acids line the selectivity filter, which is a short and narrow region of the pore. An important turning point in the understanding of the physics of Ca^{2+} versus monovalent cations selectivity was a reduced model by Nonner et al. [4] They imagined the selectivity filter of a calcium channel as a high-density fluid where the two oxygens of each of the four COO^- groups were modeled as independent hard-sphere $O^{1/2-}$ ions (with radius 0.14 nm). When both Na^+ and Ca^{2+} ions compete for space in this "electric stew" [81], the competition is won by Ca^{2+} ions because they provide twice the charge of Na^+ ions while occupying the same volume (as they have similar Pauling radii).

This mechanism was later called "Charge-Space Competition" [5] because, while the four negative charges of the selectivity filter attract cations, the crowding of those COO^- groups and the permeating ions inside the very small selectivity filter imposes entropic and energetic penalties for permeating ions (Figure 1). In this scheme, there is a competition between entropic and enthalpic components, creating an advantage for small and/or high-valence cations over large and/or low-valence cations. This effect is amplified when the dielectric constant of the protein surrounding the pore is lower than the dielectric constant of the selectivity filter lumen [13].

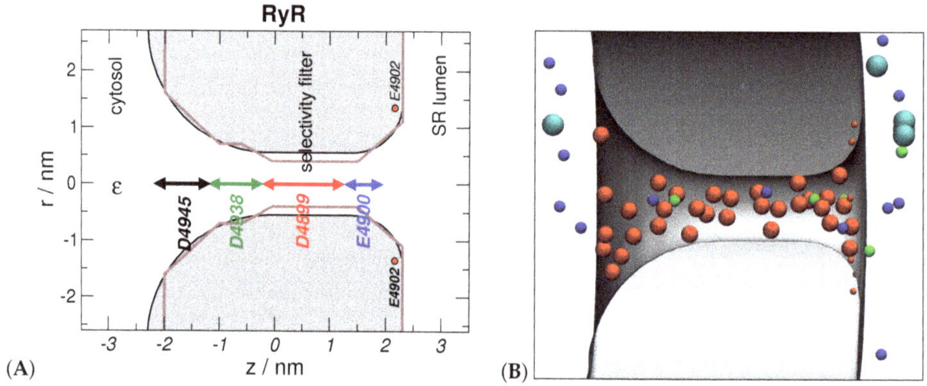

Figure 1. Model of the RyR channel [37]. (**A**) The 3D model is obtained by rotating the shaded gray area about the *z*-axis (the models have rotational symmetry). The arrows indicate the regions into which the 8 $O^{1/2-}$ ions representing the respective amino acids are confined. The charges of the E4902 residues of the RyR channel are modeled by eight point charges on a ring. The dielectric constants is $\epsilon_w = 78.5$ in the whole system. The entire simulation cell is enclosed in a large cylinder. The geometry for the NP+LEMC calculations can be found in Figure 1 of Reference [39]. The brown line indicates the countour of the 1D model of Gillespie [16]. (**B**) A snapshot of the simulation. The blue, green, light blue, and red spheres represent Na^+, Ca^{2+}, Cl^-, and $O^{1/2-}$ ions, respectively. This figure was prepared with vmd [82].

When this model of the L-type selectivity filter was incorporated into a pore and studied with GCMC simulations, the model was successful in reproducing the micromolar Ca^{2+} selectivity of the L-type calcium channel (EEEE locus). Specifically, it reproduced the seminal experiment of Almers and McCleskey [83] where, in 32 mM NaCl, 1 μM Ca^{2+} in the bath blocks Na^+ current, reducing it to half that in the absence of Ca^{2+}. The block works because Ca^{2+} ions displace Na^+ ions in the selectivity filter even though they are present in the bath at much smaller concentrations than the Na^+ ions. The model also reproduced [17,21,24] other mole fraction experiments (e.g., Ca^{2+} vs. Ba^{2+} [84–86], Li^+ vs. Na^+ [87]) and Gd^{3+}-block of ionic current [88]. Lastly, we were able to interpret [14] the experiments of Heinemann et al. [89] where a DEKA→DEEA mutation converted a sodium channel without a Ca^{2+} blockade into a calcium channel with 10^{-4} M affinity.

Concurrent to this work on the physics of L-type calcium channel selectivity, one of us (DG) created a 1D reduced model of RyR using DFT based on the Nonner et al. independent-$O^{1/2-}$ model of the COO^- groups [11]. Here, we focus on a second, improved version of this 1D DFT model [16], as it included more charged amino acids that are outside of the selectivity filter yet play an important role in cation permeation [90] (following the second principle of reduced models). The D4945, D4938, D4899, and E4900 amino acids (four copies of each of them due to the homotetrameric RyR structure) were modeled by confining eight half charged oxygen ions, $O^{1/2-}$ (with radius 0.14 nm), in the regions indicated by arrows in Figure 1. The E4902 amino acids were placed in a ring at the luminal entrance of the pore.

The purpose of this RyR model was to determine whether a reduced model of this channel could reproduce and predict experimental data. (RyR is more useful for this than L-type calcium channel because of the vast amounts of IV data available for RyR.) Both the model [16] and its subsequent applications [19,20,32,35] showed that this is indeed the case, reproducing all the available IV data from the labs of Gerhard Meissner (University of North Carolina, Chapel Hill) and Michael Fill (Rush University Medical Center, Chicago). Moreover, in these papers the model predicted (before confirming experiments were done) a number of counterintuitive and nonlinear selectivity phenomena in RyR.

Later, a 3D reduced model of RyR was created by Boda et al. [37,41]. The purpose of this model was partly to understand the success of the 1D model, trying to define the effects of radial ion distributions that are ignored in the 1D model (which assumed homogeneity in the radial direction). The profile of the pore radius is indicated by the gray shaded area in Figure 1. Here, we focus on the 3D model because it has been less well analyzed in detail and because it uses the same NP+LEMC simulation technique that is also used for the nanopores, described later, that serve as different case studies of reduced models.

Both the 1D and 3D models reproduce dozens of IV curves, some shown in the Supplementary Information for the 3D model. This indicates that both models seem to capture the basics of the RyR device physics in the axial direction. Therefore, we will discuss how each of the principles of reduced models for nanopores works in these RyR models.

3.1.1. Ionic Concentrations and Current

How the ionic profiles determine the species current has several interesting subtleties in RyR. First, given that the 3D model performs equally well as the 1D model, it seems that any radial ion packing effects do not contribute significantly to the current. Figure 2 shows examples for Na^+ and Ca^{2+}. The profiles are monotonic in the radial dimension, so the cross-section averaged axial concentration profiles are the main determinants of current. This explains the success of the 1D model.

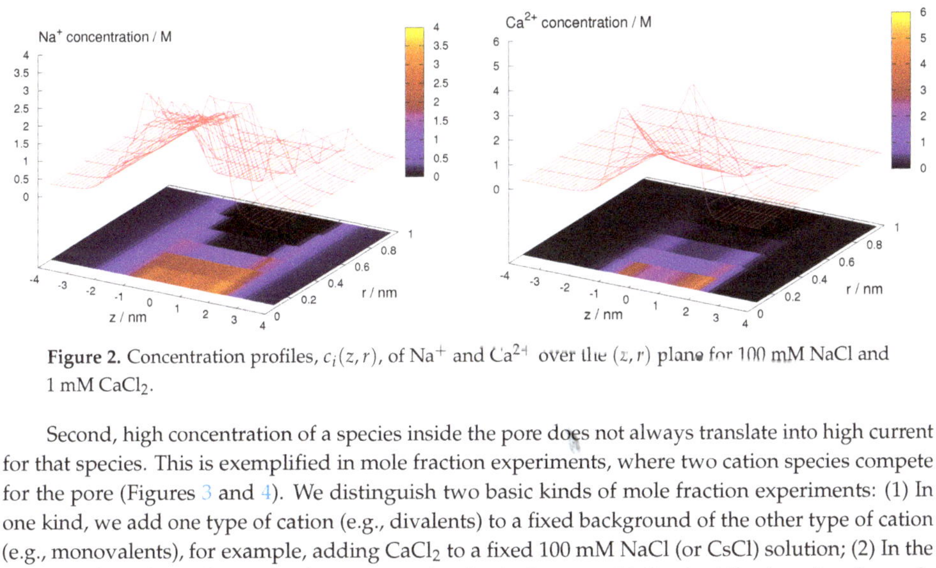

Figure 2. Concentration profiles, $c_i(z,r)$, of Na^+ and Ca^{2+} over the (z,r) plane for 100 mM NaCl and 1 mM $CaCl_2$.

Second, high concentration of a species inside the pore does not always translate into high current for that species. This is exemplified in mole fraction experiments, where two cation species compete for the pore (Figures 3 and 4). We distinguish two basic kinds of mole fraction experiments: (1) In one kind, we add one type of cation (e.g., divalents) to a fixed background of the other type of cation (e.g., monovalents), for example, adding $CaCl_2$ to a fixed 100 mM NaCl (or CsCl) solution; (2) In the other kind, we keep the total salt concentration (or ionic strength) fixed while changing the mole fraction of the two salt, for example, a NaCl/CsCl mixture at 250 mM total concentration.

Total current, I, or chord conductance, $G = I/U$ (U is the applied voltage), can be considered a primary device function in the case of ion channels. But, currents carried by the ionic species are also interesting, and we show those as well. Figure 3 shows the currents as functions of composition expressed either as $\lg[CaCl_2]$ for the added-salt experiment or the mole fraction of Na^+ ([NaCl] + [CsCl] = 250 mM) for the mole fraction experiment.

In the added-Ca^{2+} experiment with Na^+, it is seen that 10^{-3} M Ca^{2+} affects the current; against Na^+, RyR has millimolar Ca^{2+} selectivity. This $[Ca^{2+}]$ is smaller for Cs^+ because Ca^{2+} can compete more easily with the larger Cs^+. In both cases, the total current has a minimum, called the anomalous mole fraction effect (AMFE), for experiments (gray spheres), the 1D DFT RyR model (magenta lines), and the 3D NP+LEMC model (green triangles). There is also an AMFE for mixtures of Na^+ and Cs^+.

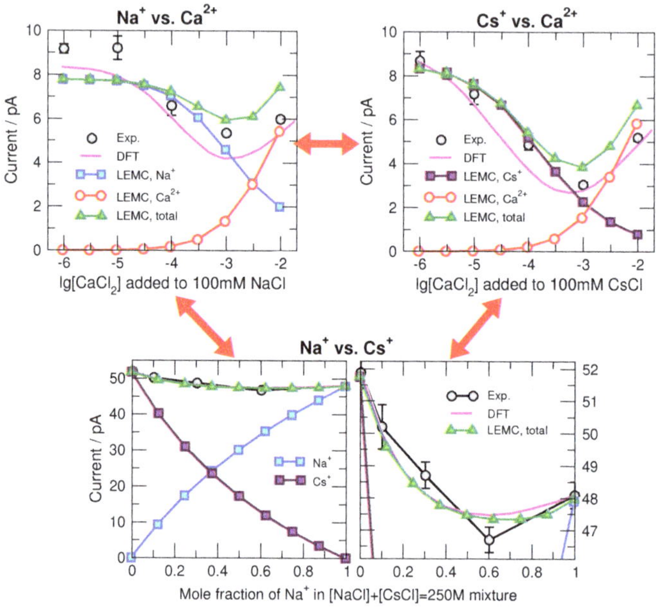

Figure 3. Added-salt experiments for Ca^{2+} vs. Na^+ and Ca^{2+} vs. Cs^+ competition (**top**), and mole-fraction experiments for Na^+ vs. Cs^+ competition (**bottom**). Details are in the main text.

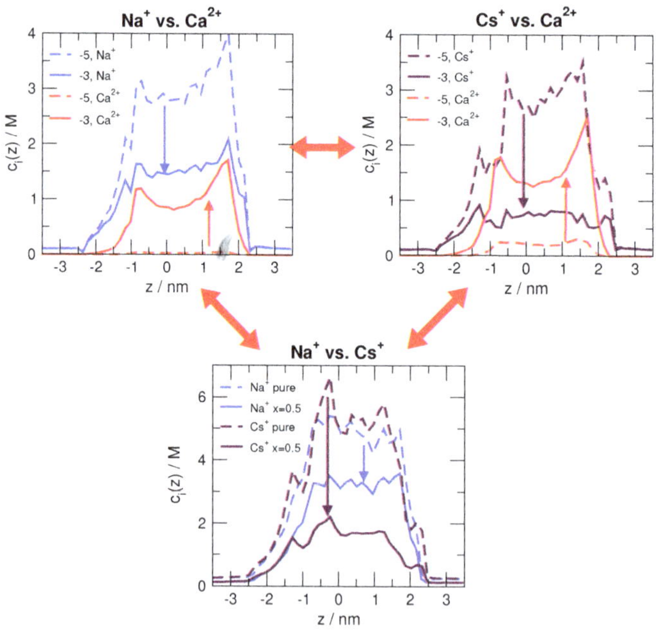

Figure 4. Axial concentration profiles for Ca^{2+} vs. Na^+, Ca^{2+} vs. Cs^+, and Na^+ vs. Cs^+ competition. In the top row, Ca^{2+} is added to either Na^+ or Cs^+. Profiles are for 10^{-5} M and 10^{-3} M added Ca^{2+}. In the bottom row, the Na^+/Cs^+ mole fraction profiles are shown for 0, 0.5, and 1 Na^+ mole fractions.

To understand the origin of the minimum in current, we first note that at the extremes all the single-species currents are very similar: the all-Na^+ current (and all Cs^+ current) at 10^{-6} M Ca^{2+} or 0

Cs^+ (0 Na^+) mole fraction is similar to the all-Ca^{2+} current at 10^{-2} M Ca^{2+}. Why then does current decrease with added Ca^{2+} (added-salt experiment) or added Na^+ (mole fraction experiment) and then increase?

Part of the answer lies in the axial concentration profiles for select Ca^{2+} concentrations and Na^+ mole fractions shown in Figure 4. In the added-salt experiment (top row of figure panels), Ca^{2+} displaces Na^+ and Cs^+ throughout the pore. Ca^{2+} has a much stronger effect on Cs^+, indicating that RyR has a higher preference for Ca^{2+} than Cs^+ (compared to Ca^{2+} versus Na^+). Specifically, 10^{-3} M Ca^{2+} displaces almost half the Na^+ in the pore and two-thirds of the Cs^+. Interestingly, however, the Ca^{2+} current in Figure 3 (top row) is below the Na^+ current and equal to the Cs^+ current at $[Ca^{2+}] = 1$ mM. Recall that all single-species currents are nearly identical. Therefore, just because Ca^{2+} has a large (or even the largest) concentration in the pore, it does not produce as much current as would be predicted from those intra-pore concentrations.

In previous work [17,20], we traced this anomaly to the fact that Ca^{2+} is at low concentration in the baths, even though it is extremely high (relatively) in the selectivity filter. This produces the counterintuitive result that the bath has a high resistance to Ca^{2+} flowing, while the selectivity filter has a low resistance. Usually it is the opposite. Only when the bath Ca^{2+} concentration is relatively high is there an appreciable amount of Ca^{2+} current. This is physiologically relevant, as resting luminal SR Ca^{2+} concentration is between 0.5 to 1 mM, and during contractions this is Ca^{2+} depleted to $\sim 50\,\%$ levels in cardiac myocytes and even lower in skeletal myocytes. The physiological cardiac ion species currents are described in Reference [32].

This is an extreme example of the depletion zones we will discuss for the nanopores later. A depletion zone (a place where ions are absent for the most part) can have as large an effect on current as the regions of high concentration. This is because the axial direction for current flow is made of several regions, the bath, the access region (at the mouth of the channel or pore), the pore, another access region, and another bath. Each of these has a resistance to current flow and the highest resistance element can dominate. In a channel this is usually the selectivity filter because it is commonly physically narrow. However, if it is highly charged, then it will always have ions in it at high concentration and so the bath resistance may dominate the current. In general, the absence of ions in a region can be as consequential as high concentrations.

3.1.2. Accurate Representation of Pore Charges is Important for Reproducing Device Function

As stated above, the first 1D RyR model [11] did not include all the charged groups that the second one [16] does. In fact, it originally included only the two then-known charged groups (Asp-4899 and Glu-4900). But, no parameters could be found to make the computed IV curves resemble, even qualitatively, the experimental curves. Only by hypothesizing the existence of a region of negative charge on the cytosolic side of the selectivity filter did the curves begin to match up. Later, it was determined that two other aspartate groups (Asp-4945 and Asp-4938) also significantly affect ion permeation and selectivity [90]. Only with the explicit addition of these and another charged group (Glu-4902) did the model reproduce all the experimental data and predict even more (which were later confirmed by experiments [16,19,20,32,35]).

3.1.3. Important versus Unimportant Degrees of Freedom

The results of both the 1D and 3D models indicate that the essential important degrees of freedom were captured. One that was left out was ion dehydration. This is crucial for the physiological function of potassium channels [77,91] and excludes Mg^{2+} from many other calcium channels [92]. However, in RyR it does not seem to play a role, as indicated by both experiments and the models. In experiments, Mg^{2+} (which has a very large ion dehydration energy compared to the otherwise similar Ca^{2+}) permeates RyR equally as well as Ca^{2+}, indicating no large energetic barrier for Mg^{2+} entry by stripping off waters. In the two models, missing an important piece of physics ought to result in (large) deviations from the experimental data, especially in Mg^{2+} versus monovalent cation competition experiments. That this was not seen implies (but does not prove) that ion dehydration is not significant for RyR.

One degree of freedom we have in both the 1D and 3D models that may be superfluous is the flexibility of the $O^{1/2-}$ to move within their regions of confinement. Our previous work on the L-type calcium channel [27] indicates that their movement in response to other ions being nearby is unimportant for selectivity. Specifically, for that model pore the selectivity behavior of the channel does not change much if we fix the positions of the $O^{1/2-}$ ions. Seemingly, the important characteristics is the density of the $O^{1/2-}$ ions inside the pore, while their exact position is secondary. We continue to include the flexibility because it is easy to include and extensive studies would be needed to verify that it is indeed superfluous.

3.1.4. Transferability of Parameters

The main parameters we had that were not based on known RyR structure and that had to be fitted to data were the ionic diffusion coefficients. For both the 1D and 3D models, after these were fit, they were never changed. Therefore, they were used at low and high ionic bath concentrations, low/high and negative/positive applied voltages, and in ionic mixtures. This indicates that they truly are transferable and independent of external conditions.

The one caveat to that statement relates to one of the differences in constructing the 1D and 3D models. In the 3D model, we used only one adjustable D_i^{pore} value in the selectivity filter and interpolated in the vestibules to the bulk. (Values are shown in Table 1.) In the 1D model, on the other hand, there were fitted diffusion coefficients not only in the selectivity filter, but also in the vestibules on either side, in the D4938 and E4900 regions (Figure 1). These were fit for K^+ based on data of RyR in symmetric 0.25 M KCl for native RyR (i.e., fully charged) and two charge-neutralizing mutations (D4938N and E4900Q). With these, the 1D model reproduces the nonlinear IV curve of another charge-neutralizing mutation (D4899N) that was not used in fitting the diffusion coefficients. This further shows the transferability of the diffusion coefficients. (All non-K^+ cation species were fitted with one experimental data point for the selectivity filter diffusion coefficient and the vestibule values were determined from ratios of the K^+ diffusion coefficients in different areas of the pore.)

Table 1. Parameters of ions as used in the NP+Local Equilibrium Monte Carlo (LEMC) simulations. The last column shows the density functional theory (DFT) value D_i^{pore}, the diffusion coefficient in the selectivity filter, for comparison; the values for the vestibules are found in Reference [16]. [a] This value was not fitted due to the fact that the channel does not let Cl^- through.

Ion	R_i (Pauling)	D_i^{bulk}	D_i^{pore} (LEMC)	D_i^{pore} (DFT)
	nm		10^{-9} m^2s^{-1}	
Na^+	0.095	1.334	0.141	0.0365
Cs^+	0.169	2.056	0.193	0.0418
Ca^{2+}	0.99	0.792	0.0243	0.0041
Cl^-	1.81	2.032	0.25 [a]	0.02

The 3D model, on the other hand, does not reproduce these charge-neutralizing experiments (data not shown). Therefore, its diffusion coefficients are not as robust against changes to external conditions (although such mutations are large perturbations). This indicates that caution is always in order when interpreting a reduced model outside its established (i.e., tested against experiments) range of external conditions.

3.2. *Nanopores of Different Device Functions from Different Charge Patterns*

In a recent work [46], we considered synthetic nanopores with varying charge patterns on their walls along the z-axis (Figure 5). Although our rules of thumb were not formulated explicitly back then, we practically organized that study along the lines of the four rules of thumb:

- We studied how different charge patterns influence concentration profiles, and, through those, device functions (rules of thumb #1 and #2).
- We performed simulations with models of different resolutions and studied the performance of reduced models compared to all-atom MD simulations. Special attention was given to whether water molecules could be treated implicitly, that is, whether they proved to be "unimportant" degrees of freedom (rule of thumb #3).
- We fit the diffusion coefficients in the pore to MD data and investigated their transferability over varying charge patterns (rule of thumb #4).

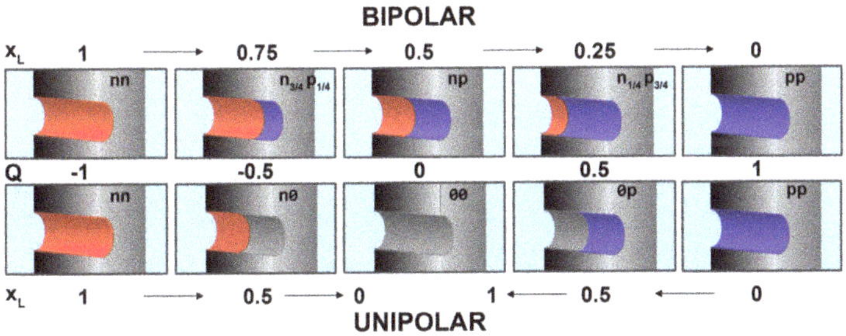

Figure 5. Schematics of the cylindrical nanopores with different charge patterns. There are two regions of lengths H_L and H_R carrying σ_L and σ_R surface charges. We consider either bipolar (top row) or unipolar (bottom row) nanopores. In the bipolar cases, the left-hand region is always negative ($\sigma_L = -\sigma_0$ with $\sigma_0 = 0.4835\ e/\text{nm}^2$), while the right-hand region is positive ($\sigma_R = \sigma_0$). In the unipolar cases, the same is true, but the other side is neutral. The dimensionless net charge, Q, increases from left to right, while the fraction of the left region, x_L (Equation (3)), changes as indicated by the arrows.

A cylindrical nanopore was considered with radius $R_{\text{pore}} = 0.97$ nm and length $H = 6.4$ nm. The wall of the pore was divided into two regions along the z-axis: a left (L) region of length H_L carrying σ_L surface charge, and a right (R) region of length $H_R = H - H_L$ carrying σ_R surface charge. The geometry can be characterized by the dimensionless parameter $x_L = H_L/H$. We gradually increased H_L, while keeping the total length, H, fixed, so we increased x_L from 0 to 1. We performed two series of calculations.

Bipolar pores: The H_L region was negative (red in Figure 5), $\sigma_L = -\sigma_0$, where $\sigma_0 = 0.4835\ e/\text{nm}^2$, while the H_R region was positive (blue in Figure 5), $\sigma_R = \sigma_0$. The limiting cases are the fully negatively ('nn') and positively ('pp') charged pores for $x_L = 0$ and 1, respectively, while we talk about bipolar pores in between ('np').

Unipolar pores: In the other series, one of the regions was neutral (grey in Figure 5) in the intermediate cases. These are actually two series of experiments. Starting from the 'nn' limiting case (from left to right in Figure 5), through unipolar 'n0' charge patterns, we reach the '00' limiting case (neutral pore) as x_L changes from 1 to 0. Starting from the 'pp' limiting case (from right to left in Figure 5), through unipolar '0p' charge patterns, we reach the '00' limiting case (neutral pore) as x_L changes from 0 to 1. The 'n0' ('0p') pore, where $\sigma_L = -\sigma_0$ and $\sigma_R = 0$ ($\sigma_L = 0$ and $\sigma_R = \sigma_0$) exhibits rectification due to the asymmetric charge pattern.

In order to characterize charge pattern, we introduced a dimensionless net charge, Q, ranging from -1 to 1, defined as

$$Q = x_L \frac{\sigma_L}{\sigma_0} + (1 - x_L)\frac{\sigma_R}{\sigma_0}. \tag{3}$$

This value is uniquely related to x_L in the cases depicted in Figure 5. Its value is -1 for the 'nn' pore, 1 for the 'pp' pore, 0 for the 'np' pore, -0.5 for the 'n0' pore, 0.5 for the '0p' pore, and 0 for the 'np' and '00' pores. We found that the pore's basic behavior is correlated with this parameter (Figure 6).

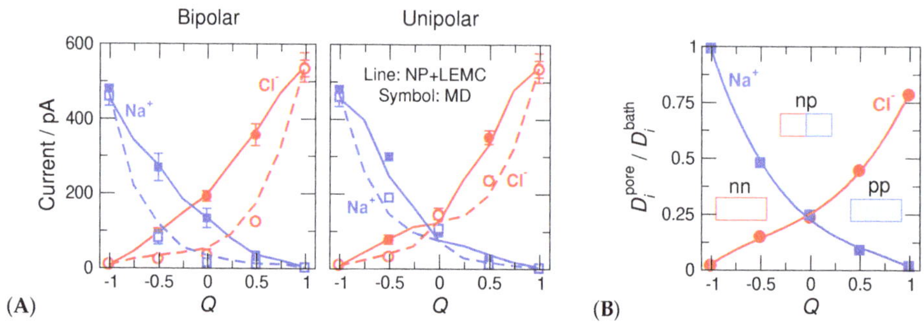

Figure 6. (**A**) Ionic currents as functions of Q for the bipolar (left panel) and unipolar (right panel) nanopores. Blue and red colors correspond to Na^+ and Cl^-, respectively. Symbols and lines correspond to MD and NP+LEMC results, respectively. Filled symbold and solid lines refer to the ON state (200 mV), while open symbols and dashed lines refer to the OFF state (−200 mV). (**B**) Diffusion coefficients in the pore, D_i^{pore}, normalized by the bulk values, D_i^{bath}, fitted to MD currents in the ON states of the bipolar pore. The fit was done for every Q separately.

In order to relate our implicit-water NP+LEMC simulations to explicit-water MD simulations, we constructed an all-atom version of the model. While we did our best in building the all-atom model that is, apart from the treatment of water, is as similar to the reduced model as possible, there are differences:

- Water is explicit (SPC) in MD, while it is implicit in LEMC.
- The ions have Lennard-Jones cores in MD, while they have hard-sphere cores in LEMC.
- The pore wall is a carbon nanotube (CNT) in MD, while it is a hard wall in LEMC.
- The membrane is confined by carbon nanosheets (CNS) in MD, while with hard walls in LEMC.
- The interior of the membrane is empty (a vacuum) in MD, while it is an $\epsilon = 78.45$ region in LEMC.
- The MD simulation cell applies periodic boundary conditions, while the LEMC simulation cell is finite (a cylinder).

The most serious difference between the two systems is the treatment of water, so we consider this study as a test of the implicit-water approximation for this nanopore system.

A continuous surface charge was mimicked by placing partial point charges at the carbon atoms of the CNT. The CNT consisted of hexagons of side width 0.142 nm. There were 1682 partial charges of strength 0.0112 e on the grid for the 'pp' pore. These same partial charges were used in the NP+LEMC calculations. This fine resolution of the pore charges was necessary, because we also compared to the PNP theory in Reference [46] (PNP results are not shown here).

The electrolyte was NaCl (for the ionic parameters see Reference [46]) at bulk concentrations 1 M. The asymmetric pores were rectifying when we applied voltages 200 and −200 mV (ON and OFF states, respectively).

3.2.1. Concentration Profiles and Device Functions

The MD simulation results are our gold standard, so we fit the diffusion coefficients inside the pore, D_i^{pore}, to MD current data for the bipolar pore in the ON state (Figure 6B). Because we decided to use only one adjustable parameter (D_i^{pore}), it was necessary to make its value Q-dependent, because the pore's behavior is severely different at different Q parameters as also shown by the concentration profiles (Figure 7).

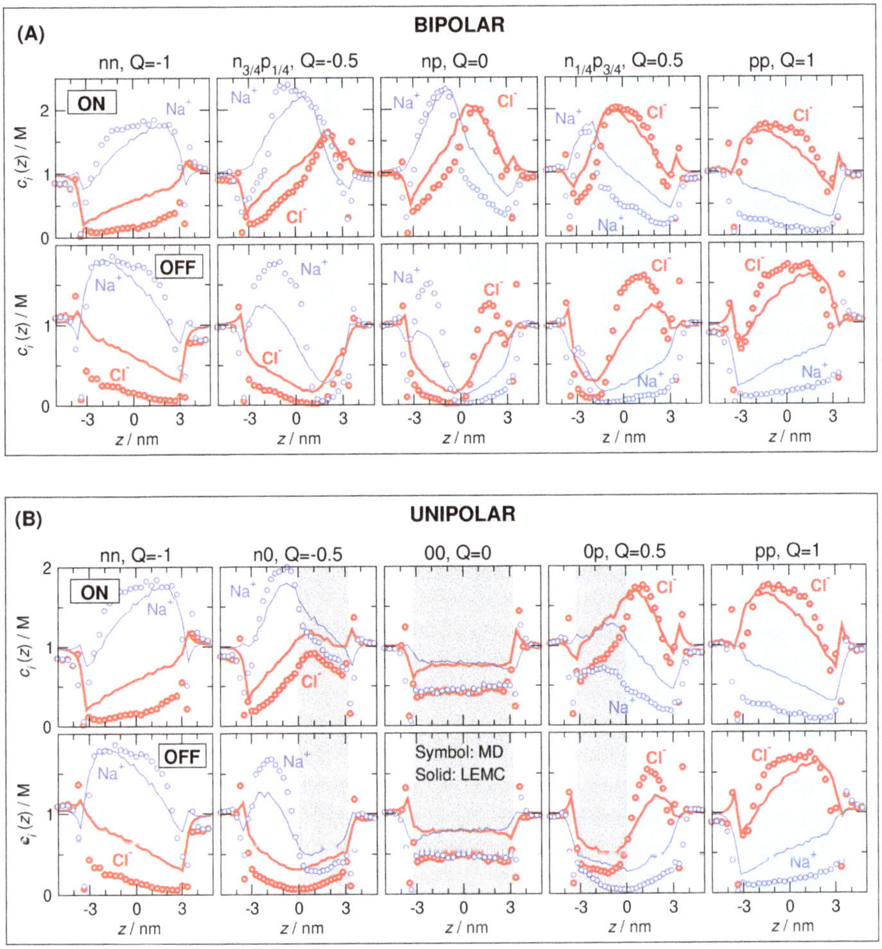

Figure 7. Cross-section averaged axial concentration profiles of Na^+ (blue) and Cl^- (red) ions for (**A**) the bipolar and (**B**) unipolar cases. In each case, top row and bottom row show the ON and OFF states, respectively. Symbols and solid lines refer to molecular dynamics (MD) and NP+LEMC results, respectively.

As the pore charge, Q, increases, Na^+ current decreases and Cl^- current increases (Figure 6A). One of the device functions, selectivity, changes with Q, with the pore being non-selective at $Q = 0$. When the charge pattern is asymmetrical, the pore rectifies, namely, the ON current is larger than the OFF current (Figure 6A). Rectification (the other device function) has a maximum at $Q \approx 0$ in the bipolar case, while it has maxima between $Q = -1$ and 0 as well as between $Q = 0$ and 1 in the unipolar case. The selectivity and rectification curves as functions of Q are shown in Reference [46] (their Figure 7).

The axial concentration profiles (Figure 7) determine the current, as in the case of the RyR ion channel. The major difference compared to the RyR channel is that the depletion zones have decisive roles inside the pore here, not only in the access regions as in the case of the RyR. Briefly, if an ionic species has a depletion zone somewhere inside the pore along the z-axis, its current is suppressed. This statement is intuitive if we imagine the pore as a collection of layers along the z-axis that, in turn, are imagined as resistors connected in series. If any of the resistors has a large resistance due to a depletion zone in that layer, the whole circuit has a large resistance.

We can also support our statement with a quantitative analysis. In Appendix A, we outline our slope-conductance approach that shows that the resistance of the pore is related to the integral of $c_i^{-1}(z)$ (Equation (A6)). Depletion zones give large contributions to that integral, and, therefore, to resistance.

3.2.2. Charge Pattern Determines Device Behavior

The decisive effect of pore charge pattern does not require special verification here; the studies of Reference [46] shown in Figure 5 were devised for the purpose of studying that effect. Figure 6A for the current and Figure 7 for the concentration profiles clearly show that the charge pattern characterized by the Q parameter squarely determines device behavior.

When electrostatic attractions and repulsions play the primary role in forming the shape of the ionic concentration profiles—namely, defining which are the coions and which are the counterions to define where depletion zones and peaks are formed—it is not a surprise that charge pattern dominates over other factors.

3.2.3. Water Molecules as Unimportant Degrees of Freedom

The decisive roles of Coulomb interactions and charge patterns also explain why water molecules can be smeared into a continuum background. Both the axial concentration profiles (Figure 7) and currents (Figure 6A) show that the device works qualitatively the same way in the case of the explicit-water (MD) and implicit-water (NP+LEMC) models.

We devoted a whole paper to this question [42], so we summarize the results of that paper. We showed that the implicit-water and explicit-water models produced qualitatively similar behavior of the current for different voltages and model parameters. Looking at the details of concentration and potential profiles, we found profound differences between the two models. However, these differences did not influence the basic behavior of the model as a device because they do not influence the z-dependence of the concentration profiles, which we found are the main determinants of current. Therefore, our simulations showed that reduced models can still capture the overall device physics correctly because they included the physics that is necessary from the point of view of device function. This is despite the fact that they get some important aspects of the molecular-scale physics quite wrong (e.g., radial ion packing produced by the structure of the water molecules).

3.2.4. Transferability of the Fitted Diffusion Coefficient

We emphasized that it is the *qualitative* behavior that is the same on the two modeling levels. If we want *quantitative* agreement, we need to fit the parameter(s) of the reduced model to MD or to experimental data. In general, we can say that if we observe an overall *qualitative* agreement, the reduced model does its job and there is a good chance that our response function that replaces the smeared degrees of freedom is transferable. The question is what transferability means. What are the external conditions that influence the response function and what are those that do not?

This question has been already touched on with the RyR ion channel, where we stated that our choice of a single adjustable parameter (the diffusion coefficient in the selectivity filter, D_i^{pore}) does not make it possible to create a response function that is transferable over charge patterns, namely, over mutations. It was, however, transferable over voltages, concentrations, and electrolyte compositions. The situation here is the same. We attempted to create a diffusion coefficient profile that is independent of Q, but due to uncertainties in MD simulations and computational demand of NP+LEMC simulations, we abandoned these efforts. Instead, we realized that the difference between the MD and NP+LEMC concentration profiles (Figure 7) depends on Q systematically. For example, as Q increases, the Na^{+} profiles as obtained from MD and NP+LEMC become increasingly different. (At the same time, Cl^{-} profiles become increasingly similar.) Exactly this difference is what must be balanced by the diffusion coefficient in the pore.

Therefore, we decided to use a single $D_i^{\mathrm{pore}}(Q)$ value all along the pore that is allowed to vary with Q. We fitted D_i^{pore} to one case (bipolar/ON), and investigated transferability for the remaining three cases (bipolar/OFF, unipolar/ON, unipolar/OFF). So, we fixed the $D_i^{\text{pore}}(Q)$ values fitted to the bipolar/ON case and used them at other cases for the same Q. These values are shown in Figure 6B as functions of Q.

As an example, let us consider Q values close to -1. This is close to the 'nn' geometry, namely, a cation selective pore. Cl^- ions have depletion zones in this case in both models, but they are deeper in MD than in NP+LEMC. We need, therefore, a very small $D_{Cl^-}^{\text{pore}}$ value to bring the NP+LEMC profiles (and, therefore, currents) down to the values yielded by MD. As Q increases, the difference between the Cl^- profiles decreases and they are pretty similar for $Q = 1$, namely, for the 'pp' pore. In general, we can state that the implicit-water approximation works better (compared to MD) in the case of peaks than in the case of depletion zones.

To summarize, one job of the $D_i^{\text{pore}}(Q)$ function is to take the differences in the explicit and implicit water models into account. The diffusion coefficient in the pore, therefore, is more than a transport coefficient that, in principle, could be calculated from autocorrelations functions or mean square displacements. It carries more information that stems from differences between the reduced model and the more realistic experimental data or MD simulations. Eventually, it is an adjustable parameter of the reduced model as a whole.

3.3. Selectivity Inversion Due to Charge Inversion

In the two case studies so far radial profiles were relatively unimportant. The narrow RyR pore was a crowded high density region (Figure 2) but without layering (oscillatory concentration profiles) in the radial dimension. The case of the wider nanopore in Section 3.2, however, is much more complex. The radial distribution of the ions is important because it determines the behavior of the axial profiles. This was discussed in detailed in our recent studies [50,52].

In the first study [50], we showed that bipolar nanopores exhibit a scaling behavior for a fixed $\sigma = \pm 1\ e/\text{nm}^2$. Specifically, we constructed a scaling parameter, $\xi = R_{\text{pore}}/\lambda\sqrt{z_+|z_-|}$, where λ is the characteristic screening length of the electrolyte computer either as the Debye length (for a point-ion model) or the Mean Spherical Approximation screening length (for Primitive Model ion). (Note that screening works differently near surfaces of different curvatures (flat, concave, convex). Different equations for the capacitance can be given with an unchanged value of the Debye length. [93]) We found that for different pore sizes and different electrolyte concentrations that had the same ξ the device function (this time rectification) was the same; that is, for a given z_+:z_- electrolyte, the relationship of R_{pore} and λ determines device behavior. If $R_{\text{pore}} \ll \lambda$, the double layers formed at the nanopore's wall in the radial dimension overlap. In that case, the counterions will be at high concentration in the middle of the pore, while coions will be at relatively low concentration. This forms depletion zones for the excluded coions. If $R_{\text{pore}} \gg \lambda$, the double layers do not overlap, a bulk electrolyte is present in the pore's center line, and depletion zones are not formed. Depletion zones are necessary for selectivity and rectification. In Section 3.2, this was not discussed because R_{pore} and λ (concentration) were fixed. Double layer overlap was present.

In the second study, [52] we considered the dependence of bipolar nanopores on σ for different electrolytes (1:1, 2:2, 2:1, 3:1). If multivalent ions are present, a deviation from the above scaling behavior (basically a mean-field phenomenon) appears because strong ionic correlations cause peculiar phenomena such as overcharging (overcharging means that more counterions are attracted to the surface than necessary to compensate the surface charge) and charge inversion [78] (charge inversion is the appearance of a layer of excess coions that produces a change in the sign of the electrical potential in this layer). Specifically, these correlations cause an increase in coion concentration in the second layer of ions behind the dense counterion first layer near the charged wall. Consequently, the electrostatic potential can change sign (relative to the potential at the charged wall). We showed that this accumulation of coions (anions) produces an anion leakage current, and this causes non-monotonic behavior in the device function (rectification) as σ increases. Charge inversion always manifests itself in the dimension perpendicular to the charged wall, which for pores is the radial dimension.

In this section, we present new results for the phenomenon of selectivity inversion in a negatively charged nanopore ($\sigma = -1\ e/\text{nm}^2$) as the electrolyte is changed from 1:1 through 2:1 to 3:1. This phenomenon was shown experimentally in our paper with the group of Zuzanna Siwy [22] and interpreted with the help of GCMC simulations. It was observed that while the pore is cation selective for a 1:1 electrolyte (KCl), it becomes anion selective for a 3:1 electrolyte ($CoSepCl_3$). The GCMC simulations supported the idea that the basic reason of this selectivity inversion is charge inversion. The trivalent cations stick to the negatively charged surface, overcharge it, and remain paralyzed; they do not contribute to the current significantly because their mobility near the pore is reduced by being trapped in an energy well.

Here, we show that this behavior can be reproduced without explicitly changing the mobilities of the ions (i.e., decreasing $D_i(r)$ near the wall) by using localized charges instead of a surface charge that is smeared over the surface relatively uniformly as it was in Section 3.2. In fact, this model is much closer to the experimental reality, because the negative charges are localized in chemical groups on the surface of an insulator, specifically, in COO^- groups for the PET nanopores used by Siwy et al.

Here, we show that adopting this idea can produce strong charge inversion around the binding sites now both in the z and r dimensions. The nanopore is practically the same as the one in Section 3.2: it is a cylindrical pore with $R_{\text{pore}} = 1$ nm and $H = 6$ nm with $c = 0.1$ M electrolytes on both sides (ionic radii are $R_+ = R_- = 0.15$ nm). We place fractional point charges on a rectangular grid on the pore's surface of width Δz in a way that the surface charge density is kept constant at $\sigma = -1\ e/\text{nm}^2$. Having $\Delta z = 1$ nm, where $-e$ point charges are sitting on the grid, corresponds to the experimental situation.

3.3.1. Axial Concentration Profiles Determine Selectivity

Cation selectivity defined as $I_+/(I_+ + I_-)$ is shown in the bottom panel of Figure 8A as a function of Δz for different electrolytes (1:1, 2:1, and 3:1) for a constant R_{pore}. The top panel shows the ionic currents from which selectivity was computed. While cation selectivity is insensitive to the fineness of the grid (the degree of localization of surface charge) in the 1:1 case, cation current (and cation selectivity with it) quickly drops as Δz increases above 0.8 nm in the 3:1 case (thick red lines).

The explanation follows from the axial cationic concentration profiles (i.e., cross-sectionally averaged concentrations) in Figure 8B. For a fine grid similar to that used in Section 3.2 and in our earlier studies ($\Delta z = 0.2$ nm), [42–52] the cation profiles are practically constant inside the pore for all the electrolytes from 1:1 to 3:1. For the case of localized charges ($\Delta z = 1$ nm), depletion zones appear along the z-dimension that are much deeper in the case of 2:1 and, especially, 3:1 electrolytes. As the axial depletion zones get deeper, cation currents decrease as Δz increases.

Anion currents, on the other hand, do not change significantly as Δz changes because the anion profiles do not change (Figure 8B). This statement is valid for the anion profiles too (see Figure 8B). This is not a surprise because the anions are far from the charged surface on average, so their distribution is less influenced by the localization of the pore charges. This indicates that it is the behavior of the cations that is responsible for selectivity inversion.

3.3.2. Charge Localization Is an Important Degree of Freedom

The appearance of those depletion zones, however, can be fully understood only if we take into account both the z- and r-dependence of the ionic distributions. Although the statement that current primarily depends on the axial profiles remains true (Equation (A6)), understanding why the axis profiles look the way they look requires the complete picture.

Figure 9A shows the $c_{3+}(z,r)$ concentration profiles for trivalent cations and $\Delta z = 1$ nm. The figure shows the large peaks near the localized pore charges and deep depletion zones between the peaks (note the logarithmic scale). Also, the cationic concentration profiles decline as $r \to 0$, namely, approaching the centerline of the pore.

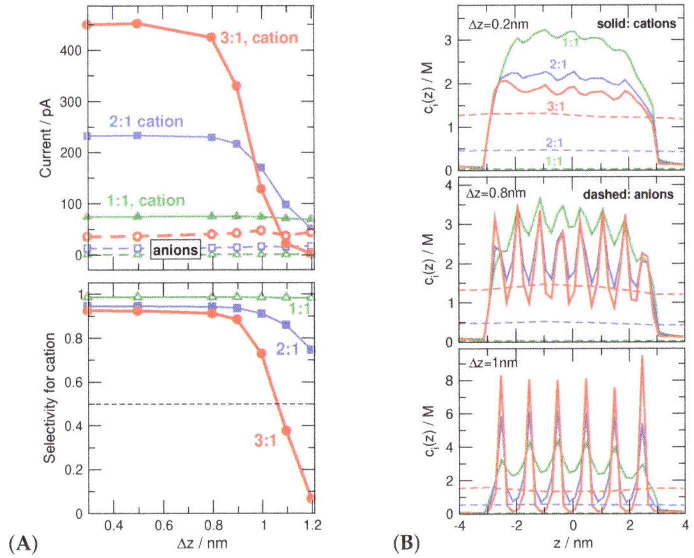

Figure 8. (**A**) The top panel shows ionic currents as functions of Δz for various electrolytes (green, blue, and red color refer to 1:1, 2:1, and 3:1, electrolytes). Solid and dashed lines refer to cations and anions, respectively. The bottom panel shows the cation selectivities computed as $I_+/(I_+ + I_-)$. Values above and below 0.5 correspond to cation and anion selectivities, respectively. (**B**) Axial concentration profiles of cations (solid lines) and anions (dashed lines) in the three electrolytes. Different panels refer to different values of Δz (0.2, 0.8, and 1 nm from top to bottom). Colors have the same meaning as in Figure 8A.

These phenomena can be observed better if we plot the radial profiles for fixed z values that correspond to either a peak (red) or a depletion region (blue). (For the actual values of z, see the caption of Figure 9B.) The left panel of Figure 9B shows radial profiles for $\Delta z = 0.8$ nm; the corresponding axial profiles were shown in the middle panel of Figure 8B. The important thing to note is that the radial profiles do not differ much for different values of z. Depletion zones, therefore, are not formed in this case (see solid blue line with filled squared). It is important to point out that charge inversion is present in this case in the radial profiles; the anion profiles are larger in and around the center line of the pore. It is, however, only present in the radial direction.

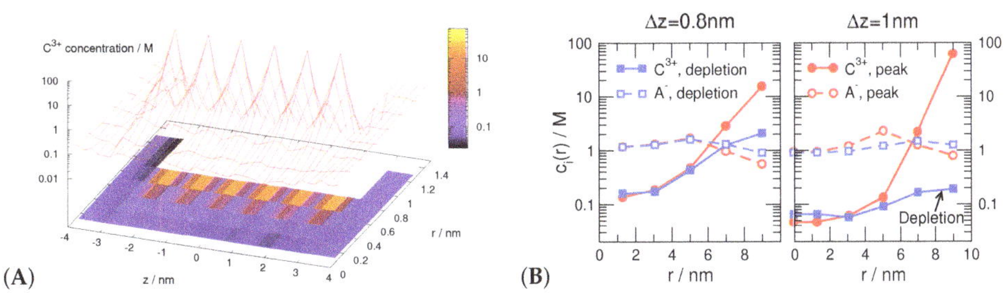

Figure 9. (**A**) The $c_{3+}(z,r)$ concentration profiles of the trivalent cations for the $\Delta z = 1$ nm case. (**B**) Radial concentration profiles of trivalent cations (C^{3+}, filled symbols with solid lines) and monovalent anions (A^-, open symbols with dashed lines) for selected values of z. In the left panel ($\Delta z = 0.8$ nm), the values $z = 0.3$ nm and $z = 0.7$ nm correspond to a peak and a depletion region, respectively. In the right panel ($\Delta z = 1$ nm), the values $z = 0.5$ nm and $z = 0.9$ nm correspond to a peak and a depletion region, respectively.

The right panel of Figure 9B shows the radial profiles for $\Delta z = 1$ nm. This small difference in Δz results in a significant change in the behavior of the ions. The cation profiles show the large peaks for $z = 0.5$ nm (solid red line with filled circles), while they exhibit depletion zone for $z = 0.9$ nm (solid blue line with filled squares). This different behavior at $z = 0.5$ nm and $z = 0.9$ nm produces the oscillating axial concentration profiles with the axial depletion zones of Figure 8B. In this case, therefore, we have charge inversion in both the radial and axial directions.

Taken together, these results show that the way we place the pore charges on the wall matters from the point of view of reproducing device function (pore selectivity and specifically its change due to charge inversion). Specifically, modelers probably need to step beyond the continuous surface charge distribution and to build localized pore charges into the reduced model. Counterion interactions with pore charges depend on the distance from a local binding site in all directions. In the case here, charge inversion around a local binding site produced important axial depletion zones. However, even when not considering cases with charge inversion, different ion correlations around localized pore charges can potentially produce similar important axial effects that are missed with a uniformly charged wall.

3.3.3. Future Work

While it is clear that the location and discreteness of pore charges are an important degree of freedom, whether we need to use explicit particles to model the atoms of the COO^- groups is a subject of ongoing research. We suspect it is not vital since the charge inversion at the core of the device behavior is a product of charge itself, not the shape or mobility of the atoms producing the charge.

Also, we do not know whether we need to change the diffusion coefficient, $D_i(z,r)$, in the radial dimension in order to fit to experiments or to dynamic simulations. Work is currently underway with all-atom MD simulations to determine this.

4. Conclusions

In reduced models, some degrees of freedom (the important ones) are modeled explicitly, while the rest (the unimportant ones) are taken into account implicitly in some way, via response functions, for example. Before the age of computers, all models were reduced. When MD simulations became an everyday computational tool, atomic models became the new standard in certain areas of chemistry, physics, and biology. While understanding nanoscale physics is vital, we believe that the ease of use of MD has sometimes caused the baby to be thrown out with bath water. Rather, we think that what is needed are clever models that are necessarily reduced to some degree to be computationally feasible.

Modeling of ion channels and synthetic nanopores is a case in point. This modeling is inherently difficult as nanoscale interactions and physics directly translate into measurable phenomena (what we call device functions). By simplifying the physics to be modeled, reduced models have a number of advantages over all-atom simulations. However, building such models is in many ways more art than science. Here, we have taken both old and new data from our simulations of ion channels and nanopores and distilled from them four rules of thumb (principles) for constructing reduced models for nanopores. These are

1. The current carried by an ionic species is mainly determined by the axial concentration profile of that species inside the pore.
2. Care must be taken to model the pore charges since they produce the local ion concentrations, and, consequently, device function.
3. The important degrees of freedom that must be included in the model are those that depend on the input parameters of the device (voltage and concentration), while those that do not can be replaced by response functions.
4. Having the parameters within a response function not depend on external conditions (or at least have minimal dependence) makes those parameters transferable to other conditions, and this makes it possible for the model to make predictions that can be tested.

Our goal is to offer insights into how to think about reduced model, but also to point out the subtleties and consequences of the choices a modeler might make. Specifically, for each rule of thumb we showed that its interpretation is not as straightforward as it might seem. For example, while large ion concentrations are important, so are areas with small concentrations which act as large resistors that can dominate the current. Also, charged groups seemingly far from the key locations (e.g., the selectivity filter of an ion channel) can grossly change current/voltage curves. Overall, testing and probing to find the important degrees of freedom that capture the axial-direction physics is the key to reproducing device function and understanding the physics behind the device function; for example, using uniform versus discrete pore charges can have measurable consequences. Once these have been identified, approximating other physics as response functions is a lot easier.

Lastly, we note that while reduced models are important to understand these devices, they are only one part of the continuum of modeling levels that are possible. All-atom and even quantum mechanical simulations play key roles as well in defining the physics of nanopores at the atomic and molecular levels. The role of reduced models is on a larger scale, namely to identify the physics of the device as a whole using the nanoscale physics defined at more detailed levels of modeling. They are the last step to couple atoms to experimental measurements.

Supplementary Materials: The following are available online at http://www.mdpi.com/1099-4300/22/11/1259/s1.

Author Contributions: Conceptualization, D.B. and D.G.; Methodology, D.B. and D.G.; Software, D.B. and D.G.; Validation, D.B. and M.V. and D.G.; Investigation, D.B. and M.V. and D.G.; Writing—Original Draft Preparation, D.B.; Writing—Review & Editing, D.B. and D.G.; Visualization, D.B. and M.V.; All authors have read and agreed to the published version of the manuscript.

Funding: This research was funded by the National Research, Development and Innovation Office (NKFIH) grant number K124353.

Acknowledgments: Present article was published in the frame of the project GINOP-2.3.2-15-2016-00053. The authors acknowledge the contributions of many talented colleagues with whom they worked together during the last 20 years: Douglas Henderson, Wolfgang Nonner, Bob Eisenberg, Tamás Kristóf, Zuzanna Siwy, Zoltán Ható, Claudio Berti, Simone Furini, Attila Malasics, Eszter Mádai, Dávid Fertig, Róbert Kovács, Janhavi Giri, Michael Fill, Gerhard Meissner.

Conflicts of Interest: The authors declare no conflict of interest.

Appendix A. Slope Conductance Theory

Let us assume rotational symmetry, so quantities depend on the variables z, r. Let assume, furthermore, that μ does not depend on r; results (not shown) support this assumption. Let us integrate the NP equation (Equation (1)) over the cross section:

$$I_i = -z_i e \int_{A(z)} j_i(z,r)da = \frac{z_i e}{kT}\left[\int_{A(z)} D_i(z,r)c_i(z,r)da\right]\frac{d\mu_i(z)}{dz} = \frac{z_i e}{kT} N_i(z)\frac{d\mu_i(z)}{dz} \tag{A1}$$

for any z inside the pore with

$$N_i(z) = \int_{A(z)} D_i(z,r)c_i(z,r)da.$$

Let us rearrange and integrate over the pore

$$I_i \int_{H_1}^{H_2} \frac{dz}{N_i(z)} = \frac{z_i e}{kT}\int_{H_1}^{H_2} d\mu_i(z) = \frac{z_i e}{kT}\Delta\mu_i. \tag{A2}$$

If we assume that bulk concentrations are the same on the two sides of the membrane, the electrochemical difference is

$$\Delta\mu_i = z_i e U, \tag{A3}$$

where U is the voltage across the pore. Substituting into Equation (A2), we obtain that

$$I_i \int_{H_1}^{H_2} \frac{dz}{N_i(z)} = \frac{z_i^2 e^2}{kT} U \tag{A4}$$

from which the resistance (the reciprocal of conductance) is obtained as

$$\frac{1}{G_i} = \frac{U}{I_i} = \frac{kT}{z_i^2 e^2} \int_{H_1}^{H_2} \frac{dz}{N_i(z)}. \tag{A5}$$

If we assume that $D_i^{\text{pore}}(z)$ does not depend on r inside the pore, we can write that

$$\frac{1}{G_i} = \frac{kT}{z_i^2 e^2} \int_{H_1}^{H_2} \frac{dz}{D_i^{\text{pore}}(z) A(z) c_i(z)}, \tag{A6}$$

where $c_i(z) = \frac{1}{A(z)} \int_{A(z)} c_i(z,r) da$ is the radially-averaged concentration. If $c_i(z)$ is very small somewhere in the pore along the z-axis, the integral, the resistance, becomes large. This analysis was used in several works [17,18,21,23,24,27,46,48,52] to relate concentration profiles to currents.

Reference

1. Eisenberg, R.S. Living Transistors: A Physicist's View of Ion Channels. Available online: http://arxiv.org/abs/q-bio/0506016v2 (accessed on 12 August 2020).
2. Nonner, W.; Eisenberg, B. Ion Permeation and Glutamate Residues Linked by Poisson-Nernst-Planck Theory in L-type Calcium Channels. *Biophys. J.* **1998**, *75*, 1287–1305.
3. Nonner, W.; Chen, D.P.; Eisenberg, B. Anomalous Mole Fraction Effect, Electrostatics, and Binding in Ionic Channels. *Biophys. J.* **1998**, *74*, 2327–2334.
4. Nonner, W.; Catacuzzeno, L.; Eisenberg, B. Binding and Selectivity in L-Type Calcium Channels: A Mean Spherical Approximation. *Biophys. J.* **2000**, *79*, 1976–1992.
5. Boda, D.; Busath, D.D.; Henderson, D.; Sokołowski, S. Monte Carlo simulations of the mechanism for channel selectivity: The competition between volume exclusion and charge neutrality. *J. Phys. Chem. B* **2000**, *104*, 8903–8910.
6. Nonner, W.; Gillespie, D.; Henderson, D.; Eisenberg, B. Ion accumulation in a biological calcium channel: Effects of solvent and confining pressure. *J. Phys. Chem. B* **2001**, *105*, 6427–6436.
7. Boda, D.; Henderson, D.; Busath, D.D. Monte Carlo study of the effect of ion and channel size on the selectivity of a model calcium channel. *J. Phys. Chem. B* **2001**, *105*, 11574–11577.
8. Boda, D.; Busath, D.D.; Eisenberg, B.; Henderson, D.; Nonner, W. Monte Carlo simulations of ion selectivity in a biological Na channel: Charge-space competition. *Phys. Chem. Chem. Phys.* **2002**, *4*, 5154–5160.
9. Boda, D.; Henderson, D.; Busath, D.D. Monte Carlo study of the selectivity of calcium channels: Improved geometrical model. *Mol. Phys.* **2002**, *100*, 2361–2368.
10. Miedema, H.; Meter-Arkema, A.; Wierenga, J.; Tang, J.; Eisenberg, B.; Nonner, W.; Hektor, H.; Gillespie, D.; Meijberg, W. Permeation properties of an engineered bacterial OmpF porin containing the EEEE-locus of Ca^{2+} channels. *Biophys. J.* **2004**, *87*, 3137–3147.
11. Gillespie, D.; Xu, L.; Wang, Y.; Meissner, G. (De)constructing the Ryanodine Receptor: Modeling Ion Permeation and Selectivity of the Calcium Release Channel. *J. Phys. Chem. B* **2005**, *109*, 15598–15610.
12. Miedema, H.; Vrouenraets, M.; Wierenga, J.; Gillespie, D.; Eisenberg, B.; Meijberg, W.; Nonner, W. Ca^{2+} selectivity of a chemically modified OmpF with reduced pore volume. *Biophys. J.* **2006**, *91*, 4392–4400.
13. Boda, D.; Valiskó, M.; Eisenberg, B.; Nonner, W.; Henderson, D.; Gillespie, D. The effect of protein dielectric coefficient on the ionic selectivity of a calcium channel. *J. Chem. Phys.* **2006**, *125*, 034901.
14. Boda, D.; Nonner, W.; Valiskó, M.; Henderson, D.; Eisenberg, B.; Gillespie, D. Steric selectivity in Na channels arising from protein polarization and mobile side chains. *Biophys. J.* **2007**, *93*, 1960–1980.
15. Boda, D.; Valiskó, M.; Eisenberg, B.; Nonner, W.; Henderson, D.; Gillespie, D. Combined effect of pore radius and protein dielectric coefficient on the selectivity of a calcium channel. *Phys. Rev. Lett.* **2007**, *98*, 168102.

16. Gillespie, D. Energetics of Divalent Selectivity in a Calcium Channel: The Ryanodine Receptor Case Study. *Biophys. J.* **2008**, *94*, 1169–1184.
17. Gillespie, D.; Boda, D. The anomalous mole fraction effect in calcium channels: A measure of preferential selectivity. *Biophys. J.* **2008**, *95*, 2658–2672.
18. Gillespie, D.; Boda, D.; He, Y.; Apel, P.; Siwy, Z. Synthetic nanopores as a test case for ion channel theories: The anomalous mole fraction effect without single filing. *Biophys. J.* **2008**, *95*, 609–619.
19. Gillespie, D.; Fill, M. Intracellular Calcium Release Channels Mediate Their Own Countercurrent: The Ryanodine Receptor Case Study. *Biophys. J.* **2008**, *95*, 3706–3714.
20. Gillespie, D.; Giri, J.; Fill, M. Reinterpreting the anomalous mole fraction effect: The Ryanodine receptor case study. *Biophys. J.* **2009**, *97*, 2212–2221.
21. Boda, D.; Valiskó, M.; Henderson, D.; Eisenberg, B.; Gillespie, D.; Nonner, W. Ion selectivity in L-type calcium channels by electrostatics and hard-core repulsion. *J. Gen. Physiol.* **2009**, *133*, 497–509.
22. He, Y.; Gillespie, D.; Boda, D.; Vlassiouk, I.; Eisenberg, R.S.; Siwy, Z.S. Tuning transport properties of nanofluidic devices with local charge inversion. *JACS* **2009**, *131*, 5194–5202.
23. Malasics, A.; Gillespie, D.; Nonner, W.; Henderson, D.; Eisenberg, B.; Boda, D. Protein structure and ionic selectivity in calcium channels: Selectivity filter size, not shape, matters. *Biochim. Biophys. Acta Biomembr.* **2009**, *1788*, 2471–2480.
24. Malasics, M.; Boda, D.; Valiskó, M.; Henderson, D.; Gillespie, D. Simulations of calcium channel block by trivalent ions: Gd^{3+} competes with permeant ions for the selectivity filter. *Biochim. Biophys. Acta Biomembr.* **2010**, *1798*, 2013–2021.
25. Rutkai, G.; Boda, D.; Kristóf, T. Relating binding affinity to dynamical selectivity from dynamic Monte Carlo simulations of a model calcium channel. *J. Phys. Chem. Lett.* **2010**, *1*, 2179–2184.
26. Boda, D.; Giri, J.; Henderson, D.; Eisenberg, B.; Gillespie, D. Analyzing the components of the free energy landscape in a calcium selective ion channel by Widom's particle insertion method. *J. Chem. Phys.* **2011**, *134*, 055102.
27. Giri, J.; Fonseca, J.; Boda, D.; Henderson, D.; Eisenberg, B. Self-organized models of selectivity in calcium channels. *Phys. Biol.* **2011**, *8*, 026004.
28. Gillespie, D. High Energy Conversion Efficiency in Nanofluidic Channels. *Nano Lett.* **2012**, *12*, 1410–1416.
29. Boda, D.; Gillespie, D. Steady state electrodiffusion from the Nernst-Planck equation coupled to Local Equilibrium Monte Carlo simulations. *J. Chem. Theor. Comput.* **2012**, *8*, 824–829.
30. Csányi, E.; Boda, D.; Gillespie, D.; Kristóf, T. Current and selectivity in a model sodium channel under physiological conditions: Dynamic Monte Carlo simulations. *Biochim. Biophys. Acta Biomembr.* **2012**, *1818*, 592–600.
31. Ható, Z.; Boda, D.; Kristóf, T. Simulation of steady-state diffusion: Driving force ensured by Dual Control Volumes or Local Equilibrium Monte Carlo. *J. Chem. Phys.* **2012**, *137*, 054109.
32. Gillespie, D.; Chen, H.; Fill, M. Is ryanodine receptor a calcium or magnesium channel? Roles of K^+ and Mg^{2+} during Ca^{2+} release. *Cell Calcium* **2012**, *51*, 427–433.
33. Hoffmann, J.; Gillespie, D. Ion Correlations in Nanofluidic Channels: Effects of Ion Size, Valence, and Concentration on Voltage- and Pressure-Driven Currents. *Langmuir* **2013**, *29*, 1303–1317.
34. Boda, D.; Henderson, D.; Gillespie, D. The role of solvation in the binding selectivity of the L-type calcium channel. *J. Chem. Phys.* **2013**, *139*, 055103.
35. Gillespie, D.; Xu, L.; Meissner, G. Selecting Ions by Size in a Calcium Channel: The Ryanodine Receptor Case Study. *Biophys. J.* **2014**, *107*, 2263–2273.
36. Berti, C.; Furini, S.; Gillespie, D.; Boda, D.; Eisenberg, R.S.; Sangiorgi, E.; Fiegna, C. A 3-D Brownian Dynamics simulator for the study of ion permeation through membrane pores. *J. Chem. Theor. Comput.* **2014**, *10*, 2911–2926.
37. Boda, D. Chapter 5: Monte Carlo Simulation of Electrolyte Solutions in Biology: In and Out of Equilibrium. In *Annual Reports in Computational Chemistry*; Wheeler, R.A., Ed.; Elsevier: Amsterdam, The Netherlands, 2014; Volume 10, pp. 127–163.
38. Boda, D.; Csányi, E.; Gillespie, D.; Kristóf, T. Dynamic Monte Carlo simulation of coupled transport through a narrow multiply-occupied pore. *J. Phys. Chem. C* **2014**, *118*, 700–707.

39. Boda, D.; Kovács, R.; Gillespie, D.; Kristóf, T. Selective transport through a model calcium channel studied by Local Equilibrium Monte Carlo simulations coupled to the Nernst-Planck equation. *J. Mol. Liq.* **2014**, *189*, 100–112.
40. Ható, Z.; Boda, D.; Gillepie, D.; Vrabec, J.; Rutkai, G.; Kristóf, T. Simulation study of a rectifying bipolar ion channel: Detailed model versus reduced model. *Cond. Matt. Phys.* **2016**, *19*, 13802.
41. Fertig, D.; Mádai, E.; Valiskó, M.; Boda, D. Simulating ion transport with the NP+LEMC method. Applications to ion channels and nanopores. *Hung. J. Ind. Chem.* **2017**, *45*, 73–84.
42. Ható, Z.; Valiskó, M.; Kristóf, T.; Gillespie, D.; Boda, D. Multiscale modeling of a rectifying bipolar nanopore: Explicit-water versus implicit-water simulations. *Phys. Chem. Chem. Phys.* **2017**, *19*, 17816–17826.
43. Mádai, E.; Valiskó, M.; Dallos, A.; Boda, D. Simulation of a model nanopore sensor: Ion competition underlines device behavior. *J. Chem. Phys.* **2017**, *147*, 244702.
44. Matejczyk, B.; Valiskó, M.; Wolfram, M.T.; Pietschmann, J.F.; Boda, D. Multiscale modeling of a rectifying bipolar nanopore: Comparing Poisson-Nernst-Planck to Monte Carlo. *J. Chem. Phys.* **2017**, *146*, 124125.
45. Mádai, E.; Matejczyk, B.; Dallos, A.; Valiskó, M.; Boda, D. Controlling ion transport through nanopores: Modeling transistor behavior. *Phys. Chem. Chem. Phys.* **2018**, *20*, 24156–24167.
46. Valiskó, M.; Matejczyk, B.; Ható, Z.; Kristóf, T.; Mádai, E.; Fertig, D.; Gillespie, D.; Boda, D. Multiscale analysis of the effect of surface charge pattern on a nanopore's rectification and selectivity properties: From all-atom model to Poisson-Nernst-Planck. *J. Chem. Phys.* **2019**, *150*, 144703.
47. Mádai, E.; Valiskó, M.; Boda, D. The effect of the charge pattern on the applicability of a nanopore as a sensor. *J. Mol. Liq.* **2019**, *283*, 391–398.
48. Fertig, D.; Valiskó, M.; Boda, D. Controlling ionic current through a nanopore by tuning pH: A Local Equilibrium Monte Carlo study. *Mol. Phys.* **2019**, *117*, 2793–2801.
49. Mádai, E.; Valiskó, M.; Boda, D. Application of a bipolar nanopore as a sensor: Rectification as an additional device function. *Phys. Chem. Chem. Phys.* **2019**, *21*, 19772–19784.
50. Fertig, D.; Matejczyk, B.; Valiskó, M.; Gillespie, D.; Boda, D. Scaling Behavior of Bipolar Nanopore Rectification with Multivalent Ions. *J. Phys. Chem. C* **2019**, *123*, 28985–28996.
51. Hohl, B.; Mádai, E.; Boda, D.; Valiskó, M. Modeling of a pH-tunable dual-response nanopore sensor. *J. Mol. Liq.* **2020**, *310*, 112946.
52. Fertig, D.; Valiskó, M.; Boda, D. Rectification of bipolar nanopores in multivalent electrolytes: Effect of charge inversion and strong ionic correlations. *Phys. Chem. Chem. Phys.* **2020**, *22*, 19033–19045.
53. Eisenberg, B. Ionic Channels as Natural Nanodevices. *J. Comput. Electron.* **2002**, *1*, 331–333.
54. Eisenberg, B. Asking biological questions of physical systems: The device approach to emergent properties. *J. Mol. Liq.* **2018**, *270*, 212–217.
55. Hille, B. *Ion Channels of Excitable Membranes*, 3rd ed.; Sinauer Associates: Sunderland, MA, USA, 2001.
56. Albrecht, T.; Gibb, T.; Nuttall, P. Ion Transport in Nanopores. In *Engineered Nanopores for Bioanalytical Applications*; Elsevier BV: Amsterdam, The Netherlands, 2013; pp. 1–30.
57. Iqbal, S.M.; Bashir, R. (Eds.) *Nanopores*; Springer Science and Business Media: Amsterdam, The Netherlands, 2011.
58. Kuo, T.C.; Sloan, L.A.; Sweedler, J.V.; Bohn, P.W. Manipulating Molecular Transport through Nanoporous Membranes by Control of Electrokinetic Flow: Effect of Surface Charge Density and Debye Length. *Langmuir* **2001**, *17*, 6298–6303.
59. Ali, M.; Ramirez, P.; Mafé, S.; Neumann, R.; Ensinger, W. A pH-Tunable Nanofluidic Diode with a Broad Range of Rectifying Properties. *ACS Nano* **2009**, *3*, 603–608.
60. Vlassiouk, I.; Kozel, T.R.; Siwy, Z.S. Biosensing with Nanofluidic Diodes. *J. Am. Chem. Soc.* **2009**, *131*, 8211–8220.
61. Ali, M.; Nasir, S.; Ramirez, P.; Cervera, J.; Mafe, S.; Ensinger, W. Calcium Binding and Ionic Conduction in Single Conical Nanopores with Polyacid Chains: Model and Experiments. *ACS Nano* **2012**, *6*, 9247–9257.
62. Siwy, Z.; Trofin, L.; Kohli, P.; Baker, L.A.; Trautmann, C.; Martin, C.R. Protein Biosensors Based on Biofunctionalized Conical Gold Nanotubes. *J. Am. Chem. Soc.* **2005**, *127*, 5000–5001.
63. Gracheva, M.E.; Vidal, J.; Leburton, J.P. p-n Semiconductor membrane for electrically tunable ion current rectification and filtering. *Nano Lett.* **2007**, *7*, 1717–1722.
64. Piruska, A.; Gong, M.; Sweedler, J.V. Nanofluidics in chemical analysis. *Chem. Soc. Rev.* **2010**, *39*, 1060–1072.

65. Lepoitevin, M.; Ma, T.; Bechelany, M.; Janot, J.M.; Balme, S. Functionalization of single solid state nanopores to mimic biological ion channels: A review. *Adv. Coll. Interf.* **2017**, *250*, 195–213.
66. Daiguji, H.; Yang, P.; Majumdar, A. Ion Transport in Nanofluidic Channels. *Nano Lett.* **2004**, *4*, 137–142.
67. Howorka, S.; Siwy, Z. Nanopores: Generation, Engineering, and Singly-Molecule Applications. In *Handbook of Single-Molecule Biophysics*; Hinterdorfer, P., van Oijen, A., Eds.; Advances in Chemical Physics; Springer: Amsterdam, The Netherlands, 2009; Chapter 11, pp. 293–339.
68. Eijkel, J.C.T.; van den Berg, A. Nanofluidics and the Chemical Potential Applied to Solvent and Solute Transport. *Chem. Soc. Rev.* **2010**, *39*, 957.
69. Gokel, G.W.; Negin, S. Synthetic Ion Channels: From Pores to Biological Applications. *Acc. Chem. Res.* **2013**, *46*, 2824–2833.
70. Guo, W.; Tian, Y.; Jiang, L. Asymmetric Ion Transport through Ion-Channel-Mimetic Solid-State Nanopores. *Acc. Chem. Res.* **2013**, *46*, 2834–2846.
71. Makra, I.; Gyurcsányi, R.E. Electrochemical sensing with nanopores: A mini review. *Electrochem. Commun.* **2014**, *43*, 55–59.
72. Zhang, H.; Tian, Y.; Jiang, L. Fundamental studies and practical applications of bio-inspired smart solid-state nanopores and nanochannels. *Nano Today* **2016**, *11*, 61–81.
73. Kalman, E.B.; Vlassiouk, I.; Siwy, Z.S. Nanofluidic bipolar transistors. *Adv. Mater.* **2008**, *20*, 293–297.
74. Duboué-Dijon, E.; Javanainen, M.; Delcroix, P.; Jungwirth, P.; Martinez-Seara, H. A practical guide to biologically relevant molecular simulations with charge scaling for electronic polarization. *J. Chem. Phys.* **2020**, *153*, 050901.
75. Nernst, W. Zur Kinetik der in Lösung befindlichen Körper. *Z. Phys. Chem.* **1888**, *2*.
76. Planck, M. Ueber die Erregung von Electricität und Wärme in Electrolyten. *Ann. Phys. Chem.* **1890**, *275*, 161–186.
77. Varma, S.; Rempe, S.B. Tuning Ion Coordination Architectures to Enable Selective Partitioning. *Biophys. J.* **2007**, *93*, 1093–1099.
78. Voukadinova, A.; Gillespie, D. Energetics of Counterion Adsorption in the Electrical Double Layer. *J. Chem. Phys.* **2019**, *150*, 154706.
79. Norman, G.E.; Filinov, V.S. Investigations of Phase Transitions by a Monte-Carlo Method. *High Temp. (USSR)* **1969**, *7*, 216–222.
80. Adams, D. Chemical potential of hard-sphere fluids by Monte Carlo methods. *Mol. Phys.* **1974**, *28*, 1241–1252.
81. McCleskey, E.W. Ion Channel Selectivity Using an Electric Stew. *Biophys. J.* **2000**, *79*, 1691–1692.
82. Humphrey, W.; Dalke, A.; Schulten, K. VMD: Visual molecular dynamics. *J. Mol. Graphics* **1996**, *14*, 33–38.
83. Almers, W.; McCleskey, E.W.; Palade, P.T. A non-selective cation conductance in frog muscle membrane blocked by micromolar external calcium ions. *J. Physiol.* **1984**, *353*, 565–583.
84. Hess, P.; Lansman, J.B.; Tsien, R.W. Calcium channel selectivity for divalent and monovalent cations. Voltage and concentration dependence of single channel current in ventricular heart cells. *J. Gen. Physiol.* **1986**, *88*, 293–319.
85. Friel, D.D.; Tsien, R.W. Voltage-gated calcium channels: Direct observation of the anomalous mole fraction effect at the single-channel level. *Proc. Natl. Acad. Sci. USA* **1989**, *86*, 5207–5211.
86. Yue, D.T.; Marban, E. Permeation in the dihydropyridine-sensitive calcium channel. Multi-ion occupancy but no anomalous mole-fraction effect between Ba^{2+} and Ca^{2+}. *J. Gen. Physiol.* **1990**, *95*, 911–939.
87. Prod'hom, B.; Pietrobon, D.; Hess, P. Interactions of protons with single open L-type calcium channels. Location of protonation site and dependence of proton-induced current fluctuations on concentration and species of permeant ion. *J. Gen. Physiol.* **1989**, *94*, 23–42.
88. Babich, O.; Reeves, J.; Shirokov, R. Block of CaV1.2 Channels by Gd^{3+} Reveals Preopening Transitions in the Selectivity Filter. *J. Gen. Physiol.* **2007**, *129*, 461–475.
89. Heinemann, S.H.; Terlau, H.; Stühmer, W.; Imoto, K.; Numa, S. Calcium channel characteristics conferred on the sodium channel by single mutations. *Nature* **1992**, *356*, 441–443.
90. Xu, L.; Wang, Y.; Gillespie, D.; Meissner, G. Two Rings of Negative Charges in the Cytosolic Vestibule of Type-1 Ryanodine Receptor Modulate Ion Fluxes. *Biophys. J.* **2006**, *90*, 443–453.
91. Biggin, P.; Smith, G.; Shrivastava, I.; Choe, S.; Sansom, M. Potassium and sodium ions in a potassium channel studied by molecular dynamics simulations. *Biochim. Biophys. Acta (BBA) Biomembr.* **2001**, *1510*, 1–9.

92. Lansman, J.B.; Hess, P.; Tsien, R.W. Blockade of current through single calcium channels by Cd^{2+}, Mg^{2+}, and Ca^{2+}. Voltage and concentration dependence of calcium entry into the pore. *J. Gen. Physiol.* **1986**, *88*, 321–347.
93. Shoorideh, K.; Chui, C.O. On the origin of enhanced sensitivity in nanoscale FET-based biosensors. *Proc. Natl. Acad. Sci. USA* **2014**, *111*, 5111–5116.

Publisher's Note: MDPI stays neutral with regard to jurisdictional claims in published maps and institutional affiliations.

entropy

Article

Electrophysiological Properties from Computations at a Single Voltage: Testing Theory with Stochastic Simulations

Michael A. Wilson [1,2] and Andrew Pohorille [1,3,*]

1 Exobiology Branch, MS 239-4, NASA Ames Research Center, Moffett Field, CA 94035, USA; Michael.A.Wilson@nasa.gov
2 SETI Institute, 189 Bernardo Ave, Suite 200, Mountain View, CA 94043, USA
3 Department of Pharmaceutical Chemistry, University of California, San Francisco, CA 94132, USA
* Correspondence: Andrew.Pohorille@nasa.gov; Tel.: +1-650-604-5759

Abstract: We use stochastic simulations to investigate the performance of two recently developed methods for calculating the free energy profiles of ion channels and their electrophysiological properties, such as current–voltage dependence and reversal potential, from molecular dynamics simulations at a single applied voltage. These methods require neither knowledge of the diffusivity nor simulations at multiple voltages, which greatly reduces the computational effort required to probe the electrophysiological properties of ion channels. They can be used to determine the free energy profiles from either forward or backward one-sided properties of ions in the channel, such as ion fluxes, density profiles, committor probabilities, or from their two-sided combination. By generating large sets of stochastic trajectories, which are individually designed to mimic the molecular dynamics crossing statistics of models of channels of trichotoxin, p7 from hepatitis C and a bacterial homolog of the pentameric ligand-gated ion channel, GLIC, we find that the free energy profiles obtained from stochastic simulations corresponding to molecular dynamics simulations of even a modest length are burdened with statistical errors of only 0.3 kcal/mol. Even with many crossing events, applying two-sided formulas substantially reduces statistical errors compared to one-sided formulas. With a properly chosen reference voltage, the current–voltage curves can be reproduced with good accuracy from simulations at a single voltage in a range extending for over 200 mV. If possible, the reference voltages should be chosen not simply to drive a large current in one direction, but to observe crossing events in both directions.

Keywords: computational electrophysiology; electrodiffusion model; stochastic simulations; current–voltage dependence; reversal potential; committor probabilities

check for updates

Citation: Wilson, M.A.; Pohorille, A. Electrophysiological Properties from Computations at a Single Voltage: Testing Theory with Stochastic Simulations. *Entropy* **2021**, *23*, 571. https://doi.org/10.3390/e23050571

Academic Editor: Peter V E McClintock

Received: 1 December 2020
Accepted: 28 April 2021
Published: 6 May 2021

1. Introduction

Ion channels are ubiquitous in living systems in which they mediate ion transport across cell walls [1–3]. Although all confirmed structures of ion channels are either bundles of α-helices or β-barrels organized around a transmembrane, water-filled pore lined largely with hydrophilic side chains, they markedly differ in their properties. Their activity is regulated by a variety of signals, such as voltage, ligands, pH or mechanical tension. Some channels are made of peptides that barely span the membrane, while others are among the largest protein assemblies in a cell. In terms of ionic conductance, defined as the ratio of ionic current to voltage, channels differ by more than two orders of magnitude and conductance is not correlated with size. For example, the single-channel conductance of a bacterial homolog of pentameric ligand gated ion channels (pLGICs), GLIC, which consists of 317 residues per subunit is 8 pS [4], similar to the lowest conductance level of a channel made of antimicrobial peptide, alamethicin, which is built of 20 amino acids [5]. Another channel-forming peptide trichotoxin (TTX), consisting of 7 helices, each containing 18 residues conducts ions at 850–900 pS [6], which is close to the conductance of mechanosensitve channels MscS containing 250–1100 residues [7], approximately equal to

1 nS [8]. Some channels exhibit exquisite selectivity whereas others are non-selective. Single point mutations can not only markedly affect conductance and selectivity but even render a channel inactive or constitutively open [9–11]. How variations in a common, general architecture translate to a variety of electrophysiological behaviors is of great interest not only for understanding regular biological systems but also for explaining a number of diseases associated with improper function of ion channels [12–14].

The availability of high-resolution structural models of ion channels has created opportunities to connect structure and function. Molecular dynamics (MD) computer simulations can contribute to this goal by providing mechanistic and thermodynamic descriptions of ion transport that is not readily accessible from experimental studies [15–26]. For a recent, comprehensive review, see Flood, et al. [27]. Furthermore, MD simulations can be used to validate experimentally derived structural models, which do not always correspond to the native structures of channels [28], select the native structure among several candidates [29], and predict functional effects of mutations. These simulations, however, have to be validated by demonstrating that they can be used to reproduce measured electrophysiological properties with satisfactory reliability.

Calculating electrophysiological properties from MD simulations with applied voltage can be done simply by way of computing the current across the simulation cell [15,30,31] or counting the number of ions that cross the channel [16,24,25,32]. However, this direct method, especially when applied to obtain I-V curves and reversal potentials, requires significant computational effort, as it involves MD simulations at a number of applied voltages. To obtain the same accuracy, channels with low conductance generally require longer simulations than channels with high conductance. For example, in simulations of TTX, we counted almost 200 K^+ crossing events in 900 ns at 50 mV [33], whereas only 23 Na^+ crossing events were observed in a 7.7 µs simulation of GLIC at 100 mV (Wilson and Pohorille, unpublished). Since the ionic currents from both simulations appear to obey Poisson statistics, we expect the relative errors in the conductance of K^+ in TTX and Na^+ in GLIC to be approximately 7% and 20%, respectively. To achieve the same relative errors for currents in GLIC as in TTX, a MD trajectory of over 60 µs in length would be required. This means that calculating the I-V curve might present a considerable challenge. To deal with this challenge, it has been common to improve statistics for ion crossing events by applying high voltages, sometimes substantially above their physiological values, or increasing ionic concentration in bulk solution [15–17,34–38]. This approach, however, is fraught with dangers, as it might lead to the disruption of membrane structure or saturation effects for ion entry to a channel [37]. Furthermore, if high voltages are used, I-V curves in physiologically relevant ranges are obtained via interpolation or extrapolation procedures of unknown accuracy [35].

If the motion of ions through the channel can be satisfactorily described as diffusion in the applied electric field and the potential of mean force (PMF) exerted by all other components of the system, which is assumed to be independent of voltage, then the computational effort can be markedly reduced. Several approaches not based on MD take advantage of this description. Methods based on Poisson-Nernst-Planck (PNP) theory rely on solving the electrodiffusion (ED) equation for electrical current in which the mobile ions are represented as a mean-field concentration profile whose distribution and motion is determined by electrostatic forces [39–44]. In Brownian Dynamics (BD), channel conductance is calculated by way of solving the Langevin equation in which both short-ranged interactions with a static model of the channel and long-ranged, electrostatic interactions are taken into account [40,45–48]. In both PNP and BD approaches, electrostatic forces are are obtained from the Poisson equation and the medium is represented as a continuum. From this perspective, we take an approach in which atomistic, dynamic information offered by MD is combined with the efficiency of the ED equation. Instead of carrying out a series of extensive MD calculations of a channel over a range of applied voltages, a substantially reduced set of simulations is combined with the one-dimensional ED model in a steady state. In this approach, the actual electrophysiological properties, such as the current, are calculated from the ED equation, whereas the quantities needed to solve this equation are supplied from MD simulations.

Not all channels conform to the ED model. This model cannot be applied directly if ion crossing events are not statistically independent [49,50], ion diffusion is single file rather than Fickian [51], there are strong binding sites for ions in the channel, the channel changes its structure in response to applied voltage in the range of interest [26] or ions experience saturation in the mouth of the channel. Despite these limitations, it appears that ion movement through many channels satisfies the assumptions of the ED model. A number of small, naturally occurring and synthetic channels and pLGICs belong to this category. The channels discussed in this paper were found to be well-described by the model. More generally, a number of different equations that are special cases of the ED equation, such as Goldman-Hodgkin-Katz (GHK) equation, have been extensively and successfully used as basic tools in experimental and computational electrophysiology for nearly 80 years, for example to determine ionic selectivity from the measured reversal potential [1,52]. Further, basic assumptions of the ED model, such as independence of ion crossing events or Fickian nature of diffusion, can be tested without substantial, additional effort. This was previously done for a number of channels [25].

Previously, we developed an approach to calculate electrophysiological properties from the integrated form of the ED equation [19,24,25]. Instead of MD calculations at several voltages, the system is simulated at a single voltage (or with no applied voltage) to obtain the PMF for each ion in the channel. Subsequently, markedly shorter simulations at voltages of interest are required to determine the densities of the ions near the ends of the channel. Calculating the currents from simulations at these voltages is not required. In addition, ionic diffusivity along the channel has to be determined. Both boundary density values and diffusivity obtained by any of these methods are burdened with errors, contributing to inaccuracies in the calculated currents.

Recently, we developed two formalisms for calculations of electrophysiological properties, including I-V curves and reversal potential, from a single MD simulation at one voltage [33]. From this simulation, the PMF, nonequilibrium density profiles and committor probabilities for ions in the channel are obtained and used to calculate currents at different voltages after appropriate transformations of the ED equation. Additional calculations to obtain the density boundary terms at different voltages and diffusivity are no longer needed. These formalisms were tested on a simple model of the TTX channel, comprised of 7 straight α-helices, each containing 18 amino acids [6], and were shown to perform very well. The improved efficiency of this novel approach derives from the fact that only one MD simulation instead of multiple ones is needed to obtain the I-V curve or reversal potential.

As is the case for any new approach, it is essential to establish the intrinsic accuracy of our formalism. This is the goal of this paper. Specifically, we focus on the question: how reliable are our approaches to calculating electrophysiological properties, independent of other sources of errors, such as inaccuracies in force fields and insufficient simulation times? Separating errors due to the proposed methods from other error is not simple. It cannot be done through a direct comparison with experiments because, for example, of inaccuracies due to force fields. In principle, it can be done via comparison with accurate MD simulations of the PMF and currents at several applied voltages, but, in practice, it is expensive to obtain sufficiently accurate free energies and currents. Although PMFs for ions in a number of channels were obtained from MD simulations, in most cases no estimates of errors were provided [21,23,26,53–61]. In a few cases in which errors are available by way of either direct estimates or comparisons of PMF obtained via different methods [20,24,25,62–68] they are of the order of 0.2–0.7 kcal/mol, which is similar to what is expected to be the intrinsic accuracy of the formalisms studied here. This means that if there were differences between electrophysiological properties obtained from MD simulations at several applied voltages and reconstructed from simulations at a single voltage it would not be possible to determine whether these differences were due to insufficient accuracy of the simulations or to inaccurate reconstruction from the new methods. For these reasons, we take a different approach.

We assume that the ED model describes ion transport with satisfactory accuracy and that the underlying PMF is known. Then, the ED equation is solved many times by way of stochastic simulations to ascertain how statistical errors depend on the number of stochastic trajectories. Even though the stochastic simulations employed here do not involve time explicitly, the number of trajectories considered in a stochastic dataset can be related to the MD simulation time. In a simple case of TTX, we demonstrate that this can be done consistently. For each set of stochastic trajectories, the PMF and the electrophysiological properties at different applied voltages are reconstructed by way of the new theoretical approaches considered here. Due to the stochastic nature of each solution to the ED equation and the limited number of trajectories in each set, which can be related to a specific simulation time, the quantities of interest obtained from each reconstruction differ between themselves and from the accurate values associated with the underlying PMF. Further, the calculated quantities also depend on the theoretical formalism used. If a sufficient number of trajectories has been generated, statistical errors on the quantities of interest can be estimated as a function of the number of trajectories or equivalently, simulation time and performance of each theoretical approach can be systematically assessed. For sets with a large number of trajectories, which corresponds to long simulation times in MD, the underlying PMFs will be reconstructed accurately. For sets with a smaller number of trajectories, the accuracy will not be as good and is expected to deteriorate as the number of trajectories is reduced. A similar systematic study cannot easily be done in practice by way of MD simulations because the computational effort to generate many MD trajectories of different length would have been prohibitive. Furthermore, no analytical method for error analysis exists for this problem.

In principle, this type of analysis can be carried out for any underlying PMF, even if it is unrelated to real ion channels. This is, however, not the direction that we pursue. Instead, we use the PMFs that we previously obtained from simulations of three actual channel models and, for the purpose of this study, assume that they are accurate. The models were selected such that they differ in size, pore structure, conductance and selectivity. The first model is TTX, which exhibits relatively high conductance, very little structure inside the pore and a weak selectivity for cations.

The second model is the high-resolution NMR structure proposed by OuYang et al. [69] for a hexameric channel p7 from the hepatitis C virus. Each subunit consists of 63 amino acids. The model has an unusual architecture not found in any other channel. The channel does not exist as a bundle of α-helices, which is the most common structural motif among membrane proteins, but instead forms an interlocked structure in which each subunit assumes a horseshoe conformation with each side comprised of a short, α-helical section. Because of these atypical features there have been concerns about the veracity of this model [28]. Recently, we calculated conductance and ionic selectivity of this model by way of MD simulations and showed that both properties differ significantly from those measured experimentally (Shannon et al., unpublished). Specifically, in contrast to the electrophysiological data, the model exhibits high conductance and strong selectivity for Cl^- over K^+. These results strongly suggest that the proposed model does not represent the native structure of the channel, demonstrating that computational electrophysiology can be used not only to support but also to disprove structural models of ion channels.

The third model is based on the crystal structure of a pentameric, cation-selective ion channel, GLIC, from a cyanobacterium Gloeobacter [70]. This channel is a bacterial homolog of receptors belonging to the family of pentameric ligand-gated ion channels. Its main electrophysiological characteristics are low-conductance (9.3 pS) and strong selectivity for cations [70]. Molecular models of all three channels are shown in the Supplementary Materials (SM), Section S4.

Both the p7 and GLIC models have a markedly more varied pore structure than TTX and, consequently, a more complex PMF. Although we will use the names of these three channels further in the text, we do not claim that the underlying PMFs faithfully represent the PMFs for these channels in their native open forms (this does not appear to be the case for p7) and, therefore, we do not compare the electrophysiological properties calculated from stochastic simulations with the same properties obtained experimentally. Instead, we fully concentrate on assessing the accuracy of the underlying theory.

2. Theory and Method

In this section, we briefly outline the theory behind three different approaches to calculating the PMF and electrophysiological characteristics of an ion channel. Two of them require simulations at only one applied voltage. A more detailed derivation of the basic formulas, which follows closely the earlier development [33], is provided in Supplementary Materials, Section S1. Next, we describe how the properties of interest can be obtained from stochastic simulations under the assumptions of the ED model specified in the introduction. Note that while the theory is developed in the context of MD simulations, here, we use the results of the theory to compute the PMF and I-V curves from density profiles and committor probabilities that were obtained from stochastic simulations.

2.1. Calculating the Potential of Mean Force

If the concentrations of ions on both sides of the membrane and the applied voltage remain constant in time, the system is in a steady state, which means that the flux of ions through the channel, J, is also constant in time. These are the conditions most often considered in both experiments and simulations aimed at extracting electrophysiological properties of channels. Then, the one-dimensional ED equation for a given type of ions can be written as

$$J = -D(z)\left(\frac{d\rho(z)}{dz} + \beta\rho(z)\frac{dE(z)}{dz}\right), \tag{1}$$

where $D(z)$ is the diffusivity that, in general, depends on position z along the reaction coordinate $\mathbf{z}$. For a transmembrane channel embedded in the membrane located in the x,y-plane, a convenient reaction coordinate is the position of an ion along the pore of the channel, which can be measured along the z-coordinate. $\rho(z)$ is the line density of ions, which is usually recorded as a histogram in computer simulations. $\beta = k_B T$, where k_B is the Boltzmann constant and T is temperature. $E(z)$ is given by

$$E(z) = A(z) + qV(z), \tag{2}$$

where $A(z)$ is the PMF, $V(z)$ is the applied voltage and q is ionic charge. In a constant electric field, $\mathcal{E}^{el}$, acting along $\mathbf{z}$, which is the most frequent experimental condition,

$$V(z) = \mathcal{E}^{el}(z - z_a). \tag{3}$$

Even though the electric field is applied across the whole system [15,30], it acts only between z_a and z_b in the non-polar phase, which has been identified as corresponding to the hydrophobic core of the membrane [25,33]. Thus, electric field is a boxcar function that is equal to $\mathcal{E}^{el}$ in the range $[z_a, z_b]$ and zero otherwise. This can be formally written as $\mathcal{E}^{el}[H(z - z_a) - H(z - z_b)]$, where H is the Heaviside function. Although we will not use this notation for simplicity, the range in which $\mathcal{E}^{el}$ is non-zero has to be kept in mind.

Integrated with the integrating factor $\exp[\beta E(z)]$ and resolved for J, the ED equation takes the form

$$J = \frac{\rho(z_{min})\exp[\beta E(z_{min})] - \rho(z_{max})\exp[\beta E(z_{max})]}{\int_{z_{min}}^{z_{max}} \frac{\exp[\beta E(z)]}{D(z)} dz}. \tag{4}$$

For a system in a steady state, J does not formally depend on the limits of integration z_{min} and z_{max}. This means that these limits do not have to coincide with the edges of the channel. In practice, the limited precision of MD simulations introduces some dependence on the limits of integration, as analyzed elsewhere [25].

To calculate J from this equation, $E(z)$ has to be known, which in turn requires determining $A(z)$. This can be done in equilibrium simulations in the absence of voltage. A host of methods exist for this purpose [71–73]. $A(z)$ can be also calculated from non-equilibrium simulations at an applied voltage. If the ED equation is integrated with the integrating factor $1/\rho(z)$ then

$$J = \frac{\ln \frac{\rho(z_{min})}{\rho(z_{max})} - \beta\left[A(z_{max}) - A(z_{min}) + q\mathcal{E}^{el}(z_{max} - z_{min})\right]}{\int_{z_{min}}^{z_{max}} \frac{1}{D(z)\rho(z)} dz}. \quad (5)$$

Since J is independent of the limits of integration, z_{max} can be substituted by z. After simple rearrangements, it yields a formula for the PMF relative to its value at z_{min}, $\Delta A(z, z_{min}) = A(z) - A(z_{min})$

$$\Delta A(z, z_{min}) = -k_B T\left[\ln \frac{\rho(z)}{\rho(z_{min})} + J\int_{z_{min}}^{z} \frac{1}{D(z')\rho(z')} dz'\right] - q\mathcal{E}^{el}(z - z_{min}). \quad (6)$$

We call this method for determining PMF the Integrated Electrodiffusion Equation Method (IEEM).

To solve Equations (5) and (6), $D(z)$ has to be known in the full range of z. $D(z)$ can be determined by way of calculating the mean square displacement of the ion at several points along the channel obtained from a series of short MD trajectories after subtracting the PMF [19], from the force-force autocorrelation function acting on a stationary ion at different positions in the channel [74], or by way of a Bayesian fitting method [75–77]. See Supplementary Materials, Section S5 for a discussion of how diffusivity was computed in our MD simulations.

Once $\Delta A(z_{max}, z_{min})$ and $D(z)$, which are both assumed to be independent on voltage, are known, the boundary density terms $\rho(z_{min})$ and $\rho(z_{max})$ have to be obtained from either MD or stochastic simulations at each voltage of interest. Since the full knowledge of $\rho(z)$ is not needed, these simulations can be markedly shorter than simulations to determine the PMF. Then, $\Delta A(z_{max}, z_{min})$, $D(z)$, $\rho(z_{min})$ and $\rho(z_{max})$ are used in Equation (5) to calculate J at a given voltage. Previously, we demonstrated that this method performs satisfactorily for simple channels [19,24,25].

Recently, we developed two alternative approaches to calculating the PMF and electrophysiological properties that require markedly less computational effort [33]. Both rely on separating the total ionic current, J, to currents moving in two opposite directions – from z_{min} to z_{max} and from z_{max} to z_{min}. We abbreviate them J^f and J^b and call them forward and backward currents, respectively.

$$J^f = \frac{\rho^f(z_{min})\exp[\beta E(z_{min})] - \rho^f(z)\exp[\beta E(z)]}{\int_{z_{min}}^{z} \frac{\exp[\beta E(z')]}{D(z')} dz'}, \quad (7)$$

$$J^b = \frac{\rho^b(z)\exp[\beta E(z)] - \rho^b(z_{min})\exp[\beta E(z_{mim})]}{\int_{z_{min}}^{z} \frac{\exp[\beta E(z')]}{D(z')} dz'}, \quad (8)$$

$$J = J^f - J^b. \quad (9)$$

Here, $\rho^f(z)$ and $\rho^b(z)$ are densities of ions that entered the range $[z_{min}, z_{max}]$ at z_{min} and z_{max}, respectively. We assume that both forward and backward currents are in a steady state and, therefore, their values do not depend on the limits of integration. This allows for setting the upper limit to $z_{min} < z \leq z_{max}$.

Assume that $J^f > 0$ and $J^b > 0$ and take the ratio of Equation (7) to Equation (8). This yields

$$\frac{J^f}{J^b} = \frac{\rho^f(z_{min}) - \rho^f(z)\exp[\beta\Delta E(z, z_{min})]}{\rho^b(z)\exp[\beta\Delta E(z, z_{min})] - \rho^b(z_{min})}, \tag{10}$$

where

$$\Delta E(z, z_{min}) = E(z) - E(z_{min}). \tag{11}$$

Combined with Equations (2) and (3), Equation (10) can be solved for $\Delta A(z, z_{min})$

$$\Delta A(z, z_{min}) = -k_B T \ln \frac{J^b \rho^f(z) + J^f \rho^b(z)}{J^b \rho^f(z_{min}) + J^f \rho^b(z_{min})} - q\mathcal{E}^{el}(z - z_{min}). \tag{12}$$

From this equation it follows that the PMF can be obtained from non-equilibrium simulations at applied electric field $\mathcal{E}^{el}$ simply from an average of ion densities in the forward and backward directions weighed by the backward and forward currents, respectively. We call this method for calculating the PMF the Current-Weighted Density Method (CWDM). Knowledge of diffusivity is not necessary in CWDM. The denominator in the argument of the logarithmic function sets the reference value of the PMF at z_{min}.

If we abbreviate the number of crossing events in forward and backward direction as n^f and n^b, respectively, then, assuming that crossing events are governed by the Poisson statistics, the corresponding errors will be approximately $1/\sqrt{(n^f)}$ and $1/\sqrt{(n^b)}$. This means that if n^f or n^b is small, $\Delta A(z, z_{min})$ calculated from Equation (12) may become inaccurate. Thus, we developed another, related theoretical approach for determining the PMF from non-equilibrium simulations that does not suffer from this disadvantage. Since it requires calculating committor probability, $P(z)$, we will call it the Committor Probability Method (CPM). For a diffusive process considered here, $P(z)$ referenced to the forward direction is defined as the probability that a particle (ion) in position z will reach z_{max} before it reaches z_{min}. $P(z)$ can be calculated either directly during computer simulations or in post-processing, as described in Supplementary Materials, Section S3. A general discussion of committor probabilities in more than one dimension and their application to chemical kinetics can be found elsewhere [78–81].

The PMF can be calculated from ion densities in the forward or the backward direction. The corresponding formulas are

$$\exp[\beta\Delta E(z, z_{min})] = \frac{\rho^f(z_{min})[1 - P(z)]}{\rho^f(z) - \rho^f(z_{max})}, \tag{13}$$

$$\exp[\beta\Delta E(z, z_{min})] = \exp[\beta\Delta E(z_{max}, z_{min})]\frac{\rho^b(z_{max})P(z)}{\rho^b(z) - \rho^b(z_{min})}. \tag{14}$$

Their derivation closely follows our earlier work [33] and is given in Supplementary Materials Section S1.

Both equations allow for calculating the same quantity— the PMF. Individually, each of them is not expected to be accurate in the full $[z_{min}, z_{max}]$ range of z, especially away from the entry point. Specifically, as z becomes close to z_{max} both $\rho^f(z) - \rho^f(z_{max})$ and $1 - P(z)$ approach zero. Since numerical inaccuracies in Equations (13) and (14) affect mainly the opposite sides of the $[z_{min}, z_{max}]$ range, these two equations can be profitably combined. Then, $\rho^f(z) - \rho^f(z_{max})$ and $\rho^b(z) - \rho^b(z_{min})$ can be considered as two biased distributions representing the same unbiased distribution $h(z)$. The problem of merging them to reconstruct $h(z)$ such that statistical error on $\Delta A(z, z_{min})$ is minimized can be solved by way of the Weighted Histogram Analysis Method (WHAM) [82]. This yields the following formula for reconstructing the PMF from non-equilibrium simulations:

$$\Delta A(z, z_{min}) = C - k_B T \ln\left[\frac{h(z)}{\rho^f(z_{min})(1 - P(z))P(z)}\right] - q\mathcal{E}^{el}(z - z_{min}), \tag{15}$$

where neither C, which is a constant that only shifts the energy scale, nor $\rho^f(z_{min})$, which is independent of z and is needed to ensure that the PMF at $z = z_{min}$ is equal to zero, influences the shape of $\Delta A(z, z_{min})$. Similarly to Equation (12), no knowledge of diffusivity is required.

Typically, MD simulations would be carried out on the channel system of interest at some applied voltage V. From this, the committor probability, $P(z)$ and the 1-sided density profiles, $\rho^f(z)$ and $\rho^b(z)$, and the number of forward and backward crossing events would be determined, and used to calculate the forward and backward fluxes, J^f and J^b, respectively. The PMF can be determined from either CWDM (Equation (12)) or CPM (Equation (15)). As will be discussed later, we created synthetic data sets of 10^6–10^8 stochastic trajectories. For each data set, we calculate the same quantities that would be calculated in MD, $P(z)$, $\rho^f(z)$ and $\rho^b(z)$, and then use these to calculate the PMF. Note that the free energy depends on ratios of density profiles, so the absolute normalization of the density profiles is not important. Similarly, the CWDM requires only ratios of the forward and backward currents, so the magnitudes are not required.

2.2. Calculating I-V Dependence from Simulation at a Single Voltage

If the PMF, the current, J_μ, and the density, $\rho_\mu(z)$, or the committor probability, $P_\mu(z)$, are known from simulations at an applied voltage, ΔV_μ, the current, J_ν, at a different voltage ΔV_ν can be obtained without any calculations at this voltage. This allows for reconstructing the I-V curve from simulations at a single voltage.

If Equation (1) is integrated with the same integrating factor, $\exp[\beta E_\nu(z)]$, for both voltages, ΔV_μ and ΔV_ν, we obtain

$$J_\mu = \frac{\rho_\mu(z_{min})\exp[\beta E_\nu(z_{min})] - \rho_\mu(z_{max})\exp[\beta E_\nu(z_{max})] + \beta q(\mathcal{E}_\nu^{el} - \mathcal{E}_\mu^{el})\int_{z_{min}}^{z_{max}} \rho_\mu(z)\exp[\beta E_\nu(z)]dz}{\int_{z_{min}}^{z_{max}} \frac{\exp[\beta E_\nu(z)]}{D(z)}dz}, \quad (16)$$

and

$$J_\nu = \frac{\rho_\nu(z_{min})\exp[\beta E_\nu(z_{min})] - \rho_\nu(z_{max})\exp[\beta E_\nu(z_{max})]}{\int_{z_{min}}^{z_{max}} \frac{\exp[\beta E_\nu(z)]}{D(z)}dz} \quad (17)$$

The latter but not the former equation is the standard integrated form of the ED equation, Equation (4).

If we take the ratio of currents J_μ/J_ν then, after some algebra given in Supplementary Materials, Section S2 we obtain

$$\frac{J_\mu}{J_\nu} = 1 + \beta q(\mathcal{E}_\nu^{el} - \mathcal{E}_\mu^{el})\int_{z_{min}}^{z_{max}} f_\mu(z)\exp\{\beta q[V_\nu(z) - V_\mu(z)]\}dz, \quad (18)$$

where

$$f_\mu(z) = \exp[\beta\Delta E_\mu(z, z_{min})]\frac{\rho_\mu(z)}{\rho_\mu(z_{min})}. \quad (19)$$

or

$$f_\mu(z) = 1 + \exp[\beta\Delta E_\mu(z, z_{min})]\frac{\rho_\mu(z_{max})}{\rho_\mu(z_{min})} - P_\mu(z), \quad (20)$$

depending on whether it is preferred to calculate J_ν from ion density, $\rho_\mu(z)$, or committor probability, $P_\mu(z)$. In both instances, neither diffusivity nor quantities at the applied voltage ΔV_ν are needed. Equation (19) is expected to be less accurate that Equation (20) because $\Delta E_\mu(z, z_{min})$ and $P_\mu(z)$ that enter the latter equation are estimated on the basis of both forward and backward simulations, whereas $\rho_\mu(z)$ in the former equation is a one-sided density that looses accuracy away from the entry point.

Unlike the free energy, Equation (18) gives only ratios of forward or backward currents with respect to a reference voltage. Consequently, to calculate the I-V curves, we need currents at this reference voltage.

2.3. Stochastic Simulations

The electrodiffusion equation was solved by generating trajectories on a free energy surface $E(z)$ that included the PMF and applied electric field with diffusivity $D(z)$ or average diffusion coefficient, $< D >$, at temperature T [83,84]. This allowed us to generate the channel crossing statistics, density profiles and committor probabilities for the ions for this free energy surface. As the crossing events that we have observed in MD simulations appear to obey Poisson statistics, independently for both ions, we consider the ED equation for each ion separately. Then, we calculated statistical errors in recovering the underlying PMF and the I-V curves as functions of the number of trajectories.

As above, we define the channel boundaries as z_{min} and z_{max}, and absorbing boundary conditions were located at these points. Trajectories were initiated at a point just inside the boundary at either z_{min} for forward trajectories or z_{max} for backward trajectories, and propagated until they reached either of the absorbing boundaries. Forward and backward trajectories are considered separately as we are interested in the 1-sided density profiles and committor probabilities, as well as their 2-sided combination. A trajectory that crossed from z_{min} to z_{max} is said to be a crossing trajectory in the forward direction. Similarly, trajectories that cross from z_{max} to z_{min} are crossing trajectories in the backward direction. For simplicity, these will be referred to as forward or backward crossing events. Since the trajectories are initiated near the absorbing boundaries, the majority of trajectories in either direction do not cross, but they do contribute to the 1-sided density profiles and first-passage statistics that are used to compute the committor probabilities (see Supplementary Materials, Section S3 for further details).

The number of trajectories initiated per data set were typically 10^6, 10^7 or 10^8, further abbreviated as N6, N7 and N8, respectively. These numbers were chosen because the average number of crossing events for PMFs corresponding to the models of TTX and p7 observed for $N = 10^6$ is of the same order of magnitude as the numbers of crossing events observed in our MD simulations of 0.5–2 µs. For the cation-selective GLIC channel, in which the free energy barrier to permeation of Na^+ is markedly higher, simulations with 10^6 trajectories yielded too few crossing events to be useful. In this case, the number of crossing events observed at the N7 level approximately corresponds to the number seen in MD simulations of 8 µs. See Supplementary Materials, Section S4 for details of the MD simulations.

The free energy surfaces for the stochastic simulations were obtained by adding the voltage ramp to the PMFs. We use a set of PMFs from our MD calculations. For our problem, the PMFs are the equilibrium free energy surfaces for moving an ion along the 1-dimensional reaction coordinate of the ion with respect to the center-of-mass of the protein channel, at the bulk ion densities of the MD. For this study, we used average diffusion coefficients, $< D >$ obtained by averaging the diffusivities estimated from MD (Supplementary Materials, Section S4). Strict matching of diffusivity is not necessary for the primary purpose of this study, but it provides a more realistic connection between statistical errors estimated in stochastic simulations and time scales of MD simulations. Additional details are given in Supplementary Materials, Section S7. The forward and backward ion density profiles were obtained from histograms of either the forward or backward trajectories in each data set. Committor probabilities (Supplementary Materials, Section S3) were calculated from the first passage statistics of the forward and backward trajectories. The density histograms and committor probabilities were computed for each data set, and not as an average over the individual trajectories in the data set. Averages were constructed over multiple data sets.

3. Results and Discussion

3.1. Connection with Molecular Dynamics

To compute the currents for the I-V curves from stochastic simulations, some connection to MD is required. MD simulations provide both forward and backward ion trajectories as part of the simulation, unless the channel is strongly rectifying or a large voltage is applied. The net current due to a particular ion is $J = J^f - J^b$ (Equation (9)). The total current is the sum of these net currents over all types of ions. As mentioned in the introduction, J can be obtained from MD simulations by way of combining the fluxes from forward and backward crossing events or calculating the ionic displacement currents. In stochastic simulations, only the former method can be used. Therefore, we tested whether both method yield the same results for MD and found that this was indeed the case. Both methods and the results of the tests are described in Supplementary Materials, Section S6.

In addition, the detailed balance condition connecting forward and backward crossing events has to be satisfied. In MD simulations this problem is implicitly solved: if there is no external voltage, simulations of transmembrane systems will exhibit no net current to within statistical errors, which means that the number of forward and backward crossing events is equal, again to within statistical errors. In stochastic simulations, detailed balance also has to be satisfied, which means that trajectories in both directions have to be combined with the correct weights. To determine these weights, we carried out sets of 10^8 simulations with no applied voltage to obtain the well converged, average numbers of forward and backward crossing events. From these simulations, the ratio of forward to backward trajectories that satisfies the detailed balance condition was established and subsequently used to compute the density profiles, committor probabilities and the PMFs at different voltages.

Once the ratio of forward to backward trajectories needed to satisfy the detailed balance condition is known, the average numbers of crossing events in both directions, $n^f(\Delta V)$ and $n^b(\Delta V)$, can be obtained from stochastic simulations at a given voltage ΔV. This, however, is still insufficient to determine currents; additional information about time scales is required. This can be obtained from a MD simulation of the system. We abbreviate the number of forward and backward crossing events observed in MD simulations at applied voltage, ΔV_{ref}, as m^f and m^b. Then, the length of the MD trajectory, t_{MD}, can be used to estimate a stochastic time, t_S, corresponding to the number of stochastic trajectories that produced $n^f(\Delta V_{ref})$ and $n^b(\Delta V_{ref})$ crossing events at the voltage ΔV_{ref}. A simple way to make such estimate is to use the number of crossing events in one direction. It is, of course, recommended to choose the direction that provides better statistics. Assuming that there are more forward than backward crossing events in the MD simulations,

$$t_S = t_{MD} \frac{n^f(\Delta V_{ref})}{m^f}.$$

If the backward events dominate, t_S would be estimated using $n^b(\Delta V_{ref})$ and m^b. Once t_S has been determined, the stochastic currents at voltage ΔV can be calculated:

$$J_S(\Delta V) = J_S^f(\Delta V) - J_S^b(\Delta V) = [n^f(\Delta V) - n^b(\Delta V)]/t_S.$$

where $J_S(\Delta V)$, $J_S^f(\Delta V)$ and $J_S^b(\Delta V)$ are the total, forward and backward currents at voltage ΔV.

3.2. Committor Probabilities

The committor probabilities for p7 are shown in Figure 1. The committor probabilities in Figure 1a have been calculated from Supplementary Materials, Equation S30, in which ions arriving at z from both sides are included. The statistical errors associated with $P(z)$ at different voltages are small, even at the N6 level. As can be seen in Figure 1b, this is not the case for one-sided $P(z)$, obtained using Supplementary Materials, Equations

S28 and S29. This is due to the decreasing number of ions from one direction as they approach the opposite side of the channel. The inset of Figure 1b shows the numbers of first-passage events from the forward and backward calculations as well as the combined number of events. At 140 mV, the statistics are satisfactory only in the forward direction, in which most of crossing events occur, whereas no reliable probabilities are obtained for the backward direction over the full range of z. The opposite is true for -140 mV; $P(z)$ in the forward direction is unreliable. Thus, combining information about $P(z)$ in both directions is preferable whenever possible.

As voltage changes from -140 mV to 140 mV, the position of the transition state for K^+ permeation through p7, defined as the x,y-plane at which $P(z) = 0.5$, Ref. [85] shifts substantially and systematically from 7 Å to -13 Å with respect to the center of mass of the membrane. Such large shifts, however, are not universal. As we have shown in the example of TTX [33], the position of the transition state changes markedly less with voltage if the underlying PMF is strongly peaked.

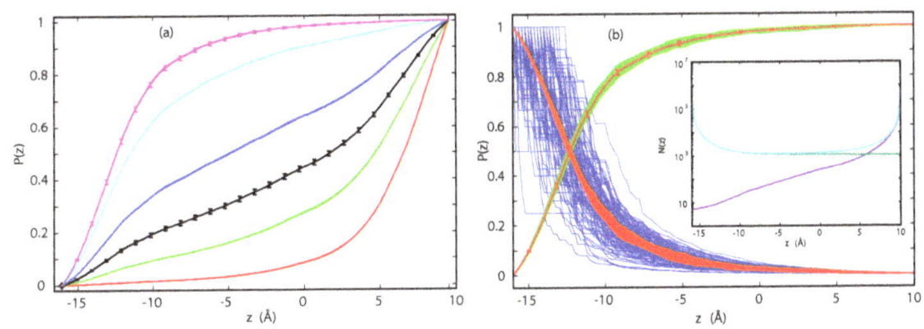

Figure 1. (**a**) Committor probabilities for Cl^- in p7 at -140 mV (red), -70 mV (green), -35 mV (black), 0 V (blue), 70 mV (cyan) and 140 mV (magenta). Error bars are shown for the N6 data sets at -35 mV and 140 mV; (**b**) Committor probabilities for p7 at 140 mV from the N6 data set for 1-sided forward (green) and backward (blue) trajectories, respectively, 2-sided data set in the backward direction (red lines), and average in the forward direction with error bars (red symbols). In the inset we show the number of first passage trajectories to reach z for one N6 data set in the forward (green) and backward (magenta) directions and the total (light blue).

Calculating $P(z)$ for GLIC is more difficult. This is a slow channel and even at the N7 level, which approximately corresponds to a MD trajectory of 8 μs in length (see Supplementary Materials, Section S4, the number of crossing events is small. At 100 mV only an average number of 0.5 forward and 29 backward crossing events were observed. In particular, N7 simulations of forward trajectories frequently produce no crossing events. At the same voltage, $P(z)$ in the backward direction is often equal to 1 over a relatively wide range of several Å near z_{max}, which means that all ions that reached this range exit the channel at z_{max}. In such circumstances, calculation of $P(z)$ from Supplementary Materials, Equation S30 is no longer possible. A different approach is needed.

Direct calculation of the committor probability requires that some number of trajectories successfully cross the channel, $N^b(z_{min}) > 0$ and $N^f(z_{max}) > 0$. If one of these conditions is not met, for example, if $N^f(z_{max}) = 0$, then $P^f(z) = N^f(z_{max})/N^f(z) = 0$. If we consider position z' ($z' < z_{max}$) at which $N^f(z') > 0$, then we can write the committor probability $P(z) = \alpha N^f(z')/N^f(z)$, where α is unknown, though formally would be equal to $N^f(z_{max})/N^f(z')$ if complete sampling of the forward direction were available. α can be determined in a self-consistent manner. Using Equation S30 from Supplementary Materials, Section S3, we can write the total committor probability in the region $z < z'$ and the backward committor probability $z > z'$:

$$P(z) = \begin{cases} \frac{\alpha N^f(z') + N^b(z) - N^b(z_{min})}{N^f(z) + N^b(z)} & \text{if } z < z' \\ 1 - \frac{N^b(z_{min})}{N^b(z)} & \text{if } z > z'. \end{cases}$$

If we require that $P(z)$ is continuous at $z = z'$, then $\alpha = 1 - N^b(z_{min})/N^b(z')$. Other ways of determining $P(z)$ for this problem are possible.

3.3. The Potential of Mean Force

Typically, the PMFs for ions in channels are calculated in simulations in the absence of electric field using enhanced sampling techniques (see the recent review by Flood, et al. [27]). In contrast, the methods outlined here allow for reconstructing PMF from steady-state simulations with an applied electric field. The underlying PMFs for p7 and GLIC used in the present study were obtained by way of this method (see Supplementary Materials, Section S1). Since TTX is a bundle of straight α-helices surrounding a featureless water pore, the PMFs for K^+ and Cl^- are quite generic, which is characteristic of several very simple channels (see Figure 2a) [24,25]. For K^+, the PMF is fairly flat over a wide range of approximately 18 Å inside the channel, which is reminiscent of classical models of ionic conductance in which it is assumed that the PMF is a step function constant inside the channel [1,86]. For Cl^-, the PMF is peaked near the center of the bilayer, which can be attributed to the Born barrier experienced by an ion permeating a rigid, featureless non-polar lamella [87]. If an ion is transferred across a membrane through a water-filled pore, the general shape of the PMF remains the same, but the barrier is substantially reduced [87,88]. For TTX, it still remains approximately 1.5 kcal/mol higher than the barrier for K^+, which is consistent with a weak selectivity of this channel toward cations.

The PMF for permeation of Cl^- in the OuYang et al. model of p7 [69] is more structured than the PMF for TTX (see Figure 3a). The barriers are low, which explains high chloride current predicted by this model [37]. In contrast to TTX, the barriers to Cl^- permeation in p7 are located near the mouths of the channel due to the presence of positively charged residues at these locations. Compared to permeation of Cl^-, the current of K^+ in this model is quite low, which indicates that the channel should be anion-selective. Both predicted selectivity and total currents are at variance with electrophysiological data [89], thus contributing to arguments that the proposed high-resolution structure [69] is not native.

The PMF representing permeation of Na^+ through GLIC is also markedly more structured than the PMF for ions in TTX (see Figure 2b). The barrier is substantially higher than in the other two channels. As a result, the conductance of this channel is relatively low [70]. This presents a challenge because the number of crossing events in both directions is small. The PMF for Cl^- in this channel is not considered because no crossing events of this ion have been observed in MD simulations.

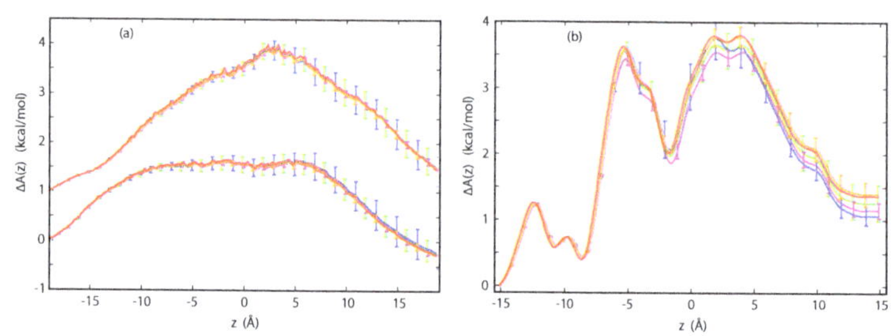

Figure 2. (**a**) PMFs for K^+ (lower curves) and Cl^- (upper curves) in TTX from stochastic simulations with an applied voltage of 50 mV. The PMFs have been reconstructed by way of CWDM at the N6 (blue) and N7 (gold) levels or by way of CPM at the N6 (green) and N7 (magenta) levels; (**b**) PMF for Na^+ in GLIC from stochastic simulations with applied voltage of 100 mV. The PMF has been reconstructed by way of CWDM at the N7 (blue) and N8 (gold) level or by way of CPM at the N7 (green) and N8 (magenta) level. In both panels, the underlying PMF is in red.

Taken together, the PMFs considered here are quite different from one another, but are typical of the variety seen in ion channels. In spite of these differences, all three PMFs were successfully reconstructed from non-equilibrium simulations by way of

Equations (12) and (15) associated, respectively, with the CWDM and CPM methods. The applied voltages were 50 mV for TTX, 140 and −35 mV for p7 and 100 mV for GLIC. For TTX and p7, reconstruction was carried out at the N6, N7 and N8 levels. For GLIC, the number of crossing events at the N6 level was quite small or equal to zero. Thus, only N7 and N8 levels were considered. At the N6 level, 50 and 100 data sets of trajectories were generated for TTX and p7, respectively. At the N7 level, 20 sets of trajectories were generated for TTX and p7, and 50 sets were generated for GLIC. At the N8 level, the number of generated sets was 4, 8 and 25 for TTX, p7 and GLIC, respectively. At each level, all reconstructed PMFs are found to be tightly clustered and their averages at each level are close to the underlying PMF, as shown in Figures 2 and 3a.

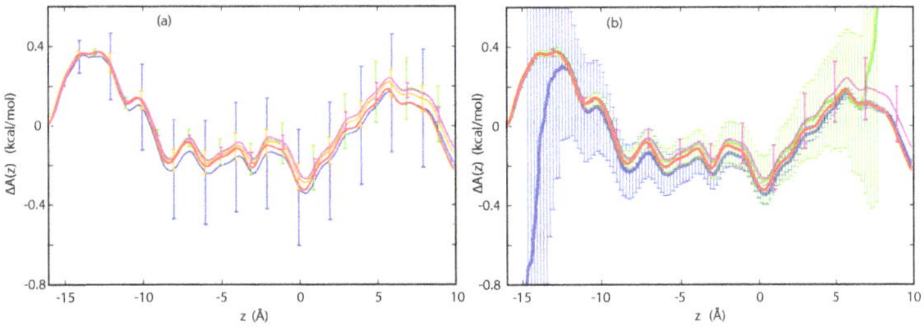

Figure 3. (**a**) PMF for Cl^- in p7 from stochastic simulations with an applied voltage of 140 mV. The PMFs have been reconstructed by way of CWDM at the N6 (blue) and N7 (gold) levels or by way of CPM at the N6 (green) and N7 (magenta) levels. The input PMF (red) is shown for reference. PMFs at the N8 level are not shown, as they coincide with the underlying PMFs and statistical errors associated with this level arequite small and are poorly visible at this scale; (**b**) PMFs for P7 reconstructed by way of one-sided forward trajectories (green) using Equation (13) and backward trajectories (blue) using Equation (14) from stochastic simulations at the N6 level with applied voltage of 140 mV. Two-sided reconstruction (magenta) and the underlying PMF (red) are shown for comparison. Note that one-sided, but not two-sided reconstructions are burdened with large errors at the ends.

In these figures, statistical errors associated with dispersion of the reconstructed PMFs are marked. For $\Delta A(z_{max}, z_{min})$, these errors are approximately ± 0.3 kcal/mol at the N6 level for TTX and p7 and at the N7 level for GLIC. As expected, they are reduced by approximately a factor of 3 with each level in which the number of sets increases by an order of magnitude. The mean PMFs obtained by way of CWDM and CPM at different levels are quite close to the underlying PMF and the corresponding statistical errors are very similar, indicating that both methods are successful in reproducing the underlying PMFs. Only for GLIC at the N7 level, does the $\Delta A(z_{max}, z_{min})$ reconstructed by way of CWDM appear to be systematically underestimated. At this level, no crossing events in one direction are observed for a considerable fraction of data sets, which makes reconstruction of the PMF from Equation (12) impossible. This systematically biases the sample in favor of sets with higher counts of crossing events and, consequently, lower $\Delta A(z_{max}, z_{min})$. From the comparison between the PMFs reconstructed for p7 from trajectories at 140 and −35 mV, it appears that precision of the reconstruction depends somewhat on applied voltage. If the forward and backward densities are well balanced, precision improves.

In CPM, the PMFs can be calculated from one-sided quantities, Equations (13) and (14), or by combining them. Here, the latter has been done by way of WHAM, Equation (15). As can be seen in Figure 3b, this approach yields improved agreement with the underlying PMF. In one-sided formulas, the densities can become quite low near the exit and, as a result, precision in this range suffers.

In summary, both CWDM and CPM provide a reliable means for reconstructing PMFs from non-equilibrium simulations. However, the relation between statistical errors obtained in stochastic and MD simulations is not straightforward. Even if the assumptions of the ED model are satisfied, precision of stochastic simulations is expected to be higher than precision in MD simulations of equivalent length. Specifically, it is usually uncertain if all degrees of freedom perpendicular to the reaction coordinate have been properly equilibrated on the time scale of the simulations. Torsional angles in the side chains of residues lining the pore or motion of whole helices are examples of degrees of freedom that might undergo slow equilibration and, by doing so, influence the calculated PMF and electrophysiological properties. The same concern applies to all other methods for calculating these quantities.

3.4. Current-Voltage Dependence

Once the PMFs for the ions permeating the channel have been reconstructed and the committor probabilities for these ions have been calculated for a reference voltage, the full current-voltage (I-V) curves can be calculated from Equations (19) and (20) without the need for additional simulations. This is the principal gain in efficiency of the method: the I-V curve can be obtained from a single MD simulation instead of multiple simulations. For example, if constructing the I-V curve required MD simulations at five different voltages in the range of interest the efficiency of our methods would be approximately five-fold. Since numerical results indicate that Equation (19) yields less accurate results than Equation (20), this equation will not be further considered. The results for TTX, p7 and GLIC are shown in Figures 4 and 5. The reference applied voltages are the voltages used for reconstructing PMFs, described in the previous subsection. For comparison, currents calculated directly from stochastic trajectories at several voltages are also shown.

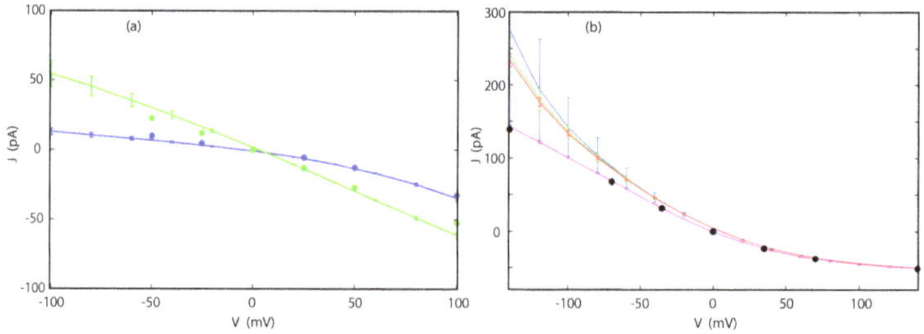

Figure 4. (**a**) I-V curves for K^+ (green) and Cl^- (blue) in TTX reconstructed from simulations at 50 mV at the N6 level. Blue and green dots are currents obtained from direct simulations at specific voltages.; (**b**) I-V curves for Cl^- in p7 reconstructed from simulations at 140 mV at the N6 (blue), N7 (green) and N8 (red) level, and for −35 mV at the N6 level (magenta). N7 and N8 curves are not shown because they are almost identical to the N6 results. Black dots are currents obtained from direct simulations at specific voltages. All reconstructions were done using the PMFs obtained by way of CPM. The results of reconstructions using the PMFs from CWDM are not displayed because they are nearly identical.

As we can see in Figure 4, the agreement between the I-V curves for both K^+ and Cl^- in TTX calculated directly and by way of Equations (18) and (20) is excellent, even at the N6 level, for the full range of voltages studied here, which extends from −100 to 100 mV. As shown in Figure 6, the I-V curves obtained for different sets of trajectories are closely clustered and deviate from each other only at the largest absolute applied voltages by no more than a few pA.

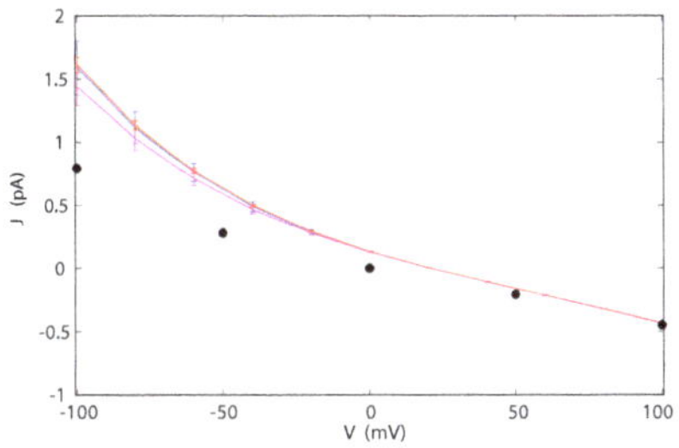

Figure 5. I-V curves for Na^+ in GLIC reconstructed from simulations at 100 mV at the N7 level with PMF from CPM (blue), at the N7 level with PMF from CWDM (magenta), and N8 with PMF from CWDM (red). N8 with CPM (not shown) is almost identical to N7 CPM. Black dots are currents obtained from direct simulations at specific voltages.

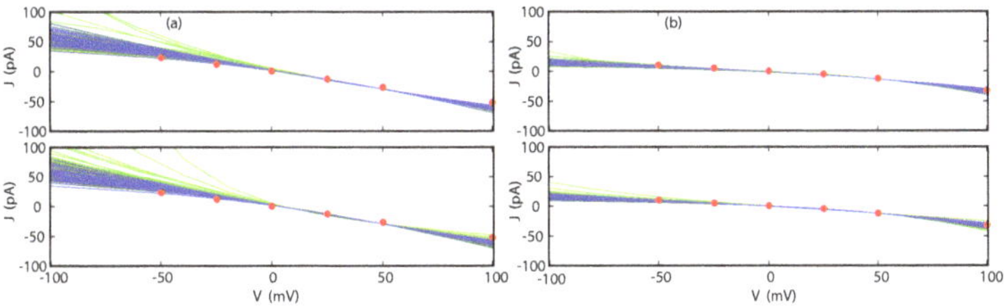

Figure 6. Reconstructions of I-V curves in TTX from individual sets of trajectories for K^+ (**a**) and Cl^- (**b**). The PMFs were obtained from CPM (upper panels) or CWDM (lower panels). The curves were calculated by way of Equation (20) (blue) or Equation (18) (green). All reconstructions were carried out from simulations at applied voltage of 50 mV at the N6 level. Note that blue curves, but not green curves, are tightly clustered together indicating that Equation (20) is more accurate than Equation (18).

For p7, the agreement is not as good if the the reference voltage of 140 mV is used for calculating the I-V curve. The Cl^- currents calculated directly and from Equation (18) agree well for positive voltages, but diverge for negative voltages, away from the reference state. The corresponding statistical errors also increase and become quite large below -50 mV. The source of this disagreement can be traced to the integrand in Equation (18). As the difference between the reference and the target voltage increases, the exponential term also increases, which magnifies inaccuracies in function $f(z)$. If the reference voltage is chosen to be -35 mV, the agreement over the full range of voltages -150 to 150 mV improves markedly, with modest deviations only at high, positive voltages (see Figure 4). A similar situation was observed for GLIC. For the reference voltage of 100 mV, the I-V curves at the N7 level satisfactorily reproduce currents calculated directly for positive voltages. For negative voltages, the performance of the method progressively deteriorates. Again, if one is interested in an I-V curve that extends to both positive and negative voltages, a different choice of reference voltage may yield significant improvements in accuracy.

As pointed out in the introduction, a number of previous studies have used unrealistically large applied voltages to increase the number of crossing events and, by doing so, improved precision of the calculated currents [15,23,35,37]. Furthermore, as discussed earlier, this may lead to electroporation of the membrane, saturation effects during the intake of ions at the mouth of the channel and involves extrapolation or interpolation to the voltages by way of *ad hoc* procedures of unknown accuracy. The approach developed here is more efficient and accurate and has a substantially stronger theoretical basis than procedures used previously, even though only calculations at the reference applied voltage are necessary. In this approach, the accuracy of the reconstructed I-V curves can be substantially improved through a judicious choice of this reference voltage. This choice depends on the range of voltage that is of interest and on several properties of a channel, in particular its rectification, which characterizes an asymmetry of currents in response to the change in direction of applied voltage. In general, maximizing total ion current through applying high voltage is not the optimal strategy. Instead, it is often better to choose a voltage that yields good statistics in both directions.

3.5. Reversal Potential

The reversal potential, ΔV_R, is the applied voltage at which there is no net current. If ionic concentrations on both sides of the membrane are equal, $\Delta V_R = 0$. Experimentally, the reversal potential is measured by maintaining different concentrations on the *cis* and *trans* side of the membrane, and then used in conjunction with the GHK equation to estimate channel selectivity [1,6]. In MD, asymmetric concentrations have to be maintained to measure directly the reversal potential [90], which markedly complicates simulations. We have only considered the situation where the concentrations of ions are the same on both sides of the membrane, as this corresponds to the conditions under which we have carried out MD. We wish to expand this to a range of concentrations.

We expect that the net number of crossing events, from which we calculate the I-V curve, depends on this concentration. If the bulk concentrations are low and the channel is not saturated, then we expect the number of crossing events, and hence the currents, to be linearly dependent on the concentrations of ions. For example, if the concentration is doubled on both sides of the membrane, the net currents will also double. Under these assumptions, we can calculate the reversal potential from our formalism. We simply need to scale the fluxes of all ion types on one side of the membrane to match the desired concentration difference.

The K^+/Cl^- selectivity of TTX obtained from the MD simulations is 2.2 [24,33]. Using currents scaled by 5:1 in the I-V reconstruction from Equation (20), we obtain a reversal potential of -9 mV. This corresponds to a GHK selectivity of 1.7, which is reasonably close to the selectivity found in MD. Note that experimentally, the reversal potential is -27 mV, corresponding to a K^+/Cl^- selectivity of 6 estimated from the GHK Equation [6]. This cannot be compared directly to our results because the actual channel structure is unknown, there are uncertainties due to force fields, and the GHK equation itself is an approximation.

4. Conclusions

Stochastic simulations were used to investigate the reliability of two new methods to calculate PMFs for ion transport across transmembrane ion channels and electrophysiological properties of these channels within the general framework of the electrodiffusion model. Both methods have the desirable features that only simulations at a single voltage are needed and information on the diffusivity is not required. In CPM, knowledge of the committor probability is required. Stochastic simulations containing 10^6 trajectories were shown to have similar numbers of crossing events for models of TTX and p7 in MD simulations of 1–2 μs in length. Analysis of 50 or 100 of such simulations indicate that errors in the free energy profiles are approximately ± 0.3 kcal/mol. For a model of a slow channel, GLIC, 10^7 trajectories, which approximately corresponds to MD simulations of 10 μs in length, are needed to achieve a similar statistical error. For both TTX and for p7 at lower applied voltages, CPM and CWDM yield similar results. In CWDM, one-sided fluxes are used directly, and for cases in which few crossing events are observed in one direction, either due to large applied voltages, such as p7 at ± 140 mV, or because the channel is rectifying, such as GLIC, CPM performs better because two-sided quantities are employed in this method. Similarly, even though one-sided CPM calculations are possible, the errors near the end of the channel become substantial because the density becomes quite small, yielding large relative errors.

Stochastic simulations were also used to investigate the reliability of a new expression to calculate the ionic currents at different voltages, ΔV, given knowledge of the PMF, committor probabilities and density profiles at a reference voltage ΔV_{ref}. We found that the I-V dependence could be reconstructed over a range of ± 100 mV, with respect to the reference voltage. Judicious choice of ΔV_{ref} can markedly improve the accuracy of the reconstruction. Specifically, the I-V reconstruction for p7 is much better for $\Delta V_{\text{ref}} = -35$ mV than for $\Delta V_{\text{ref}} = 140$ mV. Although much of the error can be attributed to the large voltage ramp for voltages away from ΔV_{ref} (3.2 kcal/mol at 140 mV), some of the error is due to the poor statistics in the direction against the field. This is also evident in the reconstruction of the I-V curve for GLIC, for which some simulations yielded no crossing events against the field.

Common goals of simulations of ion channels are to obtain the free energy profiles of ions translocating the channel and to determine electrophysiological properties of the channel. In some instances, a reliable estimate of the numbers of crossing events, from which the ionic currents can be calculated, is difficult to obtain from MD even for long simulation times. We have shown that the new methods perform very well both to obtain reliably the free energy profile across the channel and to allow for accurate determination of the I-V curves. In the latter case, it is desirable to use a reference voltage that yields good crossing statistics in both directions rather than a voltage that maximizes the total number of crossing events. In summary, if transport of ions through a channel can be satisfactorily described by the ED model, the new methods offer substantial reductions of computational effort without sacrificing accuracy. Our approach is amenable to extensions in which the advantages of MD and stochastic simulations are further combined on reliable theoretical grounds.

Supplementary Materials: The following are available at https://www.mdpi.com/article/10.3390/e23050571/s1 , Figure S1: Pictures from MD simulations of TTX, p7, and GLIC, Figure S2: Total displacement charge calculated from ion crossing statistics and displacement current.

Author Contributions: Conceptualization, A.P.; methodology, A.P. and M.A.W.; software, M.A.W.; validation, M.A.W. and A.P.; formal analysis, A.P. and M.A.W.; writing–original draft preparation, A.P.; writing–review and editing, A.P and M.A.W.; visualization, M.A.W.; project administration, A.P.; funding acquisition, A.P. All authors have read and agreed to the published version of the manuscript.

Funding: Support for this research was provided by NASA's Planetary Science Division Research Program.

Conflicts of Interest: The authors declare no conflict of interest. The funders had no role in the design of the study; in the collection, analyses, or interpretation of data; in the writing of the manuscript, or in the decision to publish the results.

Abbreviations

The following abbreviations are used in this manuscript:

BD	Brownian Dyanmics
CPM	Committor Probability Method
CWDM	Current-Weighted Density Method
ED	electrodiffusion
GHK	Goldman Hodgkin Katz (equation)
IEEM	Integrated Electrodiffusion Equation Method
MD	molecular dynamics
PNP	Poisson-Nernst-Planck
PMF	potential of mean force
TTX	trichotoxin channel
WHAM	Weighted Histogram Analysis Method

References

1. Hille, B. *Ion Channels of Excitable Membranes*, 3rd ed.; Sinauer Associates, Inc.: Sunderland, MA, USA, 2001.
2. Zheng, J.; Trudeau, M.C. *Handbook of Ion Channels*; CRC Press: Boca Raton, FL, USA, 2015.
3. Kubalski, A.; Martinac, B. *Bacterial Ion Channels and Their Eukaryotic Homologs*; ASM Press: Washington, DC, USA, 2005.
4. Bocquet, N.; Nury, H.; Baaden, M.; Le Poupon, C.; Changeux, J.P.; Delarue, M.; Corringer, P.J. X-ray structure of a pentameric ligand-gated ion channel in an apparently open conformation. *Nature* **2009**, *457*, 111–114.

5. Sakmann, B.; Boheim, G. Alamethicin-induced single channel conductance fluctuations in biological membranes. *Nature* **1979**, *282*, 336–339.
6. Duclohier, H.; Alder, G.M.; Bashford, C.L.; Bruckner, H.; Chugh, J.K.; Wallace, B.A. Conductance studies on trichotoxin_A50E and implications for channel structure. *Biophys. J.* **2004**, *87*, 1705–1710.
7. Martinac, B. Mechanosensitive ion channels: An evolutionary and scientific tour de force in mechanobiology. *Channels* **2012**, 211–213.
8. Bass, R.B.; Strop, P.; Barclay, M.; Rees, D.C. Crystal structure of Escherichia coli MscS, a voltage-modulated and mechanosensitive channel. *Science* **2002**, *298*, 1582–1587.
9. Panaghie, G.; Purtell, K.; Tai, K.K.; Abbott, G.W. Voltage-dependent C-type inactivation in a constitutively open K+ channel. *Biophys. J.* **2008**, *95*, 2759–2778.
10. Anstee, Q.M.; Knapp, S.; Maguire, E.P.; Hosie, A.M.; Thomas, P.; Mortensen, M.; Bhome, R.; Martinez, A.; Walker, S.E.; Dixon, C.I.; et al. Mutations in the Gabrb1 gene promote alcohol consumption through increased tonic inhibition. *Nat. Commun.* **2013**, *4*, 2816.
11. Hou, X.; Outhwaite, I.R.; Pedi, L.; Long, S.B. Cryo-EM structure of the calcium release-activated calcium channel Orai in an open conformation. *Elife* **2020**, *338*, 1308–1313.
12. Ashcroft, F.M. *Ion Channels and Disease*; Academic Press: Cambridge, MA, USA, 1999.
13. Bagal, S.K.; Brown, A.D.; Cox, P.J.; Omoto, K.; Owen, R.M.; Pryde, D.C.; Sidders, B.; Skerratt, S.E.; Stevens, E.B.; Storer, R.I.; et al. Ion channels as therapeutic targets: A drug discovery perspective. *J. Med. Chem.* **2013**, *56*, 593–624.
14. Garcia, M.L.; Kaczorowski, G.J. Ion channels find a pathway for therapeutic success. *Proc. Natl. Acad. Sci. USA* **2016**, *113*, 5472–5474.
15. Aksimentiev, A.; Schulten, K. Imaging α-hemolysin with molecular dynamics: Ionic conductance, osmotic permeability, and the electrostatic potential map. *Biophys. J.* **2005**, *88*, 3745–3761.
16. Sotomayor, M.; Vásquez, V.; Perozo, E.; Schulten, K. Ion conduction through MscS as determined by electrophysiology and simulation. *Biophys. J.* **2007**, *92*, 886–902.
17. Pezeshki, S.; Chimerel, C.; Bessonov, A.N.; Winterhalter, M.; Kleinekathöfer, U. Understanding ion conductance on a molecular level: An all-atom modeling of the bacterial porin OmpF. *Biophys. J.* **2009**, *97*, 1898–1906.
18. Kutzner, C.; Grubmüller, H.; De Groot, B.L.; Zachariae, U. Computational electrophysiology: The molecular dynamics of ion channel permeation and selectivity in atomistic detail. *Biophys. J.* **2011**, *101*, 809–817.
19. Wilson, M.A.; Wei, C.; Bjelkmar, P.; Wallace, B.A.; Pohorille, A. Molecular dynamics simulation of the antiamoebin ion channel: Linking structure and conductance. *Biophys. J.* **2011**, *100*, 2394–2402.
20. Zhu, F.; Hummer, G. Theory and simulation of ion conduction in the pentameric GLIC channel. *J. Chem. Theory Comput.* **2012**, *8*, 3759–3768.
21. Stock, L.; Delemotte, L.; Carnevale, V.; Treptow, W.; Klein, M.L. Conduction in a biological sodium selective channel. *J. Phys. Chem. B* **2013**, *117*, 3782–3789.
22. Jensen, M.Ø.; Jogini, V.; Eastwood, M.P.; Shaw, D.E. Atomic-level simulation of current–voltage relationships in single-file ion channels. *J. General Physiol.* **2013**, *141*, 619–632.
23. Ulmschneider, M.B.; Bagnéris, C.; McCusker, E.C.; DeCaen, P.G.; Delling, M.; Clapham, D.E.; Ulmschneider, J.P.; Wallace, B.A. Molecular dynamics of ion transport through the open conformation of a bacterial voltage-gated sodium channel. *Proc. Natl. Acad. Sci. USA* **2013**, *110*, 6364–6369.
24. Wilson, M.A.; Nguyen, T.H.; Pohorille, A. Combining molecular dynamics and an electrodiffusion model to calculate ion channel conductance. *J. Chem. Phys.* **2014**, *141*, 22D519.
25. Pohorille, A.; Wilson, M.A.; Wei, C. Validity of the electrodiffusion model for calculating conductance of simple ion channels. *J. Phys. Chem. B* **2017**, *121*, 3607–3619.
26. Callahan, K.M.; Roux, B. Molecular dynamics of ion conduction through the selectivity filter of the NaVAb sodium channel. *J. Phys. Chem. B* **2018**, *122*, 10126–10142.
27. Flood, E.; Boiteux, C.; Lev, B.; Vorobyov, I.; Allen, T.W. Atomistic Simulations of Membrane Ion Channel Conduction, Gating, and Modulation. *Chem. Rev.* **2019**, *119*, 7737–7832.
28. Oestringer, B.P.; Bolivar, J.H.; Hensen, M.; Claridge, J.K.; Chipot, C.; Dehez, F.; Holzmann, N.; Zitzmann, N.; Schnell, J.R. Re-evaluating the p7 viroporin structure. *Nature* **2018**, *562*, E8–E18.
29. Machtens, J.P.; Kortzak, D.; Lansche, C.; Leinenweber, A.; Kilian, P.; Begemann, B.; Zachariae, U.; Ewers, D.; de Groot, B.L.; Briones, R.; et al. Mechanisms of anion conduction by coupled glutamate transporters. *Cell* **2015**, *160*, 542–553.
30. Roux, B. The membrane potential and its representation by a constant electric field in computer simulations. *Biophys. J.* **2008**, *95*, 4205–4216.
31. Gumbart, J.; Khalili-Araghi, F.; Sotomayor, M.; Roux, B. Constant electric field simulations of the membrane potential illustrated with simple systems. *Biochim. Biophys. Acta* **2012**, *1818*, 294–302.
32. Faraudo, J.; Calero, C.; Aguilella-Arzo, M. Ionic partition and transport in multi-ionic channels: A molecular dynamics simulation study of the OmpF bacterial porin. *Biophys. J.* **2010**, *99*, 2107–2115.
33. Pohorille, A.; Wilson, M.A. Computational Electrophysiology from a Single Molecular Dynamics Simulation and the Electrodiffusion Model. *J. Phys. Chem. B* **2021**, *125*, 3132–3144.

34. Biró, I.; Pezeshki, S.; Weingart, H.; Winterhalter, M.; Kleinekathöfer, U. Comparing the temperature-dependent conductance of the two structurally similar E. coli porins OmpC and OmpF. *Biophys. J.* **2010**, *98*, 1830–1839.
35. Chandler, D.E.; Penin, F.; Schulten, K.; Chipot, C. The p7 protein of hepatitis C virus forms structurally plastic, minimalist ion channels. *PLoS Comput. Biol.* **2012**, *8*, e1002702.
36. Zachariae, U.; Schneider, R.; Briones, R.; Gattin, Z.; Demers, J.P.; Giller, K.; Maier, E.; Zweckstetter, M.; Griesinger, C.; Becker, S.; et al. β-Barrel mobility underlies closure of the voltage-dependent anion channel. *Structure* **2012**, *20*, 1540–1549.
37. Holzmann, N.; Chipot, C.; Penin, F.; Dehez, F. Assessing the physiological relevance of alternate architectures of the p7 protein of hepatitis C virus in different environments. *Bioorgan. Med. Chem.* **2016**, *24*, 4920–4927.
38. Wood, M.L.; Freites, J.A.; Tombola, F.; Tobias, D.J. Atomistic modeling of ion conduction through the voltage-sensing domain of the Shaker K+ ion channel. *J. Phys. Chem. B* **2017**, *121*, 3804–3812.
39. Chen, D.P.; Eisenberg, R.S. Flux, coupling, and selectivity in ionic channels of one conformation. *Biophys. J.* **1993**, *65*, 727–746.
40. Noskov, S.Y.; Im, W.; Roux, B. Ion permeation through the α-hemolysin channel: Theoretical studies based on Brownian dynamics and Poisson-Nernst-Plank electrodiffusion theory. *Biophys. J.* **2004**, *87*, 2299–2309.
41. Coalson, R.D.; Kurnikova, M.G. Poisson-Nernst-Planck theory approach to the calculation of current through biological ion channels. *IEEE Trans. Nanobiosci.* **2005**, *4*, 81–93.
42. Coalson, R.D.; Kurnikova, M.G. Poisson–Nernst–Planck theory of ion permeation through biological channels. In *Biological Membrane Ion Channels*; Springer: Berlin, Germany, 2007; pp. 449–484.
43. Liu, J.L.; Eisenberg, B. Poisson-Nernst-Planck–Fermi theory for modeling biological ion channels. *J. Chem. Physic.* **2014**, *141*, 12B640_1.
44. Liu, J.L.; Eisenberg, B. Molecular mean-field theory of ionic solutions: A Poisson-Nernst-Planck-Bikerman model. *Entropy* **2020**, *22*, 550.
45. Im, W.; Roux, B. Brownian dynamics simulations of ions channels: A general treatment of electrostatic reaction fields for molecular pores of arbitrary geometry. *J. Chem. Phys.* **2001**, *115*, 4850–4861.
46. Chung, S.H.; Kuyucak, S. Recent advances in ion channel research. *Biochim. Biophys. Acta* **2002**, *1565*, 267–286.
47. Chung, S.H.; Allen, T.W.; Kuyucak, S. Conducting-state properties of the KcsA potassium channel from molecular and Brownian dynamics simulations. *Biophys. J.* **2002**, *82*, 628–645.
48. Chung, S.H.; Krishnamurthy, V. Brownian Dynamics: Simulation for Ion Channel Permeation. In *Biological Membrane Ion Channels*; Springer: Berline, Germany, 2007; pp. 507–543.
49. Berneche, S.; Roux, B. Molecular dynamics of the KcsA K+ channel in a bilayer membrane. *Biophys. J.* **2000**, *78*, 2900–2917.
50. Köpfer, D.A.; Song, C.; Gruene, T.; Sheldrick, G.M.; Zachariae, U.; de Groot, B.L. Ion permeation in K+ channels occurs by direct Coulomb knock-on. *Science* **2014**, *346*, 352–355.
51. Allen, T.W.; Andersen, O.S.; Roux, B. Energetics of ion conduction through the gramicidin channel. *Proc. Natl. Acad. Sci. USA* **2004**, *101*, 117–122.
52. Keener, J.P.; Sneyd, J. *Mathematical Physiology*; Springer: Berlin, Germany, 1998; Volume 1.
53. Berneche, S.; Roux, B. Energetics of ion conduction through the K+ channel. *Nature* **2001**, *414*, 73–77.
54. Furini, S.; Domene, C. On conduction in a bacterial sodium channel. *PLoS Comput. Biol.* **2012**, *8*, e1002476.
55. Chakrabarti, N.; Ing, C.; Payandeh, J.; Zheng, N.; Catterall, W.A.; Pomès, R. Catalysis of Na+ permeation in the bacterial sodium channel NaVAb. *Proc. Natl. Acad. Sci. USA* **2013**, *110*, 11331–11336.
56. Boiteux, C.; Vorobyov, I.; Allen, T.W. Ion conduction and conformational flexibility of a bacterial voltage-gated sodium channel. *Proc. Natl. Acad. Sci. USA* **2014**, *111*, 3454–3459.
57. Harpole, T.J.; Grosman, C. Side-chain conformation at the selectivity filter shapes the permeation free-energy landscape of an ion channel. *Proc. Natl. Acad. Sci. USA* **2014**, *111*, E3196–E3205.
58. Finol-Urdaneta, R.K.; Wang, Y.; Al-Sabi, A.; Zhao, C.; Noskov, S.Y.; French, R.J. Sodium channel selectivity and conduction: Prokaryotes have devised their own molecular strategy. *J. General Physiol.* **2014**, *143*, 157–171.
59. Trick, J.L.; Chelvaniththilan, S.; Klesse, G.; Aryal, P.; Wallace, E.J.; Tucker, S.J.; Sansom, M.S. Functional annotation of ion channel structures by molecular simulation. *Structure* **2016**, *24*, 2207–2216.
60. Flood, E.; Boiteux, C.; Allen, T.W. Selective ion permeation involves complexation with carboxylates and lysine in a model human sodium channel. *PLoS Comput. Biol.* **2018**, *14*, e1006398.
61. Cottone, G.; Chiodo, L.; Maragliano, L. Thermodynamics and kinetics of ion permeation in wild-type and mutated open active conformation of the human α7 nicotinic receptor. *J. Chem. Inf. Model.* **2020**, *60*, 5045–5056.
62. Jensen, M.Ø.; Borhani, D.W.; Lindorff-Larsen, K.; Maragakis, P.; Jogini, V.; Eastwood, M.P.; Dror, R.O.; Shaw, D.E. Principles of conduction and hydrophobic gating in K+ channels. *Proc. Natl. Acad. Sci. USA* **2010**, *107*, 5833–5838.
63. Corry, B.; Thomas, M. Mechanism of ion permeation and selectivity in a voltage gated sodium channel. *J. Am. Chem. Soc.* **2012**, *134*, 1840–1846.
64. Modi, N.; Bárcena-Uribarri, I.; Bains, M.; Benz, R.; Hancock, R.E.; Kleinekathöfer, U. Role of the central arginine R133 toward the ion selectivity of the phosphate specific channel OprP: Effects of charge and solvation. *Biochemistry* **2013**, *52*, 5522–5532.
65. Jorgensen, C.; Furini, S.; Domene, C. Energetics of ion permeation in an open-activated TRPV1 channel. *Biophys. J.* **2016**, *111*, 1214–1222.

66. Polovinkin, L.; Hassaine, G.; Perot, J.; Neumann, E.; Jensen, A.A.; Lefebvre, S.N.; Corringer, P.J.; Neyton, J.; Chipot, C.; Dehez, F.; et al. Conformational transitions of the serotonin 5-HT 3 receptor. *Nature* **2018**, *563*, 275–279.
67. Klesse, G.; Rao, S.; Tucker, S.J.; Sansom, M.S. Induced polarization in molecular dynamics simulations of the 5-HT3 receptor channel. *J. Am. Chem. Soc.* **2020**, *142*, 9415–9427.
68. Rao, S.; Klesse, G.; Lynch, C.I.; Tucker, S.J.; Sansom, M.S. Molecular Simulations of Hydrophobic Gating of Pentameric Ligand Gated Ion Channels: Insights into Water and Ions. *J. Phys. Chem. B* **2021**, *125*, 981–994.
69. OuYang, B.; Xie, S.; Berardi, M.J.; Zhao, X.; Dev, J.; Yu, W.; Sun, B.; Chou, J.J. Unusual architecture of the p7 channel from hepatitis C virus.*Nature* **2013**, *498*, 521–525.
70. Sauguet, L.; Poitevin, F.; Murail, S.; Van Renterghem, C.; Moraga-Cid, G.; Malherbe, L.; Thompson, A.W.; Koehl, P.; Corringer, P.J.; Baaden, M.; et al. Structural basis for ion permeation mechanism in pentameric ligand-gated ion channels. *EMBO J.* **2013**, *32*, 728–741.
71. Chipot, C.; Pohorille, A. *Free Energy Calculations. Theory and Applications to Chemistry and Biology*; Springer: Berlin, Germany, 2007.
72. Pohorille, A. Free Energy Calculation for Understanding Membrane Receptors. In *Computational Biophysics of Membrane Proteins (Theoretical and Computational Chemistry, Band 10)*; The Royal Society of Chemistry: London, UK, 2017; pp. 59–106.
73. Furini, S.; Domene, C. Computational studies of transport in ion channels using metadynamics. *Biochim. Biophys. Acta* **2016**, *1858*, 1733–1740.
74. Marrink, S.J.; Berendsen, H.J.C. Simulation of water transport through a lipid membrane. *J. Phys. Chem.* **1994**, *98*, 4155–4168.
75. Hummer, G. Position-dependent diffusion coefficients and free energies from Bayesian analysis of equilibrium and replica molecular dynamics simulations. *New J. Phys.* **2005**, *7*, 34.
76. Peter, C.; Hummer, G. Ion transport through membrane-spanning nanopores studied by molecular dynamics simulations and continuum electrostatics calculations. *Biophys. J.* **2005**, *89*, 2222–2234.
77. Comer, J.; Chipot, C.; González-Nilo, F.D. Calculating position-dependent diffusivity in biased molecular dynamics simulations. *J. Chem. Theory Comput.* **2013**, *9*, 876–882.
78. Hummer, G. From transition paths to transition states and rate coefficients. *J. Chem. Phys.* **2004**, *120*, 516–523.
79. Weinan, E.; Vanden-Eijnden, E. Transition-path theory and path-finding algorithms for the study of rare events. *Ann. Rev. Phys. Chem.* **2010**, *61*, 391–420.
80. Elber, R.; Bello-Rivas, J.M.; Ma, P.; Cardenas, A.E.; Fathizadeh, A. Calculating iso-committor surfaces as optimal reaction coordinates with milestoning. *Entropy* **2017**, *19*, 219.
81. Berezhkovskii, A.M.; Szabo, A. Committors, first-passage times, fluxes, Markov states, milestones, and all that. *J. Chem. Phys.* **2019**, *150*, 054106.
82. Kumar, S.; Rosenberg, J.M.; Bouzida, D.; Swendsen, R.H.; Kollman, P.A. The weighted histogram analysis method for free-energy calculations on biomolecules. I. The method. *J. Comput. Chem.* **1992**, *13*, 1011–1021.
83. Pieprzyk, S.; Heyes, D.M.; Brańka, A.C. Spatially dependent diffusion coefficient as a model for pH sensitive microgel particles in microchannels. *Biomicrofluidics* **2016**, *10*, 054118.
84. Huber, G.A.; McCammon, J.A. Brownian Dynamics Simulations of Biological Molecules. *Trends Chem.* **2019**, *1*, 727–738.
85. Pratt, L.R. A statistical method for identifying transition states in high dimensional problems. *J. Chem. Phys.* **1986**, *85*, 5045–5048.
86. Gennis, R.B. *Biomembranes: Molecular Structure and Function*; Springer: New York, NY, USA, 1989.
87. Parsegian, V.A. Energy of an Ion crossing a Low Dielectric Membrane: Solutions to Four Relevant Electrostatic Problems. *Nature* **1969**, *221*, 844–846.
88. Neumcke, B.; Läuger, P. Nonlinear electrical effects in lipid bilayer membranes. II. *Biophys. J.* **1969**, *9*, 1160–1170.
89. Montserret, R.; Saint, N.; Vanbelle, C.; Salvay, A.G.; Simorre, J.P.; Ebel, C.; Sapay, N.; Renisio, J.G.; Bockmann, A.; Steinmann, E.; et al. NMR structure and ion channel activity of the p7 protein from hepatitis C virus. *J. Biol. Chem.* **2010**, *285*, 31446–31461.
90. Khalili-Araghi, F.; Ziervogel, B.; Gumbart, J.C.; Roux, B. Molecular dynamics simulations of membrane proteins under asymmetric ionic concentrations. *J. Gen. Physiol.* **2013**, *142*, 465–475.

entropy

Article

Maxwell Equations without a Polarization Field, Using a Paradigm from Biophysics

Robert S. Eisenberg [1,2]

[1] Department of Applied Mathematics, Illinois Institute of Technology, Chicago, IL 60616, USA; Reisenberg@iit.edu; Tel.: +1-708-932-2597
[2] Department of Physiology and Biophysics, Rush University Medical Center, Chicago, IL 60612, USA

Abstract: When forces are applied to matter, the distribution of mass changes. Similarly, when an electric field is applied to matter with charge, the distribution of charge changes. The change in the distribution of charge (when a local electric field is applied) might in general be called the induced charge. When the change in charge is simply related to the applied local electric field, the polarization field **P** is widely used to describe the induced charge. This approach does not allow electrical measurements (in themselves) to determine the structure of the polarization fields. Many polarization fields will produce the same electrical forces because only the divergence of polarization enters Maxwell's first equation, relating charge and electric forces and field. The curl of any function can be added to a polarization field **P** without changing the electric field at all. The divergence of the curl is always zero. Additional information is needed to specify the curl and thus the structure of the **P** field. When the structure of charge changes substantially with the local electric field, the induced charge is a nonlinear and time dependent function of the field and **P** is not a useful framework to describe either the electrical or structural basis-induced charge. In the nonlinear, time dependent case, models must describe the charge distribution and how it varies as the field changes. One class of models has been used widely in biophysics to describe field dependent charge, i.e., the phenomenon of nonlinear time dependent induced charge, called 'gating current' in the biophysical literature. The operational definition of gating current has worked well in biophysics for fifty years, where it has been found to makes neurons respond sensitively to voltage. Theoretical estimates of polarization computed with this definition fit experimental data. I propose that the operational definition of gating current be used to define voltage and time dependent induced charge, although other definitions may be needed as well, for example if the induced charge is fundamentally current dependent. Gating currents involve substantial changes in structure and so need to be computed from a combination of electrodynamics and mechanics because everything charged interacts with everything charged as well as most things mechanical. It may be useful to separate the classical polarization field as a component of the total induced charge, as it is in biophysics. When nothing is known about polarization, it is necessary to use an approximate representation of polarization with a dielectric constant that is a single real positive number. This approximation allows important results in some cases, e.g., design of integrated circuits in silicon semiconductors, but can be seriously misleading in other cases, e.g., ionic solutions.

Keywords: polarization; maxwell equations; gating current; dielectric constant

Citation: Eisenberg, R.S. Maxwell Equations without a Polarization Field, Using a Paradigm from Biophysics. *Entropy* **2021**, *23*, 172. https://doi.org/10.3390/e23020172

Received: 13 December 2020
Accepted: 26 January 2021
Published: 30 January 2021

1. Introduction

When forces are applied to matter, the distribution of mass changes. Similarly, when electrical forces are applied matter with charge, the distribution of charge changes.

The electric field $\mathbf{E}(x,y,z|t;\ \rho_Q(x,y,z|t;\mathbf{E}))$ changes the spatial distribution of charge $\mathbf{P}(x,y,z|t;\mathbf{E})$ producing polarization that has a central role in electrodynamics. In general, the change in charge distribution induced by the electric field will depend on time and electric field in a complex nonlinear way. We will discuss that situation later. But even when the induced charge is that of a polarization field characterized by a single dielectric constant (a real number), the actual definition of the polarization field $\mathbf{P}(x, y, z|t; \mathbf{E})$ is problematic, as major textbooks point out. Purcell and Morin [1], p. 500–507, show how the same structure can be described by different fields $\mathbf{P}(x, y, z|t)$. They conclude "The concept of polarization density **P** is more or less arbitrary" (slight paraphrase of [1], p. 507) and leads to an auxiliary variable that "is an artifice that is not, on the whole, very helpful" [1], p. 500.

Feynman shares this view. Feynman's text says (on p. 10–17 of [2]) "One more point should be emphasized. An equation like $\mathbf{D} = \varepsilon_r \varepsilon_0 \mathbf{E}$ is an attempt to describe a property of matter. But matter is extremely complicated, and the equation is in fact not correct.", as he then explains in some detail [3]. (Zangwill [2] uses quantum electrodynamics (p. 160) to deal with **P** and avoids (p. 44) the auxiliary variable **D.** He concentrates on the fundamental variable **E**, as we do here.) Neither Purcell nor Feynman propose a general explanation for the ambiguity in **P**.

The significance of the Purcell and Morin and Feynman's statements is great. If the concept of polarization is 'more or less arbitrary' (Purcell and Morin's words); and the distinction between bound and free charge is 'ambiguous', then the formulation of the Maxwell equations in textbooks is ambiguous and arbitrary.

I hope it is not necessary to say the obvious: something as important as the Maxwell equations should not be presented in a way that two Nobel Laureates (Purcell and Feynman) think is ambiguous and arbitrary (their words, not mine). It seems that " ... the conventional theory of electrodynamics inside matter needs to be redesigned": p. 13 of [4].

A general explanation is presented here following Griffiths, Ch. 4, [5]. The ambiguity in the definition of polarization arises from a mathematical property of vector fields and not from a particular physics or structure of charges. Only the divergence of the polarization field enters into the equations for the electric field **E** and so very different functions can be added to **P** without changing the observable electric field. Specifically, the curl of any function can be added to **P** without changing the electric field because the divergence of the curl of any function is zero. Thus, measurements of **E** cannot determine the polarization field **P** uniquely. Different structures of polarization charge can give the same electric field and so measurements of the electric field cannot determine the structures producing polarization or there the structures of charge itself.

A paradigm widely used in biophysics to define gating current allows resolution of this ambiguity in many cases beyond biophysics. This paradigm cannot be universally applied but when it can be applied it is very useful. The dependence of polarization on the electric field is as complicated as the motions of matter in an electric field. These motions are nearly as complicated as the motions of matter in general. It is unlikely that any single paradigm will be universal. Nonetheless, the gating current paradigm of biophysics may be generally useful and will surely make specific what is needed for paradigms in general.

The paradigm of biophysics was developed to resolve the nonlinear displacement (i.e., capacitive) current of nerve that Hodgkin and Huxley [6] suggested might be the voltage sensor of nerve. This 'gating current' was measured in nerve [7] using a paradigm developed by Schneider and Chandler [8,9] and significantly improved by Bezanilla and Armstrong [10,11] and has been studied in great detail [7,12–19] because of the insight it gives [17,18,20,21] into the physical mechanism of conformation change in a most important biological protein and process. The conformation change of the voltage sensor determines many properties of the action potential, which is the signal used by the nervous system, skeletal and cardiac muscle to send signals more than a few micrometers.

The ambiguity of **P** arises from the history of electrodynamics, in my view. Faraday and Maxwell thought all charge depends on the electric field ([3], p. 36; [22–24]. All charge would then be polarization.

Maxwell used the **D** and **P** fields as fundamental dependent variables. Charge only appeared as polarization, usually over-approximated [25–35] by a dielectric constant ε_r that is a single real positive number. Charge independent of the electric field was not included, because the electron had not been discovered: physicists at Cambridge University (UK) did not think that charge could be independent of the electric field. The electron was discovered some decades later, in Cambridge, ironically enough [36,37]. (Thomson's monograph"intended as a sequel to Professor Clerk-Maxwell's Treatise on electricity and magnetism" [38] does not mention charge, as far as I can tell. Clarendon Press: 1893. "intended as a sequel to Professor Clerk-Maxwell's Treatise on electricity and magnetism"; does not mention charge, as far as I can tell. Faraday's chemical law of electrolysis was not known and so the chemist's 'electron' postulated by Richard Laming and defined by Stoney [39] was not accepted in Cambridge as permanent charge, independent of the electric field. It is surprising that the physical unit 'the Faraday' describes a quantity of charged particles unknown to Michael Faraday. Indeed, he did not anticipate the existence or importance of permanent charge on particles or elsewhere.) It then became apparent to all that the permanent charge of an electron is a fundamental source of the electric field. The electron and permanent charge must be included in the equations defining the electric field, e.g., Equations (1) and (6) as it is in every textbook I have examined.

For physicists today, the fundamental electrical variable is the **E** field that describes the electric force on an infinitesimal test charge. **D** and **P** fields are auxiliary derived fields that many textbooks think unnecessary, at best.

2. Theory

The setup used here is described in many fine textbooks and so detail is omitted [1–5,40,41]. The specifics of the setup used to measure gating currents is described later, see Figures 1 and 2.

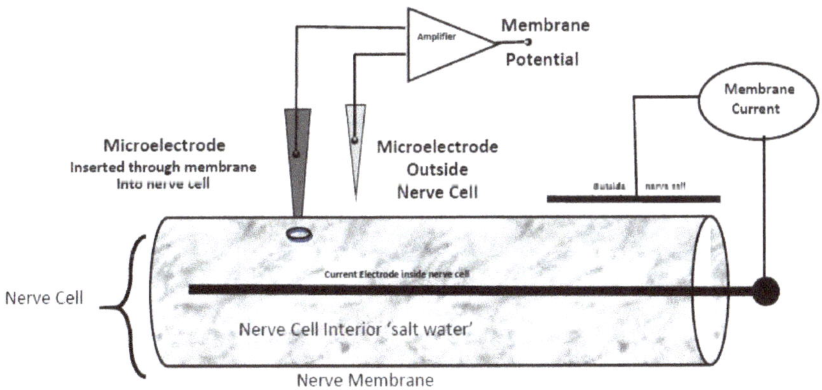

Figure 1. Setup for estimating the potential across the membrane of a cylindrical nerve cell while measuring the current through the membrane.

Maxwell's first equation for the composite variable **D** relates the 'free charge' $\rho_f(x,y,z|t)$, units cou/m^3, to the sum of the electric field **E** and polarization **P**. It is usually written as

$$\textbf{div}\,\mathbf{D}(x,y,z|t) = \rho_f(x,y,z|t) \tag{1}$$

$$\mathbf{D}(x,y,z|t;\mathbf{E}) \triangleq \varepsilon_0\,\mathbf{E}(x,y,z|t) + \mathbf{P}(x,y,z|t;\mathbf{E}) \tag{2}$$

The physical variable **E** that describes the electric field is not visible in the classical formulation Equation (1). Maxwell embedded polarization in the very definition of the dependent variable $\mathbf{D} \triangleq \varepsilon_0\,\mathbf{E} + \mathbf{P}$. ε_0 is the electrical constant, sometimes called the 'permittivity of free space'. Polarization is described by a vector field **P** with units of dipole moment per volume, cou-m/m^3, that can be misleadingly simplified to cou-m^{-2}. The charge ρ_f cannot depend on **D** or **E** in traditional formulations and so ρ_f is a permanent charge.

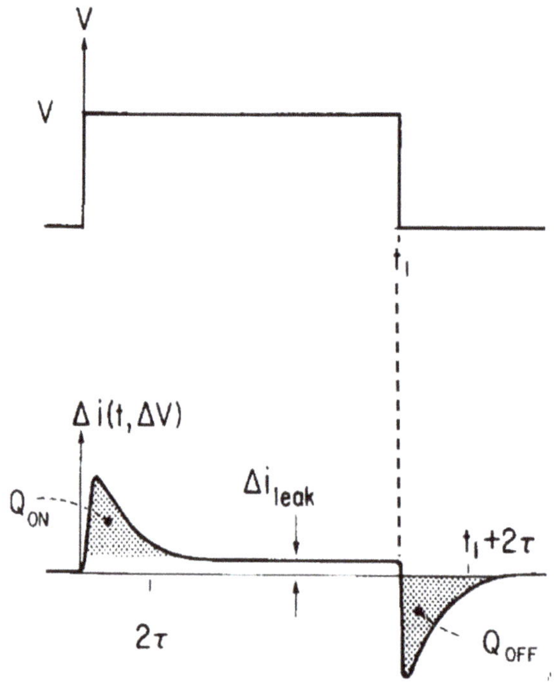

Figure 2. shows the response to a step function change in potential and the charges measured that are proposed as an operational definition of polarization.

When Maxwell's first equation is written in a style appropriate since the discovery of the electron **E** is the dependent variable, as textbooks make clear. The source terms are ρ_f and the divergence of **P**.

$$\varepsilon_0 \mathbf{div}\, \mathbf{E}(x,y,z|t) = \rho_f(x,y,z|t) - \mathbf{div}\, \mathbf{P}(x,y,z|t;\mathbf{E}) \tag{3}$$

P does not have the units of charge and should not be called the 'polarization charge'. **P** does not enter the equation by itself. Only the divergence of **P** appears on the right-hand side of Equation (3).

$\mathbf{D}(x,y,z|t)$ and the polarization $\mathbf{P}(x,y,z|t)$ are customarily over-approximated in classical presentations of Maxwell's equations: the polarization is assumed to be proportional to the electric field, independent of time.

$$\mathbf{P}(x,y,z|t) \triangleq (\varepsilon_r - 1)\varepsilon_0\, \mathbf{E}(x,y,z|t) \tag{4}$$

$$\mathbf{D}(x,y,z|t) \triangleq \varepsilon_r \varepsilon_0 \mathbf{E}(x,y,z|t) \tag{5}$$

The proportionality constant $(\varepsilon_r - 1)\varepsilon_0$ involves the dielectric constant ε_r which must be a single real positive number if the classical form of the Maxwell equations is taken as an exact mathematical statement of a system of partial differential equations. If ε_r is generalized to depend on time, or frequency, or the electric field, the form of the Maxwell equations changes. If ε_r is generalized, traditional equations cannot be taken literally as a mathematical statement of a boundary value problem. They must be changed to accommodate the generalization.

Polarization and thus ε_r—however generalized—depend on time or frequency in complex ways in all matter as documented in innumerable experiments [33–35,42–44].

Many of the most interesting applications of electrodynamics arise from the dependence of polarization and ε_r on field strength.

ε_r should be taken as a constant only when experimental estimates, or theoretical models are not available, in my view.

It is difficult to imagine a physical system in which the electric field produces a change in charge distribution independent of time (see examples shown towards the end of Discussion). The time range in which Maxwell's equations are used in the technology of our computers, smartphones, and video displays starts around 10^{-10} s. The time range in which Maxwell's equations are used in biology start around 10^{-15} s in simulations of the atoms that control protein function. The time range of the X-rays that determine protein structure is ~10^{-19} s. The time range used to design and operate the synchrotrons that generate X-rays is very much faster than that, something like 10^{-23} s. The Maxwell equations describe experiments to many significant figures over this entire range.

It is evident that a dielectric constant ε_r independent of time is an inadequate over-approximation in many cases of practical interest today, in biology, engineering, chemistry, and physics.

Maxwell's first equation for **E** is well described in many textbooks, although the inadequacies of the usual representation of polarization with a single dielectric constant are not emphasized, if mentioned at all. Students are then often unaware of the over-approximation, particularly if they have a stronger background in biology or mathematics than the physical sciences.

$$\varepsilon_r \varepsilon_0 \, \mathbf{div}\, \mathbf{E}(x,y,z|t) = \rho_f(x,y,z|t) \tag{6}$$

Polarization is particularly well described in Griffiths [5].

It is wise, in my view to combine the fields on the right-hand side of Equation (3) with the definition

$$\rho_Q(x,y,z|t;\mathbf{E}) \triangleq \rho_f(x,y,z|t) - \mathbf{div}\, \mathbf{P}(x,y,z|t;\mathbf{E}) \tag{7}$$

yielding the version of Maxwell's first law that does not involve a polarization field $\mathbf{P}(x,y,z|t)$ at all.

$$\varepsilon_0 \, \mathbf{div}\, \mathbf{E}(x,y,z|t) = \rho_Q(x,y,z|t;\mathbf{E}) \tag{8}$$

We adopt this version of Maxwell's first equation here.

3. Results

The traditional formulation of the differential equations shown in Equations (1) and (6) is ambiguous in an important way (Integral forms of the Maxwell equations show more clearly the need for boundaries. They display the charge on the surface as an integral and explicit part of the general solution of Poisson's equation for the electrical potential, for example). They do not mention the shape or boundaries of the regions in question. In fact, if **P** varies from region to region, but is constant within each region, charge is absent within each region: when **P** is constant, $\mathbf{div}\,\mathbf{P} = 0$. Charge accumulates only at the boundaries of the regions. In many situations involving dielectrics, including most of those described in classical textbooks Only the boundary charge has effects on the Maxwell Equations (1) and (6). The **P** field in the Maxwell Equation (7), and implied in Equations (1) and (6), is zero; only the boundary values of **P** are important and they are not visible in the Maxwell Equations (1) and (6) themselves.

We turn now to applications in biology where the issue of charge at boundaries is particularly important, not to say that it is unimportant in semiconductor devices as well. Dielectric boundary charges have a particular role in biological systems involving membranes or proteins. The membrane capacitance, so important in determining the electrical properties of cells, particularly cells with action potentials like nerve and muscle, is a boundary phenomenon. Boundary charges are of great importance in channel proteins that allow (nearly catalyze) ion flow through membranes, see Appendix A on Proteins and [45].

Turning back to classical electrodynamics, we remember that most of the properties of dielectric rods studied by Faraday—and predecessors going back to Benjamin Franklin, if not earlier—arise from the dielectric boundary charges. Textbooks typically spend much effort teaching why polarization charge appears on dielectric boundaries in systems with constant **P** where **div P** $= 0$ (e.g., Ch. 6 of [3]). Students wonder why regions of dielectrics without polarization charge have polarization charge on boundaries.

A general principle is at work here: a field equation in itself—like Equations (1) and (6) that are partial differential equations without boundary conditions—is altogether insufficient to specify an electric field. A model is needed that has boundary conditions. Applications of electrodynamics to biology, electrochemistry, and semiconductors are not useful until they specify models and boundary conditions that realistically describe the system of interest.

The model needs to include an explicit structure. It needs to describe the spatial variation of **P**. Indeed, the spatial variation of **P** may be a main determinant of properties [46–48] in (for example) many biological systems (e.g., channels), electrochemical systems (electrodes of batteries), and semiconductor devices. Without specifying boundary conditions (defined explicitly in specific structures), using **P** in the differential Equation (7), and implied in Equations (1) and (6), is ambiguous and confusing. Indeed, using **P** without boundary conditions is so incomplete that it might be called incorrect.

The general nature of the ambiguity in **P** becomes clear once one realizes that:

$$\text{Adding } \mathbf{curl}\ \widetilde{\mathbb{C}}(x,y,z|t) \text{to } \mathbf{P}(x,y,z|t) \text{ in Maxwell's first equation, Equation (7)} \tag{9}$$

changes nothing (Ch. 4 of [5]) because [49,50]

$$\mathbf{div}\ \mathbf{curl}\ \widetilde{\mathbb{C}}(x,y,z|t)\ \equiv 0\ ; \tag{10}$$

The ambiguity in **P** in the Maxwell differential equations means that any model $\mathbf{P}_{model}(x,y,z|t)$ of polarization can have **curl** $\widetilde{\mathbb{C}}(x,y,z|t)$ added to it, without making any change in the **div** $\mathbf{P}(x,y,z|t)$ in Maxwell's first equation (7), and implied in Equations (1) and (6).

In other words, the polarization **div** $\mathbf{P}(x,y,z|t)$ in Maxwell's first Equations (7), and implied in Equations (1) and (6), does not provide a unique structural model of polarization $\mathbf{P}_{model}(x,y,z|t)$. In particular, a model drawn from an atomic detail structure can be modified by adding a polarization $\widetilde{\mathbb{P}}(x,y,z|t) \triangleq$ **curl** $\widetilde{\mathbb{C}}(x,y,z|t)$ to its representation (i.e., 'drawing') of polarization without changing electrical properties at all: $\mathbf{div}\ \mathbf{P} \equiv \mathbf{div}\left(\mathbf{P} + \widetilde{\mathbb{P}}\right)$.

Models of the polarization $\mathbf{P}^1_{model}$ and $\mathbf{P}^2_{model}$ of the same structure written by different authors may be strikingly different but they can give the same electrical results even though the models can appear to be very different. The **curl** $\widetilde{\mathbb{C}}(x,y,z|t)$ field can be quite complex and hard to recognize in a model, particularly for structural biologists who may not be comfortable with vector calculus and its **curl** and **div** operators. The two models $\mathbf{P}^1_{model}$ and $\mathbf{P}^2_{model}$ produce the same charge distribution **div** $\mathbf{P}^1_{model}$ and div $\mathbf{P}^2_{model}$ in Maxwell's first equation Equation (11) and so they cannot be distinguished by electrical measurements.

As we have seen, the **P** field is arbitrary, as certainly has been known previously Ch. 4 of [5]. Purcell and Morin [1], see pp. 500–507, describe structural models and ways to construct different fields $\mathbf{P}(x,y,z|t)$ from the same structure as stated in the introduction to this paper. **P** fields are not unique.

Purcell and Morin are not guilty of overstatement—indeed they may be guilty of understatement—when they say "The concept of polarization density **P** is more or less arbitrary" (slight paraphrase of [1], p. 507) and the **D** field is "is an artifice that is not, on the whole, very helpful" [1], p. 500.

The classical approach criticized by Purcell and Morin [1] does not allow unique specification of a polarization field $\mathbf{P}(x,y,z|t)$ from electrical measurements.

An arbitrary artificial formulation is prone to artifact and likely to produce misunderstanding and unproductive argument: "what is the true description of a dielectric object (e.g., protein)?" is a question likely to arise and be unanswerable if the polarization field **P** is itself not unique.

The $\mathbf{P}(x, y, z|t)$ of classical theory is not a firm foundation on which to build an understanding of the structural basis of the phenomena of polarization, or the electrodynamics of matter, with problems particularly apparent in the understanding of the polarization arising from the structure of proteins (see Appendix A).

It seems clear that most formulations of electrodynamics of dielectrics in classical textbooks are "more or less arbitrary" and depend on an "artifice" (quotations from Feynman and Purcell and Morin). Because dielectrics, polarization and a dielectric constant (as a single real number) are central to the classical treatments of electrodynamics, the conclusion (p. 13) of a modern monograph on electrodynamics, using mathematics (exterior differential forms) appropriate for relativistic theories of electrodynamics, [4] quoted previously seems worth restating "We believe that the conventional theory of electrodynamics inside matter needs to be redesigned". That redesign begins with a revised treatment of polarization that reflects the ambiguity of the curl, see [5]. Ambiguity and its problems can be avoided if Maxwell's First Equation is rewritten without a polarization field $\mathbf{P}(x, y, z|t)$ as shown previously in Equation (8). The phenomena of polarization—the response of charges to an electric field—is then included in a variable $\rho_Q(x, y, z|t; \mathbf{E})$, specifically as (part of) its dependence on $\mathbf{E}$:

$$\mathbf{div}\, \varepsilon_0 \mathbf{E}(x, y, z|t) \;= \rho_Q(x, y, z|t; \mathbf{E}) \tag{11}$$

Here $\rho_Q(x, y, z|t; \mathbf{E})$ describes all charge whatsoever, no matter how fast, small or transient are their movements, including what is usually called dielectric charge and permanent charge, as well as charges driven by other fields, like convection, diffusion or temperature. The charge ρ_Q can be parsed into components in many ways (see Equations (1), (3), (6) and (8) and [43,51]). Updated formulations of the Maxwell differential equations [43,51] are needed, in my opinion, to avoid the problems produced by ambiguous **P** and over-simplified ε_r.

We turn now to a quite different property of charge matter, the flow of charges.

Most applications of electrodynamics involve flow. The most prominent application of electrodynamics is surely computational and semiconductor electronics [52–61] and that involves flow, usually described by Kirchhoff's current law. Semiconductor electronics has remade our world increasing computer power by nearly $10^9\times$ in the last seventy years [62–67]. Biology and electrochemistry (batteries) scarcely exist without flow: what physical chemists call equilibrium (no flows of any kind) is hardly worth studying in biological or electrochemical systems. Unlike thermodynamics, electrodynamics nearly always involves flow.

Thus, we study the flux of charges ρ_Q as well as their density. Maxwell's second equation describes the flow of charges, electrical current, and the magnetic field. It is understandable that Maxwell—and his Cambridge contemporaries and followers—had difficulty understanding current flow when their models did not include permanent charge, electrons or their motions.

Maxwell's extension of Ampere's law describes the special properties of current flow $\mathbf{J}_{total}$ (Equation (13) that make it so different from the flux of matter. Maxwell's field equations include the ethereal current $\varepsilon_0 \partial \mathbf{E}/\partial t$ that makes the equations resemble those of a perfectly incompressible fluid: the ethereal current always exists, whether matter is present or not, unlike the dielectric current $(\varepsilon_r - 1)\varepsilon_0 \partial \mathbf{E}/\partial t$ that exists only when matter is present.

Maxwell's field equations describe the incompressible flow $\mathbf{J}_{total}$ over the dynamic range of something like 10^{16} that is safely accessible within laboratories. The dynamic range of the Maxwell equations is much larger if one includes the interior of stars, and the core of galaxies in which light is known to follow the same equations of electrodynamics as in our laboratories.

Maxwell's field equations are different from material field equations (like the Navier–Stokes equations) because they are meaningful and valid universally [68], both in a vacuum devoid of mass and matter and within and between the atoms of matter [43].

The ethereal current $\varepsilon_0 \partial \mathbf{E}/\partial t$ responsible for the special properties of Maxwell's equations arises from the Lorentz (un)transformation of charge. Charge does not vary with velocity, unlike mass (this is the mass that determines inertia, called the 'relativistic mass' nowadays. This was the meaning of the word 'mass' in Einstein's original papers, presumably because he wanted an operational definition of 'mass' that was based on the observable properties, inertia and momentum, and that was independent of Lorentz transformations, and theoretical considerations) [69], length, and time, all of which change dramatically as velocities approach the speed of light, strange as that seems. This topic is explained in any textbook of electrodynamics that includes special relativity. Feynman's discussion of 'The Relativity of Electric and Magnetic Fields' was an unforgettable revelation to me as a student, see Section 13-6 of reference [2]: an obervers moving at the same speed as a stream of electrons sees zero current, but the forces measured by that observer are the same as the forces measured by an observed who is not moving at all. The moving observer describes the force as an electric field $\mathbf{E}(x, y, z|t)$. The unmoving observer describes the force as a magnetic field $\mathbf{B}(x, y, z|t)$. The observable forces are the same, whatever they are called, according to the principle and theory of relativity. (The principle and theory of relativity are confirmed to many significant figures every day in the GPS (global positioning systems) software of the map apps on our smartphones, and in the advanced photon sources (synchrotrons) that produce X-rays to determine the structure of proteins).

The ethereal current reveals itself in magnetic forces which have no counterpart in material fields. The ethereal current is apparent in the daylight from the sun, that fuels life on earth, and in the night light from stars that fuels our dreams as it decorates the sky. The ethereal current is the term in the Maxwell equations that produces propagating waves in a perfect vacuum like space.

Magnetism $\mathbf{B}$ is described by Maxwell's version of Ampere Law, Maxwell's Second Equation:

$$\frac{1}{\mu_0} \,\mathbf{curl}\ \mathbf{B} = \mathbf{J}_Q + \varepsilon_0 \frac{\partial \mathbf{E}}{\partial t} \tag{12}$$

$$\mathbf{J}_{total} \triangleq \mathbf{J}_Q + \varepsilon_0 \frac{\partial \mathbf{E}}{\partial t} \tag{13}$$

$$\frac{1}{\mu_0} \,\mathbf{curl}\ \mathbf{B} = \mathbf{J}_{total} \tag{14}$$

If we are interested in flux and current, we must turn to Maxwell's second equation and deal explicitly with magnetism, even if magnetic fields themselves do not carry significant energy (as in almost all biological applications). Only by dealing with Maxwell's second equation can we derive conservation of total current and compare it with the conservation of charge. Indeed, the derivation of the continuity equation used here depends on equations involving the magnetic field.

Note that $\mathbf{J}_Q$ includes the movement of all charge ρ_Q with mass, no matter how small, rapid or transient. It includes the movements of charge classically approximated as the properties of an ideal dielectric. It describes all movements of the charge described by $\rho_Q(x, y, z|t; \mathbf{E})$; ρ_f is one of the components of ρ_Q. Indeed, $\mathbf{J}_Q$ can be written in terms of $\mathbf{v}_Q$ the velocity of mass with charge. In simple cases, such as a plasma of ions each with charge $\mathbf{Q}_Q$

$$\mathbf{J}_Q = \mathbf{v}_Q \mathrm{Q}_Q \mathrm{N}_Q \tag{15}$$

where Q_Q is the charge per particle and N_Q is the number density of particles. In a mixture, sets of fluxes $\mathbf{J}_Q^i$, velocities $\mathbf{v}_Q^i$, charges Q_Q^i, number densities N_Q^i, and charge densities ρ_Q^i are needed to keep track of each elemental species i of particles. Plasmas are always mixtures because they must contain both positive and negative particles to keep electrical forces within safe bounds, as determined by (approximate) global electroneutrality.

In cases other than plasmas, the relationship of $\mathbf{J}_Q$, $\mathbf{J}_{total}$ and Q_Q to material properties is complex. The relationship often involves convection and diffusion fields and extends over a range of scales from atomic to macroscopic, in both space and time. For example, the Maxwell equations do not describe charge and current driven by other fields, like convection, diffusion, or temperature. They do not describe constraints imposed by boundary conditions and mechanical structures. Those must be specified separately. If the other fields, structures, or boundary conditions involve matter with charge, they will respond to changes in the electric field. The other fields and constraints thus contribute to the phenomena of polarization and must be included in a description of it, as we shall discuss further below in the examples shown towards the end of Discussion. The theory of complex fluids has dealt with many such cases, often with the label 'micro macro', spanning scales, connecting micro (even atomic) structures with macro phenomena.

The charge density ρ_Q and current $\mathbf{J}_{total}$ can be parsed into components in many ways, some helpful in one historical context, some in another. References [33,43,51,70–75] define and explore those representations in tedious detail. Simplifying those representations led to the treatment in this paper.

Maxwell's Ampere's law Equation (12) implies two equations of great importance and generality. First, it implies a continuity equation that describes the conservation of charge with mass. The continuity equation is the relation between the flux of charge with mass and density of charge with mass.

Derivation: Take the divergence of both sides of Equation (12), use **div curl** $= 0$ [49,50], and get

$$\mathbf{div}\,\mathbf{J}_Q = \mathbf{div}\left(-\varepsilon_0\frac{\partial \mathbf{E}}{\partial t}\right) = -\varepsilon_0\frac{\partial}{\partial t}\mathbf{div}\,\mathbf{E} \tag{16}$$

when we interchange time and spatial differentiation.

However, we have a relation between **div E** and charge ρ_Q from Maxwell's first equation, Equation (11), giving the Maxwell Continuity Equation:

$$\mathbf{div}\,\mathbf{J}_Q = -\varepsilon_0\,\varepsilon_0\frac{\partial \rho_Q}{\partial t} \tag{17}$$

$$\mathbf{div}\left(\mathbf{v}_Q Q_Q N_Q\right) = -\varepsilon_0\frac{\partial \rho_Q}{\partial t}, \tag{18}$$

for a biophysical or astrophysical plasma of ions.

Note that sets of fluxes $\mathbf{J}_Q^i$ and sets of charge densities ρ_Q^i are needed to keep track of each elemental species i of particles in a mixture, along with sets of velocities $\mathbf{v}_Q^i$, charges Q_Q^i, and number densities N_Q^i, as described near Equation (15).

Maxwell's Ampere's law Equation (12) implies a second equation of great importance. Indeed, it is this equation that allows the design of the one-dimensional branched circuits of our digital technology using the relatively simple mathematics of Kirchhoff's current law [72,74].

Derivation: Taking the divergence of both sides of Maxwell's Second law Equation (12) yields Conservation of Total Current

$$\mathbf{div}\,\mathbf{J}_{total} \triangleq \mathbf{div}\left(\mathbf{J}_Q + \varepsilon_0\frac{\partial \mathbf{E}}{\partial t}\right) = 0 \tag{19}$$

$$\mathbf{div}\,\mathbf{J}_{total} = 0 \tag{20}$$

or

$$\mathbf{div}\,\mathbf{J}_{total} \triangleq \mathbf{div}\left(\mathbf{v}_Q Q_Q N_Q \mathbf{J}_Q + \varepsilon_0\frac{\partial \mathbf{E}}{\partial t}\right) = 0 \tag{21}$$

It is easy to overlook the importance of one-dimensional systems. They may seem trivial, almost unworthy of analysis using the powerful beauty of vector calculus. However, one-dimensional systems are of great importance despite, or because of their simplicity.

Nearly all of our electronic technology occurs in one-dimensional systems, networks of branching one-dimensional conductors. Our electronic technology is driven by batteries that are one-dimensional systems. Our technology is at the hands of animals, humans in which all information transfer is done by one-dimensional circuits, unbranched in ion channels, and barely branched in nerve cells. Branched one-dimensional systems describe the metabolic pathways of biological cells that make life possible.

The importance of one-dimensional systems may come from their design. The design of one-dimensional systems is relatively easy for engineers or evolution. Design requires Kirchhoff's laws and little else. One-dimensional systems are widely used for another reason. They are reliable. The dimensionality of these circuits rules out spatial singularities. Systems are more robust when steep slopes near infinities are not present to create severe sensitivity.

Kirchhoff's laws are used to design semiconductor circuits that work over an enormous range of sizes and times, from say 10^{-10} s to many minutes, from 10^{-19} m to 10^{4} m or longer. Current flow over these ranges of time space involves a wide range of physics, described by many constitutive equations.

Current is not just the movement of point permanent charges as assumed in the textbook derivations of Kirchhoff's current law I have consulted, both in electrical engineering and electrodynamics. The derivations of Kirchhoff's current law are usually restricted to the simplest case of the long-time translation of point permanent charges, although it is very well known that is a poor model for current flow under conditions actually found in the integrated circuits of our digital technology. It is possible to show, however, that current flow in one-dimensional systems can be described accurately by a simple generalization of Kirchhoff's current law that arises naturally from the treatment of Maxwell's equations found in this paper: all the $\mathbf{J}_{total}$ that flows into a node must flow out [51,72–74]. This result seems to be rather new, although of course it seems elementary and obvious. Indeed, it is so obvious that it must exist somewhere in the literature, even though I do not know where.

Kirchhoff's current law take on simplest form in unbranched one-dimensional systems. Unbranched one-dimensional systems are important despite their utter simplicity. Indeed, the ion channels of biological systems control a wide range of biological function and are unbranched one-dimensional series systems. They cannot be considered degenerate. Nor can be the diodes of electronic technology that are also series systems. However, the greatest importance of unbranched one-dimensional systems may be the insight they give to the importance of the ethereal current $\varepsilon_0 \partial \mathbf{E}/\partial t$, as we shall soon see.

Unbranched one-dimensional systems have components in series, each with its own current voltage relation arising from its microphysics. In a series one-dimensional system, the total current $\mathbf{J}_{total}$ is equal everywhere at any time in every location no matter what the microphysics of the flux $\mathbf{J}_Q$ of charge with mass. The current through a battery is an exceedingly complicated mixture of the microphysics of electrodes, ion movement and electron flow. If that battery is connected by a wire to a vacuum capacitor, the microphysics of the vacuum capacitor $i_{capacitor} = A_{rea}\varepsilon_0 \partial \mathbf{E}/\partial t$, is as simple as the microphysics of the battery is complex, yet the total currents in the capacitor and the battery are equal at any time, in any conditions. Indeed, the microphysics of the wire linking the capacitor and the battery is totally different from the microphysics of the capacitor and battery. The microphysics of the wire actually resemble that of a waveguide at frequencies important in our digital integrated circuits. The microphysics of the wire, capacitor and battery do not change the fact that the total current through each is exactly the same, always, at every location and at every time.

How can that possibly be true? The answer is found in the Maxwell equations. They can be solved for the electric field and magnetic fields that make the total currents equal.

The solutions of Maxwell's equations ensure that the ethereal current $\varepsilon_0 \partial \mathbf{E}/\partial t$, and the other dependent variables, take on the values at every location and every time needed to make the total currents $\mathbf{J}_{total}$ equal everywhere. A practical example, not difficult to build in any laboratory, including resistor, capacitor, diode, capacitor, cylinder of salt water, and wire is described in detail near Figure 2 of [73].

There is no spatial dependence of total current in a series one-dimensional system. No spatial variable or derivative is needed to describe total current in such a system [75], although of course spatial variables are needed to describe other variables, including (1) the density of mass with charge $\mathbf{Q}_Q^i$ (2) the flux $\mathbf{J}_Q$ of charge with mass (3) the electrical current $\mathbf{J}_{total}^i$ of individual elemental species (4) the velocities, charge, and number densities $\mathbf{v}_Q$, Q_Q, ρ_Q, and N_Q.

It is important to realize that the flux of charge with mass $\mathbf{J}_Q$ is not conserved, only the total current $\mathbf{J}_{total}$ is conserved. Charges carry $\mathbf{J}_Q$ can accumulate. In fact, $\mathbf{div}\,\mathbf{J}_Q = \mathbf{div}\,(\mathbf{v}_Q Q_Q N_Q)$ supplies the flow of charge that is the current $\partial\rho_Q\,/\partial t$ necessary to change $\mathbf{div}\,(\varepsilon_0\partial\mathbf{E}/\partial t)$ as described by the following continuity equation.

$$\mathbf{div}\,\mathbf{J}_Q = \mathbf{div}\,\varepsilon_0\frac{\partial\mathbf{E}}{\partial t} = \frac{\partial}{\partial t}\mathbf{div}\,(\varepsilon_0\mathbf{E}) = \frac{\partial\rho_Q}{\partial t}. \tag{22}$$

That is to say, $\mathbf{J}_Q$ can accumulate as Q_Q. Total current $\mathbf{J}_{total}$ cannot accumulate, not at all, not anywhere, not at any time.

Because conservation of total current applies on every time and space scale, including those of thermal motion, the properties of $\mathbf{J}_Q$ differ a great deal from the properties of $\mathbf{J}_{total}$. For example, in one-dimensional channels, the material flux $\mathbf{J}_Q$ can exhibit all the complexities of a function of infinite variation, like a trajectory of a Brownian stochastic process, that reverses direction an uncountably infinite number of times in any interval. A Brownian trajectory of a Brownian stochastic process is a continuous function that does not have a (well defined) time derivative anywhere.

In marked contrast to the infinite variation of $\mathbf{J}_Q$, the electrical current $\mathbf{J}_{total}$ has no spatial variation at all. It is spatially uniform [75].

The fluctuations of $\varepsilon_0\partial\mathbf{E}/\partial t$ (in time and space) and other variables are exactly what are needed to completely smooth the infinite fluctuations of $\mathbf{J}_Q$ into the spatially uniform $\mathbf{J}_{total}$.

Maxwell's equations serve as the perfect low pass (spatial) filter converting the infinite variation of Brownian motion into a spatial constant, as strange as that seems.

These universal and exact properties of Maxwell's equations are hidden in the usual treatment of Maxwell's equations. The usual treatment includes a grossly approximate treatment of polarization as the property of a perfect dielectric. Everyone knows how bad this approximation is, so everyone understands that Maxwell's equations as usually written are not universal or exact. They are as sloppy as is the dielectric constant as a description of the polarization of matter.

ONLY when Maxwell's equations are written without a dielectric constant, with a perfectly general treatment of induced charge, does it become clear that Maxwell's equations are universal and exact independent of any property of matter.

How then is polarization included in a modified version of the Maxwell equations that does not include a dielectric constant. One needs an explicit model of polarization appropriate for the system of interest.

It is obvious that one cannot describe material flow unless one knows how matter moves in response to forces. It should be obvious that one cannot describe the flux of charges unless one knows how material charge moves in response to forces.

The use of a single real dielectric constant in Maxwell's equations is no more necessary than the use of a single spring constant (i.e., elasticity) is in material equations. But Maxwell's equations describe the total electrical current—that includes the ethereal current—not the flux of charges. Because of the ethereal current, Maxwell's equations describe light in the vacuum of space between stars.

Because of the ethereal current, Maxwell's equations are universal and exact. They describe total current as exactly as they describe anything, and their description of total current flow is entirely independent of the properties of matter. Total current flow depends on no constitutive equations, except perhaps the constitutive equation of a vacuum, more or less determined by special relativity. Electrodynamics are very different in this respect from the equations of material movement. They always depend on constitutive equations in important respects. The fundamental properties of electrodynamics do not depend on constitutive equations.

4. Discussion: From Electrodynamics to Biophysics and Back

A fundamental question arises with the updated version of Maxwell's equations. How is the phenomenon of polarization included in Equation (11) and Equation (14)?

To answer this question, we first need a general paradigm to define polarization, even when dielectrics are far from ideal, when they might be time and frequency dependent, and voltage dependent as well. We need a paradigm that describes how the charge distribution varies with the electric field in as general a system as possible, including systems with charge movement driven by forces not in the Maxwell equations at all, such as convection and diffusion.

It seems obvious that a general paradigm cannot be found. After all the motions of matter in response to a change in electric field are more or less as complex as the motions of matter itself! Nonetheless, a paradigm of that may be helpful in many cases has been in use for many years, even if it is not perfectly general.

This problem has been addressed in membrane biophysics. A community of scholars has studied the nonlinear currents that control the opening of voltage sensitive protein channels for nearly fifty years, [7,12–19] inspired by [6]. They have developed protocols that may be useful in other systems, as they have been in biophysics. Schneider and Chandler followed by Bezanilla and Armstrong are responsible for this paradigm, more than anyone else [7–9].

The basic setup used in these experiments is that of an electrochemical cell modified to deal with a cylindrical cell as shown in Figure 1. Membrane potential is measured across a biological membrane, with defined concentrations on both sides of the membrane. Current is applied through electrodes to control the potential, in the classical voltage clamp set up of Cole [76] and Hodgkin, Huxley, and Katz [77,78]. It is best to apply that current in electrodes different from those that record membrane potential using a so-called four electrode setup [79–81], like those described in textbooks of electrochemistry.

I propose using the operational definition of 'gating current' used to define nonlinear, time and voltage dependent polarization by biophysicists as a useful setup and definition of many types of polarization. Obviously, this definition is not general, but the hope is that it may be generally useful.

The basic idea is to apply a set of step functions of potential across the system—in biology across the membrane—and observe the currents that flow. The currents observed are transients that decline to a steady value, often to near zero after a reasonable (biologically relevant) time. The measured currents are perfectly reproducible. If a pulse is applied, the charge moved (the integral of the current) can be measured when the voltage step is applied. The integration goes on until t_1 when the current i_{leak} is nearly independent of time, often nearly zero. That integral is called the ON charge $\mathbf{Q_{ON}}$.

When the voltage is returned to its initial value (the value that was present before the ON pulse), another current is observed that often has quite different time course [7–9], much more so than in Figure 1. The integral of that current is the OFF charge $\mathbf{Q_{OFF}}$.

If $\mathbf{Q_{ON}} = \mathbf{Q_{OFF}}$, and the physical processes involved depend fundamentally on potential and not its time derivative, the biophysical paradigm is likely to be useful. In other cases, another paradigm is needed. If the current produced by the step in potential is in fact actually transient, the steady current will be what it was before the voltage step was applied. The transient will disappear with time as the word 'transient' implies. In that case it seems that the biophysical paradigm is not only useful but may even provide a unique definition of gating current and the corresponding polarization.

Gating current as measured in biophysical experiments depends on the membrane voltage before the step, as well as the voltage just after and during the step. It also depends separately on the voltage after the step, although Figure 1 does not illustrate the dependence documented in the literature [7–9]. The voltage and time dependence arises from the molecular motions underlying the gating current. The voltage and time dependence defines the mean molecular motions [7,16,17,19,21,82–86] and is called 'the gating current' in the biophysics literature.

If the ON charge is found experimentally to equal the OFF charge, for a variety of pulse sizes and range of experimental conditions, the current is said to arise in a nonlinear (i.e., voltage dependent) polarization capacitance and is interpreted as the movement of charged groups in the electric field. The charged groups move to one location after the ON pulse, and return to their original location following the OFF pulse. The charge is called 'gating charge', and the current that carries the charge is called 'gating current'.

The macroscale current observed in the set-up is equal to the sum of the micro (actually atomic scale) currents carried by the charged groups inside a channel protein, even though the recording electrodes are remote from the protein. Indeed, there might be 10^{18} charged atoms (ions) between the electrodes and the protein.

The currents in the electrodes and the channel protein are equal because the setup is designed to be an unbranched one-dimensional circuit with everything in series. In a one-dimensional series setup the total current is equal everywhere in the series system at any one time, even though the total current varies significantly with time. The Maxwell equations guarantee spatial uniformity of total current (including the ethereal current $\varepsilon_0 \partial \mathbf{E}/\partial t$) independent of the microphysics of movement of charge (with mass): Figure 2 of [73], and [43,75,87]. The equality of current can be checked by measuring current in different locations in the experiment. The spatial equality of current needs also to be checked in simulations as in [18,21,88] because tiny inadvertent errors in numerical procedures or coding can produce substantial deviations from spatial equality and thus misleading artifacts. Imposing periodic boundary conditions on nonperiodic systems is another possible source of such artifacts.

If the currents reach a steady value independent of time, but not equal to zero, as in Figure 1, the signal is not transient, in the strict meaning of the word. In biophysics, the steady current i_{leak} is then usually considered to flow in a resistive path that is time independent, but perhaps voltage dependent, in parallel with the path or device in which the gating charges Q_{ON} and Q_{OFF} flow. If the current does not reach a steady value, or if the areas in Figure 2 are not equal, the currents are not considered 'capacitive' and are interpreted as those through a time and voltage dependent 'resistor'. This is a biological and biophysical assumption. It is not a physical or mathematical necessity. Thus, it is important to investigate the properties of the currents through the resistive path—e.g., those that are not transient and do not return to zero and those that make $Q_{ON} \neq Q_{OFF}$ by ***independent*** methods to see if they are time independent. In biophysics, currents can be done by blocking the resistive path with drugs, or with mutations of the channel protein. If the resistive currents are not time independent, the definition of Q_{ON} and Q_{OFF} in Figure 1 needs to be changed. Indeed, experiments of another type must be designed that allow separation of polarization from conduction currents. The simplest version of the biophysics paradigm then needs to be extended.

Clearly, this approach will only work if step functions can reveal all the properties of the underlying mechanism. If the underlying mechanisms depend on the time rate of change of voltage, step functions are clearly insufficient because $\partial V/\partial t$, is zero or infinity but nothing else in a step function. In the classical language of membrane biophysics, the ionic conductances g_{Na} and g_K must not depend on the rate of change of voltage.

Much work has been conducted showing that step functions are enough to understand the voltage dependent mechanisms in the classical action potential of the squid axon [89–91], starting with [78], Figure 10 and Equation (11). Hodgkin kindly explained the significance of this issue to colleagues, including the author (around 1970). He explained the possible incompleteness of step function measurements: if sodium conductance had a significant dependence on $\partial V/\partial t$, the action potential computed from voltage clamp data would differ from experimental measurements. He mentioned that this possibility was an important motivation for Huxley's heroic hand integration [6] of the Hodgkin Huxley differential equations. Huxley confirmed this in a separate personal communication, Huxley to Eisenberg. Those computations and many papers since [89–91] have shown that voltage clamp data (in response to steps) is enough to predict the shape and propagation of the action potential in nerve and skeletal muscle. It should be clearly understood that such a result is not available for biological systems in which the influx of Ca^{++} drives the action potential and its propagation [92].

The conductance of the voltage activated calcium channel has complex dependence on the current through the channel because the concentration of Ca^{++} in the cytoplasm is so low (~10^{-8} M at rest) that the current almost always changes the local concentration in and near the channel on the cytoplasmic side. Those concentration changes, in turn, alter the gating and selectivity characteristics of the channel protein, as calcium ions are prone to do int many physical and biological systems, particularly at interfaces.

It seems unlikely that the resulting properties of voltage dependent calcium channels can be comfortably described by the same formalism [6] used for voltage-controlled sodium and potassium channels of nerve and skeletal muscle. That formalism uses variables that depend on membrane potential and not membrane current because Cole [93] and Hodgkin [94–96] guessed that neuronal action potentials were essentially voltage dependent, not current dependent. They found action potentials in 'space clamped' axons with wires down their middle [76,77,97,98] that ensured spatial uniformity of potential. These axons had very different patterns of current flow from normal axons, and so Cole and Hodgkin were confirmed in their view that the membrane processes generating the action potential were voltage dependent, much more than current dependent (personal communication Cole to Eisenberg 1960; Hodgkin to Eisenberg 1961, et al.).

Hodgkin, Huxley, Katz, and Cole did not know of action potentials driven by calcium channels [99–103], nor of the extraordinarily small concentration of calcium ions inside cells. There may of course be other reasons the formalism [6] is inadequate. In summary, experiments, theory, computations and perhaps simulations are needed to show that responses to steps of voltage allow computation of a calcium driven action potential.

The polarization protocol described here can be applied to simulations of polarization as well as experimental measurements of polarization. Indeed, the operational definition of polarization has been applied even when theories [18] or simulations are enormously complicated by atomic detail that includes the individual motions of thousands of atoms [21,88].

Another question of general interest is how does the polarization defined this way correspond to the polarization $-\mathbf{P} = (\varepsilon_r - 1)\varepsilon_0\mathbf{E}$ in the classical formulation of the Maxwell Equations (7) and implied in Equations (1) and (6)? Does the estimated polarization equal **P**?

The answer is not pleasing. Polarization cannot be defined in general. The variety of possible responses of matter to a step of potential prevents a general answer. Indeed, a main point of this paper is that polarization must be defined by a protocol in a specific setting that specifies how the local electric field changes the distribution of charge.

Polarization cannot be defined in general because there are too many possible motions of mass with charge in response to a change in the electric field. Every possible motion of mass (with charge), including rotations and translations and changes of shape and density of charge, would produce a polarization. Polarization currents can be as complicated as the motions of matter.

In mechanical systems in general these issues do not attract much attention. It seems obvious that one must have a model and theory of how a system changes shape (and distribution of mass) when forces are applied. Seeking a general treatment is silly. In electrodynamics, for illogical reasons of history, tradition, and respect for our elders, scientists have sought the general treatment that would be considered silly for mechanical systems.

Scientists, certainly including me, have used the simple electromechanical model of an ideal dielectric to describe how charge moves in response to an electric field, using the name polarization to describe the phenomena. They have tried to apply it everywhere, as is seen because that model is embedded in the traditional formulation of the Maxwell equations found universally in textbooks.

It seems to me time to abandon this forlorn hope of a general description of the response of charged matter to a change in the electric field, and to move to a more reasonable approach, in which explicit models of the response of charge to the electric field are constructed, with different models for different systems.

Insight can be developed into various kinds of polarization by constructing 'toy' models of simple systems. Those models must specify the mechanical variables $\mathbf{v}_Q$, $\mathbf{Q}_Q$, $\boldsymbol{\rho}_Q$ and $\mathbf{N}_Q$ (or their equivalent) and solve the field equations of mechanics, perhaps including diffusion, along with the Maxwell equations. The models are then studied using the operational definition of polarization, described previously (Figure 1) or other operational definitions more suitable for other systems. One can hope some of the models resemble some of the more elaborate models of polarization already in the literature [26–29,31,32,34].

Toy models might include:

(1) Simple electro-mechanical models, like a charged mass on a spring with damping.
(2) Ideal gases of permanently charged particles, i.e., biological and physical plasmas.
(3) Ideal gases of dipoles (point [104] and macroscopic), quadrupoles, and mixtures of dipoles and quadrupoles, that rotate and translate while some are attached by bonds that vibrate (see (1)). These mixtures should provide decent representations of liquid water in ionic solutions, if they include a background dielectric, even if the dielectric is over-approximated with a single dielectric constant $\varepsilon_r(H_2O) \cong 80$. Indeed, there is a substantial literature of such models, including [105,106] but one must be sure that the models include the unavoidable interactions of atoms, molecules, and structures often dominated by their electrodynamics. Atoms, molecules and structures are almost always charged and so never move independently. Their motions are correlated by the electric field, and those correlations are likely to dominate the properties of greatest interest in applications. Of course, the extensive analysis of these authors can be of great use once it is focused on issues and applications of interest and combined with experimental measurements (see (5) and (6) below).
(4) Molecular models of ionic solutions that include water as a molecule. It is best to use models that are successful in predicting the activity of solutions of diverse composition and content and include water and ions as molecules of unequal nonzero size [107].
(5) Classical models of impedance, dielectric, and molecular spectroscopy [26–29,31,32,34].
(6) Well-studied systems of complex fluids, spanning scales, connecting micro (even atomic) structures with macroscopic functions, often called 'micro-macro models' in the literature.

These examples, taken together, will help form a handbook of practical examples closely related to the classical approximations of dielectrics.

These problems have time dependent solutions except in degenerate, uninteresting cases. Time dependence poses particular problems for the classical formulations of Maxwell equations. As stated in [51] on p. 13.

"It is necessary also to reiterate that ε_r is a single, real positive constant in Maxwell's equations as he wrote them and as they have been stated in many textbooks since then, following [108–110]. If one wishes to generalize ε_r so that it more realistically describes the properties of matter, one must actually change the differential Equation (6) and the set of Maxwell's equations as a whole. If, to cite a common (but not universal) example, ε_r is to be generalized to a time dependent function, (because polarization current in this case is a time dependent solution of a linear, often constant coefficient, differential equation that depends only on the local electric field), the mathematical structure of Maxwell's equations changes".

Perhaps it is tempting to take a short cut by simply converting ε_r into a function of time $\varepsilon_r(t)$ in Maxwell's equations, as classically written. "Solving the equations with a constant ε_r and then letting ε_r become a function of time creates a mathematical chimera that is not correct. The chimera is not a solution of the equations." The full functional form, or differential equation for $\varepsilon_r(t)$ must be written and solved together with the Maxwell equations. This is a formidable task in any case, but becomes an even more formidable challenge if convection or electrodiffusion modify polarization, as well as the electric field.

If one confines oneself to sinusoidal systems (as in classical impedance or dielectric spectroscopy [27,42,111,112]), one should explicitly introduce the sinusoids into the equations and not just assume that the simplified treatment of sinusoids in elementary circuit theory [113–117] is correct. It is not at all clear that Maxwell's equations joined with constitutive equations; and boundary conditions always have steady state solutions in the sinusoidal case. The Maxwell equations joined with diffusion and convection equations (like Navier–Stokes [118–135] or PNP = Poisson Nernst Planck = drift diffusion [52,53,55,57,59,61,123,136–145]) certainly do not always have solutions that are linear functions of just the electric field [146–149]."

It seems clear that the classical Maxwell equations with the over-approximated dielectric coefficient ε_r cannot emerge in the time dependent case. Of course, the classical Maxwell equations cannot emerge when polarization has a nonlinear dependence on the electric field, or depends on the global (not local) electric field, or depends on convection or electrodiffusion.

Indeed, in my opinion, when confronted with the models of polarization listed on the previous page, the classical Maxwell equations will be useful only when knowledge of the actual properties of polarization is not available. All the models listed involve time dependence in the polarization fields that are not included in the classical Maxwell equations as usually written.

5. Conclusions

A generalization of Maxwell's **P** useful in a range of systems may emerge. The generalization would describe how the local electric field changes the distribution of charge, as one imagines that Maxwell hoped **P** and **D** would be.

Until then, one is left with:

(1) Bewilderingly complete measurements, over an enormous range of frequencies (e.g., [26–29,31,32,34,35]) of the dielectric properties and conductance of ionic solutions of varying composition and content. These measurements embarrass the theoretician with their diversity and complexity. They have not yet been captured in any formulas or programs less complicated than a look-up table of all the results.
(2) Computations of the motion of all charges on the atomic scale [21,88], described by the field equations of mechanics and electrodynamics [18].
(3) Reduced models. It is unlikely that the reduced models can be derived solely by mathematics. It is more likely that they must be 'guessed and checked' one by one, as most models are checked in science.

What should be done when little is known? Sadly, the actual properties of polarization are often unknown. Then, one is left with the over-approximated Equation (6) or nothing at all. It is almost never wise to assume polarization effects are negligible. Equation (6) is certainly better than nothing: Equation (6) can be particularly helpful if it is used gingerly: toy models can successfully represent an idealized view of a part of the real world of technological or biological importance, for example, electronic circuits or several properties of ion channels.

In some cases, the toy models can be enormously helpful. They allow the design of circuits in our analog and digital electronic technology [150–153]. They allow the understanding of selectivity [107,154–156] and current voltage relations of several important biological channel proteins in a wide range of solutions [107,157–159]. In other cases—for example, the description of ionic solutions with many components—the toy models can be too unrealistic to be useful. Experiments and experience can tell how useful the toy model actually is in a particular case: pure thought usually cannot.

Funding: This research received no external funding.

Institutional Review Board Statement: Not applicable.

Informed Consent Statement: Not applicable.

Data Availability Statement: Not available.

Acknowledgments: I thank the reviewers for suggestions and criticisms that substantially improved the paper. It is a particular pleasure to thank my friend and teacher Chun Liu for his continual encouragement and advice, and for patiently correcting mistakes in my mathematics as these ideas were developed over many years. Mistakes may remain, sad to say. All are my responsibility.

Conflicts of Interest: The author declares no conflict of interest.

Appendix A $\mathbf{P}(x,y,z|t)$ in Proteins

Ambiguities in the meaning of the polarization field $\mathbf{P}(x,y,z|t)$ can cause serious difficulties in the understanding of protein function. Understanding protein function is greatly aided by knowledge of protein structure. The protein data bank contains 173,754 structures in atomic detail today (24 January 2021) and the number is growing rapidly as cryo-electron microscopy is used more and more.

Protein structures are usually analyzed with molecular dynamics programs that assume periodic boundary conditions and chemical equilibrium, i.e., no flows. Most proteins control large flows as part of their natural biological functions. Equilibrium hardly ever occurs in living biological systems. It seems obvious that equilibrium systems cannot provide general insight into flows, any more than a nonfunctional amplifier without a power supply can show how a functional amplifier works. Proteins are not periodic in their natural setting. It seems obvious that periodic systems with flow cannot conserve total current $\mathbf{J}_{total}$ in general—or perhaps even in particular—as required by the Maxwell equations, see Equation (19). In other words, it is likely that molecular dynamics analyses of periodic structures do not satisfy the Maxwell equations, although almost all known physics does satisfy those equations.

It is also unlikely that standard programs of molecular dynamics compute electrodynamics of nonperiodic systems correctly, despite their use of Ewald sums, with various conventions, and force fields (tailored to fit macroscopic, not quantum mechanical) data. Compare the exhaustive methods used to validate results in computational electronics [61] with those in the computation of electric fields in proteins.

The electrostatic and electrodynamic properties of proteins are of great importance. Many of the atoms in a protein are assigned permanent charge greater than 0.2**e** in the force fields used in molecular dynamics, where **e** is the elementary charge, and these charges tend to cluster in locations most important for biological function, just as they cluster at high density near the electrodes of batteries and other electrochemical systems. Enormous densities of charge (>10 M, sometimes much larger) are found in and near channels of proteins [107,160–162] and in the 'catalytic active sites' [163] of enzymes. Such densities are also found near nucleic acids, DNA and (all types of) RNA and binding sites of proteins in general.

It seems likely that a hierarchy of models of different resolutions will be needed to compute the electrodynamics of proteins accurately enough to explain how the electrical properties of side chains (polarizability [21] and others) of a protein determine biological function. Analysis of gating currents suggests such an approach is feasible [17,18,20,21].

References

1. Purcell, E.M.; Morin, D.J. *Electricity and Magnetism*; Cambridge University Press: Cambridge, UK, 2013.
2. Feynman, R.P.; Leighton, R.B.; Sands, M. *The Feynman: Lectures on Physics, Mainly Electromagnetism and Matter*; Addison-Wesley Publishing Co.: New York, NY, USA, 1963; p. 592. Available online: http://www.feynmanlectures.caltech.edu/II_toc.html (accessed on 28 January 2021).
3. Zangwill, A. *Modern Electrodynamics*; Cambridge University Press: New York, NY, USA, 2013; p. 977.

4. Hehl, F.W.; Obukhov, Y.N. *Foundations of Classical Electrodynamics: Charge, Flux, and Metric*; Birkhäuser Boston: Cambridge, MA, USA, 2012.
5. Griffiths, D.J. *Introduction to Electrodynamics*, 3rd ed.; Cambridge University Press: Cambridge, UK, 2017.
6. Hodgkin, A.L.; Huxley, A.F. A quantitative description of membrane current and its application to conduction and excitation in nerve. *J. Physiol.* **1952**, *117*, 500–544.
7. Bezanilla, F. Gating currents. *J. Gen. Physiol.* **2018**, *150*, 911–932.
8. Schneider, M.F.; Chandler, W.K. Voltage Dependent Charge Movement in Skeletal Muscle: A Possible Step in Excitation-Contraction Coupling. *Nature* **1973**, *242*, 244–246.
9. Chandler, W.K.; Rakowski, R.F.; Schneider, M.F. A non-linear voltage dependent charge movement in frog skeletal muscle. *J. Physiol.* **1976**, *254*, 245–283.
10. Armstrong, C.M.; Bezanilla, F. Charge movement associated with the opening and closing of the activation gates of the Na channel. *J. Gen. Physiol.* **1974**, *63*, 533–552.
11. Armstrong, C.M.; Bezanilla, F. Currents related to movement of the gating particles of the sodium channels. *Nature* **1973**, *242*, 459–461.
12. Fernandez, J.; Bezanilla, F.; Taylor, R. Distribution and kinetics of membrane dielectric polarization. II. Frequency domain studies of gating currents. *J. Gen. Physiol.* **1982**, *79*, 41–67.
13. Taylor, R.E.; Bezanilla, F. Sodium and gating current time shifts resulting from changes in initial conditions. *J. Gen. Physiol.* **1983**, *81*, 773–784.
14. Perozo, E.; MacKinnon, R.; Bezanilla, F.; Stefani, E. Gating currents from a nonconducting mutant reveal open-closed conformations in Shaker K+ channels. *Neuron* **1993**, *11*, 353–358.
15. Sigg, D.; Stefani, E.; Bezanilla, F. Gating current noise produced by elementary transitions in Shaker potassium channels. *Science* **1994**, *264*, 578–582.
16. Bezanilla, F.; Stefani, E. Gating currents. *Methods Enzymol.* **1998**, *293*, 331–352.
17. Catacuzzeno, L.; Franciolini, F. Simulation of Gating Currents of the Shaker K Channel Using a Brownian Model of the Voltage Sensor. *Biophys. J.* **2019**, *117*, 2005–2019.
18. Horng, T.-L.; Eisenberg, R.S.; Liu, C.; Bezanilla, F. Continuum Gating Current Models Computed with Consistent Interactions. *Biophys. J.* **2019**, *116*, 270–282.
19. Lacroix, J.J.; Hyde, H.C.; Campos, F.V.; Bezanilla, F. Moving gating charges through the gating pore in a Kv channel voltage sensor. *Proc. Natl. Acad. Sci. USA* **2014**, *111*, E1950–E1959.
20. Catacuzzeno, L.; Franciolini, F.; Bezanilla, F.; Eisenberg, R. Gating current noise produced by Brownian models of a voltage sensor. *bioRxiv* **2021**.
21. Catacuzzeno, L.; Sforna, L.; Franciolini, F.; Eisenberg, R. Why are voltage gated Na channels faster than K channels: A multi-scale hierarchical model. *J. Gen. Physiol.* **2020**.
22. Buchwald, J.Z. *From Maxwell to Microphysics. Aspects of Electromagnetic Theory in the Last Quarter of the Nineteenth Century*; University of Chicago: Chicago, IL, USA, 1985.
23. Simpson, T.K. *Maxwell on the Electromagnetic Field: A Guided Study*; Rutgers University Press: Rutgers, NJ, USA, 1998; p. 441.
24. Arthur, J.W. The Evolution of Maxwell's Equations from 1862 to the Present Day. *IEEE Antennas Propag. Mag.* **2013**, *55*, 61–81.
25. Macdonald, J. Impedance spectroscopy. *Ann. Biomed. Eng.* **1992**, *20*, 289–305.
26. Banwell, C.N.; McCash, E.M. *Fundamentals of Molecular Spectroscopy*; McGraw-Hill New York: New York, NY, USA, 1994; Volume 851.
27. Kremer, F.; Schönhals, A. *Broadband Dielectric Spectroscopy*; Springer: New York, NY, USA, 2003; p. 729.
28. Barsoukov, E.; Macdonald, J.R. *Impedance Spectroscopy: Theory, Experiment, and Applications*, 2nd ed.; Wiley-Interscience: New York, NY, USA, 2005; p. 616.
29. Sindhu, P. *Fundamentals of Molecular Spectroscopy*; New Age International: New Dehli, India, 2006.
30. Angulo-Sherman, A.; Mercado-Uribe, H. Dielectric spectroscopy of water at low frequencies: The existence of an isopermitive point. *Chem. Phys. Lett.* **2011**, *503*, 327–330.
31. Rao, K.N. *Molecular Spectroscopy: Modern Research*; Elsevier: Amsterdam, The Netherlands, 2012.
32. Steinfeld, J.I. *Molecules and Radiation: An Introduction to Modern Molecular Spectroscopy*; Courier Corporation: North Chelmsford, MA, USA, 2012.
33. Eisenberg, R.S. Dielectric Dilemma. *arXiv* **2019**, arXiv:1901.10805.
34. Buchner, R.; Barthel, J. Dielectric Relaxation in Solutions. *Annu. Rep. Prog. Chem. Sect. C Phys. Chem.* **2001**, *97*, 349–382.
35. Barthel, J.; Buchner, R.; Münsterer, M. *Electrolyte Data Collection Vol. 12, Part 2: Dielectric Properties of Water and Aqueous Electrolyte Solutions*; DECHEMA: Frankfurt am Main, Germany, 1995.
36. Thomson, J.J. XL. Cathode Rays. *Lond. Edinb. Dublin Philos. Mag. J. Sci.* **1897**, *44*, 293–316.
37. Thomson, J.J. Nobel Lecture: Carriers of Negative Electricity. Nobel Media AB 2014: 1906. Available online: http://www.nobelprize.org/nobel_prizes/physics/laureates/1906/thomson-lecture.html (accessed on 28 January 2021).
38. Thomson, J.J. *Notes on Recent Researches in Electricity and Magnetism: Intended as a Sequel to Professor Clerk-Maxwell's Treatise on Electricity and Magnetism*; Clarendon Press: Oxford, UK, 1893.

39. Arabatzis, T. *Representing Electrons: A Biographical Approach to Theoretical Entities*; University of Chicago Press: Chicago, IL, USA, 2006.
40. Lorrain, P.; Corson, D. *Electromagnetic Fields and Waves*, 2nd ed.; Freeman: New York, NY, USA, 1970.
41. Kovetz, A. *Electromagnetic Theory*; Clarendon Press: Oxford, UK, 2000.
42. Barsoukov, E.; Macdonald, J.R. *Impedance Spectroscopy: Theory, Experiment, and Applications*; John Wiley & Sons: New York, NY, USA, 2018.
43. Eisenberg, B.; Oriols, X.; Ferry, D. Dynamics of Current, Charge, and Mass. *Mol. Based Math. Biol.* **2017**, *5*, 78–115.
44. Barthel, J.; Krienke, H.; Kunz, W. *Physical Chemistry of Electrolyte Solutions: Modern Aspects*; Springer: New York, NY, USA, 1998.
45. Nadler, B.; Hollerbach, U.; Eisenberg, R.S. Dielectric boundary force and its crucial role in gramicidin. *Phys. Rev. E Stat. Nonlin Soft Matter Phys.* **2003**, *68*, 021905.
46. Varsos, K.; Luntz, J.; Welsh, M.; Sarabandi, K. Electric Field-Shaping Microdevices for Manipulation of Collections of Microscale Objects. *Proc. IEEE* **2011**, *99*, 2112–2124.
47. Donnell, R.O. Prolog to Electric Field-Shaping Microdevices for Manipulation of Collections of Microscale Objects. *Proc. IEEE* **2011**, *99*, 2110–2111.
48. Fiedziuszko, S.J.; Hunter, I.C.; Itoh, T.; Kobayashi, Y.; Nishikawa, T.; Stitzer, S.; Wakino, K. Dielectric Materials, Devices, and Circuits. *IEEE Trans. Microw. Theory Tech.* **2002**, *50*, 706–720.
49. Schey, H.M.; Schey, H.M. *Div, Grad, Curl, and All That: An Informal Text on Vector Calculus*; WW Norton: New York, NY, USA, 2005.
50. Arfken, G.B.; Weber, H.J. *Mathematical Methods for Physicists*; AAPT: College Park, MD, USA, 1999.
51. Eisenberg, R.S. Updating Maxwell with Electrons, Charge, and More Realistic Polarization. *arXiv* **2019**, arXiv:1904.09695.
52. Shockley, W. *Electrons and Holes in Semiconductors to Applications in Transistor Electronics*; Van Nostrand: New York, NY, USA, 1950; p. 558.
53. Van Roosbroeck, W. Theory of flow of electrons and holes in germanium and other semiconductors. *Bell Syst. Tech. J.* **1950**, *29*, 560–607.
54. Gummel, H.K. A self-consistent iterative scheme for one-dimensional steady-state transistor calculations. *IEEE Trans. Electron Devices* **1964**, *ED-11*, 445–465.
55. Blotekjaer, K. Transport equations for electrons in two-valley semiconductors. *Electron Devices IEEE Trans.* **1970**, *17*, 38–47.
56. Sze, S.M. *Physics of Semiconductor Devices*; John Wiley & Sons: New York, NY, USA, 1981; p. 838.
57. Selberherr, S. *Analysis and Simulation of Semiconductor Devices*; Springer-Verlag: New York, NY, USA, 1984; pp. 1–293.
58. Jacoboni, C.; Lugli, P. *The Monte Carlo Method for Semiconductor Device Simulation*; Springer Verlag: New York, NY, USA, 1989; pp. 1–356.
59. Markowich, P.A.; Ringhofer, C.A.; Schmeiser, C. *Semiconductor Equations*; Springer-Verlag: New York, NY, USA, 1990; p. 248.
60. Ferry, D.K.; Goodnick, S.M.; Bird, J. *Transport in Nanostructures*; Cambridge University Press: New York, NY, USA, 2009; p. 670.
61. Vasileska, D.; Goodnick, S.M.; Klimeck, G. *Computational Electronics: Semiclassical and Quantum Device Modeling and Simulation*; CRC Press: New York, NY, USA, 2010; p. 764.
62. Moore, G.E. Lithography and the Future of Moore's Law. In Proceedings of the Integrated Circuit Metrology, Inspection, and Process Control IX, Santa Clara, CA, USA, 20–22 February 1995; pp. 2–17.
63. Moore, G.E. Cramming more components onto integrated circuits. *Electron. Mag.* **1965**, *38*, 114–117.
64. Lundstrom, M. Applied Physics Enhanced: Moore's Law Forever? *Science* **2003**, *299*, 210–211.
65. Bohr, M. A 30 year retrospective on Dennard's MOSFET scaling paper. *IEEE Solid State Circuits Soc. Newsl.* **2007**, *12*, 11–13.
66. Dennard, R.H.; Gaensslen, F.H.; Rideout, V.L.; Bassous, E.; LeBlanc, A.R. Design of ion-implanted MOSFET's with very small physical dimensions. *Solid State Circuits IEEE J.* **1974**, *9*, 256–268.
67. Dennard, R.H.; Gaensslen, F.H.; Yu, H.-N.; Rideout, V.L.; Bassous, E.; LeBlanc, A.R. Design of Ion-Implanted MOSFET's with Very Small Physical Dimensions. *Proc. IEEE* **1999**, *87*, 668–678.
68. Burns, L. Maxwell's Equations are Universal for Locally Conserved Quantities. *Adv. Appl. Clifford Algebras* **2019**, *29*.
69. Tolman, R.C. XXXIII. Non-Newtonian mechanics, the mass of a moving body. *Lond. Edinb. Dublin Philos. Mag. J. Sci.* **1912**, *23*, 375–380.
70. Eisenberg, B. Maxwell Matters. *arXiv* **2016**, arXiv:1607.06691.
71. Eisenberg, B. Conservation of Current and Conservation of Charge. *arXiv* **2016**, arXiv:1609.09175.
72. Eisenberg, B.; Gold, N.; Song, Z.; Huang, H. What Current Flows Through a Resistor? *arXiv* **2018**, arXiv:1805.04814.
73. Eisenberg, R.S. Mass Action and Conservation of Current. *Hung. J. Industry Chem.* **2016**, *44*, 1–28.
74. Eisenberg, R.S. Kirchhoff's Law can be Exact. *arXiv* **2019**, arXiv:1905.13574.
75. Eisenberg, R.S. Electrodynamics Correlates Knock-on and Knock-off: Current is Spatially Uniform in Ion Channels. *arXiv* **2020**, arXiv:2002.09012.
76. Cole, K.S. Dynamic electrical characteristics of the squid axon membrane. *Arch. Des Sci. Physiol.* **1949**, *3*, 253–258.
77. Hodgkin, A.; Huxley, A.; Katz, B. Ionic Currents underlying activity in the giant axon of the squid. *Arch. Sci. Physiol.* **1949**, *3*, 129–150.
78. Hodgkin, A.L.; Huxley, A.F.; Katz, B. Measurement of current- voltage relations in the membrane of the giant axon of *Loligo*. *J. Physiol. (Lond.)* **1952**, *116*, 424–448.
79. Cole, K.S.; Moore, J.W. Ionic current measurements in the squid giant axon membrane. *J. Gen. Physiol.* **1960**, *44*, 123–167.

80. Cole, K.S.; Moore, J.W. Potassium ion current in the squid giant axon: Dynamic characteristic. *Biophys. J.* **1960**, *1*, 1–14.
81. Taylor, R.E.; Moore, J.W.; Cole, K.S. Analysis of certain errors in squid axon voltage clamp measurements. *Biophys. J.* **1960**, *1*, 161–202.
82. Horng, T.-L.; Eisenberg, R.S.; Liu, C.; Bezanilla, F. Gating Current Models Computed with Consistent Interactions. *arXiv* **2017**, arXiv:1707.02566.
83. Bezanilla, F. How membrane proteins sense voltage. *Nat. Rev. Mol. Cell Biol.* **2008**, *9*, 323–332.
84. Bezanilla, F.; Perozo, E. The voltage sensor and the gate in ion channels. *Adv. Protein Chem.* **2003**, *63*, 211–241.
85. Bezanilla, F. Voltage Sensor Movements. *J. Gen. Physiol.* **2002**, *120*, 465–473.
86. Kubota, T.; Durek, T.; Dang, B.; Finol-Urdaneta, R.K.; Craik, D.J.; Kent, S.B.; French, R.J.; Bezanilla, F.; Correa, A.M. Mapping of voltage sensor positions in resting and inactivated mammalian sodium channels by LRET. *Proc. Natl. Acad. Sci. USA* **2017**, *114*, E1857–E1865.
87. Eisenberg, R.S. Maxwell Equations for Material Systems. 2020. Available online: https://www.preprints.org/manuscript/202011.0201/v1 (accessed on 28 January 2021).
88. Catacuzzeno, L.; Sforna, L.; Franciolini, F. Voltage-dependent gating in K channels: Experimental results and quantitative models. *Pflug. Arch. Eur. J. Physiol.* **2020**, *472*, 27–47.
89. Palti, Y.; Adelman, W.J. Measurement of axonal membrane conductances and capacity by means of a varying potential control voltage clamp. *J. Membr. Biol.* **1969**, *1*, 431–458.
90. Starzak, M.E. Analysis of clamps with time dependent voltages in the squid axon using a kinetic transport model. *J. Theor. Biol.* **1976**, *57*, 153–169.
91. Fishman, H.M. Direct and rapid description of the individual ionic currents of squid axon membrane by ramp potential control. *Biophys. J* **1970**, *10*, 799–817.
92. Dolphin, A.C. A short history of voltage-gated calcium channels. *Br. J. Pharmacol.* **2006**, *147*, S56.
93. Huxley, A.F. *Kenneth Stewart Cole 1900–1984. A biographical Memoir by Sir Andrew Huxley*; National Academies Press: Washington, DC, USA, 1996.
94. Huxley, A. Hodgkin Obituary. *Indep. (Newsp.)*. 1999. Available online: http://www.independent.co.uk/arts-entertainment/obituaries-professor-sir-alan-hodgkin-1044924.html (accessed on 28 January 2021).
95. Huxley, A.F. From overshoot to voltage clamp. *Trends Neurosci.* **2002**, *25*, 553–558.
96. Hodgkin, A.L. *Chance and Design*; Cambridge University Press: New York, NY, USA, 1992; p. 401.
97. Cole, K.S. The advance of electrical models for cells and axons. *Biophys. J.* **1962**, *2*, 101–119.
98. Marmont, G. Studies on the axon membrane. I. A new method. *J. Cell. Comp. Physiol.* **1949**, *34*, 351–382.
99. Hagiwara, S.; Naka, K.I. The Initiation of Spike Potential in Barnacle Muscle Fibers under Low Intracellular Ca++. *J. Gen. Physiol.* **1964**, *48*, 141–162.
100. Tsien, R.W.; Barrett, C.F. A brief history of calcium channel discovery. In *Voltage-Gated Calcium Channels*; Springer: New York, NY, USA, 2005; pp. 27–47.
101. Fatt, P.; Ginsborg, B.L. The ionic requirements for the production of action potentials in crustacean muscle fibres. *J. Physiol.* **1958**, *142*, 516–543.
102. Fatt, P.; Ginsborg, B.L. The production of regenerative responses in crayfish muscle fibres by the action of calcium, strontium and barium. *J. Physiol.* **1958**, *140*, 59P–60P.
103. Fatt, P.; Katz, B. The electrical properties of crustacean muscle fibres. *J. Physiol.* **1953**, *120*, 171–204.
104. Debye, P.J.W. *Polar Molecules*; Chemical Catalog Company, Incorporated: New York, NY, USA, 1929.
105. De Groot, S.R.; Suttorp, L.G. *Foundations of Electrodynamics*; North-Holland: Amsterdam, The Netherlands, 1972.
106. Robinson, F.N.H. *Macroscopic Electromagnetism*; Pergamon: Oxford, UK, 1973; Volume 57.
107. Liu, J.L.; Eisenberg, B. Molecular Mean-Field Theory of Ionic Solutions: A Poisson-Nernst-Planck-Bikerman Model. *Entropy* **2020**, *22*, 550.
108. Jeans, J.H. The Mathematical Theory of Electricity and Magnetism. *Nature* **1908**, *78*, 537–578.
109. Abraham, M.; Becker, R. *The Classical Theory of Electricity and Magnetism*; Blackie and Subsequent Dover Reprints: Glasgow, UK, 1932; p. 303.
110. Abraham, M.; Föppl, A. *Theorie der Elektrizität: Bd. Elektromagnetische Theorie der Strahlung*; BG Teubner: Wiesbaden, Germany, 1905; Volume 2.
111. Ciucci, F. Modeling Electrochemical Impedance Spectroscopy. *Curr. Opin. Electrochem.* **2018**.
112. Raicu, V.; Feldman, Y. *Dielectric Relaxation in Biological Systems: Physical Principles, Methods, and Applications*; Oxford University Press: Oxford, MA, USA, 2015.
113. Ghausi, M.S.; Kelly, J.J. *Introduction to Distributed-Parameter Networks: With Application to Integrated Circuits*; Holt, Rinehart and Winston: New York, NY, USA, 1968.
114. Guillemin, E.A. *Introductory Circuit Theory*; Wiley: New York, NY, USA, 1958.
115. Lorrain, P.; Corson, D.R.; Lorrain, F. *Electromagnetic Fields and Waves: Including Electric Circuits*; Freeman: New York, NY, USA, 1988.
116. Weinberg, L. *Network Analysis and Synthesis*; Krieger Pub. Co.: Huntington, NY, USA, 1975.
117. Tuttle, D.F. *Network Synthesis*; Wiley: New York, NY, USA, 1958; Volume 1.

118. Brannick, J.; Liu, C.; Qian, T.; Sun, H. Diffuse interface methods for multiple phase materials: An energetic variational approach. *Numer. Math. Theory Methods Appl.* **2015**, *8*, 220–236.
119. Horng, T.L.; Lin, T.C.; Liu, C.; Eisenberg, B. PNP equations with steric effects: A model of ion flow through channels. *J. Phys. Chem.* **2012**, *116*, 11422–11441.
120. Wang, Y.; Liu, C.; Tan, Z. A Generalized Poisson–Nernst–Planck–Navier–Stokes Model on the Fluid with the Crowded Charged Particles: Derivation and Its Well-Posedness. *Siam J. Math. Anal.* **2016**, 3191–3235.
121. Ryham, R.J. An Energetic Variational Approach to Mathematical Moldeling of Charged Fluids, Charge Phases, Simulation and Well Posedness. Ph.D. Thesis, The Pennsylvania State University, State College, PA, USA, 2006.
122. Eisenberg, B.; Hyon, Y.; Liu, C. Energy Variational Analysis EnVarA of Ions in Water and Channels: Field Theory for Primitive Models of Complex Ionic Fluids. *J. Chem. Phys.* **2010**, *133*, 104104.
123. Johannesson, B. Development of a Generalized Version of the Poisson– Nernst–Planck Equations Using the Hybrid Mixture Theory: Presentation of 2D Numerical Examples. *Transp. Porous Media* **2010**, *85*, 565–592.
124. Doi, M. Onsager's variational principle in soft matter. *J. Phys. Condens Matter* **2011**, *23*, 284118.
125. Mori, Y.; Liu, C.; Eisenberg, R.S. A model of electrodiffusion and osmotic water flow and its energetic structure. *Phys. D Nonlinear Phenom.* **2011**, *240*, 1835–1852.
126. Boda, D.; Gillespie, D. Steady-State Electrodiffusion from the Nernst–Planck Equation Coupled to Local Equilibrium Monte Carlo Simulations. *J. Chem. Theory Comput.* **2012**, *8*, 824–829.
127. Hsieh, C.-y.; Hyon, Y.; Lee, H.; Lin, T.-C.; Liu, C. Transport of charged particles: Entropy production and maximum dissipation principle. *arXiv* **2014**, arXiv:1407.8245v1.
128. Wu, H.; Lin, T.-C.; Liu, C. On transport of ionic solutions: From kinetic laws to continuum descriptions. *arXiv* **2014**, arXiv:1306.3053v2.
129. Xu, S.; Sheng, P.; Liu, C. An energetic variational approach to ion transport. *Commun. Math. Sci.* **2014**, *12*, 779–789.
130. Wu, H.; Lin, T.-C.; Liu, C. Diffusion limit of kinetic equations for multiple species charged particles. *Arch. Ration. Mech. Anal.* **2015**, *215*, 419–441.
131. Metti, M.S.; Xu, J.; Liu, C. Energetically stable discretizations for charge transport and electrokinetic models. *J. Comput. Phys.* **2016**, *306*, 1–18.
132. Giga, M.-H.; Kirshtein, A.; Liu, C. Variational Modeling and Complex Fluids. In *Handbook of Mathematical Analysis in Mechanics of Viscous Fluids*; Giga, Y., Novotny, A., Eds.; Springer International Publishing: Cham, Switzerland, 2017; pp. 1–41.
133. Gao, Y.; Huang, J.; Liu, Y.; Chen, S. Charge transport in confined concentrated solutions: A minireview. *Curr. Opin. Electrochem.* **2018**.
134. Jiang, J.; Ginzburg, V.; Wang, Z.-G. Density Functional Theory for Charged Fluids. *Soft Matter* **2018**.
135. Zhao, Q.; Liu, P.; Xu, Z. A Fast Method for Evaluating Green's Function in Irregular Domains with Application to Charge Interaction in a Nanopore. *Commun. Comput. Phys.* **2018**, *24*, 1214–1258.
136. Macdonald, J.R. Theory of ac Space-Charge Polarization Effects in Photoconductors, Semiconductors, and Electrolytes. *Phys. Rev.* **1953**, *92*, 4–17.
137. Rubinstein, I. *Electro-Diffusion of Ions*; SIAM: Philadelphia, PA, USA, 1990; p. 254.
138. Mason, E.; McDaniel, E. *Transport Properties of Ions in Gases*; John Wiley and Sons: New York, NY, USA, 1988; p. 560.
139. Eisenberg, R. PNP what is in a name july 25-1 2019. *Engrxiv. August* **2019**.
140. Barcilon, V. Ion flow through narrow membrane channels: Part I. *Siam J. Appl. Math* **1992**, *52*, 1391–1404.
141. Bazant, M.Z.; Thornton, K.; Ajdari, A. Diffuse-charge dynamics in electrochemical systems. *Phys. Rev. E* **2004**, *70*, 021506.
142. Kurnikova, M.G.; Coalson, R.D.; Graf, P.; Nitzan, A. A Lattice Relaxation Algorithm for 3D Poisson-Nernst-Planck Theory with Application to Ion Transport Through the Gramicidin A Channel. *Biophys. J.* **1999**, *76*, 642–656.
143. Liu, W.; Wang, B. Poisson-Nernst-Planck systems for narrow tubular-like membrane channels. *J. Dynam. Differ. Equ.* **2010**, *22*, 413–437.
144. Boda, D.; Csányi, É.; Gillespie, D.; Kristóf, T. Dynamic Monte Carlo Simulation of Coupled Transport through a Narrow Multiply-Occupied Pore. *J. Phys. Chem. C* **2013**, *118*, 700–707.
145. Schuss, Z.; Nadler, B.; Eisenberg, R.S. Derivation of Poisson and Nernst-Planck equations in a bath and channel from a molecular model. *Phys. Rev. E Stat. Nonlin Soft Matter Phys.* **2001**, *64*, 036116.
146. Boyd, R.W. *Nonlinear Optics*, 3rd ed.; Academic Press: New York, NY, USA, 2008; p. 640.
147. Sutherland, R.L. *Handbook of Nonlinear Optics*; CRC Press: New York, NY, USA, 2003.
148. Wegener, M. *Extreme Nonlinear Optics: An Introduction*; Springer Science & Business Media: New York, NY, USA, 2005.
149. Zheng, B.; Madni, H.A.; Hao, R.; Zhang, X.; Liu, X.; Li, E.; Chen, H. Concealing arbitrary objects remotely with multi-folded transformation optics. *Light Sci. Appl.* **2016**, *5*, e16177.
150. Horowitz, P.; Hill, W. *The Art of Electronics*, 3rd ed.; Cambridge University Press: Cambridge, MA, USA, 2015; p. 1224.
151. Howe, R.T.; Sodini, C.G. *Microelectronics: An Integrated Approach*; Prentice Hall: Upper Saddle River, NJ, USA, 1997; p. 908.
152. Muller, R.S.; Chan, M.; Kamins, T.I. *Device Electronics for Integrated Circuits*, 3rd ed.; Wiley India Pvt. Limited: New Delhi, India, 2003.
153. Scherz, P.; Monk, S. *Practical Electronics for Inventors*; McGraw-Hill, Inc.: New York, NY, USA, 2006; p. 1056.

154. Nonner, W.; Catacuzzeno, L.; Eisenberg, B. Binding and selectivity in L-type calcium channels: A mean spherical approximation. *Biophys. J.* **2000**, *79*, 1976–1992.
155. Boda, D.; Nonner, W.; Valisko, M.; Henderson, D.; Eisenberg, B.; Gillespie, D. Steric selectivity in Na channels arising from protein polarization and mobile side chains. *Biophys. J.* **2007**, *93*, 1960–1980.
156. Boda, D.; Valisko, M.; Henderson, D.; Eisenberg, B.; Gillespie, D.; Nonner, W. Ionic selectivity in L-type calcium channels by electrostatics and hard-core repulsion. *J. Gen. Physiol.* **2009**, *133*, 497–509.
157. Gillespie, D. A review of steric interactions of ions: Why some theories succeed and others fail to account for ion size. *Microfluid. Nanofluidics* **2015**, *18*, 717–738.
158. Boda, D.; Kovacs, R.; Gillespie, D.; Kristof, T. Selective transport through a model calcium channel studied by Local Equilibrium Monte Carlo simulations coupled to the Nernst-Planck equation. *J. Mol. Liq.* **2014**, *189*, 100–112.
159. Gillespie, D. Energetics of divalent selectivity in a calcium channel: The ryanodine receptor case study. *Biophys. J.* **2008**, *94*, 1169–1184.
160. Eisenberg, R.S. Computing the field in proteins and channels. *J. Membr. Biol.* **1996**, *150*, 1–25.
161. Eisenberg, R.S. Atomic Biology, Electrostatics and Ionic Channels. In *New Developments and Theoretical Studies of Proteins*; Elber, R., Ed.; World Scientific: Philadelphia, PA, USA, 1996; Volume 7, pp. 269–357.
162. Eisenberg, B. Living Transistors: A Physicist's View of Ion Channels (version 2). *arXiv* **2005**, arXiv:q-bio/0506016v2.
163. Jimenez-Morales, D.; Liang, J.; Eisenberg, B. Active Sites of Enzymes are Crowded with Charge. *Biophys. J.* **2011**, *100*, 218a.

entropy

Article

Unraveling of a Strongly Correlated Dynamical Network of Residues Controlling the Permeation of Potassium in KcsA Ion Channel

Salvatore M. Cosseddu †, Eunju Julia Choe and Igor A. Khovanov *

School of Engineering, University of Warwick, Coventry CV4 7AL, UK; salvatore.cosseddu@gmail.com (S.M.C.); e.j.choe.00@gmail.com (E.J.C.)
* Correspondence: i.khovanov@warwick.ac.uk
† Current address: Viseca Payment Services SA, Hagenholzstrasse 56, 8050 Zürich, Switzeland.

Abstract: The complicated patterns of the single-channel currents in potassium ion channel KcsA are governed by the structural variability of the selectivity filter. A comparative analysis of the dynamics of the wild type KcsA channel and several of its mutants showing different conducting patterns was performed. A strongly correlated dynamical network of interacting residues is found to play a key role in regulating the state of the wild type channel. The network is centered on the aspartate D80 which plays the role of a hub by strong interacting via hydrogen bonds with residues E71, R64, R89, and W67. Residue D80 also affects the selectivity filter via its backbones. This network further compromises ions and water molecules located inside the channel that results in the mutual influence: the permeation depends on the configuration of residues in the network, and the dynamics of network's residues depends on locations of ions and water molecules inside the selectivity filter. Some features of the network provide a further understanding of experimental results describing the KcsA activity. In particular, the necessity of anionic lipids to be present for functioning the channel is explained by the interaction between the lipids and the arginine residues R64 and R89 that prevents destabilizing the structure of the selectivity filter.

Keywords: ion channels; protein dynamics; molecular dynamics

Citation: Cosseddu, S.M.; Choe, E.J.; Khovanov, I.A. Unraveling of a Strongly Correlated Dynamical Network of Residues Controlling the Permeation of Potassium in KcsA Ion Channel. *Entropy* **2021**, *23*, 72. https://doi.org/10.3390/e23010072

Received: 10 November 2020
Accepted: 2 January 2021
Published: 6 January 2021

1. Introduction

Over the last few decades, the bacterial K^+ ion channel KcsA [1] found in *Streptomyces lividans* has been widely studied in order to understand the structural and functional features of potassium ion channels. It continues to be of interest [2–8] in part due to its sequence similarity to eukaryotic K^+ channels, and in part because of its role as an archetype for ion permeation, selectivity, and the complex interplay of the different "gates" which governs a variety of current patterns observed experimentally in the K^+ channel superfamily [9–11]. These patterns are defined by small structural rearrangements of the pore region once the inner gate is opened [4,10–13]. The local rearrangements are mostly obscure as current experimental techniques are unable to provide the combination of spatial and temporal resolution needed to identify the underlying atomistic-level mechanisms. Structural studies showed that the current patterns depend on a number of residues, some of which are located relatively far from the pathway of K^+ permeation, and physiological recordings revealed the strong influence of the K^+ concentration in the outer bulk on the patterns [1,10,14–19]. The anionic phospholipids modulate the function of the channel [20–22] and the addition of phosphatidic acid lipid significantly affects the permeation [23].

As with most of the K^+ ion channels, KcsA contains a highly conserved amino acid sequence motif TXXTXGYGD known as the *signature sequence*, which corresponds to residues 75 to 79 in the reference X-ray structure 1K4C [1], where "X" in position 76 is replaced by valine. The whole quaternary structure of KcsA is divided into three functional regions: the selectivity filter, a water-filled cavity, and an inner gate associated with large movements in the transmembrane helices for opening the channel [16,24–27]. The selectivity filter (SF) is the narrowest part of the pore. The SF consists of five well-defined binding sites for K^+ ions by exposing the backbone carbonyl groups of the residues toward the channel axis [28]. These sites are commonly labeled as S0 (below T75), S1 (between T75 and V76), S2 (between V76 and G77), S3 (between G77 and Y78), and S4 (between Y78 and G79). The permeation is forced to occur in a single file fashion as a hopping of an ion from one site to another site.

The filter plays a role in both ion selectivity and modulating the current. The latter corresponds to random-like switching (gating) between zero and finite values of the current. Once the inner gate is opened, the current is regulated by small structural rearrangements. They are responsible for different gating processes, such as the *C-type inactivation* and the *modal-gating*, from which complex patterns of ion current arise [13,14,16,17,28–30]. The C-type inactivation corresponds to very long inactive (zero current) time intervals under steady-state conditions. The modal gating is associated with three different modes of the single-channel currents in KcsA. Two modes correspond to a high and low probability of the pore to be in the conducting (active) state, respectively. Third mode is a high-frequency flicker mode representing in bursts of fast switching back and forth between active and inactive states [13,14].

The inactivation in the KcsA channel is a common feature in functioning potassium channels, including eukaryotic ones [10]. Therefore, the C-type inactivation has been extensively studied using a variety of different experimental techniques such as crystallography, NMR, ssNMR, fluorescence measurements, and computational studies, leading to several hypotheses reflected in the recent detailed review [6]. A combination of structural (X-ray) studies and physiological measurements of the wild type (WT) of KcsA and its different mutants suggests that several residues behind the SF could be involved in filter's structural rearrangements during the inactivation [9,14,15,18,19,29–33]. These studies led to the suggestion of four channel's states with an open or closed inner gate and a conducting or non-conducting SF [4,11]. One of the hypotheses [6,11,34] suggests that the activation by opening the inner gate simultaneously alters the SF via allosteric coupling [35,36]. This coupling leads to a slow (on a time scale of seconds) collapse of the SF to a non-conducting configuration. Although structures corresponding to an inactive channel with closed and open inner gate were reported [11,28,30], a structure of an active channel with an open gate and a conducting SF is still missing. Note that the canonical structure 1K4C [28] with a conductive configuration of the SF has a closed inner gate. Another set of experiments used mutagenesis of residues in the SF and demonstrated that the ion occupancy in specific sites controls the inactivation [37–39]. This result leads to the second hypothesis that the SF alone could play the role of an "inactivating gate" without the involvement of the inner gate [6]. This hypothesis tightly links to experimental observations that the SF's conformational dynamics in the WT KcsA and its mutants govern gating properties in the KcsA channel [14,40,41]. Although these two hypotheses are sometimes considered controversial [6,42], they could coexist and reflect the complexity of the KcsA channel.

An additional complication to this gating-permeation picture is the dependence of the K^+ current and the filter rearrangements on the extracellular K^+ concentrations, common among numerous K^+ channels [9,10,38,43,44]. The probability of the inactivation grows with decreasing K^+ concentration. This effect has been suggested to link to a "foot-in-the-door" mechanism in which an ion resident in the filter stabilizes the conductive conformation and reduces the inactivation probability [9,16,38]. The exact location of the binding site responsible for the effect is unknown. However, it is suggested such site can be located either at the extracellular mouth or in the central region of the selectivity filter [10,38,45].

In the majority of these studies static (crystallographic) X-ray structures were used for describing the function. However, these static pictures do not provide details of the essentially dynamical picture of the inactivation. Therefore, general mechanistic knowledge of the gating behavior, which comprises transitions between various states, remains obscure [4,10,11,13]. Recent applications of solid-state NMR [41], 2D IT spectroscopy [2], and florescences measurements [46] for analyzing channel dynamics could address the uncertainties in functional relevance of crystallographic structures. However, a mechanistic picture of the filter's rearrangements with simultaneous dynamical analysis of ions and water molecules is beyond the current experimental techniques. Molecular dynamics (MD) simulations offer valuable tools for exploring dynamical properties at the atomic level [47–49]. For example, MD helped discriminate between "knock-on" and "snug-fit" mechanisms of the permeation in the KcsA channel [50]. In turn, the structural study [51] recently resolved some controversy in MD simulations [2,3] on water involvement in knock-on mechanisms.

The inactivation hypotheses were also discussed by applying MD approaches [34,52]. These computational studies concluded that the activation via opening the inner gate affects the low site S0 in the SF by enhancing the permeation [34] and controlling the SF's stability [52]. The latter result leads to new perspectives [53] for the inactivation mechanism as a process tightly controlled by the inner gate, which could be in different partially open states [52]. The dynamics near SF becomes less important in this picture. A quick collapse of the SF in the case of a widely open inner gate was observed [52]. The collapse happens on much shorter than broader time scales of the inactivation, which can be the order of seconds. The same time scale for collapsing the SF was reported in recent unbiased simulations of a similar open structure [54]. The former result [34] partly supports these new perspectives as it shows that configurations of the inner gate affect the permeation. However, a collapsed SF has not been reported for the performed biased MD simulations for the open inner gate [34]. As mentioned above, a crystallographic structure of the KcsA channel with a conducting SF and an open inner gate is not available, so such a structure was created in silico [34,52] using combinations of the reported structures [55]. Differences in structures used for creating proteins with an open inner gate could explain some contradictions in those MD approaches [34,52].

Heer et al. [34] also reported that the permeation barrier in the canonical (a conducting SF and a closed inner gate) structure 1K4C [1] is too high to consider its SF configuration as conducting. This conclusion was derived from biased simulations using the umbrella sampling method [56]. The obtained barrier was found to be too high for observing the permeation rate according to experimental recordings [57]. This result is in line with the work reported earlier by Fowler et al. [58]. In contrast, other unbiased MD simulations [59–61] confirmed the conducting state of SF. Note that the SF of structure 1K4C was used in the majority of the simulations mentioned above. Two major factors could explain such discrepancies. The first factor is the use of either biased or unbiased MD approaches. The second factor is defined by differences between obtained in silico structures with an open inner gate. While generating a new structure in silico applied a tight control of SF backbones and ions' and water molecules' locations, other residues were not over-sighted. In biased approaches, just one or two so-called collective variables (typically ions locations) were considered assuming that the dynamics of all other variables (water molecules and residues) can be averaged out. Yet, in unbiased approaches some constraints are applied on the protein during MD simulations.

Thus, conformations and behavior of many residues, especially in the region of the SF, were kept out of the consideration despite the experimental studies that identified a number of residues strongly altering the inactivation and gating [9,14,15,18,19,29–33]. A series of papers by Cordero et al. [9,14,29] suggested that the stability of the SF depends on a hydrogen-bond (H-bond) network formed by the triad of residues E71-D80-W67. In particular, the substitution of glutamate E71 with alanine A71 suppresses the inactivation, and the conduction is observed even in low K^+ concentrations [14]. Therefore, there is a gap in understanding how states of this triad are linked to the permeation. In this manuscript, we aim to provide a mechanistic picture of rearrangements in the WT KcsA protein and discuss the mechanisms by which residues behind the SF interacts with the backbones of the SF, and ions and water molecules within. This picture is an essential piece of the inactivation puzzle and in addressing issues of MD biased simulations.

A large number of residues in the SF region of the KcsA protein means that a brute force (combinatoric) consideration of all possible combinations of different residues states is unrealistic. The state-of-the-art microseconds MD simulations [52,54] show that structure 1KC4 adapts one of the multistable states and no rearrangements of residues behind the SF were reported. In the present work, therefore, we first conduct a comparative analysis between the WT protein and different mutants (E71A, Y82A, R64A, and L81A) (see Figure 1) where key residues are replaced by the short, weakly interacting alanine. The selection of the mutated structures is based on previous experiments [9,14,15,18,19,29–33] which reported different probability of the inactivation. MD simulations were combined with biased free-energy methods, well-tempered metadynamics [62], and statistical analysis. The biased simulations introduce additional perturbations into the protein and, therefore, verify the stability and thermodynamics of different states of the SF. The results of MD simulations are critically assessed against published experimental and computational investigations. The study was designed to unveil the complex dynamics that underlie the permeation path in the WT KcsA protein and has allowed us to identify a cooperative network of dynamically interacting residues located near the SF. Note that preliminary results of this study were reported in work [63].

In this paper, first, an analysis of residues' dynamics in mutated structure E71A is presented. The relationship between conformational changes at the SF and rearrangements of residue D80, located at the channel's outer entrance, is explored. Second, a network of residues, which affect the ion permeation, is identified by comparing the dynamics of proteins WT, Y82A, R64A and L81A. Third, a thorough description of the network dynamics, including energetics of transitions in the network, and its influence on the filter structure and the ion permeation is presented.

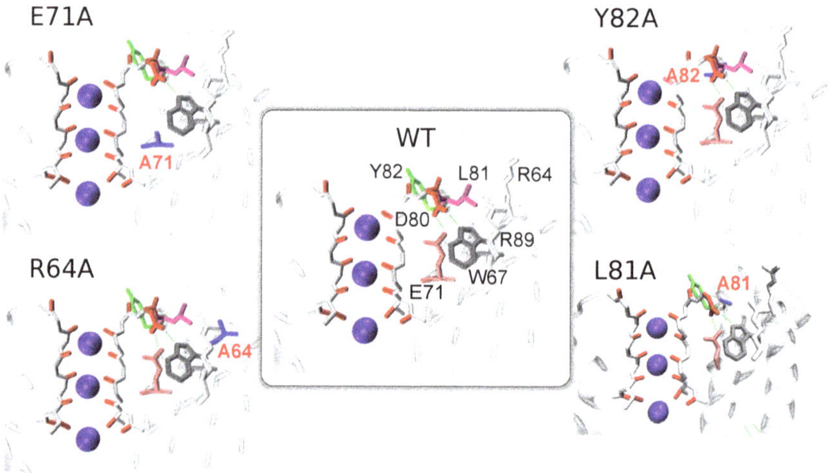

Figure 1. A region near the SF in the different proteins: WT, E71A, Y82A, R64A, and L81A, is shown. With the exception of the mutated residues, the other residues are in the X-ray conformation [28]. Ions are shown as purple spheres interacting with oxygen atoms (red color) of residues in the SF. The key residues are highlighted by different colors, mutated residues are shown in blue.

2. Methods

2.1. Setup of the Simulations

The simulations were performed using NAMD 2.8 and 2.9 [64] in the NPT ensemble with pressure 1.01 bar and temperature 310 K. A multiple timestep algorithm was used [65,66]. In the case of unbiased simulations the integration step size was 1 fs, nonbonded nonelectrostatic interactions were calculated every 2 fs, and electrostatic forces [67] every 4 fs. In biased simulations, the step size was 2 fs, nonbonded nonelectrostatic interactions were calculated every 2 fs, and electrostatic forces every 6 fs. The CHARMM27 force field (FF) was used for the protein, with a modification in the Lennard–Jones term to represent the interaction between K^+ and the carbonyl oxygens of the protein, CHARMM36 for the lipids, and TIP3P for water were applied [50,68–72]. The system was prepared by embedding the X-ray structure (pdb code 1K4C; solved at 2 Å resolution [28]) with 2 K^+ in the SF and 1 K^+ in the cavity, in a membrane patch of 222 molecules of 1-palmitoyl-2-oleoylphosphatidylcholine (POPC), and solvated by 17740 water molecules [73–75]. A potassium concentration in the aqueous phase of 0.2 M was obtained with 63 K^+ ions, and the system was neutralized by 75 Cl^- ions. The ions were distributed over the whole simulation box. Relaxation of the system and preparation of the mutants is described in Supplementary Materials.

Coordinates, if not otherwise stated, were considered every 2 ps, ignoring an initial equilibration period of 1 ns.

2.2. Collective Variables and Order Parameters

Collective variables (order parameters) used in this work are defined as follows. (i) Variables ψ_{76} and ψ_{81} are the ψ dihedral angles measured for residues indicated in the subscripts, and they follow the standard definition. (ii) Variable $\chi1_{81}$ is the $\chi1$ dihedral angle of the L81 residue, which follows the standard definition as well. (iii) Variable SC_{80} is the position of the D80 side chain considered as the distance between C_γ atom of D80 and a reference atom, C_α of A73. Note that the latter residue shows the lowest fluctuations in RMSD analysis. (iv) The distance D80–R89 is between C_γ atom of D80 and the C_ζ atom of the closest R89 residue in the quaternary structure. (v) SF length, the length of the TVGYG sequence, is measured as the distance between the C_α atoms of residues T75 and G79. (vi) The distance R64–SF is measured between C_ζ atom of R64 and the center of mass (COM) of the selectivity filter. (vii) The distance E71–D80 is between C_γ atom of D80 and the H-bond donor oxygen of E71. (viii) The coordinates z_{K1} and z_{K2} are the z coordinates of the K^+ ions bound to the filter (ions labeled as K1 and K2 in Figure 5); the coordinate system has been centered with respect of the COM of the SF, in order to remove the components associated with the protein diffusion in the membrane.

The COM of the SF was defined by the atoms N, C_α, and C of residues from 74 to 78 of all four subunits.

2.3. Free Energy Calculations—Metadynamics

Different approaches are used to enhance the sampling when high energetic barriers between states do not allow an appropriate sampling for the investigation of rare events and the reconstruction of the free energies. These are often based on non-Boltzmann sampling.

Well-tempered metadynamics (wt-metaD) is a non-Boltzmann sampling method based on a history-dependent bias potential, created as a sum of Gaussians centered along the trajectory of specified collective variables (CVs) [62,76,77]. In wt-metaD technique, the height of Gaussians added is history-dependent, and this dependence is associated with a parameter ΔT having the dimension of temperature. This parameter was adjusted for each simulation. The NAMD package [64] includes module *colvar* for performing wt-metaD. Additional details of the implementation of wt-metaD and the selection of the relevant parameters are reported in the Supplementary Materials, section "Well-tempered Metadynamics".

2.4. Initialization of WT-R64D80 Simulation

For the simulation denoted as WT-R64D80, an equilibrated conformation of WT KcsA simulated for 6 ns was used. During first 20 ps of the relaxation, residues L81 and R64 were restrained. Every residue L81 was restrained towards the flipped state by the harmonic potential with a spring constant of 24 kcal/mol degree^2 and centered on 185°. The harmonic potential (spring constant 20 kcal/mol degree^2 centered on $-160°$) was applied on $\chi 1$ dihedral angle of each R64. Note that the latter restraints were added to speed up the calculation, but are not strictly necessary to obtain the desired configuration. A further 25 ps of relaxation were performed without any restraint.

2.5. Statistical Analysis

The statistical analyses were performed using VMD 1.9 [78] and R software environment [79]. Several packages for R were used in addition to the core functions: bio3d, ggplot2, car, and MASS [80–84].

All the free-energy surfaces (FES) presented in this work were smoothed via cubic smoothing spline (grid length 80) and thin plate spline methods (grid sizes 80 × 80) which are implemented in R packages stats v2.15.3 [79] and fields v6.7.6 [85], respectively.

3. Comparative Analysis of Dynamics of WT and Mutated Proteins

3.1. Considered Proteins

The simulations commenced from relaxed systems, prepared from the X-ray structure solved at 2 Å resolution [28], as explained in the previous section "Methods". The KcsA channel has a tetrameric structure, and the four subunits of the KcsA are referred by capital letters A, B, C, and D. The SF is described as a five-site pore [24,25] through which ions and water molecules move in a single-file fashion. The standard notation of the sites is used: S0 to S4 starting from the outer site. The configurations of the SF are described by a five-character string (from S0 to S4), where a "K" represents a K^+ ion, "w" a water molecule, "0" a vacancy; when a K^+ is present in the cavity a "K" is appended, separated from the filter occupancy by sign "+". For example, the configuration wKwKw+K means the presence of K^+ ions in S1, S3, and the cavity separated by water molecules. In comparison, KwK0K implies the presence of K^+ ions in S0, S2, and S4, a water molecule in S1, a vacancy in S3 and no a K^+ ion in the cavity. Consistent to the previous literature [14,86], the results are described by considering the extracellular region as an outer region and "up" in the frame of reference, while the intracellular region is considered as inner and "down".

Among the numerous mutants, which differ from the WT in the gating behavior, three proteins have been considered: (i) E71A is be resistant to the inactivation, (ii) R64A shows a sharp reduction of the inactivation, and (iii) Y82A demonstrates an enhancement of the rate and extent of the inactivation [14]. It is, therefore, possible to specify a trend in the inactivation probability of these proteins: E71A $<$ R64A $<$ WT $<$ Y82A. An additional mutant, L81A, was created for testing the roles of residues L81 and R64, and their coupled motions.

3.2. Dynamics of Mutant E71A

The link between residues E71 and D80 is considered to be an important one for KcsA functioning. A special patch in the force fields was introduced to tune the link for observing ions' conduction [86]. However, the mutation of glutamate (E71) to alanine (A71) does not affect the conductivity and, moreover, it suppresses the inactivation. This observation means that other residues play an essential role to keep the SF in a conducting configuration.

The mutation by replacing glutamate E in position 71 by alanine A results in the structure E71A which was studied experimentally by Cordero et al. [14]. The authors demonstrated that the permeation path undergoes large conformational rearrangements in the non-inactivating mutant E71A. The rearrangements primarily occur in the region of V76 residue. Additionally, the authors [14] reported a strong upward movement of residue D80 relative to its position in the WT structure, leading to the "flipped E71A" structure.

For understanding the influence of the mutation on the dynamics and interactions of residues, an unbiased MD simulation of length 24 ns was performed. Several rearrangements in the permeation path were observed during the simulation. The most noticeable changes happened among the residues of the TVGYG sequence in the subunit B. Rotation of the V76-G77 peptide group occurred at 9 ns, and the rotation remained stable until the end of the simulation. Similar transitions have been reported in the literature for both WT and E71A. Many hypotheses [30,58,87–90] have been proposed for explaining the transitions which are usually referred to as "flipping of V76". However, understanding the significance and origin of the transitions is still missing.

The flipping of V76 has been suggested by different authors to be able to generate non-conductive conformations associated with the C-type inactivation or flicker mode [30,88]. We performed various permeation tests on V76 flipped conformations of E71A and WT by performing unbiased simulations with two ions in the cavity (see Supplementary Materials, section "Permeation in the V76 flipped configurations of E71A and WT"). The simulations revealed that reverse transitions of V76 occurring easily in the case of K^+ permeation. This result supports the hypothesis of Domene et al. [89] that flipping of V76 is not responsible for the C-type inactivation. Furthermore, the observed conductivity suggests that the flipping of V76 alone is not sufficient even for short-living inactive states, which are associated to the modal-gating, and that additional conformational readjustments are necessary for generating meta-stable non-conductive states.

The simulation of E71A showed that interactions between D80 and an arginine nearby (R89) could trigger structural rearrangements of the filter. D80 side chains, which are negatively charged, demonstrated relatively large fluctuations towards the extracellular region (see Supplementary Materials, Figure S2). These fluctuations are promoted by strong inter-domain electrostatic interactions with positively charged arginines R89. The interaction between the residues D80 and R89, and the corresponding rearrangements of the SF, are illustrated in Figure 2. The conformational space in Figure 2a is defined by the three order parameters (conformational changes for subunit B only are shown): (i) the dihedral angle ψ of V76 (ψ_{76}); (ii) the position of the D80 side chain (SC_{80}); (iii) the distance between D80 and nearest R89 (D80–R89). Initially, the dynamics of D80 and R89 appear uncorrelated (blue clouds). Some correlations arose (light blue cloud) as the time advanced because of an intermittent creation of a H-bond (D80–R89 distance is around 3 Å) between D80 and R89. Note that similar H-bonds between D80 and R89 have been reported in the literature also occurring for the WT structure [91]. The presence of the D80–R89 H-bond in E71A protein is associated with a small drift in the position of D80, SC_{80} is changed from 13.5 to 13.8 Å (see Supplementary Materials, Figures S3 and S4, for more details). The temporary strengthening and stabilization of the H-bond was accompanied by a distortion of the filter structure (in Figure 2a clouds blue to green, and in Figure 2b structure green to colored). Residue V76 assumed a *partially flipped* conformation ($\psi_{76} \approx 50°$) in the distorted structure. This observation is an important result since it demonstrates that the backbone structure of the sequence GYGD is rigid enough for delivering a perturbation from D80 to the V76-G77 peptide group. It is shown below that the rigidity of the GYGD backbone strongly affects the SF flexibility.

Time series reported in Supplementary Materials (Figure S3) further demonstrated that, in turn, V76 partial flipping affected the permeating K^+ ions, causing an inward shift of the outermost ion K1. Thus, ions' dynamics are linked with the dynamics of residues behind the filter, D80 and R89. The partially flipped conformation of V76 appears to be unstable and evolved into a complete flipping of V76. The D80–R89 H-bond caused additional small transitions in the TVGYGD sequence until a slight movement of the D80 towards the extracellular side (see Figure S3 in Supplementary Materials and Figure 2b colored to yellow) restored the initial uncorrelated motions of the D80 and R89 (red clouds in Figure 2a) causing a breakage of the H-bond.

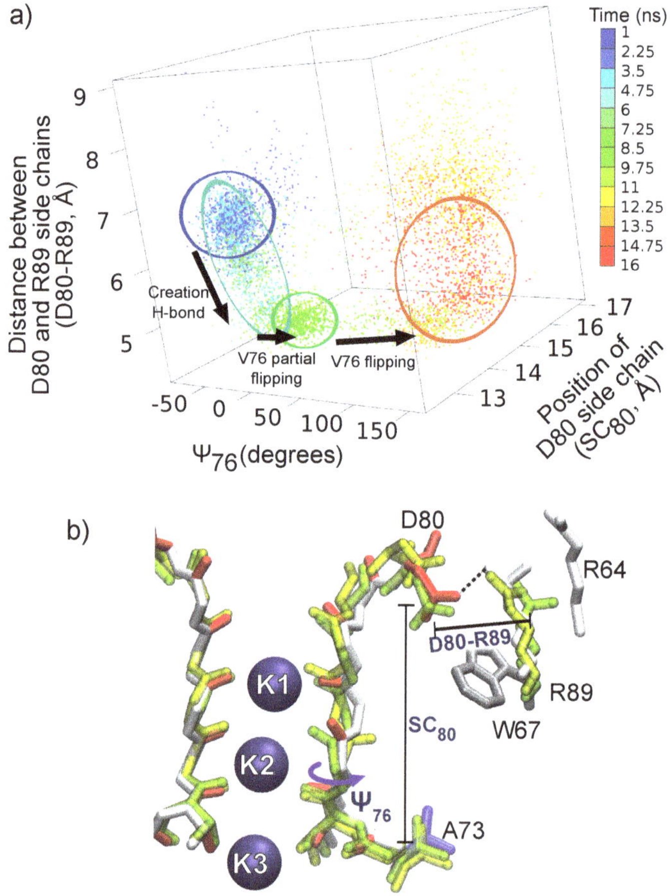

Figure 2. The influence of residues D80 and R89 on structural rearrangements in mutant E71A. (**a**) The evolution of the system (subunit B, initial 1 ns ignored as the relaxation interval) revealed that the stress induced by the D80–R89 H-bond led to rearrangements in the filter structure (the flipping of V76) and to an outward transition of D80. The conformational space is defined by a set of the order parameters (see "Methods"): (i) SC_{80} the position of side chain D80; (ii) D80–R89 distance, where residue R89 belongs to the neighboring subunit; and (iii) ψ_{76}. Time evolution of the system in the conformational space is coded by color scale shown in the colorbar. (**b**) Superposition of snapshots from the simulation of E71A: an initial configuration (green drawing); a configuration with the D80–R89 H-bond and partially flipped V76 (colored drawing); and a configuration at the end of the simulation (yellow drawing).

Although the described path is one among many available toward a V76 flipped configuration in protein E71A, these results demonstrate that the creation of H-bonds with residue D80 can trigger structural rearrangements which propagate to the filter because of the relative rigidity of the GYGD sequence backbone. The arginine R89 is able to promote the triggering transitions by creating a strong H-bond with residue D80. In the following sections, further evidence is presented for confirming that all the residues which can form H-bonds with D80 play a significant role in conformational rearrangements of the permeation path.

3.3. Correlated Dynamics of L81 and R64 Residues

The results of the previous subsection indicate that residue E71 plays an essential role in the inactivation and, therefore, in the dynamics of the WT protein. In mutant E71A, alanine in position 71 does not form bonds with D80 and residue D80 is very flexible. In contrast, in the WT protein, residue D80 is restrained by a strong link between D80 and E71. For the identification of residues that affect the permeation path, we performed a comparative analysis of three different proteins in which E71 is present. The selected proteins are the WT protein and mutants Y82A and R64A. These mutants show distinct behaviors for the inactivation: Y82A has significantly higher, and R64A has reduced the inactivation probability in comparison to WT. All three structure were simulated starting with the same initial configuration (excluding mutated residues) for different but comparable intervals: 38 ns, 28.5 ns and 23 ns for WT, Y82A and R64A respectively. Note that in the WT protein, residue R64 directly interacts with L81, which is a neighbor of residue Y82 (Figure 1).

The root mean square displacements (RMSDs) of the backbone atoms of each residue reveal residues which showed different behaviors across the three selected proteins; the X-ray structure of WT was used as the reference [28]. The results are reported in Figure S5 in the Supplementary Materials. The RMSDs analysis shows that fluctuations of the arginine R89 are wider in the proteins with a higher probability of inactivation, WT and Y82A, than in R64A. This observation additionally supports the hypothesis of a particular role of this arginine in the conformational variability of the pore. However, residue R89 in proteins WT and Y82A show similar RMSDs, and the difference in RMSDs of R89 in R64A and WT structures is relatively small. These facts imply that the dynamics of R69 by itself cannot account for the substantial diversity in the inactivation between these three proteins.

A closer inspection reveals the importance of second arginine residue, R64, which has relatively large RMSDs in WT and Y82A. The mutation of this arginine with alanine in structure R64A leads to a significant reduction of the RMSDs of the residue in position 64. In WT and Y82A proteins, arginine R64 can approach and interact with D80 and create strong H-bonds similarly to R89 in mutant E71A (Figure 3a). The possibility of a H-bond between R64 and D80 is important considering that R64 is located relatively far from D80 in the static structure provided by X-ray experiments [28] (D80-R64 distance = 9.3 Å). Residue R64 fluctuated over wide ranges and, more importantly, it can destabilize linkages between the triad of E71-D80-W67 via the interaction with residue D80 (Figure 3a). This interaction occurs more prominently in mutant Y82A, the simulation of which ended with a broken triad E71-D80-W67 in two subunits. As a result of the R64–D80 interaction, residue D80 can rotate around the dihedral angle $\chi 1$ and such rotations were observed a few times during simulations (see Figure S6 in the Supplementary Materials). In Y82A and WT proteins, the flexibility of D80 promoted by R64 leads to several multistable configurations, one of which includes a broken E71–D80 link. Note that this link is stable though the whole simulation of mutant R64A. Thus, residues R64 in WT and Y82A proteins play the destabilizing role.

In both WT and Y82A proteins, arginine R64 can interact with D80, but these structures demonstrate different inactivation behavior. Our simulations indicate that the difference in the inactivation has a dynamical origin. Residue R64 moves faster and creates quicker a H-bond with D80 in mutant Y82A than in WT. The rate of H-bond creation depends on the conformation of the leucine in position 81 (L81). This rate primarily controls by the rotation of L81 side chain, which can open by *flipping*, when angle $\chi 1_{81}$ changes from $-63°$ to $185°$, or obstruct, when residue L81 is in that conformation found in the crystallographic structure, the path toward forming the D80–R64 H-bond (Figure 3a). Conformational changes of L81 have, therefore, a critical regulatory role in the dynamics of residue R64.

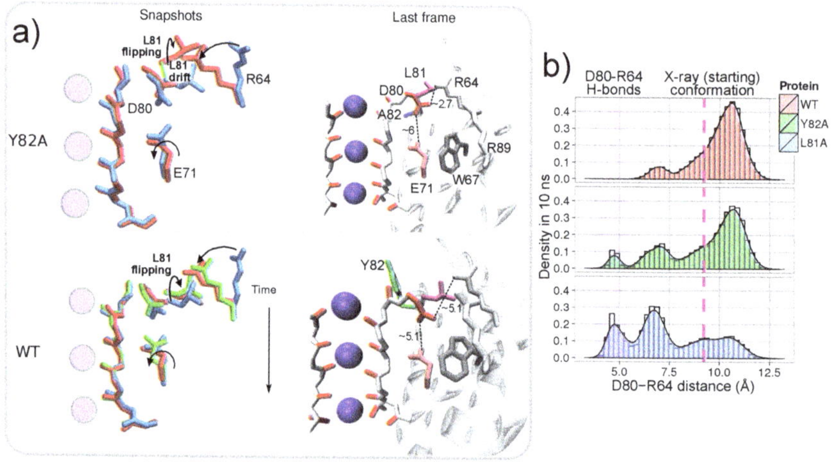

Figure 3. (**a**) The disruption of E71–D80–W67 linkages in Y82A and WT, promoted by the D80–R64 interaction. Snapshots and the final configurations for simulations of Y82A and WT are reported. Residues in the snapshots are superimposed with respect of the heavy atoms of the SF. The color sequence in the snapshots is (i) blue, an initial state; (ii) green, a transition state (distinguished only for the WT and characterized by the E71 $\chi 1$ angle of 120 degrees); and (iii) red, E71-D80-W71 linkages disrupted. Distances in the figures are reported in Å. (**b**) Comparison of the probability density of the D80–R64 distance in WT, Y82A, and L81A calculated for 10 ns of simulation. The initial D80–R64 distance is 9.3 Å and shown with the dashed magenta line. Distance D80–R64 which is less than 5.2 Å indicates the formation of the D80–R64 H-bond. Data from all the four subunits were used.

In turn, the dynamics of L81 are associated with additional readjustments in the amino acid sequence L81-X82-P83-V84, roughly definable as pivoting around the residue in position 82 (*X82-pivoting* (Figure 3a). Collective motions of this sequence can promote the flipping of L81 and a small drift of its backbone. The X82-pivoting is different in three considered proteins. The main differences are reflected in the RMSDs of residues surrounding residue 82 (X82) (Figure S5 in the Supplementary Materials). In all the three proteins, the RMSDs are similar for X82 (where X is tyrosine Y in the WT and R64A and alanine A in Y82A). In contrast, the RMSDs of the surrounding residues (L81, P83, and V84) correspond to the inactivation probabilities R64A < WT < Y82A, that is, the RMSDs are larger for Y82A and smaller for R64A than for WT. The lowest RMSDs for mutant R64A are due to the absence of a residue in position 64, which is capable of interacting with L81 via X82-pivoting. In WT and Y82A proteins, the R64–L81 interaction is controlled by bulky tyrosine Y82 and non-bulky alanine A82, respectively. In the WT protein, therefore, the motion of L81 is slower and more limited than in mutant Y82A, while in Y82A, the dynamics of L81 are faster and accompanied by a noticeable backbone drift (Figure 3a). Note that the described X82-pivoting can furthermore explain the conformational rearrangements of Y82 suggested in the experimental investigation of the C-type inactivation [17].

Thus, the mutation in position 82 changes the dynamics of residues close to the filter region, mainly affecting the conformation of L81. The enhancement of L81 transitions in the deep inactivating mutant Y82A causes the promotion of D80–R64 interactions because the dynamics of L81 and R64 are strongly coupled. Note that the X82-pivoting also alters the dynamics of residues V84 which can access D80 in a similar manner as R64. A comparable influence of V84 on D80, therefore, can be hypothesized. However, if such influence exists, it was masked by a stronger R64–D80 interaction.

For verifying the regulatory role of L81, mutant L81A (Figure 1) was additionally considered. The probability of the creation of the D80–R64 H-bond was compared for three proteins: WT, Y82A, and L81A. All the proteins were simulated with the same initial configuration, and the probability was calculated for the same time interval 10 ns. Consistently with the presented above results, the simulations confirm that the rate of the H-bond creation depends on residues L81, with a trend in the probability L81A > Y82A > WT (Figure 3b). The probability is larger in Y82A with respect to the WT by the enhanced fluctuations of L81. The probability becomes even larger in L81A when L81 is directly substituted by the small alanine which interferes less with R64 motion.

The results of this subsection show that proteins WT, R64A, and Y82A differ from each other in the dynamics of a few residues: primarily arginines R64 and R89, and leucine L81 that regulates the D80–R64 interaction. The cooperative dynamics of these three residues have a destabilizing effect on the triad E71-D80-W67 and, therefore, affect the pore region. Note that glutamate in position 71 (E71) has a non-trivial influence on the inactivation. In the absence of E71, mutant E71A is very flexible, but the inactivation is suppressed entirely. The presence of E71 is, therefore, essential to observe the inactivation, as E71 limits the flexibility of the pore region and strongly affects the motion of D80.

4. The Interactions of Residues and Ions in the WT Protein

4.1. Influence of Arginines R64 and R89 on D80, the SF and Ions

For understanding the action of arginines R64 and R89 in the WT protein, conformations with residue R64 close to residue D80 were investigated by a simulation started with a particular initial configuration. In order to enhance the probability of the R64-D80 interaction, an unbiased simulation (denoted *WT-R64D80*, duration of 45 ns) of the KcsA WT protein was commenced from a conformation with residues R64 were near D80 in all the four subunits. Details of how the initial configuration was obtained are given in the section "Methods". Figure 4b shows the initial configuration characterized by the filter occupancy wKwKw+K; the flipped state of L81 and R64 is close to D80 in all subunits. Note that in subunit C, residue R64 forms a H-bond with D80 during a short equilibration in a preparation stage (see Figure S7 in the Supplementary Materials, starting point).

Two positively charged arginines R64 and R69 can exert a sufficiently strong combined upward force on negatively charged residue D80 to overcome the strong downward attraction toward E71. From the beginning of simulation WT-R64D80, this force resulted in a large mobility of the pore region. A long breakage of E71-D80-W67 linkages (for 17 ns) occurred in subunit C as well as brief disruptions of the linkages in other subunits were observed. Conformational rearrangements in residues and content (K^+ ions and water molecules) of the SF accompanied these disruption events. Representative snapshots of changes in the SF are shown in the Supplementary Materials, Figures S7 and S8. The rearrangements observed in the subunit C were analyzed using three order parameters: angle ψ_{76}, distance SC_{80} and the length of the TVGYG sequence of the subunit, SF length. Figure 4 shows the trajectory, which reflects the time evolution of the system, in the conformational space defined by these three order parameters for the first 22.5 ns of the simulation. Initially, several transitions of residues V76 (angle ψ_{76} switches back and forth between $-50°$ and $145°$) were observed. These transitions demonstrate the inherent flexibility of the V76/G77 peptide group, which is sensitive to changes in the SF. The trajectory also shows that after 5 ns the E71-D80-W67 triad broke and residues D80 moved outward (distance SC_{80} changes from 13.5 Å to 15.5 Å). All these changes were promoted by residue R64. Residue D80 accommodated an upward state with H-bonds formed between either D80 and R64, or D80 and R89, or D80 and both arginines (see Figure S8 in Supplementary Materials). This upward state of D80 caused stretching of the TVGYGD sequence of the SF (SF length increases), and residue V76 switched to a meta-stable flipped state. The described changes correspond to the transition from the state **A** to the state **B** in the conformational space (Figure 2). The two-dimensional density for distance SC_{80} and angle ψ_{76} shown in Figure 2b emphasizes a meta-stable character of the distorted state **B** and its dependence on the position of D80 side chain.

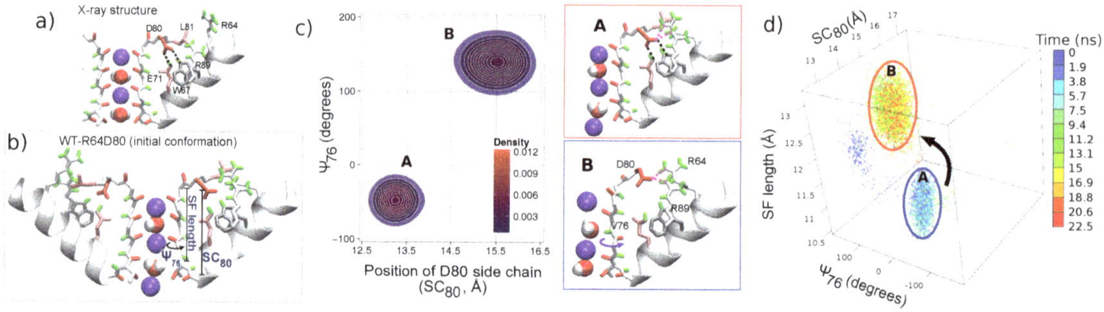

Figure 4. Panels (**a**,**b**) compare of the X-ray structure of the WT protein (a) and the initial configuration for simulation WT-R64D80 (b). Panel (**c**) shows the correlation between the states of D80 and the SF by means of the two-dimensional probability density for position SC_{80} and angle ψ_{76} (see "Methods") during first 22.5 ns of simulation WT-R64D80, i.e., before ion configuration KwK0K was reached. Two meta-stable states are denoted by letters **A** and **B**. Snapshots corresponding to each of the two states are shown on the right side panels. State **A** is the initial state in which E71-D80-W67 linkages were present. State **B** is characterized by an outer movement of D80 which followed by the break of E71-D80-W67 linkages and TVGYGD rearrangements. Panel (**d**) reports time evolution of subunit C in the conformational space defined by (i) ψ_{76}, (ii) SC_{80}, and (iii) the length of the TVGYG sequence, SF length. The trajectory in the conformational space is coded by color scale shown in the colorbar. Letters **A** and **B** indicate the same states as in panel (c).

Figure 5 and Figure S9 in Supplementary Materials show significant consequences of the distortions in subunit C on the elements bound to the filter, and in particular, on the permeating K^+ ions, which facilitate in spreading the distortions among the other three subunits. The changes in the permeation path can be characterized by the correlation between the positions of K^+ ions in the SF. Let us stress that strongly correlated motion of ions was considered as being the fundamental feature of the knock-on mechanism of the permeation in previous works [92,93]. Simulations started from the X-ray configuration demonstrate the presence of such correlated dynamics of ions in the SF (see Figure 5a): Pearson's coefficient is large (around 0.75) and the positions of ions K1 and K2 are linearly correlated. In the distorted state observed in simulation WT-R64D80 (state **B** in Figure 2b), the correlation between K^+ ions is lost; Pearson's coefficient is close to zero (Figure 5a). Ions in the SF become more flexible in the binding sites, that leads to weakening in the spatial definition of the K^+ sites (Supplementary Materials, Figure S9). An unexpected transition of the innermost ion (K2) toward the intracellular side was observed (wKwKw+K $\rightleftarrows$ wKw0K+K, Figure 5b). This transition occurred in the reverse direction with respect to the permeation path. Note that such a transition was not observed in simulations started from the X-ray conformation of the WT protein. The observation of the inverse transition is particularly important because it reveals the influence of protein distortions on single K^+ permeation events. This influence additionally can explain the different free-energy barriers obtained for the permeation path using biased approaches which induce distortions of some parts of the KcsA protein [58,86,90,94,95].

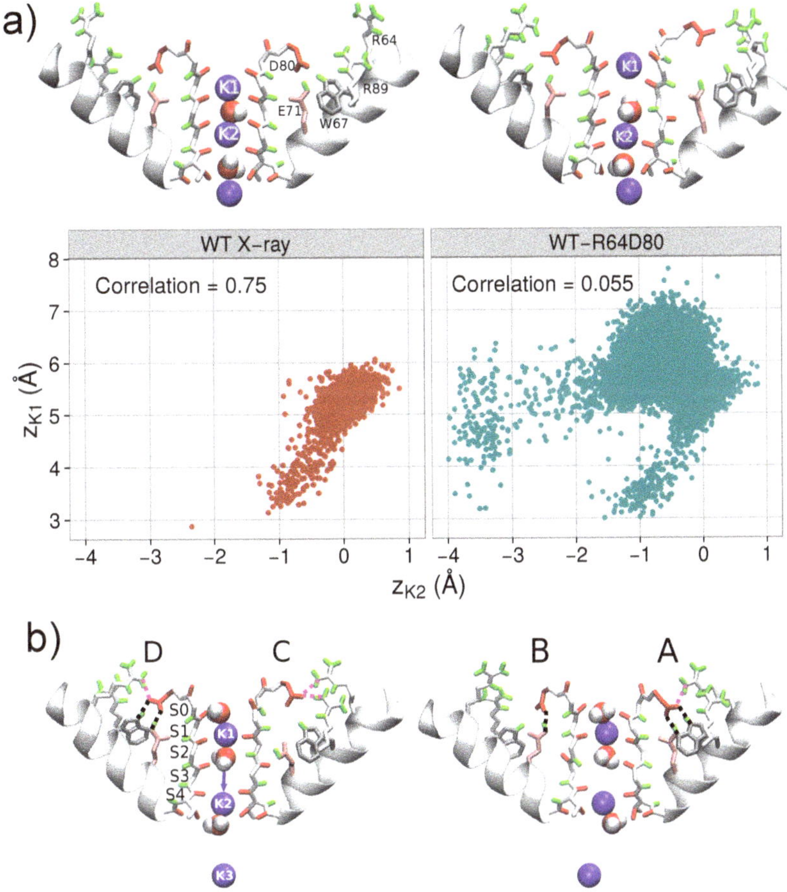

Figure 5. Panel (**a**) illustrates the correlation of the position of permeating K^+ ions (K1 and K2) for two different conformations of WT: (i) the X-ray conformation in which R64 was far from D80 in all subunits; and (ii) the confirmation used in simulation WT-R64D80 with R64 near D80 in all the subunits. Only a part (durations of 7 ns for (i) and 15 ns for (ii), respectively) of the simulations with identical filter occupancy wKwKw+K were considered. For simulation WT-R64D80 the part corresponds to broken E71-D80-W67 H-bonds. Representative snapshots and z-positions of two ions in the SF on the state plane of z_{K1} and z_{K2} for each configuration are shown. Panel (**b**) depicts a configuration of the channel after the inward transition of the innermost ion K2 (wKwKw+K $\rightleftarrows$ wKw0K+K) occurred. All four subunits denoted by letters A–D are shown.

The described distorted state of the SF is observed during the initial part of simulation WT-R64D80. In the later stage, ions underwent several further rearrangements. One of the rearrangements is the ion (K3) from the cavity enters the SF (wKwKw+K $\longrightarrow$ wKwKK) that leads to the re-establishment of E71-D80-W67 linkages. Then, the transition of the outermost ion (K1) to the site S0 (wKwKK $\longrightarrow$ KwK0K) led to a configuration close to those observed in the conductive state of the X-ray structure. The latter result suggests that the conformation in which a K^+ ion is bound to site S0 stabilizes the filter structure.

Thus, simulation WT-R64D80 demonstrates that conformational changes of the SF are dependent on a strongly correlated network of residues, in which aspartate D80 plays the central role. States of D80 with the broken E71-D80-W67 triad are promoted by the combined action of arginines R64 and R89. Furthermore, these states of D80 can destabilize the SF and cause filter's distortions. The latter affects the dynamics of ions and can prevent the permeation of ions. In turn, ions permeation through the SF can either enhance or suppress the destabilization effect.

4.2. Energetics of the Arginine Motions

Simulation WT-R64D80 was started from a particular initial configuration, and the observed changes in the SF are transient. In this subsection, therefore, the energetics of the changes are studied using a technique called well-tempered metadynamics (wt-metaD). Wt-metaD (see Supplementary Materials for further details) is a theoretical method which belongs to the family of the biased methods and has been successfully applied for both to accelerate the observation of rare events and to reconstruct free energy surfaces (FES) [62,76].

The analysis of the dynamics of mutants and simulation WT-R64D80 demonstrate that the interactions of two arginines R64 and R89 with D80 can trigger rearrangements which change the shape of the channel pore, hence alter the ion permeation in KcsA. The dynamics of R64 is strongly coupled with leucine L81 which regulates the D80–R64 interaction. Therefore, the motion of R64 needs to be analyzed together with the motion of L81. Arginines R89 is not directly controlled by neighboring residues and can be studied alone. Energetics of the motion of R69 are described in the Supplementary Materials (section "Energetics of the arginine motions").

For characterizing the dynamics of R64 and L81 the two-dimensional FES was calculated for the following order parameters: (i) the distance between R64 side chain and the center of mass of the SF (*R64–SF*), and (ii) angle $\chi 1_{81}$ (more details in section "Methods" and Supplementary Materials). The total sampling length of the wt-metaD simulation was 122 ns. The computed FES, shown in Figure 6, confirms the interplay between R64 and L81, and the regulatory role of the latter. There are several multi-stable states on the two-dimensional FES. State S_a is with non-flipped residue L81 ($\chi 1_{81} \approx 297^\circ$) and residue R64 which is far from the SF (R64–SF > 19 Å). This state is close to the X-ray structure of KcsA. It demonstrates that when L81 is in the non-flipped conformation, R64 tends to be away from the filter and D80. On the other hand, when L81 is in the flipped conformation ($\chi 1_{81} \approx 185^\circ$) residue R64 can approach closer to the SF (states S_b and S_c, R64–SF < 18 Å). Residues D80 and R64 form a H-bond in the state S_c. Two minimal-energy paths $S_a \to S_c$ are shown by dotted lines in the FES plane. The first path, highlighted by the magenta line, consists of an initial flipping of the L81 side chain ($\chi 1_{81}$ from ~297° to ~185°) followed by the subsequent movement of R64 towards the SF along a downward gradient. The second path, highlighted by the black line, involves the creation of an initially relatively unstable D80–R64 H-bond which is lately stabilized by the flipping of the L81 side chain. Both paths have a similar energy barrier (5 kcal/mol).

Note that the energy barrier for the inverse transition $S_c \to S_a$ is significantly higher (13–15 kcal/mol) than for $S_a \to S_c$. It means that state S_c corresponds to the global minimum of the FES and the configuration with a H-bound between R64 and D80 and with L81 in the flipped conformation should be observed in X-ray structural studies [28]. An analogous result, with the R89–D80 H-bond in the most probable state, was also obtained by free-energy calculations of the R89 motion for two out of three configurations of ions (see Supplementary Materials, section "Energetics of the arginine motions").

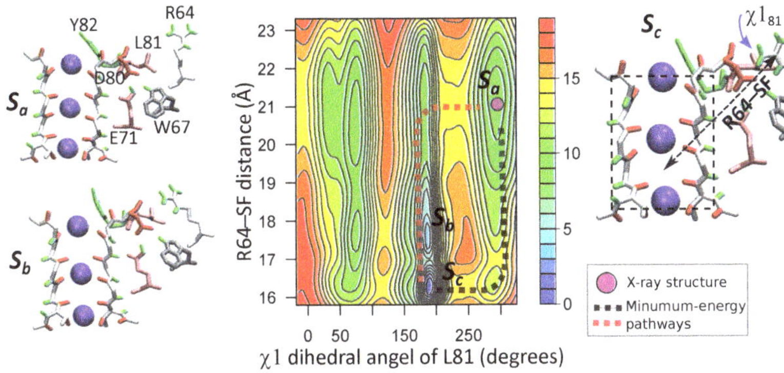

Figure 6. In the middle, two-dimensional FES computed via the wt-metaD approach is shown with respect to the distance between R64 and the SF (R64–SF) and angle $\chi 1_{81}$. In order to aid the visualization, angle $\chi 1_{81}$ is reported in the range $(0, 360)$, instead of the standard $(-180, 180)$. The FES is shown in kcal/mol, lines in the contour plot are drawn every 1 kcal/mol. Configurations of residues for three different multi-stable states: S_a, S_b and S_c, are shown on the sides of the FES plot. These three states are denoted on the FES. State S_a is with non-flipped residue L81 and close to the X-ray structure of KcsA. L81 is in the flipped conformation for states S_b and S_c. Residue R64 forms a H-bound to D80 in the state S_c. The global minimum of the FES is state S_c.

Thus, the most probable positions of arginines R64 and R89 observed in the wt-metaD calculations are different from those in the X-ray structure [28]. These positions difference can be explained by interactions between the protein and surrounding lipids. In fact, numerous experiments indicate that in common with other K^+ channels, KcsA channel is stabilized in the conductive state by the presence of the anionic lipids. In contrast, the channel is primarily non-conductive for the non-anionic lipids [20,33]. Deol et al. [96] revealed, by means of molecular dynamics simulations, that R64 and R89 can form strong, long-lived H-bonds with the head groups of the anionic lipids. Later, this result was experimentally confirmed [33]. This arginine lipid interaction could bring the arginines in positions close to those determined by the X-ray experiment [28].

Our wt-metaD simulations were performed in the absence of anionic lipids, using neutral POPC lipids which as shown experimentally have no specific interaction with KcsA [33]. However, the radial distribution function that characterizes the interaction of Cl^- ions in the bulk with residues R64 and R89, confirms the strong affinity between the arginines and negatively charged species (see Figure S11 in the Supplementary Materials). Because of this affinity, the computed FESs (Figure 6 and Figure S10 in Supplementary Materials) show that the most stable position of R64 and R89 are located in proximity to the negatively charged D80. The presence of the anionic lipids would make this position less probable by additional interactions between the arginines and these lipids. Another factor affecting the arginines is locations of ions in the SF. For example, when ions occupy sites S0, S2 and S4, the probabilities of finding R89, respectively, in proximity to D80 and far from D80 are equal (see Figure S10 in Supplementary Materials). The influence of ions' configuration on the dynamics of R64 is considered in the next subsection.

4.3. Opposite Influence of R64 and a K^+ Ion Bound to S0 on the E71–D80 H-bond

For characterizing the simultaneous action of arginine R64 and ions in the SF on the strong H-bond between E71 and D80, we calculated the FES for the interaction of E71 and D80 in different configurations of the SF. The distance between E71 and D80 residues is selected as the order parameter for the FES. The calculated FESs for arginine R89 (see Supplementary Materials, Figure S10) suggest that a K^+ ion in site S0 stabilizes the E71–D80 H-bond by reducing the probability of the R89–D80 interaction. Free energy calculations were, therefore, performed for two ions conformations: one is "KwK0K" with an ion bound to S0, and the other is "wwK0K" without an ion in S0. Two different positions of R64 with respect to D80, near and far away respectively, were additionally considered. Thus, the FES were calculated for four different configurations of R64 and ions in the SF (see Figure 7). Further details of the FES calculations are reported in Supplementary Materials (Section 7). Note that configurations with a water molecule occupied site S3 ("KwKwK" and "wwKwK") were also considered and the corresponding results are reported in Supplementary Materials (Figure S12). These results are consistent with those presented below.

The FESs in Figure 7 demonstrate the strong mutual influence between the filter occupancy and the position of R64 on the E71–D80 H-bond. The interaction between E71 and D80, therefore, does not merely depend on the nature of the residues and the nearby solvent molecules (water). Still, it originates from many different elements which constitute a strongly interacted (correlated) system.

If an ion is absent in site S0 and simultaneously R64 is far from D80 (Figure 7b), the E71–D80 H-bond is the only stable state in the FES. However, the proximity of R64 to D80 makes breaking the E71–D80 H-bond possible and leads to new meta-stable states without the bond (Figure 7d). The energetic barrier for the breaking the H-bond is relatively small (around 2.5 kcal/mol) and slightly higher (by 0.2 kcal/mol) than the barrier for re-establishing the H-bond. These new states without the interaction between E71 and D80 are close to those that led to distorted configurations in the SF observed during simulation WT-R64D80.

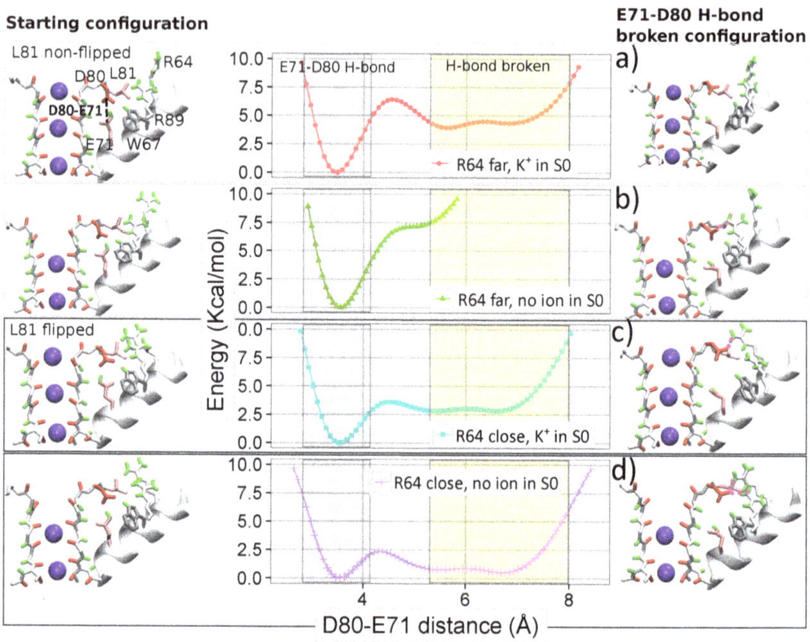

Figure 7. Graphs in the middle show the FESs for the distance between E71 and D80 in different cases: (**a**) R64 is far from D80 and ions configuration "KwK0K", (**b**) R64 is far from D80 and ions configuration "wwK0K", (**c**) R64 is close to D80 and ions configuration "KwK0K", and (**d**) R64 is close to D80 and ions configuration "wwK0K". A starting configuration for each wt-metaD simulation is shown on the left side of the figure. Examples of a configuration with a broken H-bond between E71 and D80 are shown on the right side of the figure for each corresponding initial configuration.

The presence of a K^+ ion in site S0 changes the observed picture. In the case of R64 located far from D80, the presence of an ion leads to new states with the broken

E71–D80 H-bond (compare Figure 7b and Figure 7a). However, the new states are less stable than the state with the H-bond, and the energetic barrier for the bond breaking is high (around 6.5 kcal/mol). When R64 is close to D80, the occupation of site S0 increases the energetic barrier for the breaking the H-bond and makes states without the H-bond significantly less stable (compare Figure 7d and Figure 7c). In this case, the barrier for the bond breaking is around 4 kcal/mol, and for the re-establishing, the barrier is four times less (Figure 7c). Thus, a K^+ ion occupied site S0 opposes the destabilizing influence of arginine R64, favoring the presence of the E71-D80 H-bond. The absence of an ion in site S0 induces a widening of the site that facilitates the approach of R64 to D80 and destabilizing the E71–D80 H-bond.

These results demonstrate broad cooperation between residues and ions in controlling the dynamics of the pore region. Note that the described role of the occupation of site S0 by an ion provides a mechanistic and energetic insight to the hypothesis of a 'foot-in-the-door' mechanism, widely discussed in the literature for interpreting some experimental results [9,16,38]. In particular, the strong dependence of the current on the extracellular K^+ concentration was observed experimentally [9,10,38,43,44]. For explaining this strong dependence, different authors have hypothesized that ion's occupancy in the SF rises for the high concentration of ions and an ion resident in the filter stabilizes the conductive conformation. This hypothesis was supported further by the evidence that ions with a longer occupancy (Rb^+, Cs^+, and NH_4^+) slow down the switching of the ion channel into the inactivated state [9,38]. According to our results, an ion in site S0 appears as the most valuable candidate for playing the role of the "foot-in-the-door".

As previously mentioned, Cordero et al. [14] reported a flipped structure in mutant E71A, where the replacement of glutamate E71 by alanine A71 effectively remove the E71–D80 H-bond that leads to broad outward movement of D80 and large rearrangements in the V76 region. In all our simulations, the WT protein never adopted a similar configuration, even for states with considerable free energies. It implies that residue E71 consistently plays a dual role in shaping the WT ion channel through the strong electrostatic interaction between E71 and D80 and through a steric hindrance of large rearrangements in the region of V76.

5. Conclusions

In this work, a comparative analysis of the dynamics of the WT KcsA ion channel and mutants E71A, Y82A, R64A, and L81A was conducted using molecular dynamics simulations. This analysis helped us to identify a set of residues which control the state of the SF. The interactions between the identified residues and the interdependence between the residues and ions in the SF were characterized by free-energy calculations using well-tempered metadynamics [62]. A detailed description was provided for the residues which most prominently outlined the state of the SF and the influence of the ion permeation path. Our investigations revealed that the permeation path is regulated by a strongly interconnected dynamical system. The system is centered on aspartate D80, which is linked to neighboring H-bond donors, includes ions in the SF and residues located far from the SF. Key features of this interconnected system were described, that provides a consistent and unifying picture for some experimental results on the regulation of KcsA activities. These features are highlighted below.

First, the highly conserved aspartate D80 plays the critical role in changing the structure of the SF by translating broader dynamics of the protein to the filter structure because of the relatively rigid backbone of the conserved sequence GYGD of the SF. Thus, movements of residue D80 can trigger significant rearrangements of the whole pore.

Second, two arginines (R64 and R89) can strongly interact with D80 via H-bonding. This interaction facilitates movements of D80 that triggers the changes in the protein pore. While the D80–R89 interaction was previously described in the literature [91], the possibility of the D80–R64 H-bond and the destabilizing consequences of the combined action of these two arginines on D80 were described for the first time in this work. Between the two arginines, R64 was found to exert the strongest influence on D80, and thus on the ion flow.

Third, the local dynamics of the region behind the filter is regulated by conformational changes of leucine residue L81. These changes, in turn, are linked to collective motions of the amino acid sequence L81-Y82-P83-V84, in particular to a pivoting action on residue Y82. Additionally, the simulations provided the unambiguous evidence for the regulatory role of L81: the flipping of the L81 side chain facilitates the establishment of the D80–R64 H-bond.

Fourth, the destabilization effect of arginines R64 and R89 on states D80 is reduced by the presence of a K^+ ion in the outermost binding site (S0) of the filter since the resulting electrostatic interactions stabilize the conductive structure.

We showed that the interactions between the two arginines (R64 and R89) and D80 induces the breaking the E71–D80 H-bond that could lead to a non-conducting state of the pore. This result provides an explanation of the necessity of the anionic lipids for observing the current in KcsA channel as the lipids can interact with both arginines [96], and this interaction reduces the probability of breaking the E71–D80 H-bond. Additionally, we showed that the occupancy of site S0 by an ion also stabilize the E71–D80 H-bond. The stabilizing influence of the ion bound to S0 offers an important insight into the "foot-in-the-door" mechanism proposed by various authors for explaining the influence of the extracellular K^+ concentration in stabilizing the conductive state [9,38,43].

Our comparison between the dynamics of the WT protein and mutant E71A revealed a vital role of glutamate residue E71 in response to perturbations of the pore region. In the WT protein, the residue E71 participates in E71-D80-W67 linkages, which are considered as being an essential factor driving the filter toward non-conductive conformations [9,29]. Our results demonstrated that these linkages represent just a part of the more extensive strongly correlated network which dynamically and collectively participates in determining the state of the SF. The mutation of E71 with alanine in mutant E71A generated a non-inactivating pore with freely moving D80 [14]. We showed that in mutant E71A, residue D80 interacts with arginines R64 and R89. This interaction induces the strain on the SF, which adapts and relieves the perturbation through a flipping of V76 and a transition of D80 toward the extracellular (outer) region. As a result of this adaptation, the filter remains in a conducting state. The presence of E71 in the WT protein prevents such adaptation when D80 interacts with the two arginines. This interaction, therefore, leads to distorted configurations with complicated dynamics. The resulting complex picture is defined by ions and water molecules in the filter as well as by residues interacting or controlling the interaction between the arginines and D80.

The summarized complex picture provided by this research can be represented as a network of weighted nodes which affect the permeation path (Figure 8). The sizes of the nodes are weighted according to the number of edges connecting each node. This figure reveals the primary importance of the residue D80, being the main hub. It forms the core of the network with the neighbouring H-bond donors E71, W67, and arginines R64 and R89 which mutual dynamical influence defines states of D80. The collaborative dynamics of the residues result either in the stabilization of the conductive conformation or in distorted states of the TVGYGD sequence of the SF. Note that the sequence belongs to the highly conserved signature sequence TXXTXGYGD observed in many potassium channels [16,29,31,97]. In these channels, the aspartate residue D, similar to D80 in KcsA, is surrounded by different H-bond donors. Thus, the existence of similar complex network might be a general feature in the regulation of the current in the K^+ ion channels.

The significant mutual influence between the residues behind the SF and the ion occupancy in specific sites means that perturbations imposed on either residues or ions affect the KcsA channel's state. It is reasonable to expect that numerous networks' states have distinct permeation properties. Recent experiments [7,23] with modified phospholipids showed that the interaction of arginines R64 and R89 with added phosphatidic acid lipid enhances the conduction in the KcsA channel. These experiments confirm the results of

Section 4, which describes the particular role of these arginines in regulating the network (Figure 8). In turn, the change of ion locations by an additional artificial force in biased MD simulations [34,58] could alter the residues behind the SF resulting in a non-conducting state with a high permeation barrier. It means that all the network components (Figure 8) should be included as collective variables in a biased MD simulation. Alternatively, a set of biased MD calculations for different states of the networks should be considered. The self-organized dynamics of the whole network define conducting or non-conducting states of the KcsA channel and considering the SF and ions only is not sufficient.

This dynamical network (Figure 8) is identified for the canonical structure 1KC4 with a closed inner gate. It is shown [35,36] that opening the inner gate leads to the perturbations on the backbones of the SF. Therefore, the inner gate should be included in this dynamical network as well. However, the possibility of an opposite influence of ions and residues near the SF on the inner gate is an open question. Several structures [34,52,54,59,60] with a conducting SF and an open inner gate were generated in silico by combining different crystallographic structures. Creating such structures should include a slow adaptation of the whole network to changes in the gate. The applied constraints on the SF backbones and ions only does not guarantee a realistic configuration of the SF. In this context, MD simulations of a transition of the inner gate from closed to open state are an essential missing link for clarifying the influence of the inner gate on the whole channel.

Current physical models (see, for example, the recent work in [98] and references therein) of the ion permeation in the KcsA channel consider a part of this network: ions and their interaction with the residues in the SF. Incorporating the whole network in physical models would lead to a more complex model, for example, the Markov state type, but a more realistic one. The representation of the protein's complexity via this network would lead to a comprehensive description of complicated patterns of currents observed experimentally.

Results of Section 4.1 show that one of network's states is non-conducting, and the channel in that state is inactive. This observation means that the inactivation can result from the dynamics of this network alone without the involvement of the gate residues. Future work will address the role of the network in the C-type inactivation.

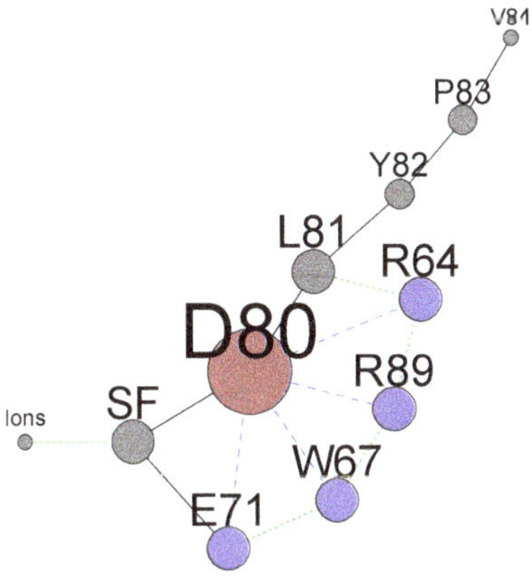

Figure 8. The network of residues that are determinant for the permeation path is drawn following certain rules: (i) blue-dashed lines represent non-bonded electrostatic interactions that can eventually lead to strong H-bonds; (ii) black lines represent connections through the backbones of the WT protein; and (iii) green dotted lines represent all the remaining non-bonded interactions, such as steric interactions or repulsions between positive or partially-positive charged groups. The sizes of the nodes are weighted according to the number of edges connecting each node. The label "SF" indicates the selectivity filter. The network was created using software package Gephi [99].

Supplementary Materials: The Supplementary Materials containing details of the methods, additional results and figures are available online at https://www.mdpi.com/1099-4300/23/1/72/s1.

Author Contributions: Conceptualization, S.M.C., E.J.C., and I.A.K.; methodology, S.M.C. and I.A.K.; formal analysis, investigation, and writing—original draft preparation, S.M.C.; writing—review and editing, S.M.C., E.J.C., and I.A.K. All authors have read and agreed to the published version of the manuscript.

Funding: The work has been supported by the Engineering and Physical Sciences Research Council (UK) under grants No. EP/G070660/1 and EP/M016889/1. Computational facilities were provided by the MidPlus Regional Centre of Excellence for Computational Science, Engineering and Mathemtatics, under EPSRC grant No. EP/K000128/1.

Data Availability Statement: Data related to this research are openly available from the University of Warwick archive at https://wrap.warwick.ac.uk/143572.

Acknowledgments: The authors thank M. Sansom, S. Takahama, D. Quigley, and M. Allen for useful discussions and critical comments. We dedicate the paper to the memory of Mark P Rodger. We are greatly indebted to him for contributions to our research.

Conflicts of Interest: The authors declare no conflict of interest.

References

1. Zhou, M.; Morais-Cabral, J.H.; Mann, S.; MacKinnon, R. Potassium channel receptor site for the inactivation gate and quaternary amine inhibitors. *Nature* **2001**, *411*, 657–661.
2. Kratochvil, H.T.; Carr, J.K.; Matulef, K.; Annen, A.W.; Li, H.; Maj, M.; Ostmeyer, J.; Serrano, A.L.; Raghuraman, H.; Moran, S.D.; et al. Instantaneous ion configurations in the K+ ion channel selectivity filter revealed by 2D IR spectroscopy. *Science* **2016**, *353*, 1040–1044.
3. Kopec, W.; Koepfer, D.A.; Vickery, O.N.; Bondarenko, A.S.; Jansen, T.L.C.; de Groot, B.L.; Zachariae, U. Direct knock-on of desolvated ions governs strict ion selectivity in K+ channels. *Nat. Chem.* **2018**, *10*, 813–820.
4. DeMarco, K.R.; Bekker, S.; Vorobyov, I. Challenges and advances in atomistic simulations of potassium and sodium ion channel gating and permeation. *J. Physiol.* **2019**, *597*, 679–698.
5. Sumikama, T.; Oiki, S. Queueing arrival and release mechanism for K+ permeation through a potassium channel. *J. Physiol. Sci.* **2019**, *69*, 919–930.
6. Xu, Y.; McDermott, A.E. Inactivation in the potassium channel KcsA. *J. Struct. Biol. X* **2019**, *3*, 100009.
7. Renart, M.; Giudici, A.; Diaz-Garcia, C.; Molina, M.; Morales, A.; Gonzalez-Ros, J.; Poveda, J. Modulation of Function, Structure and Clustering of K+ Channels by Lipids: Lessons Learnt from KcsA. *J. Mol. Sci.* **2020**, *21*, 2554.
8. Strong, S.E.; Hestand, N.J.; Kananenka, A.A.; Zanni, M.T.; Skinner, J.L. IR Spectroscopy Can Reveal the Mechanism of K+ Transport in Ion Channels. *Biophys. J.* **2020**, *118*, 254–261.
9. Cordero-Morales, J.F.; Jogini, V.; Lewis, A.; Vasquez, V.; Cortes, D.M.; Roux, B.; Perozo, E. Molecular driving forces determining potassium channel slow inactivation. *Nat. Struct. Mol. Biol.* **2007**, *14*, 1062–1069.
10. Hoshi, T.; Armstrong, C.M. C-type inactivation of voltage-gated K^+ channels: Pore constriction or dilation? *J. Gen. Physiol.* **2013**, *141*, 151–160.
11. Cuello, L.G.; Cortes, D.M.; Perozo, E. The gating cycle of a K+ channel at atomic resolution. *Elife* **2017**, *6*.
12. Chakrapani, S.; Cordero-Morales, J.F.; Perozo, E. A Quantitative Description of KcsA Gating I: Macroscopic Currents. *J. Gen. Physiol.* **2007**, *130*, 465–478.
13. Chakrapani, S.; Cordero-Morales, J.F.; Perozo, E. A Quantitative Description of KcsA Gating II: Single-Channel Currents. *J. Gen. Physiol.* **2007**, *130*, 479–496.
14. Cordero-Morales, J.F.; Cuello, L.G.; Zhao, Y.; Jogini, V.; Cortes, D.M.; Roux, B.; Perozo, E. Molecular determinants of gating at the potassium-channel selectivity filter. *Nat. Struct. Mol. Biol.* **2006**, *13*, 311–318.
15. Cordero-Morales, J.F.; Cuello, L.G.; Perozo, E. Voltage-dependent gating at the KcsA selectivity filter. *Nat. Struct. Mol. Biol.* **2006**, *13*, 319–322.
16. Kurata, H.T.; Fedida, D. A structural interpretation of voltage-gated potassium channel inactivation. *Prog. Biophys. Mol. Biol.* **2006**, *92*, 185–208.
17. Raghuraman, H.; Islam, S.M.; Mukherjee, S.; Roux, B.; Perozo, E. Dynamics transitions at the outer vestibule of the KcsA potassium channel during gating. *Proc. Natl. Acad. Sci. USA* **2014**, *111*, 1831–1836.
18. Lockless, S.W.; Zhou, M.; MacKinnon, R. Structural and Thermodynamic Properties of Selective Ion Binding in a K+ Channel. *PLoS Biol.* **2007**, *5*, e121.
19. Renart, M.L.; Montoya, E.; Fernández, A.M.; Molina, M.L.; Poveda, J.A.; Encinar, J.A.; Ayala, J.L.; Ferrer-Montiel, A.V.; Gómez, J.; Morales, A.; et al. Contribution of Ion Binding Affinity to Ion Selectivity and Permeation in KcsA, a Model Potassium Channel. *Biochemistry* **2012**, *51*, 3891–3900.

20. Marius, P.; Zagnoni, M.; Sandison, M.E.; East, J.M.; Morgan, H.; Lee, A.G. Binding of Anionic Lipids to at Least Three Nonannular Sites on the Potassium Channel KcsA is Required for Channel Opening. *Biophys. J.* **2008**, *94*, 1689–1698.
21. van der Cruijsen, E.A.W.; Nand, D.; Weingarth, M.; Prokofyev, A.; Hornig, S.; Cukkemane, A.A.; Bonvin, A.M.J.J.; Becker, S.; Hulse, R.E.; et al. Importance of lipid–pore loop interface for potassium channel structure and function. *Proc. Natl. Acad. Sci. USA* **2013**, *110*, 13008–13013.
22. Radda, R.; Kim, D.; Nimigean, C.; Andersen, O. Regulation of Ion Channel Function by the Host Lipid Bilayer Examined by a Stopped-Flow Spectrofluorometric Assay. *Biophys. J.* **2014**, *106*, 1070–1078.
23. Poveda, J.; Giudici, A.M.; Renart, M.L.; Millet, O.; Morales, A.; Gonzalez-Ros, J.M.; Oakes, V.; Furini, S.; Domene, C. Modulation of the potassium channel KcsA by anionic phospholipids: Role of arginines at the non-annular lipid binding sites. *Biochim. Biophys. Acta (BBA) Biomembr.* **2019**, *1861*, 183029.
24. Hille, B. *Ion Channels of Excitable Membranes*; Sinauer Associates: Sunderland, MA , USA, 2001.
25. Doyle, D.A.; Cabral, J.M.; Pfuetzner, R.A.; Kuo, A.; Gulbis, J.M.; Cohen, S.L.; Chait, B.T.; MacKinnon, R. The Structure of the Potassium Channel: Molecular Basis of K^+ Conduction and Selectivity. *Science* **1998**, *280*, 69–77.
26. Perozo, E.; Marien, C.D.; Cuello, L.G. Structural Rearrangements Underlying K^+-Channel Activation Gating. *Science* **1999**, *285*, 73–78.
27. Takeuchi, K.; Takahashi, H.; Kawano, S.; Shimada, I. Identification and Characterization of the Slowly Exchanging pH-dependent Conformational Rearrangement in KcsA. *J. Biol. Chem.* **2007**, *282*, 15179–15186.
28. Zhou, Y.; Morais-Cabral, J.H.; Kaufman, A.; MacKinnon, R. Chemistry of ion coordination and hydration revealed by a K^+ channel-Fab complex at 2.0 A resolution. *Nature* **2001**, *414*, 43–48.
29. Cordero-Morales, J.F.; Jogini, V.; Chakrapani, S.; Perozo, E. A Multipoint Hydrogen-Bond Network Underlying KcsA C-Type Inactivation. *Biophys. J.* **2011**, *100*, 2387–2393.
30. Chakrapani, S.; Cordero-Morales, J.F.; Jogini, V.; Pan, A.C.; Cortes, D.M.; Roux, B.; Perozo, E. On the structural basis of modal gating behavior in K^+ channels. *Nat. Struct. Mol. Biol.* **2010**, *18*, 67–74.
31. Chapman, M.L.; Krovetz, H.S.; VanDongen, A.M.J. GYGD pore motifs in neighbouring potassium channel subunits interact to determine ion selectivity. *J. Physiol.* **2001**, *530*, 21–33.
32. Cheng, W.W.L.; McCoy, J.G.; Thompson, A.N.; Nichols, C.G.; Nimigean, C.M. Mechanism for selectivity-inactivation coupling in KcsA potassium channels. *Proc. Natl. Acad. Sci. USA* **2011**, *108*, 5272–5277.
33. Marius, P.; de Planque, M.R.R.; Williamson, P.T.F. Probing the interaction of lipids with the non-annular binding sites of the potassium channel KcsA by magic-angle spinning NMR. *Biochim. Biophys. Acta (BBA) Biomembr.* **2012**, *1818*, 90–96.
34. Florian, T.H.; David, J.P.; Wojciech, W.N.; Crina, M.N.; Berneche, S. Mechanism of activation at the selectivity filter of the KcsA K+ channel. *eLife* **2017**, *6*, e25844.
35. Labro, A.J.; Cortes, D.M.; Tilegenova, C.; Cuello, L.G. Inverted allosteric coupling between activation and inactivation gates in K+ channels. *Proc. Natl. Acad. Sci. USA* **2018**, *115*, 5426–5431.
36. Sun, Z., Xu, Y., Zhang, D.; McDermott, A.F. Probing allosteric coupling in a constitutively open mutant of the ion channel KcsA using solid-state NMR. *Proc. Natl. Acad. Sci. USA* **2020**, *117*, 7171–7175.
37. Devaraneni, P.K.; Komarov, A.G.; Costantino, C.A.; Devereaux, J.J.; Matulef, K.; Valiyaveetil, F.I. Semisynthetic K+ channels show that the constricted conformation of the selectivity filter is not the C-type inactivated state. *Proc. Natl. Acad. Sci. USA* **2013**, *110*, 15698–15703.
38. Matulef, K.; Komarov, A.G.; Costantino, C.A.; Valiyaveetil, F.I. Using protein backbone mutagenesis to dissect the link between ion occupancy and C-type inactivation in K+ channels. *Proc. Natl. Acad. Sci. USA* **2013**, *110*, 17886–17891.
39. Matulef, K.; Annen, A.; Nix, J.; Valiyaveetil, F. Individual Ion Binding Sites in the K+ Channel Play Distinct Roles in C-type Inactivation and in Recovery from Inactivation. *Structure* **2016**, *24*, 750–761.
40. Baker, K.A.; Tzitzilonis, C.; Kwiatkowski, W.; Choe, S.; Riek, R. Conformational dynamics of the KcsA potassium channel governs gating properties. *Nat. Struct. Mol. Biol.* **2007**, *14*, 1089–1095.
41. Jekhmane, S.; Medeiros-Silva, J.; Li, J.; Kummerer, F.; Muller-Hermes, C.; Baldus, M.; Roux, B.; Weingarth, M. Shifts in the selectivity filter dynamics cause modal gating in K+ channels. *Nat. Commun.* **2019**, *10*, 123.
42. Li, J.; Ostmeyer, J.; Boulanger, E.; Rui, H.; Perozo, E.; Roux, B. Chemical substitutions in the selectivity filter of potassium channels do not rule out constricted-like conformations for C-type inactivation. *Proc. Natl. Acad. Sci. USA* **2017**, *114*, 11145–11150.
43. Swenson, R.P.; Armstrong, C.M. K+ channels close more slowly in the presence of external K+ and Rb+. *Nature* **1981**, *291*, 427–429.
44. Piasta, K.N.; Theobald, D.L.; Miller, C. Potassium-selective block of barium permeation through single KcsA channels. *J. Gen. Phys.* **2011**, *138*, 421–436.
45. Liu, S.; Focke, P.J.; Matulef, K.; Bian, X.; Moënne-Loccoz, P.; Valiyaveetil, F.I.; Lockless, S.W. Ion-binding properties of a K+ channel selectivity filter in different conformations. *Proc. Natl. Acad. Sci. USA* **2015**, *112*, 15096–15100.
46. Renart, M.L.; Giudici, A.M.; Poveda, J.A.; Fedorov, A.; Berberan-Santos, M.N.; Prieto, M.; Diaz-Garcia, C.; Gonzalez-Ros, J.M.; Coutinho, A. Conformational plasticity in the KcsA potassium channel pore helix revealed by homo-FRET studies. *Sci. Rep.* **2019**, *9*, 6215.
47. Sansom, M.S.; Shrivastava, I.H.; Bright, J.N.; Tate, J.; Capener, C.E.; Biggin, P.C. Potassium channels: structures, models, simulations. *Biochim. Biophys. Acta (BBA) Biomembr.* **2002**, *1565*, 294–307.

48. Maffeo, C.; Bhattacharya, S.; Yoo, J.; Wells, D.; Aksimentiev, A. Modeling and Simulation of Ion Channels. *Chem. Rev.* **2012**, *112*, 6250–6284.
49. Harpole, T.J.; Delemotte, L. Conformational landscapes of membrane proteins delineated by enhanced sampling molecular dynamics simulations. *Biochim. Biophys. Acta (BBA) Biomembr.* **2018**, *1860*, 909–926.
50. Noskov, S.Y.; Bernèche, S.; Roux, B. Control of ion selectivity in potassium channels by electrostatic and dynamic properties of carbonyl ligands. *Nature* **2004**, *431*, 830–834.
51. Tilegenova, C.; Cortes, D.M.; Jahovic, N.; Hardy, E.; Hariharan, P.; Guan, L.; Cuello, L.G. Structure, function, and ion-binding properties of a K+ channel stabilized in the 2,4-ion–bound configuration. *Proc. Natl. Acad. Sci. USA* **2019**, *116*, 16829–16834.
52. Li, J.; Ostmeyer, J.; Cuello, L.G.; Perozo, E.; Roux, B. Rapid constriction of the selectivity filter underlies C-type inactivation in the KcsA potassium channel. *J. Gen. Physiol.* **2018**, *150*, 1408–1420.
53. Delemotte, L. Opening leads to closing: Allosteric crosstalk between the activation and inactivation gates in KcsA. *J. Gen. Physiol.* **2018**, *150*, 1356–1359.
54. Furini, S.; Domene, C. Critical Assessment of Common Force Fields for Molecular Dynamics Simulations of Potassium Channels. *J. Chem. Theory Comput.* **2020**, *16*, 7148–7159.
55. Cuello, L.G.; Jogini, V.; Cortes, D.M.; Pan, A.C.; Gagnon, D.G.; Dalmas, O.; Cordero-Morales, J.F.; Chakrapani, S.; Roux, B.; Perozo, E. Structural basis for the coupling between activation and inactivation gates in K+ channels. *Nature* **2010**, *466*, 272–275.
56. Wojtas-Niziurski, W.; Meng, Y.; Roux, B.; Bernache, S. Self-Learning Adaptive Umbrella Sampling Method for the Determination of Free Energy Landscapes in Multiple Dimensions. *J. Chem. Theory Comput.* **2013**, *9*, 1885–1895.
57. LeMasurier, M.; Heginbotham, L.; Miller, C. Kcsa: It's a Potassium Channel. *J. Gen. Physiol.* **2001**, *118*, 303–314.
58. Fowler, P.W.; Abad, E.; Beckstein, O.; Sansom, M.S.P. Energetics of Multi-Ion Conduction Pathways in Potassium Ion Channels. *J. Chem. Theory Comput.* **2013**, *9*, 5176–5189.
59. Jensen, M.O.; Jogini, V.; Eastwood, M.P.; Shaw, D.E. Atomic-level simulation of current-voltage relationships in single-file ion channels. *J. Gen. Physiol.* **2013**, *141*, 619–632.
60. Köpfer, D.A.; Song, C.; Gruene, T.; Sheldrick, G.M.; Zachariae, U.; de Groot, B.L. Ion permeation in K+ channels occurs by direct Coulomb knock-on. *Science* **2014**, *346*, 352–355.
61. Sumikama, T.; Oiki, S. Digitalized K+ Occupancy in the Nanocavity Holds and Releases Queues of K+ in a Channel. *J. Am. Chem. Soc.* **2016**, *138*, 10284–10292.
62. Barducci, A.; Bussi, G.; Parrinello, M. Well-Tempered Metadynamics: A Smoothly Converging and Tunable Free-Energy Method. *Phys. Rev. Lett.* **2008**, *100*, 020603.
63. Cosseddu, S.M. Structure and Dynamics of Protein in the Permeation and Gating of Potassium Ion Channels: Identifying Molecular Determinants and Developing Coarse-Grained Approaches. Ph.D. Thesis, University of Warwick, Coventry, UK, 2013.
64. Phillips, J.C.; Braun, R.; Wang, W.; Gumbart, J.; Tajkhorshid, E.; Villa, E.; Chipot, C.; Skeel, R.D.; Kalé, L.; Schulten, K. Scalable molecular dynamics with NAMD. *J. Comput. Chem.* **2005**, *26*, 1781–1802.
65. Grubmüller, H.; Heller, H.; Windemuth, A.; Schulten, K. Generalized Verlet Algorithm for Efficient Molecular Dynamics Simulations with Long-range Interactions. *Mol. Simul.* **1991**, *6*, 121–142.
66. Tuckerman, M.; Berne, B.J.; Martyna, G.J. Reversible multiple time scale molecular dynamics. *J. Chem. Phys.* **1992**, *97*, 1990–2001.
67. Essmann, U.; Perera, L.; Berkowitz, M.L.; Darden, T.; Lee, H.; Pedersen, L.G. A smooth particle mesh Ewald method. *J. Chem. Phys.* **1995**, *103*, 8577.
68. MacKerell, A.D.; Wiorkiewicz-Kuczera, J.; Karplus, M. An all-atom empirical energy function for the simulation of nucleic acids. *J. Am. Chem. Soc.* **1995**, *117*, 11946–11975.
69. Buck, M.; Bouguet-Bonnet, S.; Pastor, R.W.; MacKerell, A.D. Importance of the CMAP Correction to the CHARMM22 Protein Force Field: Dynamics of Hen Lysozyme. *Biophys. J.* **2006**, *90*, L36–L38.
70. Klauda, J.B.; Venable, R.M.; Freites, J.A.; O'Connor, J.W.; Tobias, D.J.; Mondragon-Ramirez, C.; Vorobyov, I.; MacKerell, A.D.; Pastor, R.W. Update of the CHARMM All-Atom Additive Force Field for Lipids: Validation on Six Lipid Types. *J. Phys. Chem. B* **2010**, *114*, 7830–7843.
71. Bernèche, S.; Roux, B. The Ionization State and the Conformation of Glu-71 in the KcsA K^+ Channel. *Biophys. J.* **2002**, *82*, 772–780.
72. Bucher, D.; Guidoni, L.; Rothlisberger, U. The Protonation State of the Glu-71/Asp-80 Residues in the KcsA Potassium Channel: A First-Principles QM/MM Molecular Dynamics Study. *Biophys. J.* **2007**, *93*, 2315–2324.
73. Grubmüller, H.; Groll, V. Solvate. Available online: https://www.mpibpc.mpg.de/grubmueller/solvate (accessed on 4 January 2021).
74. Zhang, L.; Hermans, J. Hydrophilicity of cavities in proteins. *Proteins Struct. Funct. Bioinform.* **1996**, *24*, 433–438.
75. Oostenbrink, C.; Villa, A.; Mark, A.E.; Van Gunsteren, W.F. A biomolecular force field based on the free enthalpy of hydration and solvation: The GROMOS force-field parameter sets 53A5 and 53A6. *J. Comput. Chem.* **2004**, *25*, 1656–1676.
76. Laio, A.; Parrinello, M. Escaping free-energy minima. *Proc. Natl. Acad. Sci. USA* **2002**, *99*, 12562–12566.
77. Fiorin, G.; Klein, M.L.; Hénin, J. Using collective variables to drive molecular dynamics simulations. *Mol. Phys.* **2013**, *111*, 3345–3362.
78. Humphrey, W. VMD: Visual molecular dynamics. *J. Mol. Graph.* **1996**, *14*, 33–38.
79. R Core Team. *R: A Language and Environment for Statistical Computing*; R Foundation for Statistical Computing: Vienna, Austria, 2013.

80. Grant, B.J.; Rodrigues, A.P.C.; ElSawy, K.M.; McCammon, J.A.; Caves, L.S.D. Bio3d: an R package for the comparative analysis of protein structures. *Bioinformatics* **2006**, *22*, 2695–2696.
81. Wickham, H. *ggplot2: Elegant Graphics for Data Analysis*; Springer: New York, NY, USA, 2009.
82. Adler, D.; Murdoch, D. *rgl: 3D Visualization Device System (OpenGL)*; R Package Version 0.92.892. Available online: http://r-forge.r-project.org/projects/rgl (accessed on 5 January 2021).
83. Fox, J.; Weisberg, S. *An R Companion to Applied Regression*, 2nd ed.; Sage: Thousand Oaks, CA, USA, 2011.
84. Venables, W.N.; Ripley, B.D. *Modern Applied Statistics with S*, 4th ed.; Springer: New York, NY, USA, 2002.
85. Furrer, R.; Nychka, D.; Sain, S. *Fields: Tools for Spatial Data*; R Package Version 6.7.6. Available online: https://cran.r-project.org/src/contrib/Archive/fields (accessed on 5 January 2021).
86. Berneche, S.; Roux, B. Energetics of ion conduction through the K^+ channel. *Nature* **2001**, *414*, 73–77.
87. Bernèche, S.; Roux, B. Molecular Dynamics of the KcsA K^+ Channel in a Bilayer Membrane. *Biophys. J.* **2000**, *78*, 2900–2917.
88. Bernèche, S.; Roux, B. A Gate in the Selectivity Filter of Potassium Channels. *Structure* **2005**, *13*, 591–600.
89. Domene, C.; Klein, M.L.; Branduardi, D.; Gervasio, F.L.; Parrinello, M. Conformational Changes and Gating at the Selectivity Filter of Potassium Channels. *J. Am. Chem. Soc.* **2008**, *130*, 9474–9480.
90. Piccinini, E.; Ceccarelli, M.; Affinito, F.; Brunetti, R.; Jacoboni, C. Biased Molecular Simulations for Free-Energy Mapping: A Comparison on the KcsA Channel as a Test Case. *J. Chem. Theory Comput.* **2008**, *4*, 173–183.
91. Miloshevsky, G.V.; Jordan, P.C. Conformational Changes in the Selectivity Filter of the Open-State KcsA Channel: An Energy Minimization Study. *Biophys. J.* **2008**, *95*, 3239–3251.
92. Shrivastava, I.H.; Sansom, M.S.P. Simulations of Ion Permeation Through a Potassium Channel: Molecular Dynamics of KcsA in a Phospholipid Bilayer. *Biophys. J.* **2000**, *78*, 557–570.
93. Compoint, M.; Carloni, P.; Ramseyer, C.; Girardet, C. Molecular dynamics study of the KcsA channel at 2.0-A resolution: stability and concerted motions within the pore. *Biochim. Biophys. Acta Biomem.* **2004**, *1661*, 26–39.
94. Kim, I.; Allen, T.W. On the selective ion binding hypothesis for potassium channels. *Proc. Natl. Acad. Sci. USA* **2011**, *108*, 17963–17968.
95. Furini, S.; Domene, C. Atypical mechanism of conduction in potassium channels. *Proc. Natl. Acad. Sci. USA* **2009**, *106*, 16074–16077.
96. Deol, S.S.; Domene, C.; Bond, P.J.; Sansom, M.S. Anionic Phospholipid Interactions with the Potassium Channel KcsA: Simulation Studies. *Biophys. J.* **2006**, *90*, 822–830.
97. Catterall, W.A. Ion channel voltage sensors: structure, function, and pathophysiology. *Neuron* **2010**, *67*, 915–928.
98. Gibby, W.A.T.; Barabash, M.L.; Guardiani, C.; Luchinsky, D.G.; McClintock, P.V.E. Physics of selective conduction and point mutation in biological ion channels. *arXiv* **2020**, arXiv:physics.bio-ph/2010.08450.
99. Bastian, M.; Heymann, S.; Jacomy, M. Gephi: An Open Source Software for Exploring and Manipulating Networks. In Proceedings of the International AAAI Conference on Weblogs and Social Media, San Jose, CA, USA, 17–20 May 2009.

entropy

Article

Diffusion Limitations and Translocation Barriers in Atomically Thin Biomimetic Pores

Subin Sahu [1,2,3] **and Michael Zwolak** [1,*]

1 Biophysical and Biomedical Measurement Group, Microsystems and Nanotechnology Division, Physical Measurement Laboratory, National Institute of Standards and Technology, Gaithersburg, MD 20899, USA; subin.sahu@colorado.edu
2 Institute for Research in Electronics and Applied Physics and Maryland NanoCenter, University of Maryland, College Park, MD 20742, USA
3 Department of Chemical and Biological Engineering, University of Colorado Boulder, Boulder, CO 80309, USA
* Correspondence: mpz@nist.gov

Received: 19 October 2020; Accepted: 6 November 2020; Published: 20 November 2020

Abstract: Ionic transport in nano- to sub-nano-scale pores is highly dependent on translocation barriers and potential wells. These features in the free-energy landscape are primarily the result of ion dehydration and electrostatic interactions. For pores in atomically thin membranes, such as graphene, other factors come into play. Ion dynamics both inside and outside the geometric volume of the pore can be critical in determining the transport properties of the channel due to several commensurate length scales, such as the effective membrane thickness, radii of the first and the second hydration layers, pore radius, and Debye length. In particular, for biomimetic pores, such as the graphene crown ether we examine here, there are regimes where transport is highly sensitive to the pore size due to the interplay of dehydration and interaction with pore charge. Picometer changes in the size, e.g., due to a minute strain, can lead to a large change in conductance. Outside of these regimes, the small pore size itself gives a large resistance, even when electrostatic factors and dehydration compensate each other to give a relatively flat—e.g., near barrierless—free energy landscape. The permeability, though, can still be large and ions will translocate rapidly after they arrive within the capture radius of the pore. This, in turn, leads to diffusion and drift effects dominating the conductance. The current thus plateaus and becomes effectively independent of pore-free energy characteristics. Measurement of this effect will give an estimate of the magnitude of kinetically limiting features, and experimentally constrain the local electromechanical conditions.

Keywords: ion transport; nanopore; graphene; crown ether

1. Introduction

Ionic transport through nano- and sub-nano-scale pores elicits a tremendous amount of interest due to its relevance in cellular processes including neurotransmission, muscle contraction, and other biological processes [1,2], as well as its application in technologies such as desalination [3–5], osmotic power generation [6–8], and bio-chemical sensing [9–13]. Ref. [14] provides a recent review of the use of pores in 2D materials for these applications. Understanding transport mechanisms, particularly in biological settings, has remained challenging due to their complexity and dependence on atomic details [15]. Furthermore, even for uncharged membranes, the region outside the pore can play a significant role in determining ionic transport, either via access resistance [16–18] or via diffusion limitations [19].

Synthetic nanopores offer the ability to study the factors that underpin transport mechanisms, such as the role of dehydration [20–24] and functional groups [25–28], and give rise to functional behavior, such as ion selectivity [27,29]. Pores in 2D membranes, in particular, have a larger access resistance, compared to their pore resistance, due to their small aspect ratio, a_p/h_p, where a_p is the effective pore radius and h_p is the effective pore length. Therefore, 2D materials provide a unique opportunity to study geometric effects in transport [30], atomic changes in area via precise control in pore fabrication and height via layering [24], and the interplay of various length scales relevant to the problem [14], which will help in the design of separation membranes [14,31,32] and delineating factors relevant to biological channels [14]. In particular, the effective thickness is not the geometric thickness of the membrane (e.g., for graphene, twice the van der Waals radius of the carbon atoms), since ion size, hydration layer radii, and even membrane charge and the build up of charge layers give rise to an effective thickness [16]. When both the bulk and pore are within the continuum drift-diffusion regime, measuring or calculating the dependence of the conductance on pore radius quantifies the effective pore length through the equation

$$R = \gamma \left(\frac{1}{2a_p} + \frac{h_p}{\pi a_p^2} \right), \tag{1}$$

which assumes a homogeneous resistivity γ and a cylindrical pore geometry. The former entails that there are no concentration gradients and that the medium in the pore has the same resistivity as the bulk (otherwise, it requires an independent determination of the pore resistivity). For graphene, this approach yields an effective pore thickness of about 1 nm, both experimentally [33] (for unknown rim functionalization) and computationally [14,16] (for unfunctionalized rims), provided that simulations properly include the influence of the bulk via the golden aspect ratio or associated scaling analysis [16,34], as well as properly determine the pore radius [16]. The effective pore length is mostly due to the van der Waals radii and first hydration layer of the ions, which are both reflected in a build up of charge layers about 0.5 nm from the membrane.

More recently, it was demonstrated that applying strain to 2D pores can elucidate the conditions under which optimal transport and ion selectivity arise by modulating the balance of dehydration and electrostatic interactions [27]. It is unclear, however, whether 2D pores, and synthetic pores more generally, offer a means to investigate diffusion limitations. These arise due to fine details of pore structure typically thought to be out of our control. We will demonstrate here that the control provided by strain can tune atomically thin biomimetic pores into a diffusion-limited regime. Finding transitions into these regimes will help delineate and probe the electromechanical environment of nanopores, and elucidate diffusion-limited phenomena in more complex, biological settings.

Ionic transport through a pore becomes diffusion-limited when the permeability of ions in the pore is large and the current is only restricted by the rate of diffusion of the ions from the bulk to the pore mouth [35,36]. In this diffusion-limited regime, the current does not increase with voltage as expected from Ohm's law. This is similar to the diffusion-limited processes in chemical reactions [37] and other transport processes [38,39]. Diffusion-limited ionic currents are a regular occurrence in biological pores [40,41], as these can have the necessary conditions: narrow channels with high permeability for specific ions [1] and the presence of "inert" ions [42]. Pores in atomically thin membranes, such as graphene, MoS_2, and hBN, also provide a very high permeability for ions due to their sub-nanoscale channel length [14]. Thus, in these membranes, the diffusion of the ions from the bulk to the mouth of the pore may become the limiting factor in the ion transport. It is the objective of this work to determine under what conditions bulk diffusion becomes the limiting factor, particularly when drift is also present.

For the transport to be diffusion-limited, however, the drift contribution to the current in the bulk has to be small. This condition is hard to realize in pores in 2D materials under electrically driven transport, since a large portion of the applied voltage drops in the bulk solution, which in general has a higher resistance—in the form of access resistance—than the pore itself [14,16,17,33]. This is also true of small aspect ratio—short and wide—biological channels, where access resistance becomes dominant at low ion concentration [18]. There are regimes, though, where diffusion limitations may appear. Sub-nanoscale pores in graphene, for instance, have a high pore resistance [23]. As a result, most of the applied potential will drop across the pore, thus diminishing the drift current in the bulk and giving an opportunity to observe the diffusion-limited transport. At this sub-nanometer length scale, the translocation barriers and potential wells due to ion dehydration and electrostatic interactions play a major role in determining transport through such pores [27]. When one or the other interaction dominates, translocation through the pore is barrier-limited.

Under the right conditions—determined by pore size and charge, dehydration energy, etc. —the permeability of the pore will be large [43,44] (e.g., ion channels with binding sites, in particular, can have an inverse relationship between permeability and conductance [45]). To understand the conditions for having a small pore conductance and high pore permeability, one can look at the continuum-limit expressions

$$G_p = q\, c_p\, \mu_p\, A_p / h_p \tag{2}$$

and

$$P_p = D_p / h_p, \tag{3}$$

where c_p is the ion concentration in the pore, q is the charge of the ion, μ_p is the pore mobility, D_p ($=\mu_p k_B T/q$) is the diffusion coefficient, A_p is the pore area, h_p is the pore length, k_B is Boltzmann's constant, and T is the temperature. We briefly note that all pore quantities (which have a subscript p) are effective quantities, as will be abundantly clear throughout this work. An overall barrier in the pore can limit the concentration by exponentially reducing the partition coefficient into the pore. When such a barrier is present, without features internal to the pore, the mobility and diffusion can be unaffected. Similarly, the internal features, and hence mobility, can be altered with very small relative changes to pore size (and vice versa, reducing the cross-sectional area of the pore can reduce the conductance but with little effect on mobility when the pore size is relatively large). Essentially, strain and voltage will give the right knobs to tune some pores into a diffusion-limited regime by modulation of c_p and μ_p, while retaining a small A_p (i.e., a large pore resistance).

Here, we will examine the 18-crown-6 pore in graphene (see Figure 1) under the influence of a homogeneous strain in the plane of the membrane and cross-membrane voltage for different local electronic pore environments. Crown ether pores were seen by scanning transmission electron microscopy of graphene membranes that were made by exfoliation of graphite [46]. Even though strain only changes the pore area by a minuscule amount, the change in the free-energy barrier and hence the ionic current is substantial [27]. We find both barrier-limited and diffusion-limited regimes depending on the strain, voltage, and local environment. In the barrier-limited regime, the conductance increases with applied voltage, as it helps ions overcome the barrier. In the diffusion-limited regime, the pore conductance decreases with voltage as the bias depletes the charge carriers in the pore and at its entrance. The transition between these regimes depends on the charge separation in the pore (e.g., the local dipole moment at pore rim), which is not experimentally known, nor is there sufficient thermodynamic or kinetic data from experiment to constrain it. Measurement of the pore conductance versus strain and voltage, therefore, gives a possible route to determining the electromechanical environment, and thus constraining the magnitude of charge separation at the pore rim.

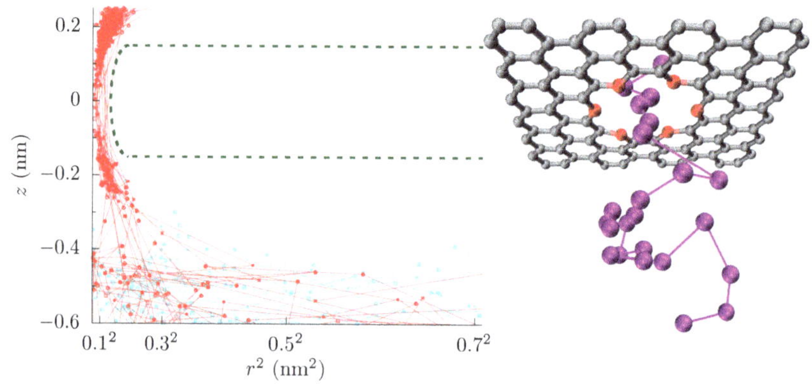

Figure 1. Ion trajectories in a graphene crown ether pore. The trajectories of K^+ ions around the pore with $q_O = -0.54\,e$ at 4% strain, with an applied bias of 0.25 V along the z-axis (making K^+ move in the positive z-direction). We plot the trajectories in z-r^2 space to keep the representation of volume constant. The red trajectories are for ions that translocate through the pore and the cyan trajectories are for ions that reflect back. The green dashed line shows the geometric boundary of the membrane determined by the van der Waals radii of the pore atoms. The effective boundary extends to $|z| \lesssim 0.5$ nm due to the size of the hydrated ions. The inset shows a portion of the graphene crown ether (red and grey spheres are oxygen and carbon atoms, respectively) and a three-dimensional trajectory of a potassium ion (purple spheres, separated in time by 10 ps, connected with purple lines) crossing the pore. Connecting lines are a guide to the eye only.

2. Methods

We performed all-atom molecular dynamics (MD) simulations using the NAMD2 simulations package [47]. The details of the simulations were the same as in Ref. [27]. We applied a voltage between 0.1 V to 1.0 V and calculated the ionic current for a 1 mol/L KCl solution by counting the ions that crossed the pore. Since the pore rim is negatively charged and sub-nanoscale in size (i.e., the electrostatic interactions are not significantly screened), only cation currents were present in all cases. We calculated the free-energy barrier using the adaptive biasing force (ABF) method [48] in a cylinder of radius 0.28 nm and height 3 nm, centered at the pore. A portion of our simulation cell and a set of ion trajectories are shown in Figure 1. We employed the golden aspect ratio method, as it is the only method that can properly capture bulk access effects [16,17], and without it, one cannot explore bulk diffusion limitations with all-atom simulations.

We calculated the ionic current and free-energy barrier at various homogeneous strains from 0% to 10% on the graphene membrane (we note that most of the features we observe occur at strains of 4% to 6%, and graphene can survive strains above 20% [49,50]). The strain was within the membrane and thus tended to enlarge the pore (albeit by small amounts) and expand in-plane distances between atoms. In the unstrained pore, the nominal pore radius (measured from the pore center to the center of the oxygen atoms) was approximately 0.29 nm and increased by about 7.5 pm for each 1% strain (this reflects a small—a factor of ≈2—geometric amplification [27]). This yielded nominal pore sizes from 0.29 nm (at 0% strain) to about 0.37 nm (at 10% strain), with a roughly linear relationship with strain. Though the change in pore size was minuscule, the energy landscape changes substantially. The landscape also depended on the pore charge, which is, however, not known. For the crown ether pore in graphene, the charge per oxygen atom (q_O) could be between $-0.2\,e$ and $-0.7\,e$ [27,51–53]. We thus used two representative test charges, $-0.24\,e$ and

$-0.54\,e$. In the former, there was an energy barrier, and in the latter, there was a potential well at the center of the unstrained pore. Each of the 12 carbon atoms—the ones adjacent to the six oxygen atoms in the pore—had charge $-q_O/2$, and the rest of the carbon atoms were neutral. When we refer to the pore charge, we are referencing the local polarization of charge from the carbon atoms of the graphene near the pore and the oxygen atoms on the pore rim. For each data point (i.e., for a particular value of q_O, strain, and voltage), we performed five parallel production runs for a total simulation between 250 ns to 500 ns. This allowed for an error estimation using the standard error, $\mathrm{SE} = \sqrt{\mathrm{var}/n_r}$, where var is the variance between the $n_r = 5$ parallel runs.

3. Results

Ionic current through a graphene crown ether pore: Figure 2 shows the potassium current, I_K, through the graphene crown ether pore versus strain at various voltages. Only potassium ions contribute to the total current, as the negatively charged pore edge does not allow any chloride ions to translocate (on the timescale of the simulations). We also plot the conductance of potassium ions, G_K, in order to demonstrate the non-Ohmic behavior of ionic current. At low voltage, the current increases by several fold for a minute strain—a couple percent strain changes the conductance by a couple hundred percent. This dramatic amplification is an example of colossal ionic mechano-conductance [27]. The current eventually maximizes around 3% strain and either decreases (for $q_O = -0.24\,e$ at small voltage) or plateaus (for all other cases). Since the pore size does not change substantially with a small strain, the colossal change in the ionic conductance is the result of a modification of the translocation barriers. The translocation landscape veers toward barrierless transport as strain tunes transport to its optimum [27]. Furthermore, the change in conductance with voltage displays non-Ohmic behavior. In some regimes, such as the colossal mechano-conductance, the conductance increases with voltage, indicating an activated process. In other regimes, the conductance decreases with voltage, indicating the diffusion-limited process. The depletion of charge carriers in the pore, see Figure 3, is also consistent with the decrease in conductance and points to a diffusion-limited regime.

Translocation barriers in sub-nanoscale pores: The single-ion energetics of transport through functionalized sub-nanoscale pores may be approximately expressed as

$$\Delta F_\nu \approx \sum_i \eta_i f_{i\nu} E_{i\nu} + \sum_{\nu'} \frac{q_\nu q_{\nu'} n_{\nu'}}{4\pi\epsilon_0 \epsilon\, r_{\nu\nu'}}, \tag{4}$$

where $f_{i\nu}$ and $E_{i\nu}$ are the fractional dehydration and energy of ith hydration layer for ion ν, $q_{\nu'}$'s and $n_{\nu'}$ are the charge and number of atom species ν' in the pore, and ϵ_0 is the vacuum permittivity. The parameter η_i is an $\mathcal{O}(1)$ factor to account for the increased binding of water molecules with the ion as dehydration increases [23], essentially giving the non-linear response of the hydration energy to the removal of water molecules. The relative permittivity of water, ϵ, under nanoscale confinement is significantly smaller than the bulk value and depends on atomic details [27,29,54]. Specifically, in the case here, when there are not intervening water molecules between the ion and charged groups in the pore, the dielectric constant is around 4 [27] and the electrostatic interaction is very large. The small dielectric constant and short distances involved give rise to the large electromechanical susceptibility of ions within the pore. The fractional dehydration $f_{\nu 1}$ also changes with the position of ion [23,24,27] and can be estimated with geometric arguments [20,22–24]. As an ion approaches the pore, the free-energy change will remain small ($<k_B T$) even at 1 nm distance from the pore, because the ion is still fully hydrated. The electrostatic interaction between the fully hydrated ion and the pore charge is weak due to the dielectric screening of the solution. However, when the ion is about 0.5 nm from the pore, it starts to dehydrate (initially in the second hydration shell and then in the first), and consequently the dehydration energy increases sharply.

Simultaneously, the electrostatic energy also rises rapidly since the ion will be significantly closer to the negatively charged oxygen atoms compared to the positively charged carbons, and the effective dielectric constant of water at this distance will be strongly diminished due to the removal of intervening molecules.

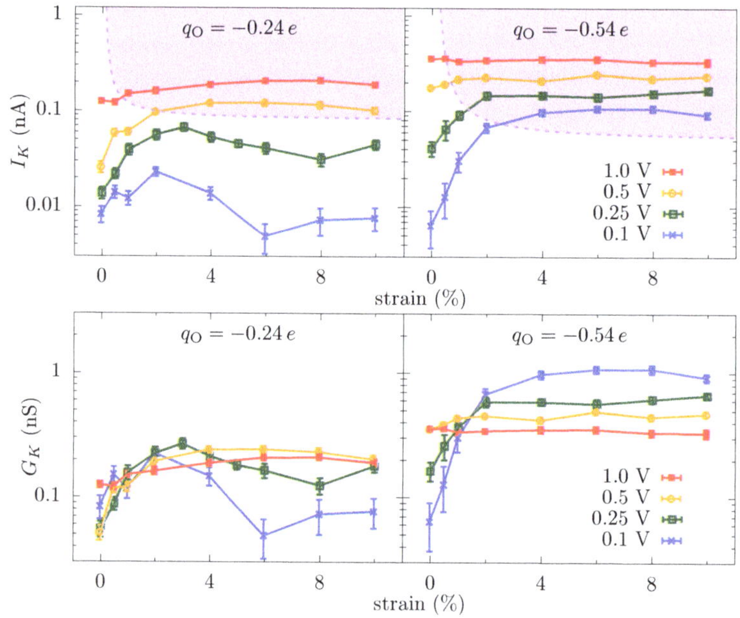

Figure 2. Colossal rise, non-monotonicity, and saturation of the ionic current. (**top panel**) Potassium current (I_K) versus strain at various voltages across the graphene crown ether pore within 1 mol/L KCl. At small voltages and minute strains, I_K increases rapidly with strain due to the large electromechanical susceptibility of the pore [27]. A further increase of the strain causes I_K to either decrease (for $q_O = -0.24\,e$) or saturate (for $q_O = -0.54\,e$), albeit the latter will also decrease when the electrostatic well disappears and dehydration begins to control the current. At large voltage, the current becomes less sensitive to strain because the applied bias dominates over the energy landscape of the pore, self-consistently washing out relevant features—ones that are contributing to resistance—of the landscape. As voltage increases further still, the current saturates at a smaller strain where the relevant free energy features are commensurate with the voltage drop (**bottom panel**). The conductance versus strain shows that for $q_O = -0.24\,e$ there is an intricate interplay of voltage and strain, indicating that the variation of free energy features with these two parameters is playing a defining role. At larger strain (greater than about 6%), the conductance tends to increase with the voltage (i.e., superlinear behavior). This is a telltale sign of an activated process, where the voltage helps overcome an overall barrier, but does not yet wash it out. In this particular case, this is due to a reduction in electrostatic compensation of dehydration as strain pulls away the counteracting negatively charged oxygen atoms. For $q_O = -0.54\,e$, the conductance increases with the voltage at smaller strain (superlinear behavior) and decreases with voltage at larger strain (sublinear behavior). The superlinear behavior indicates barrier-limited transport and sublinear behavior diffusion-limited. The error bars are plus/minus one SE from five parallel runs. Connecting lines are a guide to the eye only. Purple shaded regions in the upper panels approximately delineate the region where bulk limitations control the current.

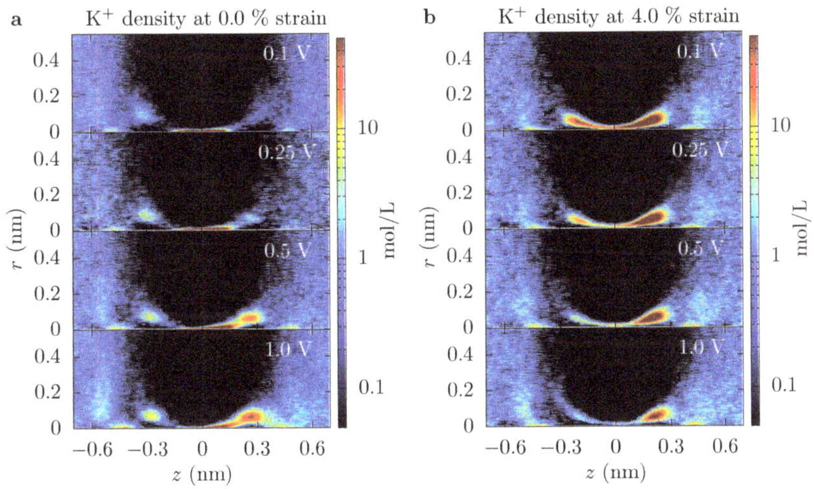

Figure 3. Voltage dependence of the ion concentration. Concentration of potassium ions near a graphene crown ether pore with $q_O = -0.54\,e$ at (**a**) 0% and (**b**) 4% strain for various voltages. For 4% strain, where we see diffusion limitations of the ionic current, we also see the depletion of ions in the pore as the voltage increases. See the Supplemental Material (SM) for additional plots with different parameter values.

Equation (4) gives these qualitative features of the energy landscape and helps to understand why electrostatics can play such a strong role even in high salt solutions. Still, we use all-atom molecular dynamics (MD) simulation for the calculation of quantitative landscape—see the Methods and later discussion about an additional entropic barrier to move into the ABF constriction. The free-energy profiles from MD are shown in Figure 4. Since we are driving the ionic current through a nanopore by an external voltage, we also calculate the energy landscape of ion transport with an applied bias. The equilibrium free-energy barrier alone does not fully represent the energy landscape of ion transport, especially when the applied bias is large compared to the features in the free energy. We note, of course, that even the energy landscape with the bias does not fully capture the current due to kinetic prefactors and averaging effects.

The equilibrium free-energy profiles (blue lines in Figure 4) exhibit a potential barrier for $q_O = -0.24\,e$ and potential well for $q_O = -0.54\,e$ at the center of the pore in the unstrained membrane. In the former, the electrostatic energy (between the ion and the pore charges) is less than the dehydration barrier, while the opposite is true for the latter. Additionally, there can be small potential wells just outside the pore where the ion maintains a larger hydration yet stays close to the negatively charged oxygen atoms of the pore. The energy landscape changes markedly with strain, which is primarily due to the change in the electrostatic interactions within the pore and dehydration outside of the pore. An increase in the pore size—by picometers—due to the strain causes the attractive electrostatic energy to decrease rapidly. The dehydration energy penalty in the pore also decreases with strain but, for small strain, it does not change as rapidly as the electrostatic energy. Consequently, there is a net increase in the energy of the ion at the center of the pore. As a result, the barrier in $q_O = -0.24\,e$ increases, and the potential well in $q_O = -0.54\,e$ flattens and then disappears at large strain (a dehydration-based barrier does appear in the middle of the $q_O = -0.54\,e$ pore, a feature which is already present in the $q_O = -0.24\,e$ pore at 0% strain due to the lower electrostatic compensation). In contrast, the effect of the strain on the free-energy outside the pore is in the opposite direction. The barrier outside the pore decreases with strain as ion can hydrate better with reduced hindrance from the pore oxygen atoms. The electrostatic energy, however, does not change as rapidly as in the center of the pore. The basic mechanism behind these large changes in free energies is that at the 0.1 nm to 0.5 nm scale; picometer changes in atomic configuration result in large changes in electrostatic and dehydration energies [27]. Dielectric screening (from the solution), in particular, is not that effective at this length scale.

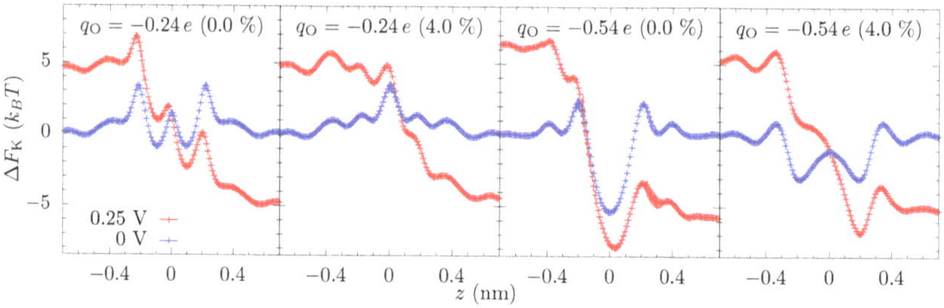

Figure 4. Equilibrium and voltage-dependent landscape for ion transport. The free-energy profile of K^+ going through a graphene crown ether pore at 0% and 4% strain for equilibrium and non-equilibrium (V_{ext} = 0.25 V) cases. The charge of the oxygen atoms of the crown ether is either $-0.24\,e$ or $-0.54\,e$ (and adjacent carbon atoms have half this charge). The potential wells and barriers are mainly the result of competition between the electrostatic attraction and the dehydration. The applied voltage reduces the features (barriers and wells) in the energy landscape but some sharp features still remain, either due to the barriers' size or due to their *irrelevance*, a term we use operationally, see the main text and Figure 5. Irrelevance of barriers occurs since the influence of a barrier on the current is both a kinetic and thermodynamic effect, and other bottlenecks (e.g., diffusion limitations) can exist, i.e., these are the relevant processes at a given voltage and strain. In particular, for 4% strain and $q_O = -0.54\,e$, the supply of ions from bulk has a much larger influence on the ionic transport than the dissociation from the well at 0.2 nm. Hence, that barrier remains roughly unchanged. Error bars are plus/minus one SE from five parallel runs.

The free-energy landscape explains many of the features seen in the ionic current versus strain, Figure 2. At small voltages, the current changes significantly with strain because of the change in the energy landscape of the pore. For $q_O = -0.24\,e$, the entrance barrier, just outside the pore, initially decreases with strain and the current increases rapidly. Eventually, the increase in the energy at the center of the pore will negate the decrease in the outer barrier, and the current subsequently decreases, thus giving a turnover behavior with the optimal current around 3% strain. At very large strains, ion hydration will increase in the pore, and thus the energy barrier at the center will start to disappear. For $q_O = -0.54\,e$, the potential well at the center of the pore becomes shallow with strain, making it easier for ions to dissociate from the pore and contribute to the increase in the current. A common principle for the colossal mechano-conductance change is that the free energy veers toward a barrierless landscape for both these example pore environments. At intermediate strains (4% to 10%), while strain does influence barriers, bulk limitations have kicked in and the barrier change will not be manifest in the current versus strain. Even at small strain, the current will become flat if the applied voltage is large enough, since the larger voltage can wash out larger free energy features. There are, however, irrelevant free energy features—ones that are not rate limiting—that remain even as voltage increases, for which we introduce discrete-barrier and one-way rate analyses below that help identify relevant and irrelevant features. The evolution of features under strain can also suggest their relevance (i.e., if the current is constant versus strain, yet a large feature disappears, that feature is likely—but not guaranteed, since other factors can conspire together—to be irrelevant).

To elucidate the effect of the applied bias on the energy landscape, and hence the current, we calculated the free-energy of the potassium ion in the presence of an external voltage ($V_{\text{ext}} = 0.25$ V). Figure 4 shows that the applied voltage raises the potential on one side of the membrane and decreases on the opposite side and the overall potential roughly drops over $|z| < 0.5$ nm (we also see this drop in the calculation of the electrostatic potential). Note that although the graphene is only 0.3 nm thick, the double layer of cations and anions on the opposite side of the membrane will be separated by a distance of about 1 nm due to their hydrated radii. Nonetheless, even a small voltage will result in a large electric field in the pore which can suppress the energetic features. Yet, some sharp features have a spatial variation larger than the applied field and are still prominent in the free energy landscape with an applied voltage.

To capture how the features in the equilibrium free-energy profile change with applied voltage, we plot in Figure 5 the discrete gradients from each energy minimum, i, to the next maximum in positive z-direction, i.e., $(\Delta F^i_{\text{max}} - \Delta F^i_{\text{min}})/(z^i_{\text{max}} - z^i_{\text{min}})$. The gradient of applied voltage, which is in the opposite direction to these gradients, reduces the barrier to transport (we do not plot the gradients in negative z-direction, which assist rather than hinder the ion translocation). Figure 5 shows that some of the gradients are larger than the electric field from the applied bias ($V_{\text{ext}} = 0.25$ V), and thus these barriers are still present in the energy landscape with applied bias. More importantly, though, the examination of how these discrete gradients change with voltage enables one to identify rather large features that remain unchanged at finite voltages, such as the well at 0.2 nm for the $q_O = -0.54\,e$ and 4% strain case. This well (and associated barrier) is not a limiting factor in transport at this strain and thus the applied bias does not self-consistently remove it. This type of plot (and a related plot we will examine later) give a clear depiction of what features are influencing transport, including indirectly the influence of kinetic prefactors. We further note that the largest gradient for the unstrained pore at $q_O = -0.54\,e$ is about 40 k_BT/nm and thus will require $V_{\text{ext}} \approx 1$ V to effectively wash it out, which is reflected in Figure 2. Once the applied voltage produces local fields larger than the relevant discrete gradients, the ionic current will have little dependence on the equilibrium landscape of the pore, which explains the saturation of the ionic current across all values of strain for large voltages, as we see in Figure 2. Saturation at smaller voltage is a combination of this same washing out plus the presence of irrelevant features due to high kinetic rates (compared to other rates, such as diffusion and entrance-side feeding; see the Supplemental Material (SM) for additional plots of the equilibrium and non-equilibrium free energy barriers).

Some features in the energy-landscape, though, are beyond 0.5 nm from the pore, albeit they are small. These features can survive to large applied voltages. Thus, while they matter little for smaller voltages, they eventually can become important when their energy- and kinetic-scales are commensurate with the other renormalized features. Thus, ions will eventually have to overcome additional entrance barriers. These barriers will directly affect the rate at which ions can enter and exit the pore and thus influence the saturation current through the pore. Conversely, the barrier on the exit side, though significant, has a smaller influence due to the larger dissociation rate, which we will discuss later when examining the interpretation of the rate constants within the model.

We note that since the K^+ ion is confined to a cylindrical region during the ABF calculations, the free energy we present does not include the entropic, 'constriction' barrier to move an ion from bulk to the ABF cylindrical constriction of radius $r_{\text{ABF}} = 0.28$ nm, and vice versa (on the exit side). The ABF constriction allows other ions (both coions and counterions) to be in the volume. The ratio of accessible states is thus approximately $\Omega_{\text{ABF}} = \pi r^2_{\text{ABF}} l / l^3$, where l is the typical distance between co-ions in bulk ($\approx$1.2 nm at 1 mol/L KCl). Thus, the contribution to the free energy of this constriction penalty is $-k_BT \ln \Omega_{\text{ABF}} \approx 1.7\,k_BT$. From within the ABF constriction, the entropic penalty to then go into the pore is included within the ABF calculation. For comparison, this contribution can also be estimated as follows: The geometric pore radius is $r_p = 0.137$ nm, taken as the pore center to oxygen center, 0.29 nm, minus oxygen's van der Waals radius, 0.152 nm. A typical approach to estimate the entropic penalty is the formula, $-k_BT\ln\left(1-(r_{K^+}/a_p)^2\right)$. However, $r_{K^+} \approx a_p$ (see, e.g., Ref. [55] for ionic sizes) and this approach will lead to large errors and, in fact, does not include important physical processes, such as the movement of oxygen atoms at the pore rim. A better approach is to estimate the entropy from the actual trajectories of ions going through the pore. Potassium ions cross the pore within a radius of about $r_c \approx 0.02$ nm from the origin. Assuming that the ions are not localized in a well, but still are locally in equilibrium, the entropic penalty is approximately $-k_BT\ln\left(r_c^2/r_{\mathrm{ABF}}^2\right) \approx 5\,k_BT$. The presence of a well of size $l_W \approx 0.3$ nm in some cases gives an additional contribution $-k_BT\ln\left(l_W/l\right) \approx 1\,k_BT$ to $2\,k_BT$.

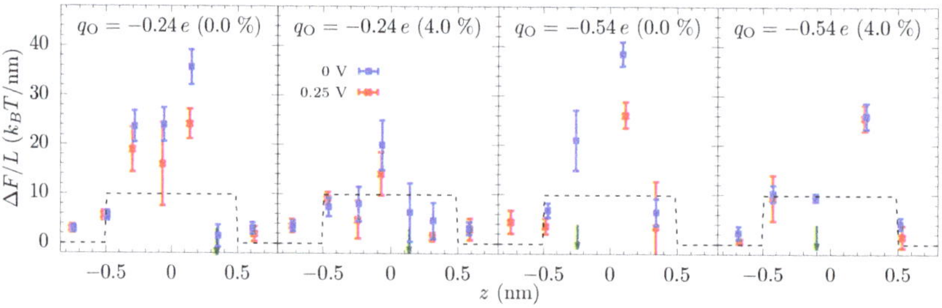

Figure 5. Discrete gradients and the renormalization of energy barriers. The discrete energy gradients encountered by a K^+ ion in graphene crown ether pores at 0% and 4% strain for 0 V (blue) and 0.25 V (red) applied bias. The gradients are between each local minimum and the next maximum in the positive z-direction in Figure 4 and we plot them against the mean position of the barrier. The gradient from the applied bias ($V_{\mathrm{ext}} = 0.25$ V), ideally about 10 k_BT/nm between $z = \pm 0.5$ nm (shown with black, dashed line), reduces the translocation barrier, completely eliminating it in some cases (shown with a green arrow). To the first approximation (in particular, ignoring the self-consistent development of the potential drop), the sharp features that are larger than the ideal electric field will have a significant influence on the ionic current, as they are still present (though reduced) when the voltage is applied. This ideal behavior is approximately occurring in the $q_O = -0.24\,e$ pore, as for both values of strain shown the gradients within the membrane region are being collectively diminished. For $q_O = -0.54\,e$, more complex behavior is occurring, with some features changing more than others. Examining the change in discrete energy gradients upon application of a voltage gives a clear indication of the presence of irrelevant features—these barriers do not change, as they are not rate limiting and do not create a self-consistent potential drop around themselves. The errors are due to the uncertainties in both the position and the magnitude of minima and maxima.

Radius of the pore: Before moving forward, we address an issue that permeates the whole field of transport in sub-nanoscale pores and is apparent in the proceeding paragraph—that of the pore radius. For smooth, uncharged pores, the radius or open area (when not circular) can be rigorously defined with all-atom simulation: One samples the trajectories of ion crossings and takes a weighted average of discrete area elements (see Ref. [16], where the current density was roughly uniform, enabling a direct and intuitive treatment). However, when pore charge is present or the pore has structure, whether steric or energetic, along its length, there is clearly no simple answer for pore radius. The effective radius that defines access resistance, for instance, will not be the same as the geometric radius of the pore mouth. This is easy to see when charge is present at the pore mouth, since the effective opening within a continuum approximation will increase by about a Debye length due to electrostatic attraction of counterions [14]. The fact that effective sites are present will change this picture further. For instance, there are association-side sites that form a staging area from around $z = -0.4$ nm to -0.3 nm with a spread r_s of about 0.1 nm, which is related to a_p but can be influenced by other factors (their numerical values here are the same). It is this region that has to be "accessed."

Moreover, if the pore has a conical shape (e.g., even for this graphene crown ether pore, ions seem to follow a coarse canonical shape, see Figure 3), what radius is relevant to defining the "open area" of the pore, especially when energetic features are present? When variation in size or energy is large on the scale of inter-ion separation and the ion mean-free path, this issue can be handled simply by assuming local equilibrium and appropriately averaging. The graphene pore examined here, as well as other pores in 2D membranes and biological channels, do not have such a simple separation of scales. Fortunately, here, the important length scales that define size fall within the range $r_c \approx 0.02$ nm (spread of trajectories of ion crossings) to $r_s \approx 0.1$ nm (spread of association-side sites) to $r_p = 0.137$ nm (geometric radius) to $\lambda_D \approx 0.3$ nm (Debye length). We will take the effective pore radius as $a_p \approx 0.1$ nm. This value is in the middle of this range and thus, except for a few particular quantities such as the entropic barrier, it gives a reasonable starting point for estimating values of different pore characteristics.

Incoming rates: Before discussing the modeling of these pores, we first introduce a simple tool to further assess the influence of different energetic features. Figure 1 shows the trajectories of K^+ ions moving toward the graphene crown ether pore. The trajectories of ions that eventually translocate through the pore are shown with a red line, and others are shown in cyan lines. Only a few non-translocating trajectories go into the range of $z = -0.3$ nm. This becomes more apparent by plotting the trajectories near the pore (within the radial distance of 0.6 nm from the center of the pore) versus z and time, as seen in Figure 6 upper panel. Information regarding the rejection of ions would thus be helpful. In Figure 6, we thus also plot the incoming rate J_{in} of ions crossing a z-plane versus the z-distance at various applied voltage. Initially, J_{in} drops rapidly with z, as ions have to go through a diffusion constriction and also get reflected by the entrance barrier. At a certain location, J_{in} becomes flat, indicating all ions that made it to that distance will complete the translocation. For $q_O = -0.24\,e$, for example, the rate drop sharply going from $z = -0.5$ nm to $z = -0.2$ nm in the unstrained pore due to the presence of an occupation barrier. The rate then becomes flat, as ions cannot go back (we note that we do see some ion crossing events that go backward, up the potential gradient. These are few and far between, but the small gap in Figure 6 for some cases quantifies this magnitude of these events). For 4% strain, the rate continues to drop until $z = 0$, as there is a large barrier at the center of the pore.

Similar observations can be made for $q_O = -0.54\,e$. For the unstrained pore at small voltage, we see a large drop in the incoming rate between $z = -0.5$ nm and $z = -0.2$ nm due to the repulsion from the ion already in the pore. There is a smaller drop due to dissociation of the ion from the pore. Importantly, both of these drops are due to dissociation, with the former due to a blockade (many-body) effect and the latter being actual ion dissociation. We note that many-body and single-ion effects can be unraveled by comparing the free energies at finite concentration to the free energy of a single ion pair in solution [27], which shows that the satellite barriers for the unstrained, $q_O = -0.54\,e$ pore are due to the presence of an ion in the pore. For the 4% strain (and for the unstrained pore at larger voltage), the drop in the rate is small and it essentially saturates at $z = -0.3$ nm. This means that ions do not feel a significant barrier going through the pore and the total current is only limited by the rate at which ions arrive at the mouth of the pore. As with the discrete barrier gradients, the plot of the one way ion rate allows for the identification of what features matter. For $q_O = -0.54\,e$ at strain at about 4% and above, the reduction in ion flux at the entrance side is due to the diffusion constriction and entrance barriers. These incoming ion rate plots thus provide both qualitative and quantitative information. We will use this to motivate the modeling choices below (specifically, the use of a staging site and the assumption of one-way current flow in the pore).

Reaction rate model: The 18-crown-6 pore in graphene can only fit a single ion at a time. It is thus intuitive to analyze the ionic transport process using rate theory [42]: Ions arrive at the pore at a certain rate and depart at a certain rate, which together provide the ionic current.

The simplest case would be to assume a single site and that ions only move in one direction. The latter takes into account that the bias is sufficiently large that ions cannot move backward, up the potential gradient (this is a reasonable assumption for the voltages in this work, as we saw above, but cannot correctly reproduce equilibrium conditions). In that case, the ionic current through the pore is given as $I/q = (1/k_a + 1/k_d)^{-1}$, where k_a (k_d) is the rate constant for association (dissociation) of ions into (from) the pore. In terms of the site occupancy (equivalently, probability of being occupied), $I/q = k_a(1-P) = k_d P$ and $P = k_a/(k_a + k_d)$. These latter equations make it clear that, with a strain independent association rate k_a, the current will linearly depend on occupancy and thus cannot plateau, as seen in Figure 2, until P is effectively zero. This limit, $k_a \ll k_d$, gives $I = q\,k_a$, in which case the ionic current is fully determined by the incoming rate.

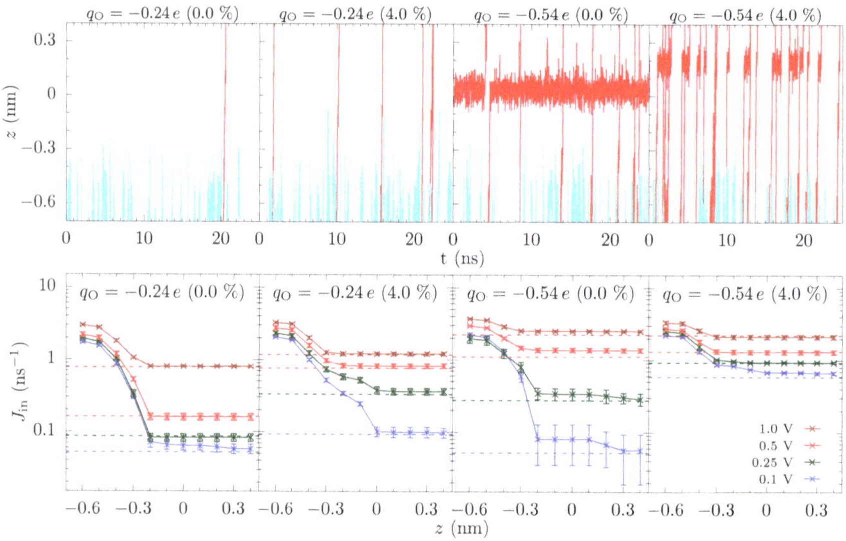

Figure 6. Translocation events and one-way rates. (**Top panel**) Time trace of the z-position of potassium ions that translocate (red) through the graphene crown ether pore and reflect (cyan) after coming within 0.6 nm of the center of the pore. For $q_O = -0.24\,e$, ions cross the pore very quickly and the association rate is the primary determinant of the current. In the unstrained pore with $q_O = -0.54\,e$, the ions spend a significant time in the pore, and thus the dissociation rate determines the current. (**Bottom panel**) The inward rate of K^+ ions versus z-distance at different applied voltages. The dashed horizontal lines gives the net rate. For small voltage, J_{in} near the pore is much smaller than the bulk diffusion rate, and thus the current is limited by the barriers to transport. Error bars are plus/minus one SE. Connecting lines are a guide to the eye only.

We will see that P is still substantial on some of the plateau. Thus, while the fit to a single-site model is reasonable when allowing k_a to have some voltage dependence (i.e., $k_a = k_{a0} + \kappa_a V$), the model is not qualitatively consistent with the data, as the model current still increases when the actual current has leveled off. This assessment of the single-site model is the same regardless of whether only one way motion is assumed or not: Allowing fluctuations in and out of the pore on both sides of the membrane still gives a linear dependence on P with a similar coefficient.

Instead, we examine a three-site model, despite the fact that the channel is atomically thin. The data in Figure 3 show that there are multiple localized regions of enhanced K^+ density. On the association side (left side of the figure), there are candidate sites—staging sites—at about -0.3 nm and about -0.4 nm (offset from each other also in the radial direction), and similarly on the dissociation side. That is, there are 4 or 5 candidate sites in the parameter regimes of that figure (ambiguity results from the fact that the candidate sites on the dissociation side are not fully disconnected—there is a non-negligible probability to find an ion in between some locations). These sites are due to ripples in the free energy, which extend outside the pore, as discussed above and seen in Figure 4. Due to the proximity of the association-side staging sites, we will assume they are the same and employ a three site model. Moreover, the one-way rate data in Figure 6 supports this view of the pore, as well as the assumption that current (mostly) flows in one direction at the pore binding site.

The kinetic equations for the three site system are

$$\dot{P}_1 = k_b(1-P_1) - k_b' P_1 - k_a P_1(1-P_2) \quad (5)$$

$$\dot{P}_2 = k_a P_1(1-P_2) - k_d P_2(1-P_3) \quad (6)$$

$$\dot{P}_3 = k_d P_2(1-P_3) + k_{b'} P_3 - k_{b'}'(1-P_3), \quad (7)$$

where P_i is the occupancy of the site i =1, 2, and 3, k_b (k_b') is the incoming (outgoing) rate from bulk on the association side, and $k_{b'}$ ($k_{b'}'$) the dissociation side. Again, the set of equations assume only one way motion into (the association side), and out of (the dissociation side), the internal pore site $i = 2$. Backward fluctuations can easily be included, but this adds extra parameters to be fitted and will only influence the fit in a minor way. We will apply this model only to the behavior of the $q_O = -0.54\,e$ pore, since the $q_O = -0.24\,e$ pore has more intricate behavior that would ultimately require association rates that are strain-dependent, i.e., that depend on the variation of the free energy landscape. We have discussed the $q_O = -0.24\,e$ pore extensively already in Ref. [27], including the origin and scale of the free energy variation. We only note here that, as seen in Figure 2, the 0.5 V and 1 V biases for $q_O = -0.24\,e$ also give an entrance-limited region. The magnitude of the currents in this region are lower than $q_O = -0.54\,e$ by only an order one factor for the same voltages and strain. The similarities in current are expected for bulk-limited behavior. The fact that they are lower by a small amount is likely due to the increased capture effectiveness of the higher charge pore. The specific estimates for parameters will thus apply in this case, albeit with some small modifications of effective radii and rates.

Even with the assumptions regarding one-way rates at the $i = 2$ site and the symmetry of bulk rates, there are a number of parameters. Instead of direct fitting, we can employ main pore site (P_2) occupancy data from MD to reduce the number of parameters and see if a consistent model results. Considering only Equation (5) and setting $\dot{P}_1 = 0$ yields the occupancy of the first site

$$P_1 = \frac{k_b}{k_b + k_b' + k_a(1-P_2)}. \quad (8)$$

This site and the third site are the least well-defined, and thus eliminating them from the expressions is key to reducing mathematical and computational acrobatics in defining and fitting the quantities in the model. The particle current is given by the last term in Equation (5) (or, equivalently in the steady state, the sum of the first two terms),

$$I/q = k_a P_1(1-P_2) = \frac{k_b k_a (1-P_2)}{k_b + k_b' + k_a(1-P_2)} \quad (9)$$

$$= \left(\frac{1}{k_b} + \frac{1}{\tilde{k}_a(1-P_2)}\right)^{-1}, \quad (10)$$

where $\tilde{k}_a = k_a P_1^{\mathrm{eq}}$ is the effective association rate and $P_1^{\mathrm{eq}} = k_b/(k_b + k_b')$ is the equilibrium density of site 1 in the absence of its connection to the main pore site (in this absence, we can examine equilibrium of P_1). We do not have to separately determine or fit P_1^{eq}, since we can examine solely $\tilde{k}_a$ for association and only k_b to give the influence of bulk. Note that Equation (10) has made no assumptions regarding the relative magnitude of the dissociation rate, or, for that matter, the influence of any of the factors that appear in Equations (6) and (7), other than the $k_a P_1(1 - P_2)$ term common with Equation (5). Thus, the model can be thought of as just Equation (5), which has only the assumption that there is a negligible backward rate from the site 2 to site 1, which as we have seen is justified for much of the parameter ranges examined for $q_O = -0.54\,e$. The form of the bulk rates on the dissociation side and the lack of backward processes on that side is thus inconsequential. Moreover, whether the model is two or three sites is also irrelevant due to our approach. The inclusion of $(1 - P_2)$ in the model, which will be directly extracted from MD, captures the influence of all potential processes on the dissociation side, whether included in Equations (6) and (7) or not.

Figure 7 shows the occupancy of the pore, P_2, and the model results overlaid with the current data. Note that we only fit the model for select points (the one for which occupancy data is shown). Since the conductance depends on voltage, we let $k_b = k_{b0} + \kappa_b V$, which together with $\tilde{k}_a$ gives a three parameter fit. The resulting fit parameters are $k_{b0} = (0.50 \pm 0.03) \times 10^9$ ion/s, $\kappa_b = (2.2 \pm 0.2) \times 10^9$ ions/(V·s), and $\tilde{k}_a = (1.2 \pm 0.3) \times 10^{10}$ ion/s, with uncertainties given by the standard error of the fit. We will discuss these parameters shortly, including their agreement with back-of-the-envelope estimates, as well as providing a quantification of diffusion limitations versus drift-supplied ions.

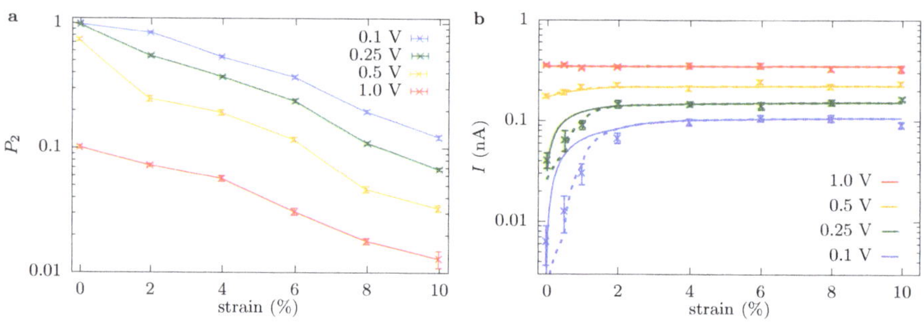

Figure 7. Pore occupancy versus strain and model fit to the current for the $q_O = -0.54\,e$ pore. (**a**). Pore occupancy versus strain for the four voltages indicated. The occupancy is decreasing exponentially with strain and voltage, with some additional, minor features and an apparent threshold behavior with voltage at zero strain. These data are reproduced in the SM along with data for the $q_O = -0.24\,e$ pore. (**b**). Current versus strain at the four voltages labeled, along with the model. The latter was fitted using current-voltage and P_2 data at 0% to 10% strain at 2% increments (i.e., the P_2 data shown in **a**). The continuous model plot is found by linearly interpolating the P_2 data. The model is very good when accounting for diffusion and access limitations. When using the interpolated P_2 data from 2% increments (solid line), there is some deviation at 0.5% and 1% strain. However, using the interpolated P_2 data including those two additional points (dashed line) shows that the issue is that P_2 has features not captured by interpolation at 2% increments (see the SM for the additional P_2 data). The R^2 and adjusted R^2 for the fit are 0.998 and 0.997, respectively (for data from 2% increments). The step-like features are solely due to employing (linear) interpolation to create a continuous curve. Error bars are plus/minus one SE. Connecting lines are a guide to the eye only.

When the model and data are viewed in tandem, the physical behavior is apparent. When the current plateaus versus strain, it is due to combined diffusion and entrance/access limitations, for which without some component of the latter, the current would not increase substantially with voltage (the voltage could only decrease local ion density, increasing and eventually saturating the diffusive contribution in the process). For smaller strains and voltages, the current is dominated by the $\tilde{k}_a(1 - P_2)$ component. That is, the current is dictated by a many-body effect: localization of a K^+ ion prevents current flow until that ion dissociates, in which case an effective particle current of $\tilde{k}_a$ flows while the pore is empty (i.e., in more concrete terms, this regime can be thought of as a current of zero flowing, while the $i = 2$ site is occupied and $\tilde{k}_a$ otherwise, giving $I/q = 0 \cdot P_2 + \tilde{k}_a(1 - P_2)$ and considering P_2, which is between 0 and 1, to be a probability). The many-body nature of transport in this regime is further supported by a decreasing ionic current versus concentration, which shows the saturating nature of the process; see the concentration figures in the SM.

The expression in Equation (10) quantitatively captures the current versus strain and voltage behavior for most of the data. Where it gives the least fidelity to the full simulation result (small strain and low voltage), it still qualitatively captures the trend in the current. For small strain and/or low voltages, this is precisely where backward motion that was neglected in the model is most important, as well as the fact that it is where the sites (except the main binding site) are the least well-defined, see Figure 3. We have seen from molecular dynamics simulations, as well, that there is a small, backward moving current, even at quite large voltage drops. Despite neglecting these effects, we still conclude here that Equation (10) is sufficient to understand and capture ionic transport through the graphene crown ether pore at $q_O = -0.54\,e$, as well as $q_O = -0.24\,e$ at 0.5 V and higher (with a slight modification of rates).

Rate constants: The rate constants k_b, k_a, and k_d depend on attempt frequencies and free-energy barriers that ions encounter during the translocation from one side of the membrane to the other [42,56]. We will consider k_b to have separate diffusion and drift components and for the other two rates to have explicit barriers. When U_a and U_d are the barriers to enter the pore and exit the pore, respectively, then $k_a = k_a^0 e^{-U_a/k_BT}$ and $k_d = k_d^0 e^{-U_d/k_BT}$, where k_a^0 and k_d^0 are the rate constants for barrierless transport.

Bulk rate constant—We first consider the rate constant from bulk, $k_b = k_{b0} + \kappa_b V$, to the sites on the association side. For pure diffusion, the standard result is to take a capture radius equal to the pore radius a_p and solid angle Θ [1], which would give

$$k_{b0} \approx \Theta\, D\, c\, a_p, \tag{11}$$

where D and c are the diffusion coefficient and the bulk ion concentration. We note that here one could argue that we should take r_s (the spread of staging sites) or some modification depending on the Debye length. However, r_s and a_p are related and, indeed, they are equal in this work (as discussed above). The influence of electrostatic interactions is even less clear, as the pore rim is charge neutral on the scale of the Debye length. Thus, we consider a_p only, but there could be further refinement of the estimates of the model and parameters. The solid angle, Θ, is generally taken to be 4π in chemical reactions [37,57] and 2π for transport through pores [38,39]. Instead, considering the pore to be a circular disc rather than a sphere, one obtains $\Theta = 4$ [58,59]. These estimates of Θ implicitly assume that the particle size is negligible compared to the capture radius. A similar estimation for sub-nanoscale pores is difficult due to the commensurate length scales involved: pore size, hydrated ion size, and Debye length are all similar. For example, Läuger pointed out that the effective capture radius of the pore can be as small as the difference between the geometric radius and the ion radius [35], which, of course, would give rise to similar issues that we discussed in the context of the pore radius.

Assuming that a region of radius a_p mimics the capture of a circular disc gives $k_{b0} \approx 0.5\ \text{ns}^{-1}$ for one molar concentration. Different assumptions about the capture geometry yield only an order one deviation in this estimate. Thus, this is in excellent agreement with that found by fitting the current data to the model, employing the pore occupancy directly from MD, which gives 0.5 ns^{-1} also. We note that the same scale of k_b is used for biological ion channels [42].

The driven component of the incoming rate from bulk is also inline with heuristic expectations: Ignoring diffusion and when the current is determined predominantly by the pore itself, see Equation (1), the current will be

$$I \approx \frac{V}{\gamma_p h_p / (\pi a_p^2)} \tag{12}$$

in the continuum limit with pore resistivity γ_p. Again ignoring diffusion, the bulk drift has to supply this same amount of current. Converting to a rate, this gives

$$\kappa_b V \approx \pi a_p^2 V / (q \gamma_p h_p). \tag{13}$$

Alternatively, one can think of this scenario as one where the voltage drop in bulk on one side of the membrane is $V_b \approx \pi a_p \gamma_b V / (4 \gamma_p h_p)$, which comes from taking the exact—assuming a continuum with bulk resistivity γ_b—voltage drop on one side of the bulk $I \cdot R_a$ and approximating the current as in Equation (12). This partial voltage drop then supplies ions at a rate determined by its bulk, access resistance, V_b/R_a (note that here $R_a = \gamma_b/(4a_p)$ as we are dealing with one side of the membrane). This yields a bulk rate identical to Equation (13). The bulk resistivity is $\gamma_b = 0.071\,\Omega{\cdot}\text{m}$ for 1 mol/L KCl in rigid TIP3P water [14]. Putting in approximate values $h_p \approx 1$ nm (see, e.g., Ref. [16]) and $a_p \approx 0.1$ nm yields either 1.4×10^9 ions/(V·s) when using just the K^+ resistivity ($\gamma_p \approx 2\gamma_b$) in Equation (12) or 2.8×10^9 ions/(V·s) when using the KCl resistivity as the pore resistivity (in the bulk, we use γ_b), which is in reasonable agreement with the extracted value of $(2.2 \pm 0.2) \times 10^9$ ions/(V·s). Again, some parameters, such as a_p and h_p, may be different, including when one is looking at different characteristics (access versus pore resistance), but at most this will give an order one change—for instance, employing $a_p \approx 0.13$ nm and $\gamma_p \approx 2\gamma_b$ would give 2.3×10^9 ions/(V·s). In these estimates, we do allow $\gamma_p \neq \gamma_b$, but this is imposed from above rather than a consequence of free energy barriers or concentration gradients, both of which have more complex repercussions. Computing these even within a continuum picture would require a self-consistent solution, including without local electroneutrality. Barriers in the pore, though, are easy to incorporate, they lower the current and thus lower the drift-induced feeding κ_b (equivalently, they reduce V_b). The proximity of the estimates, though, suggests that the pore in the plateau regime is similar to that of a small open pore—"open" meaning no free-energy features. While the pore does have energetic features for smaller strain (i.e., 2% to 6%, see the SM), this entails that those features are irrelevant in the sense developed above. For larger strain (8% and 10%), the pore is basically barrierless, even in a more strict sense (see the SM). The agreement between treating the drift rate as that in response to a small but otherwise open pore ($h_p \approx 1$ nm and $a_p \approx 0.1$ nm) may be coincidental, however, as the ion crossings happen at a smaller scale in the middle of the pore (≈0.02 nm). We will discuss this further below.

We note also that the values for k_{b0} and κ_b are in rough agreement with the one way rates shown in Figure 6. Those rates seemingly would suggest a k_{b0} about 4 times higher. However, these are one-way rates to cross a whole z-plane. Therefore, they will be larger than the rate to go into the staging sites. There are simply more fluctuations in both directions across a z-plane far from the pore when one is in the charge layers that maintain the potential drop. If one instead looks at the rates crossing a hemispherical surface (see the SM), the magnitudes are about a factor of two different than k_{b0}. This agreement is thus still only approximate, but it does suggest consistency of the model and MD data. The agreement with κ_b is also reasonable – the increase of the one-way rates when voltage goes from 0.1 V to 1 V is about 10^9 ion/s to 2×10^9 ion/s, which agrees with the extracted $\kappa_b \approx 2.2 \times 10^9$ ion/s.

Finally, we discuss an alternative potential interpretation (pun intended): Above, we considered rates given separately by bulk diffusion and bulk drift. However, it could be that small entrance barriers, specifically into the association-side staging sites, are giving a weakly-activated process, and hence the voltage enters through the exponent, i.e., $k_b = k_{b0}e^{\beta V/k_B T}$. There are small features in the free energy around which ions would associate into the staging sites, as well as depleted ion density there, see Figures 3 and 4 (and similar figures in the SM). Using this as a fitting form also results in a reasonable fit, albeit slightly worse than the form we use above, especially at low voltage. The resulting fit parameters are $k_{b0} = (0.79 \pm 0.06) \times 10^9$ ion/s, $\beta = (1.3 \pm 0.1)k_B T/\text{V}$ (at room temperature), and $\tilde{k}_a = (1.0 \pm 0.4) \times 10^{10}$ ion/s, with uncertainties given by the standard error of the fit. All these numbers are inline with the heuristic estimates.

The major difference between these two interpretations is the behavior at small voltage and that the drift-based interpretation better captures the data at the smallest voltage we examine (0.1 V). Otherwise, it will be difficult to discern the exact diffusion/entrance mechanism: The expected one-sided access-induced potential drop is 1.5 $k_B T$ at room temperature when the total applied voltage is 1 V (i.e., about 40 $k_B T$). From a homogeneous drift theory [58], about half of this is expected to drop within a distance a_p from the pore (i.e., within the "Hille" hemisphere [19])—that is, one would have to dissect small changes in the free energy and potential with voltage in the same spatial vicinity. The precision to which the calculations would have to be performed is astounding—not just statistical precision, which can be made smaller than this, but non-scaling finite-size effects would have to be nearly completely removed [16,17]. Clearly, studying the temperature dependence can help further delineate these two interpretations by revealing activation energies (provided that the temperature dependence of other factors, such as the resistivity, can be accounted for), as can an even more comprehensive study including smaller voltages (where activation will be more clearly visible) and larger simulation cells (to completely—to within more than $k_B T$—remove non-scaling finite-size effects [16,17]). While the data here favor the drift-based interpretation, it is not conclusive but it does not affect the main findings of a diffusion-limited regime around 0.1 V. Yet another alternative model is to just retain κ_b in k_b (i.e., $k_{b0} = 0$). This assumes that just drift is feeding ions to the pore. However, this model gives a poorer fit and is not consistent with the data. Thus, diffusive contributions from the bulk are present.

Association rate constant—We next consider k_a. The model fit did not directly give us this parameter, but instead the effective association rate $\tilde{k}_a = k_a P_1^{\text{eq}} = (1.2 \pm 0.3) \times 10^{10}$ ion/s. This rate is three times larger than k_b at 1 V, and 30 times larger than k_b at 0.1 V. Thus, the only time this component of the resistance matters is when the factor, $1 - P_2$, multiplying it in Equation (10), is small, which occurs only when both strain and voltage are small. It is difficult to estimate this parameter a priori without sufficient gymnastics as to obscure the truth of the matter. However, there are two qualitative features that support its magnitude. The first is that P_1^{eq} is relatively small, which can be seen from Figure 3, probably around 1/10 or smaller. This means that k_a is an order of magnitude or more larger. From Figure 4, a large value of k_a is expected. There is only a small barrier around −0.4 to −0.3 nm for the $q_O = -0.54\,e$ pore, and then the ion will be driven downhill into the pore binding site. That is, we do expect a large k_a for an ion already in the staging area.

Dissociation rate constant—We next consider k_d. This parameter does not participate in the model fitting at all, since we instead used the computationally determined P_2, which enabled us to only employ the first of the three equations, Equation (5), in the model. However, we can employ the outgoing current from the second site, i.e., the second term in Equation (6), to estimate k_d. We can do this by noticing that P_3 is also small (just as P_1 is). Since here, $1 - P_3$ is present, the estimate assuming P_3 is small will be less sensitive to this assumption compared to P_1^{eq} and k_a. On the plateau, this entails that $k_d P_2 = I/q =$ constant. Examining this relation, data point by data point, gives k_d estimates between 10^9 ion/s to 10^{11} ion/s, with well defined trends versus strain and voltage. For instance, at 1 V, one obtains $k_d = 2.0 \times 10^{10} e^{(0.20 \pm 0.01)s}$ ion/s, where s gives the strain in percent and the confidence interval of the prefactor is $[1.9 \times 10^{10}, 2.2 \times 10^{10}]$ ion/s. Thus, the k_d varies from 2.0×10^{10} ion/s at 0% strain to 1.5×10^{11} ion/s at 10% strain. For completeness, the remaining voltages give fits for k_d of $2.7 \times 10^9 e^{(0.28 \pm 0.03)s}$ ion/s, $8.3 \times 10^8 e^{(0.29 \pm 0.02)s}$ ion/s, and $4.3 \times 10^8 e^{(0.24 \pm 0.02)s}$ ion/s for 0.5 V, 0.25 V, and 0.1 V, respectively, with corresponding prefactor confidence intervals $[2.2 \times 10^9, 3.3 \times 10^9]$ ion/s, $[7.1 \times 10^8, 9.7 \times 10^8]$ ion/s, and $[3.7 \times 10^8, 5.0 \times 10^8]$ ion/s. A reasonable fit to all the plateau data (versus voltage and strain) is $3.2 \times 10^9 V e^{(0.043 \pm 0.003) qV/k_B T + (0.22 \pm 0.01)s}$ ions/(V · s) with confidence interval $[2.8 \times 10^9, 3.6 \times 10^9]$ ions/(V·s) (note that V is a prefactor out front, as well as in the exponent, since ions are driven across the pore). One takeaway from this is not only the order of magnitude of k_d, but that strain and voltage in this regime change barriers in the pore region by $\mathcal{O}(k_B T)$. This is in agreement with estimates of how barriers change due to strain, see Equation (4) and Ref. [27]. However, it is somewhat surprising that voltages, that are 10s of $k_B T$, do not change the barriers by more. The reason that this is not the case is that the pore barriers simultaneously are not playing a strong role in the resistance (i.e., they are irrelevant in the language above and thus do not self-consistently get removed by the voltage) and they occur on a scale of 0.1 nm. This means that what is relevant is the ΔV on this scale. For 1 V that evenly drops over 1 nm, this is only $4\, k_B T$, i.e., only about a factor of 2 to 4 (assuming 0.2 nm over which the barrier occurs) above the actual change found in the fitted form above. We do not expect back-of-the-envelope estimates to do much better.

To give an independent estimate of the dissociation constant for comparison, we assume that only one ion occupies the pore at a time and the translocation is driven by a constant electric field, E_p, across the internal pore site of length Δ_p. Thus, the rate constant for exit from the pore may be estimated from the drift velocity-like picture (with an effective diffusion coefficient as an attempt frequency times an Arrhenius factor) as

$$k_d = \frac{v_d}{\Delta_p} = \frac{q\, D\, e^{-U_d/k_B T}\, E_p}{k_B T\, \Delta_p} = \frac{q\, P_p\, E_p}{k_B T}, \tag{14}$$

where $P_p = D\, e^{-U_d/k_B T} / \Delta_p = D_p/\Delta_p$ is the permeability of the ion. Assuming that $E_p = V/h_p$ with $h_p = 1$ nm (i.e., a potential drop over the effective membrane thickness that is larger than the internal pore site), we obtain $k_d^0 \approx 76\,\mathrm{ns}^{-1}$ (for $\Delta_p = h_p$) to $190\,\mathrm{ns}^{-1}$ (for $\Delta_p = 0.4$ nm, which is more represented of the P_2 site) for 1 V applied voltage and the potassium mobility $\mu_K = qD/k_B T = 7.62 \times 10^{-8}\ \mathrm{m}^2/(\mathrm{V{\cdot}s})$. This is in reasonable agreement with the $k_d \approx 150\,\mathrm{ns}^{-1}$ found above for 1 V and 10% strain, where the latter has the smallest influence of barriers and is thus most similar to the barrier-free estimate here. Notably, however, this estimate decreases linearly with voltage. For 0.1 V, the estimate is 10 times too high compared with the one found from the MD data. However, it is clear from the P_2 data that occupancy is dropping faster than exponentially with voltage, meaning that k_d increases faster than exponentially (note that the pore conductance, proportional to $k_d P_2$ decreases with voltage, in line with diffusion-limited expectations). The form fitted above for all plateau data assumed a form Ve^{vV}, with v as a positive constant. This form performs well and indicates that Equation (14) is only reasonable where the voltage is not modifying the energetic landscape at all.

Finally, we comment on the magnitude of k_d compared to the bulk rate constant k_b. Even taking into account the effect of the potential well in the pore, the dissociation rate constant is still larger than the association rate constant ($\sim$0.5 ns^{-1}) for most cases. For plateau data, the smallest the dissociation constant becomes a factor of two larger, but for almost all data, it is an order of magnitude larger or more. Only in the unstrained pore with $q_O = -0.54\, e$, where there is a large exit barrier, are the two rate constants comparable at a small voltage. Thus, ionic transport in the crown ether pore is—outside of the colossal mechano-conductance regime—generally controlled by the rate at which ions arrive in the pore, i.e., the diffusion and drift rates, possibly with some reflection at the pore mouth due to a small entrance barrier. The latter includes an overall barrier for pore occupation when extended to $q_O = -0.24\,e$, see Figure 4.

Diffusion-limited currents: The observation of diffusion-limited currents requires both that current-carrying ions spend little time in the pore and that drift component of feeding ions to the pore is small. In terms of the model, Equation (10), we need $k_b \ll \tilde{k}_a(1-P_2)$ and that $\kappa_b V_b \ll k_{b0}$. To meet the former condition requires that $k_b \ll \tilde{k}_a$ (i.e., entrance barriers should not be large, or otherwise $\tilde{k}_a$ will be small) and $\tilde{k}_a \ll k_d$ (to ensure that P_2 is not close to one), which combines to the chain of inequalities

$$k_b \ll \tilde{k}_a \ll k_d. \tag{15}$$

In other words, both entrance and exit barriers should be small (i.e., transport in a near barrierless regime, where "near" is defined in terms of how fast ions arrive at the pore from bulk and thus, under realistic conditions, even barriers in the range of 5 k_BT can be "near" barrierless for this pore, but what quantifies "near" depends on pore characteristics).

When $k_b \ll \tilde{k}_a(1-P_2)$ in Equation (10), we get $I = q\,k_b$, in which case the ionic current is fully determined by the incoming rate from bulk and is independent of pore conditions, as seen in Figure 2. Albeit, one has to compare k_b to the association rate, which not only can have a free energy barrier associated with it, but also the equilibrium occupancy of the stating site $P_1^{\rm eq}$, and thus $\tilde{k}_a$ can be quite small itself.

The conditions above can in turn be employed to put conditions on the voltage. First consider a lower bound: Equation (14) gives the rate at which ions cross the pore, including both the drift (due to the local electric field) and the dissociation from a pore well (if present). This rate is proportional to the voltage. Considering the chain inequality above and considering k_d at $U_d = 0$ (i.e., the time spent in the pore without a well needs to be much greater than the bulk feeding—the presence of a well will only push this inequality toward not being satisfied):

$$k_d \gg k_b \approx k_{b0} \implies E_p \approx V/h_p \gg \Theta\, k_BT\, c\, a_p\, h_p/q, \tag{16}$$

where we can take $k_b \approx k_{b0}$, since we are interested in the regime where diffusion dominates over drift. We also take $\Delta_p \approx h_p$ (this only drops an order one factor). This relation indicates that the diffusion-limited current is likely to be observed in short and narrow pores, provided that entrance and exit barriers are small. Note that, although many biological ion channels are not necessarily short compared to their width, ions can move in a single-file concerted motion via "knock-on" mechanisms [27,60], which diminishes the effective length of the channel.

An approximate upper bound on the voltage to observe diffusion limitations is, as already noted, for there to be little voltage drop in the bulk. For instance, Läuger points out that the presence of excess impermeable, or "inert", electrolyte increases the impact of diffusion limitations [35], a fact that occurs in our pore (i.e., Cl^- is inert). This is due to the fact that an impermeable electrolyte shifts the balance of pore and bulk resistance, making the former much larger relative to the latter. Assuming a cylindrical pore, the pore resistance is dominant if $h_p/\pi a_p \gg \gamma_b/2\gamma_p$, where γ_b (γ_p) is the resistivity in bulk (pore). For larger graphene pores, and even some nanoscale ones, this condition is unlikely to be true. In fact, access resistance is larger than the pore resistance for most sizes of graphene pores, and thus the majority of the voltage will drop in the bulk. In such a case, the current will be limited, not by diffusion but mostly by drift.

In the graphene crown ether pore, however, the effective pore radius is around 0.1 nm and, when under strain, near barrierless in the sense used above (there may be barriers and wells, but the prefactors—the transition rates or attempt frequencies—are still determining the hierarchy of rate scales).

Using the same estimate to find κ_b as above, where we assume a homogeneous, continuum medium both inside and outside the pore, with the pore resistance the dominant factor, we obtain

$$\kappa_b V \approx \pi a_p^2 V/(q\gamma_p h_p) = \pi a_p^2 V c_p \mu_p / h_p \ll k_{b0} \approx \Theta D c a_p. \tag{17}$$

This upper bound can be derived directly from the steady state Nernst–Planck equation, assuming hemispherical symmetry (i.e., with only a radial component) and a homogeneous medium. There, one wants

$$\frac{\partial c}{\partial r} \gg \frac{qc}{k_B T}\frac{\partial \Phi}{\partial r}, \tag{18}$$

to have the diffusion contribution much larger than the drift, where Φ is the electric potential. Taking the pore mouth to be a hemisphere with radius a_p. The RHS is $qca_pV_b/(r^2k_BT)$. The LHS is $\Delta ca_p/r^2$, with Δc the concentration bias between the bulk (infinitely far from the pore) and the hemispherical pore mouth. This presumes that the diffusive and drift components are decoupled.

Assuming further that the staging site has zero occupancy (and thus zero concentration), this gives the maximum diffusion contribution and Equation (18) results in

$$qV_b \ll k_B T. \tag{19}$$

This relation is interesting in itself. Its simplicity is due to Einstein's relation of mobility and diffusion coefficients, which results in additional factors dropping out, and due to comparing a maximum diffusive current occurring at the largest concentration gradient with the maximum drift current occurring at zero concentration gradient. Equation (19) indicates that for drift to be negligible, the voltage drop in bulk has to be less than the thermal energy. The latter "drives" the diffusion. It should be larger than the drive of the drift current from bulk to the pore. Plugging in the form of V_b assuming a bulk potential drop in the presence of a dominant pore resistance (see just below Equation (13)) gives the same inequality as Equation (17) up to order one factors.

Rewriting Equation (17) together with Equation (16), assuming $c_p = c$ and $\mu_p = \mu$ (i.e., that these two quantities are equal to their bulk), to obtain a two-sided inequality for V yields

$$\Theta k_B T c a_p h_p^2 \ll qV \ll \Theta k_B T h_p/(\pi a_p). \tag{20}$$

This foundational relation gives one of the main predictions of this paper: Diffusion-limited currents appear within a sweet spot when free energy features are irrelevant. For the graphene crown ether pore, the voltage should be between about 6 mV and 300 mV. At voltages higher than this range, drift will be important and, below this range, ions will not be removed from the pore region fast enough to create a concentration gradient (and free energy features will also become relevant). The simulations and modeling validate the upper bound (at 0.25 V, the bulk drift and diffusion contributions are roughly equal), but they do not address the lower bound (in any case, free energy barriers will likely be relevant at 6 mV for the $q_O = -0.54\,e$ pore, as they are with the $q_O = -0.24\,e$ pore still at 100 mV and 250 mV, the relevance of which is inconsistent with the assumptions leading to Equation (20)). This range includes the voltage, 0.1 V, that we see the strongest diffusion limitations, whereas at higher voltages, drift starts to determine the current. The pore charge is important, as it determines when free energy features are irrelevant (e.g., at 0.1 V but 0% strain, the free energy landscape is dominant). Around 0.1 V is a typical value for graphene pore experiments, small enough to not degrade the membrane (i.e., 0.5 V and higher will start to see membrane degradation), but large enough that typical currents are in the 10 s of picoampere or more (while the time-resolution is irrelevant to measuring the dc conductance, we do note that pin hole leaks or other factors can set a baseline resolution of the current, around 0.5 pA at 0.1 V (see Ref. [33] where such currents in "as-grown" membranes could vary by an order of magnitude from membrane to membrane). We note that the rate model that we are developing cannot be used at very small voltages, as it includes only one way currents in the pore which cannot capture the approach to equilibrium as $V \to 0$.

In any case, there should be a very small drift current in the bulk when the applied voltage is around 0.1 V and it should start to become comparable to the diffusive component at about 0.25 V and dominant for higher voltage. We can make this quantitative using the fit to the model in Equation (10). For instance, at 0.1 V, the unstrained membrane has an effective pore associate rate, $\tilde{k}_a(1 - P_2)$ of 12/µs due to the presence of a localized ion that creates a many-body blocking effect (i.e., $P_2 \approx 1$). This effective rate increases to 11/ns for 10% strain. Meanwhile, the diffusive rate is 0.5/ns and the drift rate is 0.2/ns. Thus, diffusion supplies ions over drift by more than a factor of 2 over the whole range of strains, and already at 2%, the strain is smaller (though comparable) to the pore association rate (about 1.9/ns). At 0.25 V, the diffusive and drift components are comparable at 0.5/ns and 0.55/ns, respectively. These values are slightly more than the 0.25/ns effective pore association rate at 0% strain, but are the controlling factors for essentially all strains at 2% and above.

Since the smallest voltage we consider, 0.1 V, has smallest bulk drift contribution, we can postulate that the plateau resistance is the closest to $\gamma_p h_p/(\pi a_p^2)$ (i.e., without any access component). Employing $\gamma_p/2 = \gamma_b = 0.071\ \Omega{\cdot}\mathrm{m}$ (γ_b is the resistivity of 1 mol/L KCl in TIP3P water, see Ref. [14]) and $h_p \approx 1$ nm, this gives $a_p \approx 0.11$ nm, in agreement with the effective pore radius. This is unexpected, since the potassium ions cannot make use of the full pore area for transport and there are diffusion limitations. There may be several factors that conspire to give this agreement. One is that the pore rim is not fixed but can instead move, so that a_p can be bigger than a priori expectations. This does not, however, seem to be the case, since the density plots show that ions are translocating closer to the origin than 0.1 nm. Another factor is the role of h_p. The effective thickness may be smaller than 1 nm (its value for unfunctionalized graphene pores [16]). Moreover, while transport veers toward barrierless transport, the pores are not becoming barrierless in the strict sense for either $q_O = -0.54\,e$ (until high strain, see the SM) or $q_O = -0.24\,e$ (see Ref. [27] for the discussion of the latter case). However, localized binding sites can give a rate that is similar in magnitude to free diffusion through the pore constriction, or even a higher rate, because, while ions have to jump out of the well and the barrier height thus suppresses the rate, there is still a large prefactor, since the ion is fluctuating rapidly. The enhanced density can push the currents higher than expected based on just an open area. Whether we should think about the pore as an open pore of radius 0.1 nm or whether it is a pore of radius 0.02 nm with an enhanced density due to binding, is an interesting question. Evidence—specifically the higher concentration in a smaller spatial region—suggests the latter. However, we only point out that there is still broad agreement between these two perspectives and they only inform us how we should dissect the pore resistance R_p into component pieces (meaning, the utility of the perspectives is limited).

It is to be noted that the access resistance in an MD simulation (or any other method) depends on the simulation cell size: one can make it arbitrarily small (using a wide and short cell) or large (using tall and narrow cell) [16]. Therefore, in order to match experimental conditions, which effectively has an infinite bulk, one has to exert great care. In our simulation, we chose the simulation cell aspect ratio to be the golden aspect ratio [17], which ensures that the access resistance represents the infinite, balanced bulk resistance. Without taking this approach, one could not examine the bulk diffusion and access limitations. In the SM, we show results of simulations for several different voltages and strains, showing that the golden aspect ratio gives converged currents, ones where the bulk is properly included.

4. Conclusions

Diffusion-limited ionic currents are commonly observed in biological channels because they can provide the necessary conditions: a large pore resistance (compared to access resistance) due to the small pore radius, but also a high permeability of ions due to the functional groups that facilitate the transport of ions [61]. However, diffusion limitations have not been studied systematically in synthetic nanopores, since it is difficult to replicate the permeability of biological ion channels. In this regard, strained synthetic pores may provide a platform, not only to investigate the competition of dehydration and electrostatic interactions within precision atomic constructions that lead to optimal transport characteristics [27], but to investigate diffusion and entrance effects in ionic transport.

We have shown that there is broad agreement between a simple, many-body model developed here for the $q_O = -0.54\,e$ pore and the all-atom simulations, encompassing not only the residuals and current fit, but also with the fit parameters themselves and independent estimates. This agreement suggests that, with strain, this pore transitions from a barrier-limited pore with current dictated by many-body mechanisms (i.e., a well with a localized ion that blocks the pore), to one equivalent to an open tiny pore. The pore still has a free energy structure, but this structure is irrelevant in the plateau regime: At small voltages, current-limiting regions of the landscape—dictated both by the barrier scale and the kinetics—will appear. Larger voltages will start to self-consistently remove those limiting regions by the counteracting local voltage drop. Other regions of the landscape will start to be limiting, and those regions will subsequently be washed out. For a given strain and voltage, though, the type of behavior observed depends on the pore charge and ion dehydration energy. In the particular pore here, the extent of the bulk-limited region reflects whether the unstrained pore has an internal (dehydration-dominated) barrier or (electrostatically stabilized) well.

Therefore, the transition from the barrier-limited to the diffusion-limited regime gives the opportunity to experimentally delineate and constrain the electromechanical environment of the pore, thus pushing further the limits of employing synthetic pores to understand complex mechanisms in sub-nanoscale ion transport. Functionalized pores in two dimensional membranes are thus simultaneously complicated enough to display a wide-range of ionic phenomena seen in biological pores and simple enough to be amenable to direct modeling. Moreover, if other information can be experimentally determined, such as the (equilibrium) pore occupancy (P_2 here), then measurement will enable the extraction of kinetic rates and barriers via modeling. In other words, the graphene crown ether pore is about as simple a sub-nanoscale pore as possible. Yet, it displays a wide variety of behavior: single versus many-body ion competition, optimality, diffusion limitations, relevant versus irrelevant features, etc. Its behavior, for instance, will enable quantifying aspects of transport, such as the role of precision atomic placement and charge in biological systems, and a theoretical understanding of what "near barrierless" entails in particular pores. This area is vast, and pores in 2D membranes will provide the landscape for a systematic experimental exploration and validation of theoretical models of sub-nanoscale pores and biological channels.

Supplementary Materials: The supplemental material (MS) is available online at http://www.mdpi.com/1099-4300/22/11/1326/s1.

Author Contributions: S.S. performed the numerical calculations. Both authors modeled and analyzed data, wrote the manuscript, and clarified the ideas. All authors have read and agreed to the published version of the manuscript.

Funding: This research received no external funding.

Acknowledgments: The authors thank David P. Hoogerheide, James Alexander Liddle, Jacob Majikes, and Justin Elenewski for comments on the manuscript. S.S. acknowledges support under the Cooperative Research Agreement between the University of Maryland and the National Institute of Standards and Technology Physical Measurement Laboratory, Award 70NANB14H209, through the University of Maryland.

Conflicts of Interest: The authors declare no conflict of interest.

Abbreviations

The following abbreviations are used in this manuscript:

MD	Molecular Dynamics
ABF	Adaptive Biasing Method
SE	Standard Error

References

1. Hille, B. *Ion Channels of Excitable Membranes*; Sinauer: Sunderland, MA, USA, 2001.
2. Zheng, J.; Trudeau, M.C. *Handbook of Ion Channels*; CRC Press: Boca Raton, FL, USA, 2015.
3. Abraham, J.; Vasu, K.S.; Williams, C.D.; Gopinadhan, K.; Su, Y.; Cherian, C.T.; Dix, J.; Prestat, E.; Haigh, S.J.; Grigorieva, I.V.; et al. Tunable sieving of ions using graphene oxide membranes. *Nat. Nanotechnol.* **2017**, *12*, 546–550.
4. Surwade, S.P.; Smirnov, S.N.; Vlassiouk, I.V.; Unocic, R.R.; Veith, G.M.; Dai, S.; Mahurin, S.M. Water desalination using nanoporous single-layer graphene. *Nat. Nanotechnol.* **2015**, *10*, 459–464.
5. Cohen-Tanugi, D.; Grossman, J.C. Water Desalination across Nanoporous Graphene. *Nano Lett.* **2012**, *12*, 3602–3608.
6. Feng, J.; Graf, M.; Liu, K.; Ovchinnikov, D.; Dumcenco, D.; Heiranian, M.; Nandigana, V.; Aluru, N.R.; Kis, A.; Radenovic, A. Single-layer MoS 2 nanopores as nanopower generators. *Nature* **2016**, *536*, 197.
7. Walker, M.I.; Ubych, K.; Saraswat, V.; Chalklen, E.A.; Braeuninger-Weimer, P.; Caneva, S.; Weatherup, R.S.; Hofmann, S.; Keyser, U.F. Extrinsic Cation Selectivity of 2D Membranes. *ACS Nano* **2017**, *11*, 1340–1346.
8. Graf, M.; Lihter, M.; Unuchek, D.; Sarathy, A.; Leburton, J.P.; Kis, A.; Radenovic, A. Light-enhanced blue energy generation using MoS2 nanopores. *Joule* **2019**, *3*, 1549–1564.
9. Zwolak, M.; Di Ventra, M. Colloquium: Physical approaches to DNA sequencing and detection. *Rev. Mod. Phys.* **2008**, *80*, 141–165.
10. Heerema, S.J.; Dekker, C. Graphene nanodevices for DNA sequencing. *Nat. Nanotechnol.* **2016**, *11*, 127–136.
11. Deamer, D.; Akeson, M.; Branton, D. Three decades of nanopore sequencing. *Nat. Biotechnol.* **2016**, *34*, 518.
12. Si, W.; Aksimentiev, A. Nanopore sensing of protein folding. *ACS Nano* **2017**, *11*, 7091–7100.
13. Danda, G.; Drndić, M. Two-dimensional nanopores and nanoporous membranes for ion and molecule transport. *Curr. Opin. Biotechnol.* **2019**, *55*, 124–133.
14. Sahu, S.; Zwolak, M. Colloquium: Ionic phenomena in nanoscale pores through 2D materials. *Rev. Mod. Phys.* **2019**, *91*, 021004.
15. Faucher, S.; Aluru, N.; Bazant, M.Z.; Blankschtein, D.; Brozena, A.H.; Cumings, J.; Pedro de Souza, J.; Elimelech, M.; Epsztein, R.; Fourkas, J.T.; et al. Critical knowledge gaps in mass transport through single-digit nanopores: A review and perspective. *J. Phys. Chem. C* **2019**, *123*, 21309–21326.
16. Sahu, S.; Zwolak, M. Maxwell-Hall access resistance in graphene nanopores. *Phys. Chem. Chem. Phys.* **2018**, *20*, 4646–4651.
17. Sahu, S.; Zwolak, M. Golden aspect ratio for ion transport simulation in nanopores. *Phys. Rev. E* **2018**, *98*, 012404.
18. Aguilella-Arzo, M.; Aguilella, V.M. Access resistance in protein nanopores. A structure based computational approach. *Bioelectrochemistry* **2020**, *131*, 107371.
19. Hille, B. Ionic channels in nerve membranes. *Prog. Biophys. Mol. Biol.* **1970**, *21*, 1–32.
20. Zwolak, M.; Lagerqvist, J.; Di Ventra, M. Quantized ionic conductance in nanopores. *Phys. Rev. Lett.* **2009**, *103*, 128102.
21. Song, C.; Corry, B. Intrinsic ion selectivity of narrow hydrophobic pores. *J. Phys. Chem. B* **2009**, *113*, 7642–7649.
22. Zwolak, M.; Wilson, J.; Di Ventra, M. Dehydration and ionic conductance quantization in nanopores. *J. Phys. Condens. Matter* **2010**, *22*, 454126.

23. Sahu, S.; Di Ventra, M.; Zwolak, M. Dehydration as a Universal Mechanism for Ion Selectivity in Graphene and Other Atomically Thin Pores. *Nano Lett.* **2017**, *17*, 4719–4724.
24. Sahu, S.; Zwolak, M. Ionic selectivity and filtration from fragmented dehydration in multilayer graphene nanopores. *Nanoscale* **2017**, *9*, 11424–11428.
25. Sint, K.; Wang, B.; Král, P. Selective ion passage through functionalized graphene nanopores. *J. Am. Chem. Soc.* **2008**, *130*, 16448–16449.
26. He, Z.; Zhou, J.; Lu, X.; Corry, B. Bioinspired graphene nanopores with voltage-tunable ion selectivity for Na^+ and K^+. *ACS Nano* **2013**, *7*, 10148–10157.
27. Sahu, S.; Elenewski, J.; Rohmann, C.; Zwolak, M. Optimal transport and colossal ionic mechano–resistance in graphene crown ethers. *Sci. Adv.* **2019**, *5*, eaaw5478.
28. Guardiani, C.; Gibby, W.A.; Barabash, M.L.; Luchinsky, D.G.; McClintock, P.V. Exploring the pore charge dependence of K+ and Cl- permeation across a graphene monolayer: a molecular dynamics study. *RSC Adv.* **2019**, *9*, 20402–20414.
29. Gibby, W.A.T.; Barabash, M.; Guardiani, C.; Luchinsky, D.; McClintock, P.V.E. The role of noise in determining selective ionic conduction through nano-pores. In Proceedings of the 2018 IEEE 13th Nanotechnology Materials and Devices Conference (NMDC), Portland, OR, USA, 14–17 October 2018; pp. 1–6.
30. Liu, K.; Lihter, M.; Sarathy, A.; Caneva, S.; Qiu, H.; Deiana, D.; Tileli, V.; Alexander, D.T.; Hofmann, S.; Dumcenco, D.; et al. Geometrical effect in 2D nanopores. *Nano Lett.* **2017**, *17*, 4223–4230.
31. Bocquet, L. Nanofluidics coming of age. *Nat. Mater.* **2020**, *19*, 254–256.
32. Epsztein, R.; DuChanois, R.M.; Ritt, C.L.; Noy, A.; Elimelech, M. Towards single-species selectivity of membranes with subnanometre pores. *Nat. Nanotechnol.* **2020**, *15*, 426–436.
33. Garaj, S.; Hubbard, W.; Reina, A.; Kong, J.; Branton, D.; Golovchenko, J. Graphene as a subnanometre trans-electrode membrane. *Nature* **2010**, *467*, 190–193.
34. Sahu, S.; Zwolak, M. The golden aspect ratio for ion transport. *arXiv* **2017**, arXiv:1711.00472.
35. Läuger, P. Diffusion-limited ion flow through pores. *Biochim. Biophys. Acta* **1976**, *455*, 493–509.
36. Gates, P.Y.; Cooper, K.E.; Eisenberg, R.S. Analytical diffusion models for membrane channels. In *Ion Channels*; Springer: Boston MA, USA, 1990; pp. 223–281.
37. Collins, F.C.; Kimball, G.E. Diffusion-controlled reaction rates. *J. Colloid Sci.* **1949**, *4*, 425–437.
38. Behr, J.P.; Kirch, M.; Lehn, J.M. Carrier-mediated transport through bulk liquid membranes: dependence of transport rates and selectivity on carrier properties in a diffusion-limited process. *J. Am. Chem. Soc.* **1985**, *107*, 241–246.
39. Wanunu, M.; Morrison, W.; Rabin, Y.; Grosberg, A.Y.; Meller, A. Electrostatic focusing of unlabelled DNA into nanoscale pores using a salt gradient. *Nat. Nanotechnol.* **2010**, *5*, 160.
40. Brelidze, T.I.; Magleby, K.L. Probing the geometry of the inner vestibule of BK channels with sugars. *J. Gen. Physiol.* **2005**, *126*, 105–121.
41. Schroeder, I.; Hansen, U.P. Saturation and microsecond gating of current indicate depletion-induced instability of the MaxiK selectivity filter. *J. Gen. Physiol.* **2007**, *130*, 83–97.
42. Läuger, P. Ion transport through pores: a rate-theory analysis. *Biochim. Biophys. Acta-Biomembr.* **1973**, *311*, 423–441.
43. Berezhkovskii, A.M.; Bezrukov, S.M. Optimizing transport of metabolites through large channels: molecular sieves with and without binding. *Biophys. J.* **2005**, *88*, L17–L19.
44. Kasianowicz, J.J.; Nguyen, T.L.; Stanford, V.M. Enhancing molecular flux through nanopores by means of attractive interactions. *Proc. Natl. Acad. Sci. USA* **2006**, *103*, 11431–11432.
45. Bormann, J.; Hamill, O.P.; Sakmann, B. Mechanism of anion permeation through channels gated by glycine and gamma-aminobutyric acid in mouse cultured spinal neurones. *J. Physiol.* **1987**, *385*, 243–286.
46. Guo, J.; Lee, J.; Contescu, C.I.; Gallego, N.C.; Pantelides, S.T.; Pennycook, S.J.; Moyer, B.A.; Chisholm, M.F. Crown ethers in graphene. *Nat. Commun.* **2014**, *5*, 5389.
47. Phillips, J.C.; Braun, R.; Wang, W.; Gumbart, J.; Tajkhorshid, E.; Villa, E.; Chipot, C.; Skeel, R.D.; Kale, L.; Schulten, K. Scalable molecular dynamics with NAMD. *J. Comput. Chem.* **2005**, *26*, 1781–1802.

48. Darve, E.; Rodríguez-Gómez, D.; Pohorille, A. Adaptive biasing force method for scalar and vector free energy calculations. *J. Chem. Phys.* **2008**, *128*, 144120.
49. Lee, C.; Wei, X.; Kysar, J.W.; Hone, J. Measurement of the elastic properties and intrinsic strength of monolayer graphene. *Science* **2008**, *321*, 385–388.
50. Pereira, V.M.; Neto, A.C.; Peres, N. Tight-binding approach to uniaxial strain in graphene. *Phys. Rev. B* **2009**, *80*, 045401.
51. Heath, J.J.; Kuroda, M.A. First principles studies of the interactions between alkali metal elements and oxygen-passivated nanopores in graphene. *Phys. Chem. Chem. Phys.* **2018**, *20*, 25822–25828.
52. Glendening, E.D.; Feller, D.; Thompson, M.A. An ab initio investigation of the structure and alkali metal cation selectivity of 18-crown-6. *J. Am. Chem. Soc.* **1994**, *116*, 10657–10669.
53. Fang, A.; Kroenlein, K.; Riccardi, D.; Smolyanitsky, A. Highly mechanosensitive ion channels from graphene-embedded crown ethers. *Nat. Mater.* **2019**, *18*, 76–81.
54. Fumagalli, L.; Esfandiar, A.; Fabregas, R.; Hu, S.; Ares, P.; Janardanan, A.; Yang, Q.; Radha, B.; Taniguchi, T.; Watanabe, K.; et al. Anomalously low dielectric constant of confined water. *Science* **2018**, *360*, 1339–1342.
55. Mähler, J.; Persson, I. A study of the hydration of the alkali metal ions in aqueous solution. *Inorg. Chem.* **2012**, *51*, 425–438.
56. Nelson, P.H. A permeation theory for single-file ion channels: One-and two-step models. *J. Chem. Phys.* **2011**, *134*, 04B615.
57. Berg, H.C.; Purcell, E.M. Physics of chemoreception. *Biophys. J.* **1977**, *20*, 193–219.
58. Hall, J.E. Access Reistance of a Small Circular Pore. *J. Gen. Physiol.* **1975**, *66*, 531–532.
59. Hill, T.L. Effect of rotation on the diffusion-controlled rate of ligand-protein association. *Proc. Natl. Acad. Sci. USA* **1975**, *72*, 4918–4922.
60. Kopec, W.; Köpfer, D.A.; Vickery, O.N.; Bondarenko, A.S.; Jansen, T.L.; de Groot, B.L.; Zachariae, U. Direct knock-on of desolvated ions governs strict ion selectivity in K+ channels. *Nat. Chem.* **2018**, *10*, 813.
61. Doyle, D.A.; Cabral, J.A.M.; Pfuetzner, R.A.; Kuo, A.; Gulbis, J.M.; Cohen, S.L.; Chait, B.T.; MacKinnon, R. The Structure of the Potassium Channel: Molecular Basis of K+ Conduction and Selectivity. *Science* **1998**, *280*, 69–77.

entropy

MDPI

Article

Review and Modification of Entropy Modeling for Steric Effects in the Poisson-Boltzmann Equation

Tzyy-Leng Horng

Department of Applied Mathematics, Feng Chia University, Taichung 40724, Taiwan; tlhorng@fcu.edu.tw

Received: 11 May 2020; Accepted: 4 June 2020; Published: 8 June 2020

Abstract: The classical Poisson-Boltzmann model can only work when ion concentrations are very dilute, which often does not match the experimental conditions. Researchers have been working on the modification of the model to include the steric effect of ions, which is non-negligible when the ion concentrations are not dilute. Generally the steric effect was modeled to correct the Helmholtz free energy either through its internal energy or entropy, and an overview is given here. The Bikerman model, based on adding solvent entropy to the free energy through the concept of volume exclusion, is a rather popular steric-effect model nowadays. However, ion sizes are treated as identical in the Bikerman model, making an extension of the Bikerman model to include specific ion sizes desirable. Directly replacing the ions of non-specific size by specific ones in the model seems natural and has been accepted by many researchers in this field. However, this straightforward modification does not have a free energy formula to support it. Here modifications of the Bikerman model to include specific ion sizes have been developed iteratively, and such a model is achieved with a guarantee that: (1) it can approach Boltzmann distribution at diluteness; (2) it can reach saturation limit as the reciprocal of specific ion size under extreme electrostatic conditions; (3) its entropy can be derived by mean-field lattice gas model.

Keywords: steric effect; Poisson-Boltzmann model; Bikerman model; entropy; specific ion size

1. Introduction

One of the major limitations of the Poisson-Boltzmann (PB) and Poisson-Nernst-Planck (PNP) models is the assumption of point-like ions without considering their sizes. These models based on mean field theories work well for dilute electrolytes, but break down when the concentration is high and ions are crowded in it. A high concentration would generally cause steric repulsions and additional electrostatic correlations among ions, that cannot be described by classical PB/PNP models [1]. For example, the concentration of counter-ions, predicted by PB, can be unrealistically high near the electrode surface, when the electrode voltage is large. Another example occurs at the selectivity filter in a potassium channel, where potassium ions are strongly attracted into this extremely narrow filter by the strong negative charges of oxygens on the backbone of the filter. Employing classical PB/PNP models would overestimate the density of potassium inside the filter and give incorrect channel current predictions. Therefore, many researchers have worked on the modification of PB/PNP to include the steric effect of ions.

Steric effect has long been approached in modeling by modifying either the internal energy or entropy in the Helmholtz free energy. Through internal energy, the steric effect has been featured as excess hard-sphere energy either by density functional theory (DFT) [2,3] or Lennard-Jones potential [4]. These energies were all formulated using non-local potentials and cause the resultant modified PB/PNP to produce a series of complicated integro-differential equations, which are hard to compute in higher dimensions. For practical implementations, localization of hard-sphere potential and simplifying integro-differential equations into pure differential equations has been conducted in [5–8] for DFT and [9] for Lennard-Jones potential.

Through an entropy approach, Bikerman modified the classical Boltzmann distribution by adjusting bulk and local ion concentrations via the excluded volume concept [10]. Borukhov et al. [11] rigorously derived the same formula independently by adding solvent entropy through excluded volume into the Helmholtz free energy. Although the localized hard-sphere model-based DFT [5–8] also captures this solvent entropy as one of the terms accounting for excess hard-sphere chemical potential, the Bikerman model [10,11] has been a more popular steric model due to its easiness of application and qualitatively good agreement with experiments [12–15].

In order to obtain the potential and further derive a neat modified PB equation from free energy, Borukhov et al. [11] treated all ions as having identical size, which has been long criticized for neglecting specific ion sizes. Researchers have tried to address this shortcoming with specific ion sizes, and many of them simply extended original Bikerman model by replacing the identical ion size with specific ones without any rigorous justification. Although the resultant model has a better agreement with experiments than the original Bikerman model [16–18], it does not have a Helmholtz free energy to support it. Here modifications of the Bikerman model to include specific ion sizes have been developed iteratively in Sections 4–6, preceded by derivations of classical PB in Section 2, and the original Bikerman model in Section 3. Finally, in the Discussion and Conclusions section a specific-ion-size Bikerman model is presented with a guarantee that: (1) it can approach Boltzmann distribution at diluteness; (2) it can reach the saturation limit as the reciprocal of specific ion size under extreme electrostatic conditions; (3) its entropy can be derived by a mean-field lattice gas model.

2. Classical Poisson-Boltzmann Model

Though the classical PB model is well known, we still derive the model here for review and comparison with its modified versions discussed later. Starting by stating the Helmholtz free energy, internal energy and entropy, we have:

$$F = U - TS \tag{1}$$

$$U = \int \left[-\frac{\varepsilon}{2} |\nabla\phi|^2 + z_p e p\phi + z_n e n\phi + q\phi + pW_{sol,p} + nW_{sol,n} \right] dV, \tag{2}$$

$$-TS = \int k_B T \left[p \, \log\frac{p}{c_0} - p + n \, \log\frac{n}{c_0} - n \right] dV, \tag{3}$$

where F is Helmholtz free energy; U is internal energy; T is temperature; S is entropy; ϕ is electric potential. p, n denote cation/anion concentrations, and z_p, z_n denote their valence, respectively. e denotes elementary charge. q denotes permanent charge. Permittivity $\varepsilon = \varepsilon_0\varepsilon_r$ with ε_0 being the permittivity for vacuum and ε_r being the relative permittivity or dielectric constant. c_0 is some reference concentration such as bulk concentration of electrolyte. $W_{sol,p}$ and $W_{sol,n}$ denote the solvation energies for cations and anions, respectively. Although the traditional PB model generally does not include solvation energy in the expression, it is important when modeling some electrolyte systems involving hydration/dehydration of ions and is therefore it is explicitly included in the energy here. Based on the Born model, the solvation energies for cations and anions are:

$$W_{sol,i} = \frac{z_i^2 e^2}{8\pi\varepsilon_0 r_i}\left(\frac{1}{\varepsilon_r(\boldsymbol{x})} - 1\right), \; i = p, n. \tag{4}$$

Differentiation of F with respect to ϕ gives the Poisson equation:

$$-\nabla\cdot(\varepsilon(\boldsymbol{x})\nabla\phi) = z_p e p + z_n e n + q \tag{5}$$

By doing the differentiation of F with respect to p and n, we obtain the chemical potentials for p and n, respectively:

$$\frac{\partial F}{\partial p} = \mu_p = z_p e\phi + k_B T \, \log\frac{p}{c_0} + W_{sol,p}, \tag{6}$$

$$\frac{\partial F}{\partial n} = \mu_n = z_n e\phi + k_B T \log\frac{n}{c_0} + W_{sol,n} \tag{7}$$

At equilibrium, the chemical potential is uniform everywhere and therefore the local chemical potential must be equal to its bulk value, which is usually known:

$$\mu_p = \mu_{p,b},\ \mu_n = \mu_{n,b}, \tag{8}$$

with the bulk chemical potential for cations and anions:

$$\mu_{p,b} = k_B T \log\left(\frac{p_b}{c_0}\right) + W_{sol,p,b},\ \mu_{n,b} = k_B T \log\left(\frac{n_b}{c_0}\right) + W_{sol,n,b}, \tag{9}$$

where the subscript b denotes the bulk situation. Equations (8) and (9) can be solved solve for p and n:

$$p = p_b e^{-\beta z_p e\phi} e^{-\beta \Delta W_{sol,p}} = p_b e^{-\beta E_p},\ n = n_b e^{-\beta z_n e\phi} e^{-\beta \Delta W_{sol,n}} = n_b e^{-\beta E_n}, \tag{10}$$

where $\beta = 1/k_B T$, $E_p = z_p e\phi + \Delta W_{sol,p}$, $E_n = z_n e\phi + \Delta W_{sol,n}$, and:

$$\Delta W_{sol,i} = W_{sol,i} - W_{sol,i,b} = \frac{z_i^2 e^2}{8\pi\varepsilon_0 r_i}\left(\frac{1}{\varepsilon_r(\boldsymbol{x})} - \frac{1}{\varepsilon_{r,b}}\right),\ i = p, n. \tag{11}$$

From (10), as $\phi \to -\infty$, we obtain $p \to \infty,\ n \to 0$. Likewise, as $\phi \to \infty$, we obtain $p \to 0,\ n \to \infty$. These unrealistic infinite concentrations for p and n are mainly because ions are treated as particles without size in the classical PB model. This pitfall has motivated modifications of the classical PB/PNP model to account for the finite-size effect, or so-called steric effect, of ions. In reality, the limit of p should be at most $1/v_p$, where v_p is the particle volume of p. This can be derived by considering a volume V fully occupied by cation p only, with the number of cation particles being N_p, and then:

$$p_{max} = \frac{N_p}{V} = \frac{N_p}{N_p v_p} = \frac{1}{v_p}. \tag{12}$$

Likewise, the limit of n is at most $1/v_n$, where v_n is the particle volume of n.

Substituting (10) into (4), and we obtain the classical PB equation:

$$-\nabla\cdot(\varepsilon(\boldsymbol{x})\nabla\phi) = z_p e p_b e^{-\beta E_p} + z_n e n_b e^{-\beta E_n} + q. \tag{13}$$

For z:z electrolyte without considering solvation energy, the equation above reduces to:

$$\nabla\cdot(\varepsilon(\boldsymbol{x})\nabla\phi) = 2zec_b sinh(\beta ze\phi) - q, \tag{14}$$

with $c_b = p_b = n_b$.

3. Bikerman Model

As stated earlier, the Bikerman model [10] has been a popular steric-effect model due to its easiness of application and qualitatively good agreement with experimental data. It modifies the free energy of the classical PB (1)–(3) by adding a solvent entropy term. This term also partially represents the excessive energy accounting for overcrowding of ions and solvent molecules in localized hard-sphere models based on DFT [5–8]. The free energy in the Bikerman model treats all species of ions and solvent molecules with an identical size, and is stated as follows:

$$F = U - TS, \tag{15}$$

$$U = \int \Big[-\frac{\varepsilon}{2} |\nabla\phi|^2 + z_p e p \phi + z_n e n \phi + q\phi + pW_{sol,p} + nW_{sol,n} \Big] dV, \tag{16}$$

$$-TS = \int k_B T [p \log(pv) - p + n \log(nv) - n + w \log(wv) - w] dV, \tag{17}$$

where w is concentration of solvent (such as water); v is the universal particle volume. Why are all solute and solvent particles treated as having the same size? Why are specific sizes of ions and solvent molecules not used here? This was not explained in the original model [10,11], and the justification of using identical size for all species particles will be addressed later in Section 5.

If we assume that, besides occupation of ions, the rest of space is occupied by solvent molecules (which can be taken as water here). Then:

$$w = \frac{N_w}{V} = \frac{N_w v}{Vv} = \frac{V - N_p v - N_n v}{Vv} = \frac{1}{v}(1 - pv - nv), \tag{18}$$

where V is the whole volume of electrolyte; N_p, N_n, N_w are number of cation, anion and solvent particles in an electrolyte with volume V, respectively. Equation (18) can then be rewritten as:

$$wv + pv + nv = 1, \tag{19}$$

which simply means the sum of volume fractions of water, cation and anion is one. Note that here we assume that, besides occupation of ions, the rest of space is occupied by water.

Substituting (19) into (17) we can obtain:

$$-TS = \int k_B T \Big[p \log(pv) - p + n \log(nv) - n + \tfrac{1}{v}(1 - pv - nv) \log(1 - pv - nv) - \tfrac{1}{v}(1 - pv - nv) \Big] dV. \tag{20}$$

Differentiation of F with respect to ϕ again gives the Poisson equation:

$$-\nabla \cdot (\varepsilon(\boldsymbol{x}) \nabla\phi) = z_p e p + z_n e n + q. \tag{21}$$

By doing the derivation of F with respect to p and n, we obtain the chemical potentials for p and n, respectively:

$$\frac{\partial F}{\partial p} = \mu_p = z_p e\phi + k_B T [log(pv) - log(1 - pv - nv)] + W_{sol,p}, \tag{22}$$

$$\frac{\partial F}{\partial n} = \mu_n = z_n e\phi + k_B T [log(nv) - log(1 - pv - nv)] + W_{sol,p}. \tag{23}$$

At equilibrium, the chemical potential is uniform everywhere and therefore the local chemical potential must be equal to its bulk value, which is usually known:

$$\mu_p = \mu_{p,b}, \quad \mu_n = \mu_{n,b}, \tag{24}$$

with:

$$\mu_{p,b} = k_B T [\log(p_b v) - \log(1 - p_b v - n_b v)] + W_{sol,p,b}, \tag{25}$$

$$\mu_{n,b} = k_B T [\log(n_b v) - \log(1 - p_b v - n_b v)] + W_{sol,n,b}. \tag{26}$$

By substituting (22), (23), (25) and (26) into (24), we can relate the local ion-to-solvent volume fraction ratios (denoted as γ_i, $i = p, n$.) to their counterparts in bulk solution in a Boltzmann manner for p and n, respectively:

$$\gamma_p = \frac{pv}{1 - pv - nv} = \frac{p_b v}{1 - p_b v - n_b v} e^{-\beta E_p}, \tag{27}$$

$$\gamma_n = \frac{nv}{1 - pv - nv} = \frac{n_b v}{1 - p_b v - n_b v} e^{-\beta E_n}. \tag{28}$$

Summation of (27) and (28) gives the solute-to-solvent volume fraction ratio as:

$$\frac{pv + nv}{1 - pv - nv} = \gamma_p + \gamma_n,$$

and we can further obtain the solute volume fraction:

$$pv + nv = \frac{\gamma_p + \gamma_n}{1 + \gamma_p + \gamma_n}. \tag{29}$$

From (29) and (19), we know then volume fraction for p, n and w, respectively.

$$pv = \frac{\gamma_p}{1 + \gamma_p + \gamma_n}, \; nv = \frac{\gamma_n}{1 + \gamma_p + \gamma_n}, \; wv = \frac{1}{1 + \gamma_p + \gamma_n}, \tag{30}$$

which further gives p and n in terms of their bulk values p_b, n_b, identical particle size v, and local energy E_p, E_n:

$$p = \frac{p_b e^{-\beta E_p}}{(1 - p_b v - n_b v) + p_b v e^{-\beta E_p} + n_b v e^{-\beta E_n}}, \; n = \frac{n_b e^{-\beta E_n}}{(1 - p_b v - n_b v) + p_b v e^{-\beta E_p} + n_b v e^{-\beta E_n}} \tag{31}$$

Equations (27) and (28) can be re-arranged to obtain:

$$p = p_b e^{-\beta(E_p + S^{trc})}, \quad n = n_b e^{-\beta(E_n + S^{trc})}, \tag{32}$$

with ionic steric potential S^{trc} expressed as:

$$S^{trc} = k_B T \log\left(\frac{1 - p_b v - n_b v}{1 - pv - nv}\right). \tag{33}$$

The steric potential S^{trc}, first described in [16–19], characterizes the crowding of ions and their finite-size effect by a bulk-to-local water fraction ratio. Larger local ion concentrations would have a larger steric potential.

Also, by letting $1 - pv - nv = wv$, and $1 - p_b v - n_b v = w_b v$, Equations (27) and (28) can be simplified as:

$$\frac{pv}{wv} = \frac{p_b v}{w_b v} e^{-\beta E_p} = \gamma_p, \tag{34}$$

$$\frac{nv}{wv} = \frac{n_b v}{w_b v} e^{-\beta E_n} = \gamma_n. \tag{35}$$

Therefore, by (30), we obtain:

$$z_p pv + z_n nv = \frac{z_p \frac{p_b v}{w_b v} e^{-\beta E_p} + z_n \frac{n_b v}{w_b v} e^{-\beta E_n}}{1 + \frac{p_b v}{w_b v} e^{-\beta E_p} + \frac{n_b v}{w_b v} e^{-\beta E_n}} = \frac{z_p p_b v e^{-\beta E_p} + z_n n_b v e^{-\beta E_n}}{w_b v + p_b v e^{-\beta E_p} + n_b v e^{-\beta E_n}}. \tag{36}$$

Since $\frac{p_b}{n_b} = \frac{-z_n}{z_p}$ due to electric neutrality in bulk conditions, therefore:

$$p_b v = \frac{-z_n \mu}{z_p - z_n} v, \; n_b v = \frac{z_p \mu}{z_p - z_n} v, \tag{37}$$

where $\mu = p_b + n_b$. Also:

$$w_b v = 1 - p_b v - n_b v = 1 - \mu v. \tag{38}$$

then (36) becomes:

$$z_p p + z_n n = \frac{z_p z_n \mu\left(-e^{-\beta E_p} + e^{-\beta E_n}\right)}{(1-\mu v)\left(z_p - z_n\right) + \mu v\left(z_p e^{-\beta E_n} - z_n e^{-\beta E_p}\right)}. \quad (39)$$

Substituting (39) into (21), we obtain the Bikerman-PB equation:

$$-\nabla\cdot(\varepsilon(\boldsymbol{x})\nabla\phi) = \frac{z_p z_n e\mu\left(-e^{-\beta E_p} + e^{-\beta E_n}\right)}{(1-\mu v)\left(z_p - z_n\right) + \mu v\left(z_p e^{-\beta E_n} - z_n e^{-\beta E_p}\right)} + q. \quad (40)$$

For z:z electrolyte without considering the solvation energy, the equation above becomes:

$$\nabla\cdot(\varepsilon(\boldsymbol{x})\nabla\phi) = \frac{2zec_0 sinh(\beta ze\phi)}{1 + 2r sinh^2\left(\frac{\beta ze\phi}{2}\right)} - q, \quad (41)$$

as shown in [11] with $c_0 = p_b = n_b$, $r = \mu v = 2c_0 v$.

Two important criteria need to be checked for all modified PB/PNP models accounting for steric effects:

CRITERION I: When ion concentrations p and n are dilute, will they follow the classical Boltzmann distribution?

CRITERION II: As $\phi \to \mp\infty$, will p and n approach their saturation limits $\frac{1}{v_p}$ and $\frac{1}{v_n}$, respectively?

For CRITERION I when p and n are dilute here, it means their volume fractions are negligible, and therefore $1 - pv - nv \approx 1$, and $1 - p_b v - n_b v \approx 1$. Steric potential term S^{trc} then vanishes, and by (32) $p = p_b e^{-\beta E_p}$, and $n = n_b e^{-\beta E_n}$, which follows the Boltzmann distribution.

For CRITERION II, let us consider $\phi \to -\infty$ first, and $\phi \to \infty$ can be derived similarly. As $\phi \to -\infty$, $\gamma_p \to \infty$, and $\gamma_n \to 0$. Therefore, $pv \to 1$, and $nv \to 0$ by (30), which further means $p \to \frac{1}{v_p} = \frac{1}{v}$, and $n \to 0$. Likewise, as $\phi \to \infty$, we can get $n \to \frac{1}{v_n} = \frac{1}{v}$, and $p \to 0$.

4. The Bikerman Model with Specific Ion Sizes

The shortcoming of the Bikerman model is the usage of a universal particle size, denoted by v, for cations, anions and solvents. Using specific ion and solvent sizes would be closer to reality. Taking NaCl solution as an example, the spherical diameters for Cl^-, Na^+ and water are $D_{Cl^-} = 3.62$ Å, $D_{Na^+} = 2.04$ Å, and $D_w = 2.08$ Å, and then the particle volume ratio is $v_{Na^+} : v_{Cl^-} : v_w = 1 : 5.59 : 1.06$, in which using universal particle volume would be far from reality in the case of high ion concentrations. In appearance, it seems, and many researchers did, we can just simply modify the model to include specific ion sizes by changing pv and $p_b v$ to pv_p and $p_b v_p$; similarly, nv and $n_b v$ to nv_n and $n_b v_n$ for (22) to (41). With this straightforward extension, we obtain p, n as:

$$p = \frac{p_b e^{-\beta E_p}}{\left(1 - p_b v_p - n_b v_n\right) + p_b v_p e^{-\beta E_p} + n_b v_n e^{-\beta E_n}},$$
$$n = \frac{n_b e^{-\beta E_n}}{\left(1 - p_b v_p - n_b v_n\right) + p_b v_p e^{-\beta E_p} + n_b v_n e^{-\beta E_n}} \quad (42)$$

and the specific-ion-size Bikerman-PB equation:

$$-\nabla\cdot(\varepsilon(\boldsymbol{x})\nabla\phi) = \frac{e\left(z_p p_b e^{-\beta E_p} + z_n n_b e^{-\beta E_n}\right)}{1 - \left(p_b v_p + n_b v_n\right) + (p_b + n_b)\left(z_p v_n e^{-\beta E_n} - z_n v_p e^{-\beta E_p}\right)/\left(z_p - z_n\right)} + q \quad (43)$$

For z:z electrolyte without considering solvation energy, the equation above becomes:

$$\nabla\cdot(\varepsilon(\boldsymbol{x})\nabla\phi) = \frac{2zec_0 sinh(\beta ze\phi)}{1 - c_0\left(v_p + v_n\right) + c_0\left(v_n e^{\beta ze\phi} + v_p e^{-\beta ze\phi}\right)} - q. \tag{44}$$

Let us denote (42) as the specific-ion-size Bikerman model 1 (SISBM1) for convenience of notation. However, we can not find an energy functional like (15)–(17) to support this naive extension, which means chemical potentials (22) and (23) with universal particle size replaced by specific ion sizes cannot be derived. The correct specific-ion-size energy functional and chemical potentials should be derived as follows:

$$F = U - TS, \tag{45}$$

$$U = \int\left[-\frac{\varepsilon}{2}|\nabla\phi|^2 + z_p ep\phi + z_n en\phi + q\phi + pW_{sol,p} + nW_{sol,n}\right]dV, \tag{46}$$

$$-TS = \int k_B T\left[p\ \log\left(pv_p\right) - p + n\ \log(nv_n) - n + w\ \log(wv_w) - w\right]dV. \tag{47}$$

By $wv_w = 1 - pv_p - nv_n$, (47) can be rewritten as:

$$-TS = \int k_B T\left[p\ \log\left(pv_p\right) - p + n\ \log(nv_n) - n + \tfrac{1}{v_w}\left(1 - pv_p - nv_n\right)\ \log\left(1 - pv_p - nv_n\right) - \tfrac{1}{v_w}\left(1 - pv_p - nv_n\right)\right]dV. \tag{48}$$

Differentiation of F with respect to ϕ again gives the Poisson equation:

$$-\nabla\cdot(\varepsilon(\boldsymbol{x})\nabla\phi) = z_p ep + z_n en + q. \tag{49}$$

By doing the differentiation of F with respect to p and n, we can obtain the chemical potentials for cations and anions, respectively:

$$\mu_p = z_p e\phi + k_B T\left[\log\left(pv_p\right) - \log\left(1 - pv_p - nv_n\right)^{k_p}\right] + W_{sol,p} \tag{50}$$

$$\mu_n = z_n e\phi + k_B T\left[\log(nv_n) - \log\left(1 - pv_p - nv_n\right)^{k_n}\right] + W_{sol,n} \tag{51}$$

where $k_p = \frac{v_p}{v_w}$, $k_n = \frac{v_n}{v_w}$.

At equilibrium, the chemical potential is uniform everywhere and therefore the local chemical potential must be equal to its bulk value, which is usually known:

$$\mu_p = \mu_{p,b},\ \ \mu_n = \mu_{n,b}, \tag{52}$$

with:

$$\mu_{p,b} = k_B T\left[\log\left(p_b v_p\right) - \log\left(1 - p_b v_p - n_b v_n\right)^{k_p}\right] + W_{sol,p,b}, \tag{53}$$

$$\mu_{n,b} = k_B T\left[\log(n_b v_n) - \log\left(1 - p_b v_p - n_b v_n\right)^{k_n}\right] + W_{sol,n,b}. \tag{54}$$

To solve p and n from (52), there is no closed form solution like (31) for p and n due to the nonlinearity, unless some simplified case such as $k_p = k_n = 1$ is considered, which is actually reduced to the original Bikerman model with $v_p = v_n = v_w = v$. Like (32), p and n at most can be expressed as:

$$p = p_b e^{-\beta(E_p + k_p S^{trc})},\ \ n = n_b e^{-\beta(E_n + k_n S^{trc})}, \tag{55}$$

with the steric potential S^{trc} being modified from (33) to include specific ion sizes:

$$S^{trc} = k_B T\ \log\left(\frac{1 - p_b v_p - n_b v_n}{1 - pv_p - nv_n}\right) \tag{56}$$

Let us denote (55) as the specific-ion-size Bikerman model 2 (SISBM2) for convenience. Note that a similar model was also obtained in [18,19] without a rigorous derivation.

Again, we need to check criteria I and II for this specific-ion-size model. For CRITERION I, when p and n are dilute, it again means $1 - pv_p - nv_n \approx 1$, and $1 - p_b v_p - n_b v_n \approx 1$. Therefore $S^{trc} \approx 0$ by (56), and then $p = p_b e^{-\beta E_p}$, and $n = n_b e^{-\beta E_n}$ by (55), which follows a classical Boltzmann distribution. For CRITERION II, as $\phi \to -\infty$ in (50), $k_B T\left[\log(pv_p) - \log(1 - pv_p - nv_n)^{k_p}\right]$ should approach $+\infty$ for μ_p to be finite. This can only be achieved by $n \to 0$, and $p \to \left(\frac{1}{v_p}\right)^{-}$ (saturation). Applying the same reasoning for (51), as $\phi \to \infty$, $k_B T\left[\log(nv_n) - \log(1 - pv_p - nv_n)^{k_n}\right]$ should approach $+\infty$ for μ_n to be finite. Then $p \to 0$, and $n \to \left(\frac{1}{v_n}\right)^{-}$ (saturation). This specific-ion-size model seems correct and reasonable so far, but actually there is a pitfall. That is its entropy formula (48) cannot be derived by the traditional mean-field lattice gas model. This will be explained in the next section.

5. Mixing Entropy Derivation Based on the Mean-Field Lattice Gas Model

In this section, we would like to derive the entropy in (20) by the traditional mean-field lattice gas model. Consider the entropy for an aqueous electrolyte system:

$$TS = k_B T \log W, \tag{57}$$

where W is the number of microstates at equilibrium which possess a maximum number of microstates. Mixing entropy in electrolyte studies macrostates through spherical particles' (solute and solvent) occupation of identical cubic sites is based on the mean-field lattice gas model. The necessity of using identical cubic sites provides a combinatorial basis when computing the maximum number of microstates. The most probable distribution of all solute (ions) and solvent particles, reaching maximum number of microstates for each species, is that each identical cubic site generally would be at most occupied by one solute/solvent particle as depicted in Figure 1a. This is based on the concept that the size of each species' particle is infinitesimal or finite but dilute. When the actual size for each species' particle is considered and an aqueous electrolyte is extremely concentrated as depicted in Figure 1b, the most probable distribution above may not be available. The situation in Figure 1b will be addressed in the next section.

The entropy based on the most probable distribution of K-species solute (ions) and solvent (treated as $K+1$-th species) particles, under dilute situation, over a total of $N = \sum_{j=1}^{K+1} N_j$ available identical sites in a system is:

$$\begin{aligned} W &= \prod_{j=1}^{K+1} W_j \\ &= \binom{N}{N_1}\binom{N-N_1}{N_2}\cdots\binom{N-N_1-N_2\cdots-N_{K-1}}{N_K}\binom{N-N_1-N_2\cdots-N_K}{N_{K+1}} \\ &= \frac{N!}{N_1!(N-N_1)!}\frac{(N-N_1)!}{N_2!(N-N_1-N_2)!}\cdots\frac{(N-\sum_{j=1}^{K-1}N_j)!}{N_K!(N-\sum_{j=1}^{K}N_j)!}\frac{(N-\sum_{j=1}^{K}N_j)!}{N_{K+1}!} \\ &= \frac{N!}{(\prod_{j=1}^{K}N_j!)N_{K+1}!}, \end{aligned} \tag{58}$$

where N_j, $j = 1, \cdots, K$, is the particle number of j-species ion. N_{K+1} is the particle number of solvent, so the entropy becomes:

$$TS = k_B T \log \frac{N!}{(\prod_{j=1}^{K} N_j!)N_{K+1}!}. \tag{59}$$

Using the Stirling formula $\log M! \approx M\log M - M$ with $M \gg 1$, we can rewrite the entropy as:

$$\begin{aligned} TS &= k_BT\left[N\log N - N - \sum_{j=1}^{K} N_j\log N_j + \sum_{j=1}^{K} N_j - N_{K+1}\log N_{K+1} + N_{K+1}\right] \\ &= k_BT\left[N\log N - \sum_{j=1}^{K} N_j\log N_j - N_{K+1}\log N_{K+1}\right] \\ &= k_BT\left[N\log N - \sum_{j=1}^{K} N_j\log N_j - \left(N - \sum_{j=1}^{K} N_j\right)\log\left(N - \sum_{j=1}^{K} N_j\right)\right] \\ &= k_BT\left[N\log\frac{N}{N-\sum_{j=1}^{K}N_j} - \sum_{j=1}^{K} N_j\log\frac{N_j}{N-\sum_{j=1}^{K}N_j}\right] \end{aligned} \tag{60}$$

using the following relations:

$$V = Nv_s, \; or \; \frac{N}{V} = \frac{1}{v_s}, \tag{61}$$

$$\frac{N_j}{V} = c_j, \tag{62}$$

$$\frac{N_j}{N} = \frac{N_jv_s}{Nv_s} = \frac{N_jv_s}{V} = c_jv_s, \tag{63}$$

where c_j is the concentration of j-species particle; V is the volume of system; v_s is the volume of an identical cubic site that composes the volume of system. It is naturally requested that $v_s \geq max_{1\leq j\leq K+1}v_j$, where v_j is the particle volume of j-species particle. Usually $v_s = max_{1\leq j\leq K+1}v_j$ in aqueous electrolyte system, where solute and solvent particles are generally crowded.

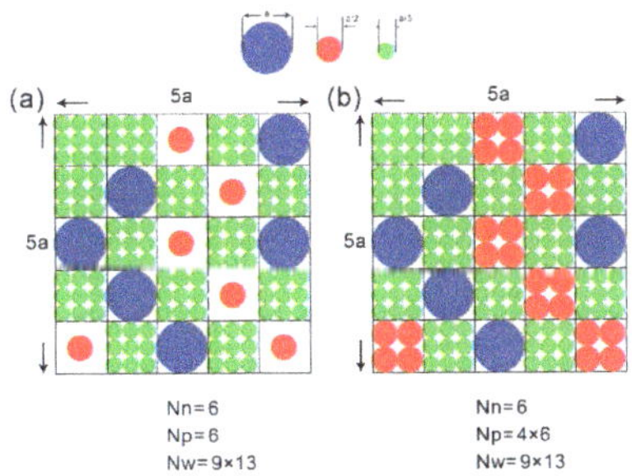

Figure 1. (**a**) Moderately concentrate situation with each solute/solvent particle only occupy one identical site. (**b**) Extremely concentrate situation with each identical site can be allowed to be occupied by multiple solute/solvent particles of the same species in order to increase packing efficiency in space.

Applying (61)–(63) to (60), the entropy density can be expressed as:

$$\frac{TS}{V} = k_BT\left[\frac{1}{v_s}\log\frac{1}{1-\sum_{j=1}^{K}c_jv_s} - \sum_{j=1}^{K}c_j\log\frac{c_jv_s}{1-\sum_{j=1}^{K}c_jv_s}\right]. \tag{64}$$

For a binary electrolyte, (64) can be expressed as:

$$TS = \int k_BT\left[\frac{1}{v_s}\log\frac{1}{1-pv_s-nv_s} - p\,\log\frac{pv_s}{1-pv_s-nv_s} - n\,\log\frac{nv_s}{1-pv_s-nv_s}\right]dV, \tag{65}$$

or:

$$-TS = \int k_BT\left[p\,\log(pv_s) + n\,\log(nv_s) + \frac{1}{v_s}(1-pv_s-nv_s)\log(1-pv_s-nv_s)\right]dV, \tag{66}$$

which, without loss of generality, can be augmented as:

$$-TS = \int k_B T\Big[p\ \log(pv_s) - p + n\ \log(nv_s) - n + \tfrac{1}{v_s}(1 - pv_s - nv_s)\log(1 - pv_s - nv_s) - \tfrac{1}{v_s}(1 - pv_s - nv_s)\Big]dV. \quad (67)$$

Equation (67) is exactly the same as (20) with:

$$v = v_s = max\Big\{v_p, v_n, v_w\Big\}. \quad (68)$$

This means the universal particle volume v in the Bikerman model is actually the volume of an identical occupation site v_s, which is limited from below by the largest particle size among all solute and solvent particles. The original Bikerman model has long suffered criticism for assuming all ions have the same size instead of using specific ion sizes in the model. The above reasoning explains why specific ion size information is left out mainly due to the need for all cubic sites to be identical in order to support the combinatorial basis demanded by the mean-field lattice gas model. Actually, information of specific ion sizes is still carried but only implicitly as shown in (68). Researchers may prefer to use SISBM2 as illustrated in Section 4, but actually its entropy formula (48) cannot be derived by the mean-field lattice gas model described above. Note that usually solute and solvent particles are treated as spheres in modeling. If a_p, a_n, a_w are diameters for p, n, and w, respectively and their maximum is a_n for example, then $v = v_s = a_n^3$ not $\frac{4\pi}{3}\left(\frac{a_n}{2}\right)^3$ since the identical occupation is cubic. This is why a^3, instead of $\frac{4\pi}{3}\left(\frac{a}{2}\right)^3$, used in [10,11].

In CRITERION II described above, as $\phi \to \mp\infty$, p and n should approach their saturation limits $\frac{1}{v_p}$ and $\frac{1}{v_n}$, respectively. Here, this would be changed to approach $\frac{1}{v_s}$ instead of $\frac{1}{v_p}$ and $\frac{1}{v_n}$ respectively, although approaching $\frac{1}{v_p}$ and $\frac{1}{v_n}$ sounds more physically correct. This paradoxical conclusion is from entropy rigorously derived by the traditional mean-field lattice gas model based on combinatorics requiring identical occupation sites. Can this be fixed to resume the limit approach to $\frac{1}{v_p}$ and $\frac{1}{v_n}$ and still holding the ground of combinatorics at the same time? An attempt at this is discussed in the next section.

6. Entropy Fixing for Electrolytes under Extreme Concentration Conditions

Here we hope to construct a steric PB model with entropy able to be derived by the mean-field lattice gas model, and at the same time showing physically correct saturation limits for ions as $\phi \to \mp\infty$. The mean-field lattice gas model is fixed here such that each identical cubic site is allowed to be occupied by more than one solute particle of the same species as illustrated in Figure 1b. Although this kind of distribution is no more a most probable distribution as stated earlier, it allows more efficient packing when space is extremely limited and size among species varies largely. Again, we consider the entropy for an aqueous electrolyte system:

$$TS = k_B T\ \log W, \quad (69)$$

Let $\widetilde{N}_j,\ j = 1, \cdots, K+1$, be the particle number of species j and $N_j,\ j = 1, \cdots, K+1$, the number of identical sites occupied by j-species particles with $\widetilde{N}_j \geq N_j$. This means that in an extremely concentrated situation an identical site can be occupied by more than one particle of the same species. If an identical cubic site, on average, can allow r_j j-species particles to occupy it, we can then relate $\widetilde{N}_j$ and N_j by $\widetilde{N}_j = N_j r_j$, or equivalently $v_s = r_j v_j$. Again, v_j is the particle volume of species j. v_s is the volume of an identical cubic site with $v_s = max_{1\leq j\leq K+1} v_j$. The entropy based on the most probable distribution of all 'grouped' species particles over a total of $N = \sum_{j=1}^{K+1} N_j$ available identical sites in a system is:

$$W = \prod_{j=1}^{K+1} W_j = \frac{N!}{N_1!(N-N_1)!}\frac{(N-N_1)!}{N_2!(N-N_1-N_2)!}\cdots\frac{\left(N-\sum_{j=1}^{K-1} N_j\right)!}{N_K!\left(N-\sum_{j=1}^{K} N_j\right)!}\frac{\left(N-\sum_{j=1}^{K} N_j\right)!}{N_{K+1}!} = \frac{N!}{\left(\prod_{j=1}^{K} N_j!\right)N_{K+1}!}. \quad (70)$$

Note that, after all the ions (in group) are distributed, there are $N - \sum_{j=1}^{K} N_j = N_{K+1}$ sites that will be filled by solvent molecules in group, so the entropy becomes:

$$TS = k_B T \log \frac{N!}{\left(\prod_{j=1}^{K} N_j!\right) N_{K+1}!}. \tag{71}$$

Using the Stirling formula $\log M! \approx M \log M - M$ with $M \gg 1$, we can rewrite the entropy as:

$$\begin{aligned} TS = k_B T \Bigg[N \log N - N - \sum_{j=1}^{K} N_j \log N_j + \sum_{j=1}^{K} N_j - \left(N - \sum_{j=1}^{K} N_j\right) \log\left(N - \sum_{j=1}^{K} N_j\right) \\ + \left(N - \sum_{j=1}^{K} N_j\right)\Bigg] = k_B T \left[N \log \frac{N}{N - \sum_{j=1}^{K} N_j} - \sum_{j=1}^{K} N_j \log \frac{N_j}{N - \sum_{j=1}^{K} N_j} \right], \end{aligned} \tag{72}$$

Using the following relations:

$$V = N v_s, \text{ or } \frac{N}{V} = \frac{1}{v_s}, \tag{73}$$

$$\frac{N_j}{V} = \frac{\frac{\widetilde{N}_j}{r_j}}{V} = \frac{c_j}{r_j}, \tag{74}$$

$$\frac{N_j}{N} = \frac{N_j v_s}{N v_s} = \frac{N_j v_s}{V} = \frac{c_j}{r_j} r_j v_j = c_j v_j, \tag{75}$$

where c_j is the concentration of species j; V is the volume of the system.

The entropy per unit volume can be expressed as:

$$\frac{TS}{V} = k_B T \left[\frac{1}{v_s} \log \frac{1}{1 - \sum_{j=1}^{K} c_j v_j} - \sum_{j=1}^{K} \frac{c_j}{r_j} \log \frac{c_j v_j}{1 - \sum_{j=1}^{K} c_j v_j} \right]. \tag{76}$$

Compared with (64), specific ion sizes can now appear explicitly in the entropy formula (76). For a binary electrolyte:

$$TS = \int k_B T \left[\frac{1}{v_s} \log \frac{1}{1 - p v_p - n v_n} - \frac{p}{r_p} \log \frac{p v_p}{1 - p v_p - n v_n} - \frac{n}{r_n} \log \frac{n v_n}{1 - p v_p - n v_n} \right] dV, \tag{77}$$

or:

$$-TS = \int k_B T \left[\frac{p}{r_p} \log(p v_p) + \frac{n}{r_n} \log(n v_n) + \frac{1}{v_s}(1 - p v_p - n v_n) \log(1 - p v_p - n v_n) \right] dV, \tag{78}$$

which, without loss of generality, can be augmented as:

$$-TS = \int k_B T \left[\frac{p}{r_p} \log(p v_p) - \frac{p}{r_p} + \frac{n}{r_n} \log(n v_n) - \frac{n}{r_n} + \frac{1}{v_s}(1 - p v_p - n v_n) \log(1 - p v_p - n v_n) - \frac{1}{v_s}(1 - p v_p - n v_n) \right] dV. \tag{79}$$

by:

$$\mu_p = \frac{\delta f}{\delta p} = z_p e \phi + W_{sol,p} + \frac{k_B T}{r_p} \log \frac{p v_p}{1 - p v_p - n v_n}, \tag{80}$$

$$\mu_n = \frac{\delta f}{\delta n} = z_n e \phi + W_{sol,n} + \frac{k_B T}{r_n} \log \frac{n v_n}{1 - p v_p - n v_n}, \tag{81}$$

Again, at equilibrium, the chemical potential is uniform everywhere and therefore the local chemical potential must be equal to its bulk value, which is usually known:

$$\mu_p = \mu_{p,b}, \ \ \mu_n = \mu_{n,b}. \tag{82}$$

Usually bulk solutions are dilute and chemical potentials under that condition can be formulated following (80) and (81) with $r_p = r_n = 1$:

$$\mu_{p,b} = k_B T \log \frac{p_b v_p}{1 - p_b v_p - n_b v_n} + W_{sol,p,b}, \tag{83}$$

$$\mu_{n,b} = k_B T \log \frac{n_b v_n}{1 - p_b v_p - n_b v_n} + W_{sol,n,b}. \tag{84}$$

By denoting $\gamma_p = e^{\frac{\mu_{p,b} - z_p e\phi - W_{sol,p}}{k_B T / r_p}}$, $\gamma_n = e^{\frac{\mu_{n,b} - z_n e\phi - W_{sol,n}}{k_B T / r_n}}$,
(82) forms

$$\frac{p v_p}{1 - p v_p - n v_n} = \gamma_p, \quad \frac{n v_n}{1 - p v_p - n v_n} = \gamma_n, \tag{85}$$

and can solve for p and n:

$$p v_p = \frac{\gamma_p}{1 + \gamma_p + \gamma_n}, \quad n v_n = \frac{\gamma_n}{1 + \gamma_p + \gamma_n}, \tag{86}$$

or:

$$p v_p = \frac{\left(\frac{p_b v_p}{1 - p_b v_p - n_b v_n}\right)^{r_p} e^{-\beta_p E_p}}{1 + \left(\frac{p_b v_p}{1 - p_b v_p - n_b v_n}\right)^{r_p} e^{-\beta_p E_p} + \left(\frac{n_b v_n}{1 - p_b v_p - n_b v_n}\right)^{r_n} e^{-\beta_n E_n}} \tag{87}$$

$$n v_n = \frac{\left(\frac{n_b v_n}{1 - p_b v_p - n_b v_n}\right)^{r_n} e^{-\beta_n E_n}}{1 + \left(\frac{p_b v_p}{1 - p_b v_p - n_b v_n}\right)^{r_p} e^{-\beta_p E_p} + \left(\frac{n_b v_n}{1 - p_b v_p - n_b v_n}\right)^{r_n} e^{-\beta_n E_n}}, \tag{88}$$

where $E_p = z_p e\phi + \Delta W_{sol,p}$, $E_n = z_n e\phi + \Delta W_{sol,n}$, $\beta_p = \left(k_B T / r_p\right)^{-1}$, $\beta_n = (k_B T / r_n)^{-1}$. Let us denote (87), (88) as the specific-ion-size Bikerman model 3 (SISBM3) for convenience.

Again, we need to check this new model with criteria I and II. For CRITERION II, we can easily deduce from (87) $\phi \to -\infty$, $n \to 0$, $p v_p \to 1$, $p \to \frac{1}{v_p}$ (saturation). Similarly, from (88), $\phi \to \infty$, $p \to 0$, $n v_n \to 1$, $n \to \frac{1}{v_n}$ (saturation). There is no constraint like $p, n \to \frac{1}{v_s}$ as $\phi \to \mp\infty$ any more, and entropy here can be derived by mean-field lattice gas model.

For CRITERION I, p and n will not approach a Boltzmann distribution $p_b e^{-\beta E_p}$ and $n_b e^{-\beta E_n}$ at diluteness unless $r_p = r_n = 1$. This violation of the Boltzmann distribution at the dilution limit is because we allow multiple ions of the same species to occupy an identical cubic site. This failure and a possible cure will be discussed in next section.

7. Discussion and Conclusions

If we wish to obtain a model for electrolytes such that: (1) it can approach a Boltzmann distribution at diluteness; (2) it can reach the saturation limit as the reciprocal of specific ion size under extreme electrostatic conditions; (3) its entropy can be derived by a mean-field lattice gas model. The only options here is SISBM3 with $r_p = r_n = 1$, since SISBM2 satisfies (1) and (2) but not (3). How can we justify $r_p = r_n = 1$ for SISBM3 here? Interpreting all ion sizes as being about the same is certainly not acceptable. Remember SISBM3 is designed for extremely high ion concentrations motivated by the more efficient packing shown in Figure 1b. Actually for situations that would give rise to extremely high ion concentrations and make the steric effect not negligible, such as the Stern layer in the electric double layer (EDL) of a charged wall (discussed next) and the selectivity filter of a K channel [20], there would be 'locally' one species only, which is the counter-ion of the local electrostatic environment, since co-ions (and even water) would be totally expelled. Taking the K channel selectivity filter as an example, its extreme narrowness and the strong negative oxygen charges inside it would definitely justify only one species being inside the selectivity filter, which is definitely potassium. This implies r_p

to be 1 locally in the filter, and we can justify $r_n = 1$ inside the filter as well since anions would be extremely dilute there due to strong electrostatic repulsion. For the rest of the K channel where ions are at most moderately concentrated, the steric effect is much less significant, and basically the original Bikerman model would be appropriate for it. Since SISBM3 with $r_p = r_n = 1$ would be a very good approximation of the original Bikerman model under mild ion concentrations, we can use SISBM3 with $r_p = r_n = 1$ globally for the whole K channel then. Notice, under $r_p = r_n = 1$, SISBM3 is actually same as SISBM1, but with a rigorous derivation now. This model has been useful and proven to fit the experimental data quite well [16–18]. Although here we just discussed steric-effect modifications for PB, modifications for PNP can be likewise derived.

Here we compare SISBM1 and PB by computing ion distributions in a 1D charged wall problem. Many researchers have used this physical model to investigate the surface differential capacitance of electrodes adjacent to electrolyte solutions [12–15]. Here (14) for PB and (44) for SISBM1 were used to calculate the ion distributions of a binary KCl electrolyte solution without considering the solvation energy and permanent charges. The associated boundary conditions are $\phi(0) = V_{wall}$, and $\phi(\infty) = 0$. The bulk concentration of KCl as $x \to \infty$ is set to be $c_0 = 100$ mM, and dielectric constant is set to 80 for the whole domain $(0, \infty)$. The Debye length, featuring the order of thickness of EDL, is $\lambda_D = \sqrt{\frac{\varepsilon_0 \varepsilon_r k_B T}{c_0 e^2}} = 13.78$ Å. The simulation result is shown in Figure 2 with Figure 2a being the distributions of $[K^+]$ for SISBM1 and PB under $V_{wall} = 0.1$ V and 2 V. Figure 2b is the counterpart plot of Figure 2a for $[Cl^-]$. In Figure 2a, the $[K^+]$ distributions for SISBM1 and PB are very close to each other and almost indistinguishable in the graph at a weak wall voltage $V_{wall} = 0.1$ V. When the wall voltage increases to $V_{wall} = 2$ V, $[K^+]$ calculated by SISBM1 reaches its saturation limit $1/v_K = 7.90 \times 10^4$ mM right adjacent to the wall, but $[K^+]$ unrealistically increases beyond the saturation limit when computed by the PB model. The main effect of SISBM1 is to offer a saturation limit for counter-ions (K here) when electrostatic attraction from electrode is strong enough, while it is very close to the result of PB when the electrostatic attraction is weak. In Figure 2b, $[Cl^-]$ distributions calculated by SISBM1 and PB are very close to each other for both strong and weak wall voltages due to the diluteness caused by electrostatic repulsion to the co-ion (Cl here) of the electrode. Note that, corresponding to a saturation layer of $[K^+]$ adjacent to wall at $V_{wall} = 2$ V (see Figure 2a), $[Cl^-]$ almost vanishes at that layer as well (see Figure 2b). This implies a total exclusion of Cl over there due to the saturation of K, and justifies the locally one-species argument above. If we use the original Bikerman model (41), in which ion sizes are universal, a similar saturating phenomenon for counter-ion concentration can still be obtained. However, specific ion sizes are particularly desired when electrolyte solutions are ternary, like a mixture of KCl and NaCl solutions since K and Na have different sizes, which would saturate at different limiting concentrations. These would be otherwise indistinguishable if using the original Bikerman model.

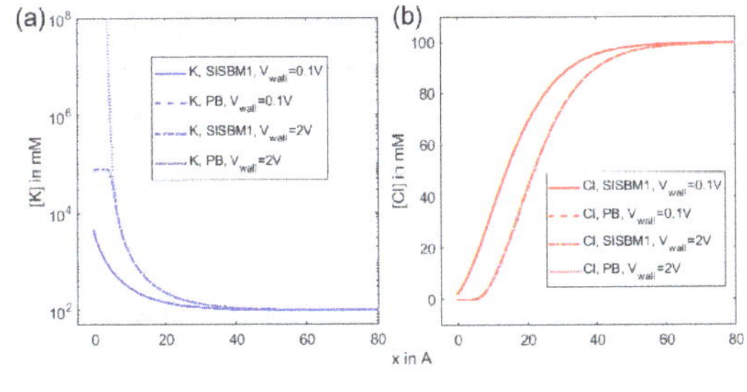

Figure 2. (**a**) K distributions in charged wall problem computed by SISBM1 and PB models under different wall voltages. (**b**) Cl distributions with conditions same as (**a**).

Above we assume the rest of space after the occupation of ions is exclusively occupied by solvent particles such as water. [9–12] have suggested that the rest of space should be occupied by solvent or void, so the $K+1$ species in (58) and (70) should be interpreted as solvent or void. This may make more sense. Taking the selectivity filter of a K channel as an example, more and more evidences have shown the selectivity filter of a K channel is exclusively occupied by potassium and voids, and water is not allowed there due to the strong solvation energy barrier. [9–12] even explicitly separate water and voids as two species in their modeling. However, that means the species transport equation (Nernst-Planck equation) of water needs to be modeled explicitly when constructing a PNP type model. This water equation is generally hard to model due to its physical complexity, especially for its electrostatic behavior since water is a dipole.

Funding: This research was funded by Ministry of Science and Technology of Taiwan under Grant Nos. MOST 108-2115-M-035 -002 -MY2 and MOST 108-2218-E-035-005.

Acknowledgments: The author would like to thank Robert Eisenberg and Jinn-Liang Liu for inspiring suggestions and discussions at Hsinchu Taiwan.

Conflicts of Interest: The author declares no conflict of interest.

References

1. Bazant, M.Z.; Storey, B.D.; Kornyshev, A. Double layer in ionic liquids: Overscreening versus crowding. *Phys. Rev. Lett.* **2012**, *109*, 149903.
2. Rosenfeld, Y. Free-energy model for the inhomogeneous hard-sphere fluid mixture and density-functional theory of freezing. *Phys. Rev. Lett.* **1989**, *63*, 980–983.
3. Rosenfeld, Y. Free energy model for the inhomogeneous fluid mixtures: Yukawa-charged hard spheres, general interactions, and plasmas. *J. Chem. Phys.* **1993**, *98*, 8126–8148.
4. Eisenberg, B.; Hyon, Y.; Liu, C.J. Energy variational analysis of ions in water and channels: Field theory for primitive models of complex ionic fluids. *J. Chem. Phys.* **2010**, *133*, 104104.
5. Ji, S.; Liu, W. Poisson-Nernst-Planck systems for ion flow with Density Functional Theory for hard-sphere potential: I-V relations and critical potentials. Part I: Analysis. *J. Dyn. Diff. Equat.* **2012**, *24*, 955–983.
6. Lin, G.; Liu, W.; Yi, Y.; Zhang, M. Poisson-Nernst-Planck systems for ion flow with a local hard-sphere potential for ion size effects. *SIAM J. Appl. Dyn. Syst.* **2013**, *12*, 1613–1648.
7. Liu, W.; Tu, X.; Zhang, M. Poisson–Nernst–Planck systems for ion flow with density functional theory for hard-sphere potential: I–V relations and critical potentials. Part II: Numerics. *J. Dyn. Diff. Equat.* **2012**, *24*, 985–1004.
8. Qiao, Y.; Lu, B.; Chen, M. A local approximation of fundamental measure theory incorporated into three dimensional Poisson-Nernst-Planck equations to account for hard sphere repulsion among ions. *J. Stat. Phys.* **2016**, *163*, 156–174.
9. Horng, T.-L.; Lin, T.-C.; Liu, C.; Eisenberg, B. PNP equations with steric effects: A model of ion flow through channels. *J. Phys. Chem. B* **2012**, *116*, 11422–11441.
10. Bikerman, J.J. Structure and capacity of electrical double layer. *Philos. Mag.* **1942**, *33*, 384–397.
11. Borukhov, I.; Andelman, D.; Orland, H. Steric effects in electrolytes: A modified Poisson-Boltzmann equation. *Phys. Rev. Lett.* **1997**, *79*, 435–438.
12. Gavish, N.; Promislow, K. Systematic interpretation of differential capacitance data. *Phys. Rev. E* **2015**, *92*, 012321.
13. López-García, J.J.; Horno, J.; Grosse, C. Differential capacitance of the diffuse double layer at electrode-electrolyte interfaces considering ions as dielectric spheres: Part I. Binary electrolyte solutions. *J. Colloid Interface Sci.* **2017**, *496*, 531–539.
14. Nakayama, Y.; Andelman, D. Differential capacitance of the electric double layer: The interplay between ion finite size and dielectric decrement. *J. Chem. Phys.* **2015**, *142*, 044706.
15. Zhao, H. Influence of nonelectrostatic ion-ion interactions on double-layer capacitance. *Phys. Rev. E* **2012**, *86*, 051502.
16. Liu, J.-L.; Eisenberg, B. Poisson-Nernst-Planck-Fermi theory for modeling biological ion channels. *J. Chem. Phys.* **2014**, *141*, 22D532.
17. Liu, J.-L.; Eisenberg, B. Poisson-Fermi model of single ion activities in aqueous solutions. *Chem. Phys. Lett.* **2015**, *637*, 1–6.

18. Liu, J.-L.; Eisenberg, B. Poisson-Fermi modeling of ion activities in aqueous single and mixed electrolyte solutions at variable temperature. *J. Chem. Phys.* **2018**, *148*, 054501.
19. Liu, J.-L.; Xie, D.; Eisenberg, B. Poisson-Fermi Formulation of Nonlocal Electrostatics in Electrolyte Solutions. *Comput. Math. Biophys.* **2017**, *5*, 116–124.
20. Song, Z.; Cao, X.; Horng, T.-L.; Huang, H. Selectivity of the KcsA potassium channel: Analysis and computation. *Phys. Rev. E* **2019**, *100*, 022406.

entropy

MDPI

Article

Electric Double Layer and Orientational Ordering of Water Dipoles in Narrow Channels within a Modified Langevin Poisson-Boltzmann Model

Mitja Drab [1,†], Ekaterina Gongadze [1,†], Veronika Kralj-Iglič [2] and Aleš Iglič [1,*]

1 Faculty of Electrical Engineering, Tržaška Cesta 25, University of Ljubljana, SI-1000 Ljubljana, Slovenia; mitja.drab@fe.uni-lj.si (M.D.); ekaterina.gongadze@fe.uni-lj.si (E.G.)

2 Faculty of Health Sciences, Zdravstvena Pot 5, University of Ljubljana, SI-1000 Ljubljana, Slovenia; kraljiglic@gmail.com

* Correspondence: ales.iglic@fe.uni-lj.si

† These authors contributed equally to this work.

Received: 28 July 2020; Accepted: 18 September 2020; Published: 21 September 2020

Abstract: The electric double layer (EDL) is an important phenomenon that arises in systems where a charged surface comes into contact with an electrolyte solution. In this work we describe the generalization of classic Poisson-Boltzmann (PB) theory for point-like ions by taking into account orientational ordering of water molecules. The modified Langevin Poisson-Boltzmann (LPB) model of EDL is derived by minimizing the corresponding Helmholtz free energy functional, which includes also orientational entropy contribution of water dipoles. The formation of EDL is important in many artificial and biological systems bound by a cylindrical geometry. We therefore numerically solve the modified LPB equation in cylindrical coordinates, determining the spatial dependencies of electric potential, relative permittivity and average orientations of water dipoles within charged tubes of different radii. Results show that for tubes of a large radius, macroscopic (net) volume charge density of coions and counterions is zero at the geometrical axis. This is attributed to effective electrolyte charge screening in the vicinity of the inner charged surface of the tube. For tubes of small radii, the screening region extends into the whole inner space of the tube, leading to non-zero net volume charge density and non-zero orientational ordering of water dipoles near the axis.

Keywords: electric double layer; orientational ordering of water dipoles; Helmholtz free energy; modified Langevin Poisson-Boltzmann model

1. Introduction

The electric double layer (EDL) is a central phenomenon found at the boundary between a charged surface and an electrolyte solution [1–8]. The counterions are accumulated close to the charged surface and the coions are depleted from this region, resulting in a non-homogeneous distribution of ions. The physical properties of the EDL are crucial in understanding colloidal systems, transport of charged molecules across biological membrane channels or binding of charged proteins to biological surfaces.

Recently, much attention is being devoted to inorganic and organic hollow cylindrical structures in the nanometer range due to their potential benefit in technology, biology and medicine [9]. Potential applications range from microelectronics to microfluidics [10]. Ion channels or pores in biological membranes and blood capillaries are also examples for cylindrical nanotubes.

In some biological systems, the walls of organic nanotubes are charged and in contact with electrolyte solution, where the primary agents of interaction are electrostatic forces, both between charged particles and polar water molecules. Due to the surface charge of the walls, counterions and coions of the electrolyte are, respectively, accumulated and depleted near the walls. At the internal surfaces concave electrical double layers of cylindrical geometry are formed [11].

Furthermore, when bound to a cylindrical geometry, the effect of curvature on EDL properties is significant on small enough scales. Such biological cylindrical channels, where EDL interactions are important, encompass axons or tunneling nanotubes [12]. When artificially made channels, for example, those found in nanoporous materials, are used in the manufacture of electrochemical nanocapacitors, their power and energy densities are dependent on EDL characteristics such as capacitance [13–15].

EDL was first modeled by Helmholtz who assumed that the charged surface attracts the surrounding point-like counterions and a single layer is formed to screen the charge [16,17]. Later, these ions have been described by a Boltzmann distribution, forming a diffuse layer extending into the bulk [18,19]. The finite size has been incorporated by Stern with the so-called distance of closest approach [20] and later developed further by numerous authors [3,21–26]. In recent decades, EDL has been the subject of numerous analytical and numerical studies from Monte-Carlo methods, DFT theories and lattice models [3,7,27–44]. Additionally, interest in nanostructured materials [45–48] requires that theoretical models of EDL are revisited [49–51], also by taking into account the possible quantum effects [52,53].

It has been shown that close to the charged surface, orientational ordering and depletion of water molecules may result in a strong decrease in the local permittivity of the electrolyte solution [54–61]. Considering the orientational ordering of water and finite size of molecules, Outhwaite and collaborators developed a modified Poisson-Boltzmann's (PB) theory of EDL composed of a mixture of hard spheres with point-like dipoles and finite-sized ions [54,62]. Later, Szalai et al. [63] published a mean spherical approximation-based theory [64] that can reproduce simulation results for the electric field dependence of the dielectric permittivity of a dipolar fluid in a saturation regime. The problem was also considered within a discrete lattice statistics model taking into account the asymmetric size of ions and orientational ordering of water dipoles [44]. Recently, ion-ion and ion-water correlations were also considered in a mean-field approach [65,66].

In the present paper, we first discuss the relative permittivity of water molecules within a cavity field model. We then go on to the derivation of a modified Langevin Poisson-Boltzmann (LPB) equation for point-like ions and water dipoles for planar geometry and then generalize the equations for arbitrary geometry. In derivation of modified LPB equation we construct a Helmholtz free energy functional and minimize it to derive the analytical expressions for ion distributions and spatial dependence of statistical averages orientations of water dipoles. The free energy expression also includes contributions from configurational entropy of ions and rotational entropy of water dipoles. In the second part of the paper the modified LPB equation and the corresponding boundary conditions, generalized for an arbitrary geometry, are utilized to present the numerical solution for a cylindrical geometry with special emphasis given to very narrow cylindrical channels (Figure 1).

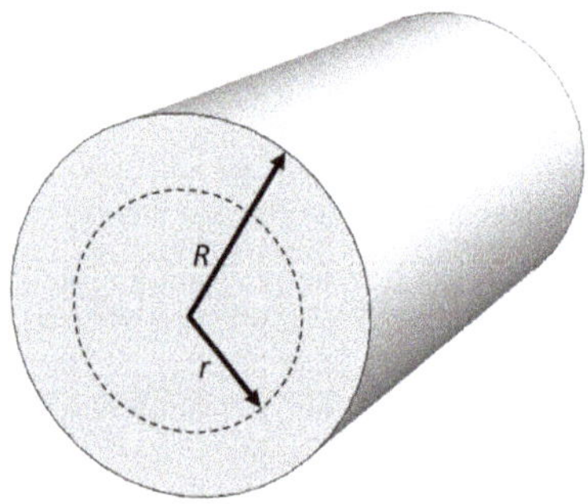

Figure 1. A schematic of a tubular structure with labeled independent coordinate r that can be at most R.

2. Relative Permittivity of Water

The dipole moment of an isolated water molecule is around 1.85 D (Debye is 3.336×10^{-30} Cm). In a solution, the dipole moment of a single water molecule differs from an isolated one since each molecule is also polarized by the electric field of the neighboring water molecules, creating an effective value of the dipole moment around 2.4 D–2.6 D [67,68]. The body of literature dealing with the dielectric permittivity of water is voluminous and comprehensive, from analytic models detailing the state of bound water molecules and water in charged crevices [69,70] to molecular dynamics simulations with nonlinear response to external electric fields [71,72].

The effect of a polarizing environment can be reproduced in the most simple way by introduction of the cavity field [61,73–75]. Cavity field is derived by solving the Poisson's equation of a model water molecule placed in an outside homogeneous electric field (for a detailed derivation, see Reference [76]). The present section deals with polarization of water dipoles that follows directly from the cavity field.

The water molecules are described within the modified Kirkwood approach [75] as point-like dipoles $\mathbf{p}$ with magnitude $|\mathbf{p}| = p$ at the centres of finite sized spheres, embedded in a medium with electric permittivity representing the ion-water solution ε_r (Figure 2) [7,61]. Within this medium, a spatially homogeneous electric field, $\mathbf{E}$, is present. Due to the built up charge at the interface between the inside and outside of the sphere, the dipole experiences the so called cavity field $\mathbf{E}_c$. The relative permittivity of water is given by $\varepsilon_r = 1 + P_{\text{tot}}/(\varepsilon_0 E)$, where P_{tot} is the total polarization of water dipoles, E is the magnitude of the spatially homogeneous electric field and ε_0 is the permittivity of vacuum. The total polarization is the sum of electronic polarization, P_e, and orientational polarization due to the permanent water dipoles P, so that $P_{\text{tot}} = P_e + P$. The electronic polarization determines the refractive index of water [51,61] $n^2 = 1 + P_e/(\varepsilon_0 E) \approx 1.8$ and ε_r can be expressed as

$$\varepsilon_r = n^2 + \frac{P}{\varepsilon_0 E}. \tag{1}$$

To find the expression for P we must take into account the constant number density of water n_w and the statistical-average orientation of water molecules in the solution [7]:

$$P = n_w p \langle \cos\theta \rangle. \tag{2}$$

Here, θ is the angle between $\mathbf{p}$ and the cavity field $\mathbf{E}_c$ acting on it (see Figure 3). Statistical averaging is labeled by $\langle ... \rangle$. To estimate $\langle \cos\theta \rangle$, we must first find the expression for $\mathbf{E}_c$. This involves solving the Poisson equation for a sphere with electric permittivity n^2 embedded in a medium with a relative permittivity ε_r described in detail in Reference [76]. Neglecting the short range interactions between dipoles, the local electric field strength at the centre of the sphere at the location of the permanent point-like dipole (Figure 2) can be expressed as [7,76]

$$\mathbf{E}_c = \frac{3\varepsilon_r}{n^2 + 2\varepsilon_r}\mathbf{E}. \tag{3}$$

When the surrounding medium has a relative permittivity much larger than the refractive index of water $\varepsilon_r \gg n^2$, it follows that

$$E_c \approx \frac{3}{2}E \quad \rightarrow \quad \mathbf{E}_c \approx \frac{3}{2}\mathbf{E}. \tag{4}$$

So far we have neglected the reaction field, which is the field of the point-like dipole at the center of the cavity itself. This reaction field is directly proportional to the strength of dipole $\mathbf{E}_{\text{react}} \propto \mathbf{p}$. In vacuum, in the case of a single isolated water molecule, the external dipole moment is also the experimentally measured dipole moment of a single water molecule $\mathbf{p}_e$ given by [7,76]:

$$\mathbf{p}_e = \frac{3}{n^2 + 2}\mathbf{p}. \tag{5}$$

Here, $\mathbf{p}$ is the permanent point-like internal water dipole at the center of the sphere. The energy of an internal point-like dipole $\mathbf{p}$ in a local field $\mathbf{E}_c$ is [61]

$$W_e = -\mathbf{p} \cdot \mathbf{E}_c. \tag{6}$$

Substituting from Equation (4) and Equation (5), we can express the dipole energy as [61]

$$W_e = -\frac{3}{2}\left(\frac{2+n^2}{3}\right) p_0 E \cos\theta, \tag{7}$$
$$W_e = \gamma p_0 E \cos\omega. \tag{8}$$

Here, p_0 is the magnitude of $\mathbf{p}_e$ and ω is supplementary to θ, as shown in Figure 3. The constant γ equals [7,61] (see Equations (7) and (8))

$$\gamma = \frac{3}{2}\left(\frac{2+n^2}{3}\right). \tag{9}$$

With this in mind, the ensemble average in Equation (2) can be calculated as:

$$\langle\cos\omega\rangle = \frac{\int \cos\omega e^{-(\beta\gamma p_0 E\cos\omega)}\, d\Omega}{\int e^{-(\beta\gamma p_0 E\cos\omega)}\, d\Omega} = -\mathcal{L}(\beta\gamma p_0 E). \tag{10}$$

Here, β is the Boltzmann's factor equal to $\beta = 1/kT$, where kT is the thermal energy. The element of solid angle is $d\Omega = 2\pi \sin\omega d\omega$, meaning that the integral runs from 0 to π with assumed azimuthal symmetry. The Langevin function is defined as $\mathcal{L}(u) = \coth u - 1/u$. By taking into account Equations (1), (2), (5) and (10), we can express the relative water permittivity as [7]:

$$\varepsilon_r = n^2 + \frac{n_w p_0}{\varepsilon_0}\left(\frac{2+n^2}{3}\right)\frac{\mathcal{L}(\beta\gamma p_0 E)}{E}. \tag{11}$$

In the limit of vanishing electric field strength ($E \to 0$), the above expression for the relative permittivity of water yields to the Onsager limit [7]

$$\varepsilon_r = n^2 + \frac{n_w p_0^2 \beta}{2\varepsilon_0}\left(\frac{2+n^2}{3}\right)^2. \tag{12}$$

For $p_0 = 3.1$ D and $n_w/N_A = 55$ mol/L, [7,44], where N_A is the Avogadro number, Equation (12) yields the value $\varepsilon_r = 78.5$ at room temperature. Returning to Equation (2), we can write the final result for the orientational polarization of water dipoles P, which will be needed for our Helmholtz free energy minimization in the following section:

$$P = -n_w p_0 \left(\frac{2+n^2}{3}\right) \mathcal{L}(\beta\gamma p_0 E). \tag{13}$$

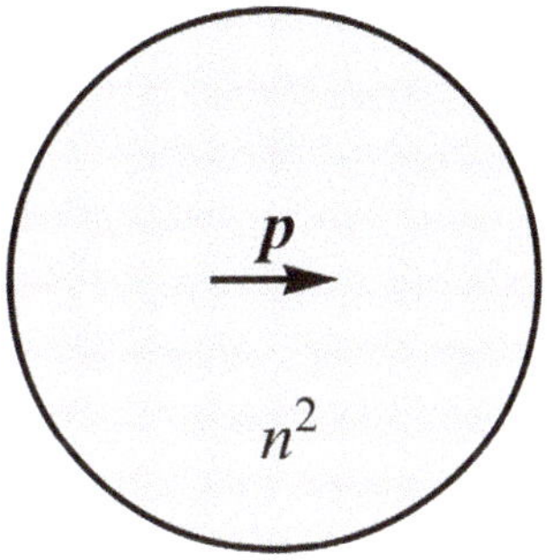

Figure 2. A single water molecule is modelled by a sphere with relative permittivity n^2, where $n = 1.33$ is the refractive index of water. A permanent point-like rigid dipole with magnitude, p, is located at the center of the sphere [61]. Due to the built up charge, the point dipole experiences the so-called cavity field $\mathbf{E}_c$.

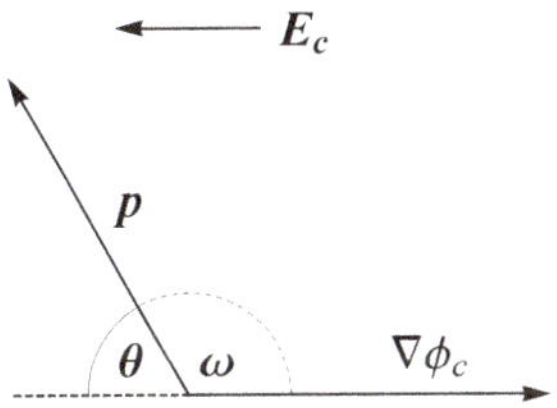

Figure 3. Relation between angles θ and ω. The water internal dipole moment is marked by $\mathbf{p}$, the local cavity field, $\mathbf{E_c}$, points in the opposite direction of $\nabla\phi_c$.

3. Derivation of the Modified LPB Equation by Minimization of Helmholtz Free Energy

Our model assumes the electrolyte solution is a mixture of point-like monovalent co- and counterions and permanent water dipoles, representing the water molecules. The expression for the spatial dependence of the solution permittivity $\varepsilon_r(x)$, arising as a direct consequence of the spontaneous ordering of water dipoles, can be obtained by using the minimization of the Helmholtz free energy in a one-dimensional setting with the charged planar surface located at $x = 0$. In the minimization procedure, the local electric field at the positions of the hydrated point-like ions in the electrolyte solution is denoted by $E(x)$, while the local cavity field at the positions of the water internal point-like dipoles is denoted by $E_c(x)$. We can write the Helmholtz free energy of the system F as (see also Reference [58]):

$$
\begin{aligned}
F = {} & \underbrace{\frac{\varepsilon_0 n^2}{2}\int E_c^2(x)\,dV}_{F_1} + \underbrace{\frac{\varepsilon_0 n^2}{2}\int E^2(x)\,dV}_{F_2} + kT\Big[\underbrace{\int\Big(n_+(x)\ln\frac{n_+(x)}{n_0} - (n_+(x)-n_0)\Big)\,dV}_{F_3} + \\
& + \underbrace{\int\Big(n_-(x)\ln\frac{n_-(x)}{n_0} - (n_-(x)-n_0)\Big)\,dV}_{F_4} + \underbrace{\int(\lambda_+ n_+(x) + \lambda_- n_-(x))\,dV}_{F_5} + \\
& + \underbrace{\int n_w\langle\mathcal{P}(x,\omega)\ln\mathcal{P}(x,\omega)\rangle_\omega\,dV}_{F_6} + \underbrace{\int n_w\eta(x)\left(\langle\mathcal{P}(x,\omega)\rangle_\omega - 1\right)\,dV}_{F_7}\Big].
\end{aligned}
\tag{14}
$$

The thermal energy is given by kT, while n is the refractive index. For greater clarity, we split the particular contributions to the free energy as marked by the underbraces in Equation (14). The mean field created by coions and counterions and the water dipoles polarization contribution are given by terms F_1 and F_2, respectively. Mixing entropy free energy contributions of point-like counterions and coions are accounted for in terms F_3 and F_4. The constraint of a constant number of ions in the system is given in F_5, where λ_+ and λ_- are the global Lagrange's multipliers for counterions and coions. The free energy that corresponds to orientational entropy of permanent water dipoles is given by F_6, while the last term, F_7, gives the local constraint for orientation of dipoles. $\mathcal{P}(x,\omega)$ is the probability that a permanent water dipole located at x is oriented at angle ω with respect to the normal to the charged surface (Figure 3). The brackets $\langle ... \rangle_\omega$ denote the average:

$$\langle \mathcal{F}(x,\omega) \rangle_\omega = \frac{1}{4\pi} \int_0^\pi \mathcal{F}(x,\omega)\, 2\pi \sin\omega d\omega. \tag{15}$$

Here, ω is the angle between the internal dipole moment vector, $\mathbf{p}$, and $\mathbf{n}_\phi = \nabla\phi_c/|\nabla\phi_c|$ (see Figure 3). We perform variation on the Helmholtz free energy, F, in Equation (14), so that $\delta F = 0$. Let us deal with the variational approach of every contribution in Equation (14) particularly, beginning with F_1 and F_2. For clarity of notation, direct spatial dependence will sometimes be omitted, so that for example, $n_+(x) \equiv n_+$.

3.1. Variation Procedure

3.1.1. Electric Fields (F_1 and F_2)

Since there are no time dependent magnetic fields, we can express the electric fields as potentials $E(x) = -\phi'(x)$, $E_c(x) = -\phi_c'(x)$, where the prime labels the spatial derivative, and perform a variation on the electrostatic term pertaining to water dipoles.

$$\delta\left(\frac{\varepsilon_0 n^2}{2}\int \phi_c'^2\, dV\right) = \frac{\varepsilon_0 n^2}{2}\int 2\phi_c'\delta(\phi_c')\, dV. \tag{16}$$

We can rearrange this term, if we consider the rules of differentiating a function product

$$\begin{aligned}(\phi_c\delta\phi_c')' &= \phi_c'\delta\phi_c' + \phi_c\delta\phi_c'', \\ \phi_c'\delta\phi_c' &= (\phi_c\delta\phi_c')' - \phi_c\delta\phi_c''.\end{aligned} \tag{17}$$

The integral in Equation (16) can be rewritten as,

$$\begin{aligned}\varepsilon_0 n^2 \int \phi_c'\delta(\phi_c')\, dV = \varepsilon_0 n^2 \Big(\underbrace{\phi_c\delta\phi_c'|_0^\infty}_{=0} - \\ - \int \phi_c\delta(\phi_c'')\, dV\Big),\end{aligned} \tag{18}$$

where the first term on the right-hand side equals 0 at infinity, since we impose the electric potential there to be constant and equal to 0. Taking into account the Poisson's equation for the water dipoles, namely $\phi_c''(x) = \boldsymbol{\nabla}\cdot\mathbf{P}/\varepsilon_0 n^2$, where $\mathbf{P}$ represents the net polarization of the permanent water dipoles, we get

$$-\varepsilon_0 n^2 \int \phi_c\delta(\phi_c'')\, dV = \int \phi_c\delta\rho_c\, dV. \tag{19}$$

Here, ρ_c corresponds to the bound charge density due to the dipoles' polarizations, which is related to net polarization $\rho_c = -\boldsymbol{\nabla}\cdot\mathbf{P}$. We observe that $\delta(\boldsymbol{\nabla}\cdot\mathbf{P}) = \boldsymbol{\nabla}\cdot\delta\mathbf{P}$. The integral in (Equation (19)) can now be rewritten:

$$\int \phi_c\delta\rho_c\, dV = -\int \phi_c(\boldsymbol{\nabla}\cdot\delta\mathbf{P})\, dV. \tag{20}$$

The product rule for divergence can be used $\boldsymbol{\nabla}\cdot(\phi_c\delta\mathbf{P}) = (\nabla\phi_c)\cdot\delta\mathbf{P} + \phi_c(\boldsymbol{\nabla}\cdot\delta\mathbf{P})$, so that the integral of Equation (20) can now be written differently again:

$$\int \phi_c(\boldsymbol{\nabla}\cdot\delta\mathbf{P})\,dV = \underbrace{\int \boldsymbol{\nabla}\cdot(\phi_c\delta\mathbf{P})\,dV}_{=0} - \int(\nabla\phi_c)\cdot\delta\mathbf{P}\,dV. \tag{21}$$

Here, the first integral on the right hand side vanishes by virtue of the divergence theorem; since the potential far away from the plates is constant and equal to zero. We therefore arrive at the final result

$$\delta F_1 = \int(\nabla\phi_c)\cdot\delta\mathbf{P}\,dV. \tag{22}$$

The polarization, $\mathbf{P}$, is related to the average orientation of all water dipoles (Equation (2)):

$$\mathbf{P}(x) = n_w\langle\mathcal{P}(x,\omega)\rangle_\omega p\mathbf{n}_\phi. \tag{23}$$

Here, n_w is the number density of water molecules in the solution, $p = |\mathbf{p}|$ is the internal point-like dipole magnitude, $\mathbf{n}_\phi = \nabla\phi_c/|\nabla\phi_c|$ is the unit vector away from the charged plate and $\langle\mathcal{P}(x,\omega)\rangle_\omega$ is defined by Equation (15) (see Figure 3). Since our case deals with a negatively charged surface ($\sigma < 0$), $\mathbf{P}$ points in the direction opposite to the direction of the x-axis and is thus negative (for details see Reference [76]). Since the variation of $\mathbf{P}$ can be written $\delta\mathbf{P}(x) = \langle n_w\mathbf{p}\delta\mathcal{P}(x,\omega)\rangle_\omega$, we arrive at the variation of F_1:

$$\delta F_1 = n_w\int\langle\delta\mathcal{P}(x,\omega)(\nabla\phi_c)\cdot\mathbf{p}\rangle_\omega\,dV. \tag{24}$$

Similarly, for F_2 by taking into account Equation (17), we get

$$\delta\left(\frac{\varepsilon_0 n^2}{2}\int\phi'^2\,dV\right) = \int\phi\delta\rho_{\text{free}}\,dV. \tag{25}$$

The Poisson equation is different for free charges (ions): $\phi''(x) = -\rho_{\text{free}}/\varepsilon_0 n^2 = e_0(n_+(x) - n_-(x))$. The variation by $\phi''(x)$ in Equation (25) can be written with macroscopic net volume charge density $\rho_{\text{free}}(x)$, which in turn is the sum of the contributions of the local net ion charges. Performing the variation on ion charge distribution $\rho_{\text{free}}(x)$ gives

$$\delta\rho_{\text{free}} = e_0(\delta n_+ - \delta n_-), \tag{26}$$

finishing the variation of the term F_2:

$$\delta F_2 = \int e_0\phi(\delta n_+ - \delta n_-)\,dV. \tag{27}$$

3.1.2. Ion Mixing (F_3, F_4 and F_5)

It makes sense to perform the variation of the ion mixing terms (F_3 and F_4), together with their Lagrange multipliers (F_5), since the variation δn_+ and δn_- will be a common term for positive and negative ions, respectively. It is easily shown from Equation (14) that

$$\delta F_3 + \delta F_4 + \delta F_5 = kT\int\delta n_+(\lambda_+ + \ln\frac{n_+}{n_0})\,dV + kT\int\delta n_-(\lambda_- + \ln\frac{n_-}{n_0})\,dV. \tag{28}$$

3.1.3. Dipole Mixing (F_6 and F_7)

Variation of the terms F_6 and F_7 is straightforward. Since the bulk water number density, n_w, is taken to be constant, the variation of F_6 is

$$\delta F_6 = kTn_w\int(\langle\delta\mathcal{P}(x,\omega)\ln\mathcal{P}(x,\omega) + \delta\mathcal{P}(x,\omega)\rangle_\omega)\,dV. \tag{29}$$

The last variation of F_7 is performed over the probability, $\mathcal{P}(x,\omega)$, and the Lagrange multiplier, $\eta(x)$. Expanding and applying the product rule, we find that

$$\delta F_7 = kTn_w \int (\delta\eta(x)\langle\mathcal{P}(x,\omega)\rangle_\omega + \eta(x)\langle\delta\mathcal{P}(x,\omega)\rangle_\omega - \delta\eta(x))\, dV. \tag{30}$$

3.2. Euler-Lagrange Equations

Combining the variations of all the integrals given in Equation (14), their sum δF gives us the variation of Helmholtz free energy. Factoring all the variation terms with respect to $n_+(x), n_-(x), \mathcal{P}(x,\omega)$ and $\eta(x)$ gives

$$\begin{aligned} \delta F = &\int dV \delta n_+(x)\Big[kT\Big(\ln\frac{n_+(x)}{n_0} + \lambda_+\Big) + \phi e_0\Big] + \int dV \delta n_-(x)\Big[kT\Big(\ln\frac{n_-(x)}{n_0} + \lambda_-\Big) - \phi e_0\Big] + \\ &+ \int dV n_w \langle \delta\mathcal{P}(x,\omega)\Big(\nabla\phi_c\cdot\mathbf{p} + \frac{\ln\mathcal{P}(x,\omega)+\eta(x)+1}{\beta}\Big)\rangle_\omega + kT\int dV n_w \delta\eta(x)\Big(\langle\mathcal{P}(x,\omega)\rangle_\omega - 1\Big). \end{aligned} \tag{31}$$

The volume differentials in a planar geometry are $dV = S\,dx$. Since the minimization condition demands $\delta F = 0$, the expressions multiplied by $\delta n_+(x)$, $\delta n_-(x)$, $\delta\mathcal{P}(x,\omega)$ and $\delta\eta(x)$ in the last equation must equal zero, resulting in a system

$$kT\Big(\ln\frac{n_+(x)}{n_0} + \lambda_+\Big) + \phi e_0 = 0, \tag{32}$$

$$kT\Big(\ln\frac{n_-(x)}{n_0} + \lambda_-\Big) - \phi e_0 = 0, \tag{33}$$

$$E_c p\cos\omega + \frac{\ln\mathcal{P}(x,\omega)+\eta(x)+1}{\beta} = 0, \tag{34}$$

$$\langle\mathcal{P}(x,\omega)\rangle_\omega - 1 = 0. \tag{35}$$

Here, we write $\beta = 1/kT$ and expand the dot product $\nabla\phi_c\cdot\mathbf{p} = E_c p\cos\omega$ (see Figure 3). Solving Equations (32) and (33), we obtain the ion spatial distributions

$$n_+(x) = n_0\exp\Big(-\beta e_0\phi - \lambda_+\Big), \tag{36}$$

$$n_-(x) = n_0\exp\Big(\beta e_0\phi - \lambda_-\Big). \tag{37}$$

The boundary conditions state that $\phi(x\to\infty) = 0$ and $n_{+,-}(x\to\infty) = n_0$, which renders $\lambda_+ = \lambda_- = 0$. We may now turn our attention to the variation of permanent water dipoles orientation. Solving for $\mathcal{P}(x,\omega)$, Equation (34) gives

$$\mathcal{P}(x,\omega) = \Lambda(x)\exp\Big(-\beta E_c p\cos\omega\Big), \tag{38}$$

where $\Lambda(x) = \exp(-\eta(x)-1)$. Substituting the cavity field E_c by E (Equation (4)) and dipole moment magnitude p by p_0 (Equation (5)) gives

$$\mathcal{P}(x,\omega) = \Lambda(x)\exp\Big(-\beta\frac{3E}{2}\Big(\frac{2+n^2}{3}\Big)p_0\cos\omega\Big), \tag{39}$$

where p_0 is the magnitude of $\mathbf{p}_e$. The final result is expressed using the constant γ defined in Equation (9):

$$\mathcal{P}(x,\omega) = \Lambda(x)\exp\Big(-\beta\gamma E p_0\cos\omega\Big). \tag{40}$$

We can now evaluate the average internal dipole moment by integrating over mean orientations (considering Equation (23)),

$$
\begin{aligned}
p\langle\cos\omega\rangle &= p_0\Big(\frac{2+n^2}{3}\Big)\langle\cos\omega\rangle \\
&= \frac{\int_0^\pi \left(\frac{2+n^2}{3}\right) p_0 \cos\omega \exp\left(-\beta\gamma E p_0 \cos\omega\right) d\Omega}{\int_0^\pi \exp\left(-\beta\gamma E p_0 \cos\omega\right) d\Omega} \\
&= -p_0\left(\frac{2+n^2}{3}\right)\mathcal{L}\left(\beta\gamma E p_0\right).
\end{aligned}
\tag{41}
$$

The orientational polarization of water is thus (see Equations (2) and (5)):

$$
\begin{aligned}
P(x) &= n_w p\langle\cos\omega\rangle \\
&= -n_w p_0\left(\frac{2+n^2}{3}\right)\mathcal{L}\left(\beta\gamma E(x) p_0\right).
\end{aligned}
\tag{42}
$$

If we insert the above result and the ion distribution functions (Equations (36) and (37)) into the average microscopic charge density equation $\rho(x) = \rho_{\text{free}}(x) - dP/dx$ [61,77], where ρ_{free} is the contribution of the net ion charges Equations (26), (36) and (37) and $P(x)$ is the polarization due to partially oriented water dipoles, we get the expression for $\rho(x)$ in a one-dimensional case:

$$
\rho(x) = -2e_0 n_0 \sinh\left(\beta e_0 \phi(x)\right) + n_w p_0\left(\frac{2+n^2}{3}\right)\frac{d}{dx}\left(\mathcal{L}\left(\beta\gamma E(x) p_0\right)\right). \tag{43}
$$

Inserting the above expression for average microscopic volume charge density $\rho(x)$ into the Poisson's equation,

$$
\phi''(x) = -\frac{\rho(x)}{n^2\varepsilon_0}, \tag{44}
$$

we get the modified LPB differential equation for the electric potential $\phi(x)$:

$$
\phi''(x) = \frac{1}{n^2\varepsilon_0}\Big[2e_0 n_0 \sinh\left(\beta e_0\phi(x)\right) - n_w p_0\left(\frac{2+n^2}{3}\right)\frac{d}{dx}\left(\mathcal{L}\left(\beta\gamma E(x)p_0\right)\right)\Big], \tag{45}
$$

where $\phi''(x)$ is the second derivative of the electric potential $\phi(x)$ with respect to x and $E(x) = -\phi'(x)$. Equation (45) can be factorized via a product rule if we take into account that the Langevin function is odd and its derivative is $\mathcal{L}'(u) = 1/u^2 - 1/\sinh^2 u$ in the following form [7]:

$$
\frac{d}{dx}\Big[\varepsilon_0\varepsilon_r(x)\phi'(x)\Big] = 2e_0 n_0 \sinh\left(\beta e_0\phi(x)\right), \tag{46}
$$

$$
\varepsilon_r(x) = n^2 + n_w\frac{p_0}{\varepsilon_0}\left(\frac{2+n^2}{3}\right)\frac{\mathcal{L}(\beta\gamma E(x)p_0)}{E(x)}, \tag{47}
$$

where $\varepsilon_r(x)$ is the relative permittivity (Equation (11)). This modified Langevin Poisson-Boltzmann (LPB) differential equation (Equation (46)) is subject to two boundary conditions. The first boundary condition arises from the electro-neutrality of the system, which assumes that the total net charge of the system is zero, hence

$$
\int \rho_{\text{free}}(x)\, dV - \sigma S = 0, \tag{48}
$$

where σ is the negative membrane surface charge density, S is the total membrane surface area and $\rho_{\text{free}}(x) = -2e_0 n_0 \sinh\left(\beta e_0\phi(x)\right)$ is the macroscopic (net) volume charge density of coions and counterions. Since the macroscopic volume charge density is only dependent on x (Equation (43)) and the differential $dV = S\,dx$, Equation (48) may be rewritten

$$
\int_0^\infty 2e_0 n_0 \sinh\left(\beta e_0\phi(x)\right) dx = -\sigma. \tag{49}
$$

If we integrate Equation (45) once over the whole system, we get

$$\phi'(x=0) = -\frac{1}{n^2\varepsilon_0}\left[\sigma + n_w p_0 \left(\frac{2+n^2}{3}\right) \cdot \mathcal{L}(\beta\gamma E p_0|_{x=0})\right]. \tag{50}$$

The second boundary condition states that the electric potential far away from the charged surface (in the bulk) is constant $\phi'(x \to \infty) = 0$, rendering $\mathcal{L}(\beta\gamma E p_0|_{x\to\infty}) = 0$. The modified LPB equation (Equation (46)) was derived in one dimension, but can be rewritten in a more general form to apply to an arbitrary three-dimensional geometry. In three dimensions, the steps are analogous and discussed in detail in a previous work [58], where a three-dimensional Lagrangian was derived for a model of finite-sized ions. With this in mind, the modified LPB equation (Equation (46)) can be rewritten:

$$\nabla \cdot \left[\varepsilon_0 n^2 \nabla\phi(\mathbf{r})\right] + n_w p_0 \left(\frac{2+n^2}{3}\right) \boldsymbol{\nabla} \cdot \left(\mathbf{n}_\phi\, \mathcal{L}\left(\beta\gamma E p_0\right)\right) = 2e_0 n_0 \sinh\left(\beta e_0 \phi(\mathbf{r})\right), \tag{51}$$

where $\mathbf{n}_\phi = \nabla\phi/|\nabla\phi| = \nabla\phi/E$. We may factor the last equation, so that

$$\nabla \cdot \left[\varepsilon_0 \left(n^2 + \frac{n_w p_0}{\varepsilon_0}\left(\frac{2+n^2}{3}\right)\frac{\mathcal{L}\left(\beta\gamma E p_0\right)}{E}\right)\nabla\phi(\mathbf{r})\right] = 2e_0 n_0 \sinh\left(\beta e_0 \phi(\mathbf{r})\right). \tag{52}$$

This modified LPB equation can be written even more compactly, considering the definition of spatially dependent permittivity $\varepsilon_r(\mathbf{r})$ given by Equation (47) (for details, see Reference [58]):

$$\nabla \cdot \left[\varepsilon_0 \varepsilon_r(\mathbf{r}) \nabla\phi(\mathbf{r})\right] = 2e_0 n_0 \sinh\left(\beta e_0 \phi(\mathbf{r})\right), \tag{53}$$

$$\varepsilon_r(\mathbf{r}) = n^2 + n_w \frac{p_0}{\varepsilon_0}\left(\frac{2+n^2}{3}\right)\frac{\mathcal{L}(\beta\gamma E(\mathbf{r}) p_0)}{E(\mathbf{r})}. \tag{54}$$

Here, $\rho_{\text{free}}(\mathbf{r})$ is the macroscopic (net) volume charge density of coions and counterions. A corresponding three-dimensional variant of the boundary condition (Equation (50)) is

$$\nabla\phi(\mathbf{r} = \mathbf{r}_{\text{surf}}) = -\frac{1}{n^2\varepsilon_0}\left[\sigma\mathbf{n}_\phi + \mathbf{n}_\phi n_w p_0 \left(\frac{2+n^2}{3}\right) \cdot \mathcal{L}(\beta\gamma E(\mathbf{r}) p_0(\mathbf{r})|_{\mathbf{r}=\text{surf}})\right]. \tag{55}$$

Rearranging, it follows that

$$\nabla\phi(\mathbf{r} = \mathbf{r}_{\text{surf}})\left[1 + \frac{\mathbf{n}_\phi}{\nabla\phi(\mathbf{r} = \mathbf{r}_{\text{surf}})}\frac{n_w p_0}{n^2\varepsilon_0}\left(\frac{2+n^2}{3}\right) \cdot \mathcal{L}(\beta\gamma E(\mathbf{r}) p_0(\mathbf{r})|_{\mathbf{r}=\text{surf}})\right] = -\frac{\sigma}{n^2\varepsilon_0}\mathbf{n}_\phi. \tag{56}$$

Evaluating the second expression on the left hand side of the last equation gives

$$\frac{\mathbf{n}_\phi}{\nabla\phi(\mathbf{r} = \mathbf{r}_{\text{surf}})} = \frac{\nabla\phi(\mathbf{r} = \mathbf{r}_{\text{surf}})}{|\nabla\phi(\mathbf{r} = \mathbf{r}_{\text{surf}})|}\frac{1}{\nabla\phi(\mathbf{r} = \mathbf{r}_{\text{surf}})} = \frac{1}{E(\mathbf{r} = \mathbf{r}_{\text{surf}})}. \tag{57}$$

Combining this simplification with Equation (42), Equation (55) becomes

$$\nabla\phi(\mathbf{r} = \mathbf{r}_{\text{surf}})\varepsilon_r(\mathbf{r} = \mathbf{r}_{\text{surf}}) = -\frac{\sigma\mathbf{n}_\phi}{\varepsilon_0}. \tag{58}$$

Here we also take into account the expression for ε_r (Equation (47)). We see that the term inside the square brackets on the left-hand side of Equation (56) is precisely the definition of the relative permittivity on the surface of charged membrane $\varepsilon_r(\mathbf{r} = \mathbf{r}_{\text{surf}})$ (Equation (54)), yielding the general result

$$\nabla\phi(\mathbf{r} = \mathbf{r}_{\text{surf}}) = -\frac{\sigma\mathbf{n}_\phi}{\varepsilon_0\varepsilon_r(\mathbf{r} = \mathbf{r}_{\text{surf}})}. \tag{59}$$

4. Results

Figure 4 shows the dependency of the calculated macroscopic (net) volume charge density of the electrolyte solution inside the nanotubes ($\rho_{\text{free}}(r)$) on the radial distance from the geometrical axis of the tube. It can be seen in the figure that for larger radii of the inner cross-sections of the nanotubes (R), the value ρ_{free} at the geometrical axis of the tube is zero, which means that the number densities of counterions and coions there are equal and the electric potential is constant, that is, zero in our case (see the right panel in Figure 5).

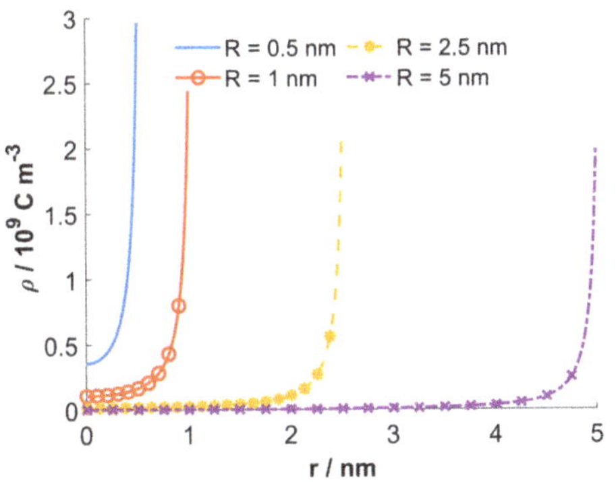

Figure 4. Macroscopic (net) volume charge density of coions and counterions (ρ_{free}) as a function of the radial distance from the geometrical axis of tube (r) calculated for 4 values of the inner tube diameter R: 0.5 nm, 1.0 nm, 2.5 nm and 5.0 nm. The bulk concentrations of counterions and coions $n_0/N_A = 0.15\,\text{mol/L}$ and $\sigma = -0.25\,\text{As/m}^2$, $T = 298\,\text{K}$, constant concentration of water $n_w/N_A = 55\,\text{mol/L}$, optical refractive index $n = 1.33$ and magnitude of external dipole moment of water $p_0 = 3.1$ Debye, where N_A is the Avogadro number.

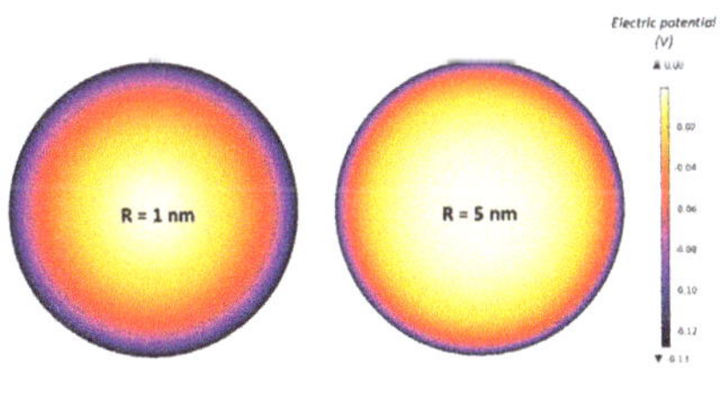

Figure 5. Space dependence of electric potential in the cross-section of the tube interior calculated for 2 values of the inner tube diameter R: 1.0 nm and 5.0 nm. The values of the model parameters are the same as given at Figure 4.

On the contrary, for smaller values of the nanotube radius R, the value of ρ_{free} at geometrical axis of the nanotube is not zero (Figure 4). Accordingly, for small values of the radius of the inner nanotube also the gradient of the electric field (Figure 6) and the electric potential at the nanotube geometrical axis are not zero (left panel in Figure 5). Hence, the bulk condition of the equal number densities of counterions and coions is fulfilled outside the interior of the nanotube.

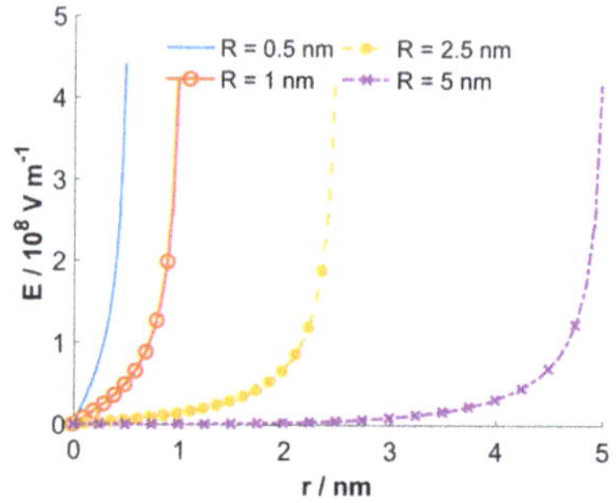

Figure 6. The magnitude of electric field strength as a function of the radial distance from the geometrical axis of tube (r), calculated for 4 values of the inner tube diameter R: 0.5 nm, 1.0 nm, 2.5 nm and 5.0 nm. The values of the model parameters are the same as given at Figure 4. The narrow nanotube before and after entrance of the nanoparticles.

Figure 7 shows the dependency of the average orientation $\langle \cos(\omega) \rangle_\omega$ and the relative permittivity ε_r on the radial distance from the geometrical axis of tube (r), calculated for four different values of nanotube inner radius R. It can be seen that for small radii, R, the average orientational of water dipoles is relatively strong also in the vicinity of geometrical axis of the tube, while for larger R the average orientation of water dipoles is strong only in the region near the charged inner surface of the tube.

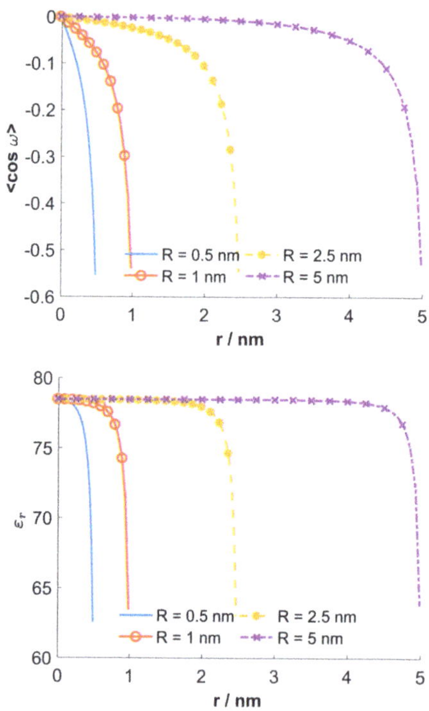

Figure 7. Average orientation $\langle \cos(\omega) \rangle_\omega$ and relative permittivity ε_r as a function of the radial distance from the geometrical axis of tube (r), calculated for 4 values of the inner tube diameter R: 0.5 nm, 1.0 nm, 2.5 nm and 5.0 nm. The values of the model parameters are the same as given at Figure 4.

5. Discussion and Conclusions

In this paper, we derived a modified Langevin Poisson-Boltzmann (LPB) model and the modified LPB equation to theoretically describe the electric double layer (EDL) for a monovalent electrolyte solution inside very narrow nanotubes with a negatively charged inner surface. In the modified LPB approach, the electronic polarization of the water is taken into account by assuming a permanent dipole embedded in the center of the sphere with a volume equal to the average volume of a water molecule. The effect of a polarizing environment is reproduced by introduction of the cavity field in the saturation regime [7,61,76]. In past EDL studies, treatments of cavity fields and structural correlations between water dipoles were limited to cases of relatively small electric field strengths, far away from the saturation limit of polarization and orientational ordering of water molecules [73–75]. High magnitudes of electric field strength were later added in several works [44,61,63,78]. A commonly oversimplified assumption when theoretically describing the EDL is the assumption of a surface charge density-independent relative permittivity in the inner (Stern) layer. Due to orientational ordering of water dipoles, the relative permittivity of the Stern layer depends on the electric field strength, that is, on the surface charge density (σ) of the electrode [51,79–82]. Furthermore, fitting the model curves with a range of free parameters to the experimental points [83] cannot prove that the Stern layer capacitance and permittivity is σ-independent. The decrease in the relative permittivity close to the charged surface (electrode) is obviously partially the consequence of orientational ordering of water dipoles close to the saturation regime or in the saturation regime as shown theoretically in References [6,27,44,54,58,59,61–64,80,82].

Within a recently presented phenomenological approach it is claimed that close to the charged surface, almost all water molecules belong to water shells around the ions, while the free water molecules are excluded [83]. The results of simulations clearly refute this fact [84] by showing increased water ordering in the direction towards the charged surface (including the region close to the charged surface) (Figure 7, upper panel) even for high salt concentrations [84], in quantitative agreement with mean-field theoretical predictions [7,82]. For example, for a magnitude of 0.16 As/m^2 surface charge density, there is practically no difference in the orientational ordering and space distribution of water dipoles close to the charged surface between water with and without NaCl (of concentration 500 mmol/L) [84]. In general, for magnitudes of surface charge density up to around 0.3 As/m^2, where the mean-field approach can still be justified [7,82], there is only a weak quantitative influence of salt on the profile of orientational ordering of water dipoles in Stern and diffuse layers, but not qualitative [84]. Note that the multi-layering of water predicted in simulations [84] cannot be predicted within our mean-field approach [44,61] as well also not in the oversimplified phenomenological models [83].

Besides the saturation in polarization/water dipole orientation at high magnitudes of the electric field strength, the important thing to consider in the EDL studies is also the saturation in the counterion concentration near the charged surface due to the finite size of ions. These steric effects were first predicted in the Wicke-Eigen's model (also called the Bikerman's model) and their modifications [3,5,22,25,27,35,85,86]. For finite sized ions, the dielectric permittivity profile in the vicinity of a charged surface is modulated by the depletion of water dipoles at the charged surface due to accumulated counterions [58,82]. In the modified LPB model [7,59], described in the present paper, these steric effects were not taken into account.

The described decrease in the relative permittivity relative to its bulk value in the present paper is the consequence of strong orientational ordering of the water dipoles in the vicinity of the charged surface (Figure 7). Contrary to our results it is claimed in Reference [87] that the relative permittivity is increased in direction to the charged surface. As pointed out in publications of different authors the predicted increase of relative permittivity near the charged surface in Reference [87] is unphysical [6,59] and defies the common wisdom in electrochemistry [56]. In addition, the experiments report just the converse as predicted in Reference [87], that is, the experiments indicated the decrease of relative permittivity near the charged surface [88,89]. The predicted substantial increase of relative permittivity in the inner part of the double layer near the charged surface in Reference [87] is due to arise in the dipole density near the surface as pointed out in Reference [56]. This unphysical result [6] is the consequence of inconsistency of so-called dipolar PB theory presented in Reference [87] as indicated in Reference [59]. Namely, it was shown [59] that the dipolar PB theory for point-like ions in Reference [87] assumes an orientationally averaged Boltzmann factor for spatial distribution function for water dipoles, which is however not compatible with the assumption of point-like ions. Energy dependent spatial distribution of water dipoles cannot be taken into account simultaneously with the assumption of point-like ions, but only if the finite size of molecules in the electrolyte solution are taken into account [35,61]. This means that the dipolar PB model presented in Reference [87] is not a self-consistent model and consequently predicts unphysical results which are not compatible with experimental results even qualitatively, as noticed in References [6,56,59] and other publications.

The other important difference between our modified LPB model and the theoretical model presented in Reference [87] is that our value for (external) water dipole moment 3.1 D [7,44,51,61] is considerably smaller than the corresponding value 4.86 D used in Reference [87]. The value 3.1 D is closer to the experimental values of the effective dipole moment of water molecules in clusters (2.7 D) and in bulk solution (2.4–2.6 D) (see for example Reference [68]). The value 4.86 D is so large in order to compensate for the cavity field [6,61,74,75,78] that is not taken into account in Reference [87], as noticed also in Reference [6], but is considered in the present modified LPB model. The model value 3.1 D can be additionally decreased by taking into account structural correlations between water dipoles [60,78]. The ion-ion and ion-water correlations were taken into account also in the mean-field models of References [8,65,66].

It has been shown that for finite-sized ions the drop in the number density of water near a charged surface results in an additional decrease of permittivity [7,58]. A further generalization of the modified LPB model with steric effects taken into account within a lattice-statistics model of a modified LPB is found in References [44,51,82]. By taking into account asymmetric finite size of ions the modified LPB equation was generalized to (modified Langevin Eigen-Wicke model) [44,51,82]:

$$\frac{d}{dx}\left[\varepsilon_0 \varepsilon_r(x) \frac{d\phi}{dx}\right] = 2e_0 n_s n_0 \frac{\sinh(\beta e_0 \phi)}{\mathcal{D}_A(\phi, E)}, \tag{60}$$

where $\varepsilon_r(x)$ is the spatial dependence of relative permittivity:

$$\varepsilon_r(x) = n^2 + n_{0w} n_s \frac{p_0}{\varepsilon_0}\left(\frac{2+n^2}{3}\right)\left(\frac{\mathcal{F}(\gamma p_0 E \beta)}{\mathcal{D}_A(\phi, E) E}\right) \tag{61}$$

and

$$\mathcal{D}_A(\phi) = \alpha_+ n_0 e^{-e_0 \phi \beta} + \alpha_- n_0 e^{+e_0 \phi \beta} + \frac{n_{0w}}{\gamma p_0 E \beta} \sinh(\gamma p_0 E \beta). \tag{62}$$

Here, the parameters α_+ and α_- are the number of lattice sites occupied by a single positive and negative hydrated ion, respectively, where a single water molecule is assumed to occupy just one lattice site. The reduced number density of lattice sites $n_s/N_A = 55$ mol/L is equal to the concentration of pure water [44,51,82]. The symbol n_{0w} stands for the bulk number density of water molecules. The function $\mathcal{F}(u)$ is defined as $\mathcal{F}(u) = \mathcal{L}(u)\sinh(u)/u$, where $\mathcal{L}(u)$ is the Langevin function.

The results of the present paper are important when considering electric fields within artificial as well as biological channels containing an electrolyte. Much attention has recently been given to understanding tunneling nanotubes (TNTs), small tubular structures that drive cell communication and spreading of pathogens [12]. Not yet fully understood, it is thought that these tubular structures initiate from local membrane bending facilitated by laterally distributed proteins or anisotropic membrane nanodomains. Further research is needed to clarify the role of EDL in the inception of these structures, since cytoplasmatic proteins and other elements are electrically charged. When such motor proteins are complemented by protruding cytoskeletal forces provided by the polymerization of f-actin, TNT formation is crucial in determining cell morphology, sometimes even leading to endovesiculation of the red blood cell membrane [90–92]. Recently, within a molecular mean-field approach and taking into account the asymmetric size of ions, polarization of water, and ion-ion and ion-water correlations, the ionic and water flows through biological ion channels was theoretically considered [65,66].

To conclude, in the present paper we started from a mean-field Helmholtz free energy functional, presented a thorough derivation of the modified LPB equation and model by minimization of the system free energy for the case of planar geometry. A special emphasis was devoted to orientational ordering of water dipoles, taken into account in the expression for the free energy by rotational entropy. Our approach provides a distinct analytical description of the interplay between mean-field electrostatic and entropic effects arising from the mixing entropy of ions and rotational entropy of water dipoles in EDL.

The derived modified LPB equation in planar geometry is then generalized for arbitrary geometry and then used to calculate numerically the average orientation of water dipoles, relative permittivity ε_r, magnitude of electric field strength, electric potential and the macroscopic (net) volume charge density of coions and counterions for a cylindrical geometry (in dependence on radial distance from the center of the tube).

Among other things it is indicated that in the saturation regime close to the charged surface, where the magnitude of electric field is very large (Figure 6), strong orientational water dipole ordering (Figure 7, upper panel) may result in a strong local decrease of permittivity (Figure 7, lower panel). The relative permittivity of the electrolyte solution decreases with increasing magnitude of the electric field strength.

Most interesting, we have shown that in the case of very narrow nanotubes the macroscopic (net) volume charge density of coions and counterions (ρ_{free}) at geometrical axis of the nanotube is not zero (Figure 4). In addition, in narrow nanotubes the water dipoles are partially oriented also close to the axis of the nanotube (Figure 7, upper panel), as schematically shown in (Figure 8). The potential importance of this phenomena for the transport through the narrow channels with the charged inner surface, specific only for very narrow nanotubes, should be investigated in the future. The channels in biological membranes can be an interesting example of such systems.

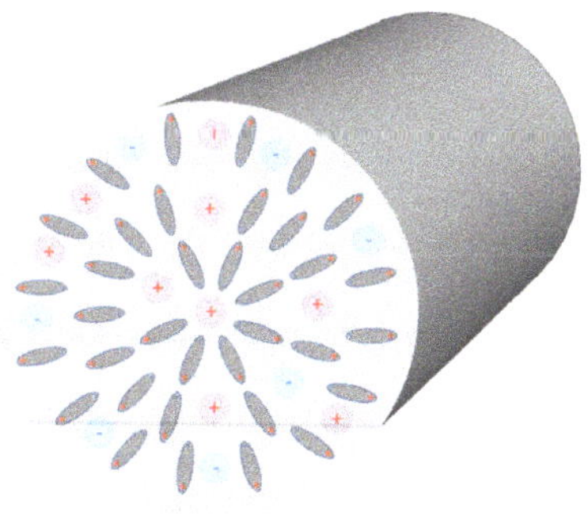

Figure 8. A schematic figure of a radial arrangement of water dipoles inside a very narrow cylindrical nanotube. The inner surface of the tube is negatively charged.

6. Materials and Methods

To solve Equation (52), a partial differential equation, we have used Comsol Multiphysics and its electrostatics stationary solver. The mesh consists of 4946 elements, the boundary condition (Equation (59)) was applied on the 2D cross-section of the nanotube and the geometry was solved for 10293 DoFs. The numerical results were solved using a stationary nonlinear solver (Automatic (Newton)), which implements a damped Newton's approach, with a minimum damping factor of 10^{-6}.

Author Contributions: M.D.: writing original draft preparation, equation derivation, writing-review and editing, visualization; E.G.: equation derivation, software, methodology, numerical calculations, calculations and preparation of the figures, editing; V.K.-I.: equation derivation, resources, editing, conceptualization, supervision, funding acquisition; A.I.: equation derivation, methodology, resources, editing, conceptualization, supervision, validation, funding acquisition. All authors have read and agreed to the published version of the manuscript.

Funding: This research was funded by the Slovenian Research Agency (ARRS), grant numbers P2-0232, P3-0388, J3-9262 and J1-9162.

Conflicts of Interest: The authors declare no conflict of interest.

Abbreviations

The following abbreviations are used in this manuscript:

EDL	Electric double layer
modified LPB	modified Langevin Poisson-Boltzmann

References

1. Kenkel, S.W.; Macdonald, J.R. A lattice model for the electrical double layer using finite-length dipoles. *J. Chem. Phys.* **1984**, *81*, 3215–3222.
2. Cevc, G. Membrane electrostatics. *Biochim. Biophys. Acta* **1990**, *1031*, 311–382.
3. Kralj-Iglič, V.; Iglič, A. A simple statistical mechanical approach to the free energy of the electric double layer including the excluded volume effect. *J. Phys. II* **1996**, *6*, 477–491.
4. Israelachvili, J.N. *Intermolecular and Surface Forces*; Academic Press: London, UK, 1997.
5. Kornyshev, A.A. Double-layer in ionic liquids: Paradigm change? *Chem. Phys. Lett.* **2007**, *111*, 5545–5557.
6. Misra, R.; Das, S.; Mitra, S. Electric double layer force between charged surfaces: Effect of solvent polarization. *J. Chem. Phys.* **2013**, *138*, 114703.
7. Gongadze, E.; Velikonja, A.; Perutkova, Š .; Kramar, P.; Maček-Lebar, A.; Kralj-Iglič, V.; Iglič, A. Ions and water molecules in an electrolyte solution in contact with charged and dipolar surfaces. *Electrochim. Acta* **2014**, *126*, 42–60.
8. Goodwin, Z.A.H.; Feng, G.; Kornyshev, A.A. Mean-field theory of electrical double layer in ionic liquids with account of short-range correlations. *Electrochim. Acta* **2017**, *225*, 190–197.
9. Kralj-Iglič, V.; Remškar, M.; Iglič. Deviatoric elasticity as a mechanism describing stable shapes of nanotubes. In *Horizons in World Physics*; Reimer, A., Ed.; Nova Science Publisher: Hauppauge, NY, USA, 2004; Volume 244, pp. 111–156.
10. Karlsson, A.; Karlsson, R.; Karlsson, M.; Cans, A.S.; Strömberg, A.; Ryttsen, F.; Orwar, O. Networks of nanotubes and containers. *Nature* **2001**, *409*, 150–152.
11. McLaughlin, S. The electrostatic properties of membranes. *Annu. Rev. Biophys. Biophys. Chem.* **1989**, *18*, 113–136.
12. Drab, M.; Stopar, D.; Kralj-Iglič, V.; Iglič, A. Inception mechanisms of tunneling nanotubes. *Cells* **2019**, *8*, 626.
13. Candelaria, S.L.; Shao, Y.; Zhou, W.; Li, X.; Xiao, J.; Zhang, J.G.; Wang, Y.; Liu, J.; Li, J.; Cao, G. Nanostructured carbon for energy storage and conversion. *Nano Energy* **2012**, *1*, 195–220.
14. Costa, R.; Pereira, C.M.; Silva, A.F. Insight on the effect of surface modification by carbon materials on the ionic liquid electric double layer charge storage properties. *Electrochim. Acta* **2015**, *176*, 880–886.
15. Signorelli, R.; Ku, D.C.; Kassakian, J.G.; Schindall, J.E. Electrochemical double-layer capacitors using carbon nanotube electrode structures. *Proc IEEE* **2009**, *97*, 1837–1847.
16. Helmholtz, H. Über einige Gesetze der Verteilung elektrischer Ströme in körperlichen Leitern mit Anwendung auf die tierelektrische Versuche. *Ann. Phys.* **1853**, *165*, 211–233.
17. Helmholtz, H. Studien uber elektrische Grenzschichten. *Ann. Phys.* **1879**, *243*, 337–382.
18. Gouy, M.G. Sur la constitution de la charge électrique à la surface d'un électrolyte. *J. Phys. Radium.* **1910**, *9*, 457–468.
19. Chapman, D.L. A contribution to the theory of electrocapillarity. *Philos. Mag.* **1913**, *6*, 475–481.
20. Stern, O. Zur Theorie der elektrolytischen Doppelschicht. *Z. Elektrochem.* **1924**, *30*, 508–516.
21. Bikerman, J. Structure and capacity of electrical double layer. *Philos. Mag.* **1942**, *33*, 384–397.
22. Wicke, E.; Eigen, M. Über den Einfluß des Raumbedarfs von Ionen in wäßriger Lösung auf ihre Verteilung in elektrischen Feld und ihre Aktivitätskoeffizienten. *Z. Elektrochem.* **1952**, *38*, 551–561.

23. Eigen, M.; Wicke, E. The thermodynamics of electrolytes at higher concentration. *J. Phys. Chem.* **1954**, *58*, 702–714.
24. Freise, V. Zur Theorie der diffusen Doppelschicht. *Z. Elektrochem.* **1952**, *56*, 822–827.
25. Borukhov, I.; Andelman, D.; Orland, H. Steric effects in electrolytes: A modified Poisson-Boltzmann equation. *Phys. Rev. Lett.* **1997**, *79*, 435.
26. Zelko, J.; Iglič, A.; Kralj-Iglič, V.; Kumar, P.B.S. Effects of counterion size on the attraction between similarly charged surfaces. *J. Chem. Phys.* **2010**, *133*, 204901.
27. Lamperski, S.; Outhwaite, C. W. Exclusion volume term in the inhomogeneous Poisson-Boltzmann theory for high surface charge. *Langmuir* **2002**, *18*, 3423–3424.
28. Bhuiyan, L.; Outhwaite, C. Comparison of exclusion volume corrections to the Poisson-Boltzmann equation for inhomogeneous electrolytes. *J. Coll. Int. Sci.* **2009**, 331, 543–547.
29. Nielaba, P.; Forstmann, F. Packing of ions near an electrolyteelectrode interface in the hnc/lmsa approximation to the rpm model. *Chem. Phys. Lett.* **1985**, *117*, 46–48.
30. Caccamo, C.; Pizzimenti, G.; Blum, L. An improved closure for the Born-Green-Yvon equation for the electric double layer. *J. Chem. Phys.* **1986**, *84*, 3327–3335.
31. Kjellander, R.; Marčelja, S. Interaction of charged surfaces in electrolyte solutions. *Chem. Phys. Lett.* **1986**, *127*, 402–407.
32. Plischke, M.; Henderson, D. Pair correlation functions and density profiles in the primitive model of the electric double layer. *J. Chem. Phys.* **1988**, *88*, 2712–2718.
33. Mier-y-Teran, L.; Suh, S.; White, H.; Davis, H. A nonlocal free energy density functional approximation for the electrical double layer. *J. Chem. Phys.* **1990**, *92*, 5087–5098.
34. Strating, P.; Wiegel, F. Effects of excluded volume on the electrolyte distribution around a charged sphere. *J. Phys. A Math. Gen.* **1993**, *26*, 3383–3391.
35. Bohinc, K.; Kralj-Iglič, V.; Iglič, A. Thickness of electrical double layer. Effect of ion size. *Electrochim. Acta* **2001**, *46*, 3033.
36. Borukhov, I. Charge renormalization of cylinders and spheres: Ion size effects. *Sci. B Polym. Phys.* **2004**, *42*, 3598–3615.
37. Torrie, G.M.; Valleau, J.P. Electrical double layers. I. Monte Carlo study of a uniformly charged surface. *J. Chem. Phys.* **1980**, *73*, 5807–5816.
38. Torrie, G.M.; Valleau, J.P. Electrical double layers. 4. Limitations of the Gouy-Chapman theory. *J. Chem. Phys.* **1982**, *86*, 3251–3257.
39. Lamperski, S.; Kłos, J. Grand canonical Monte Carlo investigations of electrical double layer in molten salts. *J. Chem. Phys.* **2008**, *129*, 164503.
40. Lamperski, S.; Outhwaite, C.W.; Bhuiyan, L. The electric double layer differential capacitance at and near zero surface charge for a restricted primitive model electrolyte. *J. Phys. Chem. B* **2009**, *113*, 8925–8929.
41. Kłos, J.; Lamperski, S. Monte Carlo study of molten salt with charge asymmetry near the electrode surface. *J. Chem. Phys.* **2014**, *140*, 054703.
42. Lian, C.; Liu, K.; Van Aken, K.L.; Gogotsi, Y.; Wesolowski, D.J.; Liu, H.L.; Jiang, D.E.; Wu, J.Z. Enhancing the capacitive performance of electric double-layer capacitors with ionic liquid mixtures. *ACS Energy Lett.* **2016**, *1*, 21–26.
43. Lee, J.W.; Nilson, R.H.; Templeton, J.A.; Griffiths, S.K.; Kung, A.; Wong, B.M. Comparison of molecular dynamics with classical density functional and poisson-boltzmann theories of the electric double layer in nanochannels. *J. Chem. Theory Comput.* **2012**, *8*, 2012–2022.
44. Gongadze, E.; Iglič, A. Asymmetric size of ions and orientational ordering of water dipoles in electric double layer model—An analytical mean-field approach. *Electrochim. Acta* **2015**, *178*, 541–545.
45. Miodek, A.; Castillo, G.; Hianik, T.; Korri-Youssoufi, H. Electrochemical aptasensor of human cellular prion based on multiwalled carbon nanotubes modified with dendrimers: A platform for connecting redox markers and aptamers. *Anal. Chem.* **2013**, *85*, 7704–7712.
46. Kabaso, D.; Gongadze, E.; Perutkova, Š .; Kralj-Iglič, V.; Matschegewski, C.; Beck, U.; van Rienen, U.; Iglič, A. Mechanics and electrostatics of the interactions between osteoblasts and titanium surface. *Comput. Methods Biomech.* **2011**, *14*, 469–482.

47. Imani, R.; Pazoki, M.; Tiwari, A.; Boschloo, G.; Turner, A.P.F.; Kralj-Iglič, V.; Iglič, A. Band edge engineering of TiO2@DNA nanohybrids and implications for capacitive energy storage devices. *Nanoscale* **2015**, *7*, 10438–10448.
48. Kulkarni, M.; Patil-Sen, Y.; Junkar, I.; Kulkarni, C.V.; Lorenzetti, M.; Iglič, A. Wettability studies of topologically distinct titanium surfaces. *Coll. Surf. B Biointerfaces* **2015**, *129*, 47–53.
49. Mohajernia, S.; Mazare, A.; Gongadze, E.; Kralj-Iglič, V.; Iglič,A.; Schmuki, P. Self-organized, free-standing TiO2 nanotube membranes: Effect of surface electrokinetic properties on flow-through membranes. *Electrochim. Acta* **2017**, *245*, 25–31.
50. Hwang, I.; Riboni, F.; Gongadze, E.; Iglič, A.; Yoo, J.; So, S.; Mazare, A.; Schmuki, P. Dye-sensitized TiO2 nanotube membranes act as a visible-light switchable diffusion gate. *Nanoscale Adv.* **2019**, *1*, 4844–4852.
51. Dubtsov, A.V.; Pasechnik, S.V.; Shmeliova, D.V.; Saidgaziev, A.S.; Gongadze, E.; Iglič, A.; Kralj, S. Liquid crystalline droplets in aqueous environments: Electrostatic effects. *Soft Matter* **2018**, *14*, 9619–9630.
52. Drab, M.; Kralj-Iglič, V. Diffuse electric double layer in planar nanostructures due to Fermi-Dirac statistics. *Electrochim. Acta* **2016**, *204*, 154–159.
53. Drab, M.; Kralj-Iglič, V. Electric double layer of electrons: Attraction between two like-charged surfaces induced by Fermi-Dirac statistics. *Phys. Lett. A* **2019**, *383*, 358–365.
54. Outhwaite, C.W. A treatment of solvent effect in the potential theory of electrolyte solution. *Mol. Phys.* **1976**, *31*, 1345–1357.
55. Butt, H.J.; Graf, K.; Kappl, M. *Physics and Chemistry of Interfaces*; Wiley-VCH: Weinheim, Germany, 2003.
56. Bazant, M.; Kilic, M.; Storey, B.; Ajdari, A. Towards an understanding of induced-charge electrokinetics at large applied voltages in concentrated solutions. *Adv. Colloid. Interface Sci.* **2009**, *152*, 48–88.
57. Nagy, T.; Henderson, D.; Boda, D. Simulation of an electrical double layer model with a low dielectric layer between the electrode and the electrolyte. *J. Phys. Chem. B* **2011**, *115*, 11409–11419.
58. Iglič, A.; Gongadze, E.; Bohinc, K. Excluded volume effect and orientational ordering near charged surface in solution of ions and Langevin dipoles. *Bioelectrochemistry* **2010**, *79*, 223–227.
59. Gongadze, E.; Rienen, van U.; Kralj-Iglič, V.; Iglič, A. Langevin Poisson-Boltzmann equation: Point-like ions and water dipoles near a charged membrane surface. *Gen. Physiol. Biophys.* **2011**, *30*, 130–137.
60. Gongadze, E.; Rienen, U. van; Kralj-Iglič, V.; Iglič, A. Spatial variation of permittivity of an electrolyte solution in contact with a charged metal surface: A mini review. *Comput. Methods Biomech.* **2013**, *16*, 463–480.
61. Gongadze, E.; Iglič, A. Decrease of permittivity of an electrolyte solution near a charged surface due to saturation and excluded volume effects. *Bioelectrochemistry* **2012**, *87*, 199–203.
62. Outhwaite, C.W. Towards a mean electrostatic potential treatment of an ion-dipole mixture or a dipolar system next to a plane wall. *Mol. Phys.* **1983**, *48*, 599–614.
63. Szalai, I.; Nagy, S.; Dietrich, S. Nonlinear dielectric effect of dipolar fluids. *J. Chem. Phys.* **2009**, *131*, 154905.
64. Szalai, I.; Dietrich, S. Magnetization and susceptibility of ferrofluids. *J. Phys. Condens. Matter* **2008**, *20*, 204122.
65. Liu, J.-L.; Eisenberg, B. Numerical methods for a Poisson-Nernst-Planck-Fermi model of biological ion channels. *Phys. Rev. E* **2015**, *92*, 012711.
66. Liu, J.-L.; Eisenberg, B. Molecular mean-field theory of ionic solutions: A Poisson-Nernst-Planck-Bikerman model. *Entropy* **2020**, *22*, 550.
67. Gregory, J.K.; Clary, D.C.; Liu, K.; Brown, M.G.; Saykally, R.J. The water dipole moment in water clusters. *Science* **1997**, *275*, 814–817.
68. Dill, K.; Bromberg, S. *Molecular Driving Forces: Statistical Thermodynamics in Chemistry and Biology*; Garland Science: New York, NY, USA, 2003.
69. Takashima, S.; Casaleggio, A.; Giuliano, F.; Morando, M.; Arrigo, P.; Ridella, S. Study of bound water of poly-adenine using high frequency dielectric measurements. *Biophys. J.* **1986**, *49*, 1003–1008.
70. Chiabrera, A.; Morro, A.; Parodi, M. Water concentration and dielectric permittivity in molecular crevices. *Il Nuovo Cimento D* **1989**, *11*, 981–992.
71. Kaatze, U. The dielectric properties of water in its different states of interaction. *J. Solution Chem.* **1997**, *26*, 1049–1112.
72. Giordano, S.; Rocchia, W. Shape-dependent effects of dielectrically nonlinear inclusions in heterogeneous media. *J. Appl. Phys.* **2005**, *98*, 104101.
73. Onsager, L. Electric moments of molecules in liquids. *J. Am. Chem. Soc.* **1936**, *58*, 1486–1493.

74. Kirkwood, J.K. The dielectric polarization of polar liquids. *J. Chem. Phys.* **1939**, *7*, 911–919.
75. Fröhlich, H. *Theory of Dielectrics*; Clarendon Press: Oxford, UK, 1964.
76. Drab, M.; Gongadze, E.; Mesarec, L.; Kralj, S.; Kralj-Iglič, V.; Iglič, A. The internal and external dipole moment of a water molecule and orientational ordering of water dipoles in an electric double layer. *Elektrotehniški Vestnik* **2017**, *84*, 221–234.
77. Jackson, J.D. *Classical Electrodynamics*; Wiley: New York, NY, USA, 1999.
78. Booth, F. The dielectric constant of water and the saturation effect. *J. Chem. Phys.* **1951**, *19*, 391–394.
79. Velikonja, A.; Gongadze, E.; Kralj-Iglič, V.; Iglič, A. Charge dependent capacitance of Stern layer and capacitance of electrode/electrolyte interface. *Int. J. Electrochem. Sci.* **2014**, *9*, 5885–5894.
80. Lorenzetti, M.; Gongadze, E.; Kulkarni, M.; Junkar, I.; Iglič, A. Electrokinetic properties of TiO2 nanotubular surfaces. *Nanoscale Res. Lett.* **2016**, *11*, 378.
81. Gongadze, E.; Mesarec, L.; Kralj-Iglič, V.; Iglič, A. Asymmetric finite size of ions and orientational ordering of water in electric double layer theory within lattice model. *Mini Rev. Med. Chem.* **2018**, *18*, 1559–1566.
82. Iglič, A.; Gongadze, E.; Kralj-Iglič, V. Differential capacitance of electric double layer—Influence of asymmetric size of ions, thickness of Stern layer and orientational ordering of water dipoles. *Acta Chim. Slov.* **2019**, *66*, 534–541.
83. Lopez-Garcia, J.J.; Hornoa, J.; Grosse, C. Differential capacitance of the diffuse double layer at electrode electrolyte interfaces considering ions as dielectric spheres: Part I. Binary electrolyte solutions. *J. Colloid Interface Sci.* **2017**, *496*, 531–539.
84. Marcovitz, A.; Naftaly, A.; Levy, Y. Water organization between oppositely charged surfaces: Implications for protein sliding along DNA. *J. Chem. Phys.* **2015**, *142*, 085102.
85. Gongadze, E.; Kralj-Iglič, V.; Iglič, A. Unequal size of ions in modified Wicke-Eigen model of electric double layer. *Gen. Phys. Biophys.* **2017**, *36*, 229–234.
86. Lopez-Garcia, J.J.; Horno, J.; Grosse, C. Influence of steric interactions on the dielectric and electrokinetic properties in colloidal suspensions. *J. Colloid Interface Sci.* **2015**, *458*, 273–283.
87. Abrashkin, A.; Andelman, D.; Orland, H. Dipolar Poisson-Boltzmann equation: Ions and dipoles close to charge interfaces. *Phys. Rev. Lett.* **2007**, *99*, 077801.
88. Teschke, O.; Ceotto, G.; de Souza, E.F. Interfacial aqueous solutions dielectric constant measurements using atomic force microscopy. *Chem. Phys. Lett.* **2000**, *326*, 328.
89. De Souza, E.F.; Ceotto, G.; Teschke, O. Dielectric constant measurements of interfacial aqueous solutions using atomic force microscopy. *J. Mol. Catal. A Chem.* **2001**, *167*, 235.
90. Fošnarič, M.; Penič, S.; Iglič, A.; Kralj-Iglič, V.; Drab, M.; Gov, N. Theoretical study of vesicle shapes driven by coupling curved proteins and active cytoskeletal forces. *Soft Matter* **2019**, *15*, 5319–5330.
91. Penič, S.; Fošnarič, M.; Mesarec, L.; Iglič, A.; Kralj-Iglič, V. Active forces of myosin motors may control endovesiculation of red blood cells. *Acta Chim. Slov.* **2020**, *67*, 674–681.
92. Penič, S.; Mesarec, L.; Fošnarič, M.; Mrowczynska, L.; Hägerstrand, H.; Kralj-Iglič, V.; Iglič, A. Budding and fission of membrane vesicles: A mini review. *Front. Phys.* **2020**, in print.

entropy

MDPI

Article

Application of a Statistical and Linear Response Theory to Multi-Ion Na^+ Conduction in NaChBac

William A. T. Gibby [1,*], Olena A. Fedorenko [2,3], Carlo Guardiani [1,4], Miraslau L. Barabash [1], Thomas Mumby [1], Stephen K. Roberts [3], Dmitry G. Luchinsky [1,5] and Peter V. E. McClintock [1,*]

1 Department of Physics, Lancaster University, Lancaster LA1 4YB, UK; carlo.guardiani@uniroma1.it (C.G.); m.barabash@lancaster.ac.uk (M.L.B.); thomas.mumby@googlemail.com (T.M.); d.luchinsky@lancaster.ac.uk (D.G.L.)

2 School of Life Sciences, University of Nottingham, Nottingham NG7 2UH, UK; olena.fedorenko@nottingham.ac.uk

3 Division of Biomedical and Life Sciences, Lancaster University, Lancaster LA1 4YQ, UK; s.k.roberts@lancaster.ac.uk

4 Department of Mechanical and Aerospace Engineering, Sapienza University, 00185 Rome, Italy

5 KBR Inc., Ames Research Center, Mountain View, CA 94035, USA

* Correspondence:w.gibby@lancaster.ac.uk (W.A.T.G.); p.v.e.mcclintock@lancaster.ac.uk (P.V.E.M.)

Abstract: Biological ion channels are fundamental to maintaining life. In this manuscript we apply our recently developed statistical and linear response theory to investigate Na^+ conduction through the prokaryotic Na^+ channel NaChBac. This work is extended theoretically by the derivation of ionic conductivity and current in an electrochemical gradient, thus enabling us to compare to a range of whole-cell data sets performed on this channel. Furthermore, we also compare the magnitudes of the currents and populations at each binding site to previously published single-channel recordings and molecular dynamics simulations respectively. In doing so, we find excellent agreement between theory and data, with predicted energy barriers at each of the four binding sites of $\sim$4, 2.9, 3.6, and 4kT.

Keywords: ion channel; statistical theory; linear response; ionic transport; NaChBac

Citation: Gibby, W.A.T.; Fedorenko, O.A.; Guardiani, C.; Barabash, M.L.; Mumby, T.; Roberts, S.K.; Luchinsky, D.G.; McClintock, P.V.E. Application of a Statistical and Linear Response Theory to Multi-Ion Na^+ Conduction in NaChBac. *Entropy* **2021**, *23*, 249. https://doi.org/10.3390/e23020249

Academic Editor: Antonio M. Scarfone

Received: 11 December 2020
Accepted: 11 February 2021
Published: 21 February 2021

1. Introduction

Biological channels are natural nanopores that passively transport ions across cellular membranes. These channels are of enormous physiological and pharmacological importance, and so investigation of their transport properties is an area of great interest and research. For example, Na^+ channels play a key role in the generation of the action potential [1–3]. Furthermore, artificial nanopores are primarily designed for their transport functionality which can be informed by our understanding of biological channels.

A primary function of these channels is their ability to discriminate effectively between ions, whilst still conducting them at high rates. An example is NaChBac from *Bacillus halodurans*, which is the first bacterial voltage-gated sodium channel (Nav) to have been characterised, and thus is a prokaryotic prototype for investigating the structure–function relationship of Nav channels [4]. It conducts ions at rates of 10^7 s^{-1} despite having permeability ratios favouring Na^+ over K^+ and over Ca^{++}. Recently we reported these values to be *at least* 10:1 and 5:1 respectively [5]. In fact from the reversal potential the Na^+/K^+ permeability ratio is found to be 25:1, which is closer in agreement but still less than [6] who found the ratio to be 170:1. This contrasts with potassium channels such as KcsA where selectivity is reversed, favouring K^+ over Na^+ at 1000:1 [7]. The channel itself is formed from several coupled subsystems, but we focus on the selectivity filter (SF), which is the primary region responsible for selectivity between ions. The SF can readily be mutated to generate a range of conducting (and non-conducting) channel types which exhibit different selectivity and conductivity properties compared to those exhibited by the wild-type (WT) channel (see [5]).

The SF has the amino acid (Here: T = threonine, L = Lecucine, E = glutamte, S = serine, W = tryptophan, A = alanine and x highlights where the sequence is not conserved and can be several possible amino acids.) sequence TLESWAS, and thus shares the TxExW sequence with eukaryotic calcium channels [6]. Unfortunately, a crystal structure of NaChBac is not available. However, Guardiani et al. [8–10] applied homology structural modelling to produce a structure of NaChBac that we will use in this publication. We conduct a variety of different Molecular Dynamics (MD) simulations (see Figure 1) to explore its properties. During simulation the SF was found to have an average radius $R_c \sim 2.8$ Å, length $L_c \sim 12$ Å and 4 binding sites for conducting Na^+ ions labelled S1–4 from the intra- to the extra-cellular side respectively. The conduction mechanism was found to involve knock-on between at least two, if not three, ions. Each binding site has a volume, as estimated in Table 1, whose sum gives the total volume of the pore V_c. The first two sites are formed at the backbone carbonyls of the threonine and leucine residues respectively. S1 is wider than the average pore radius with diameter 3.06 Å, but S2 has the average pore radius of 2.8 Å. As a result, these two sites accommodate the primary hydration shell with around 5–6 waters per ion, and thus prevent bare ion-protein interaction. S3 is of approximately the same size as S2, but the ion only interacts with four waters because it also interacts directly with the glutamate ring. The fourth site is formed on the extracellular side from the side chain of the serine residues and a sodium ion here has a 40% probability of interacting with one or two serines and a 60% probability of being fully hydrated by water. This is in stark contrast to the narrower potassium channels where K^+ ions are almost fully dehydrated as they permeate the pore. The Na^+ occupancies at each site have been determined by molecular simulation using 0.5 M bulk solutions. Both S1 and S4 have energy minima that are higher in energy than S2,3 and so are less likely to be occupied. In fact the average occupancy of S1,4 is only around half that of the most occupied site S2 (see Figure 7c).

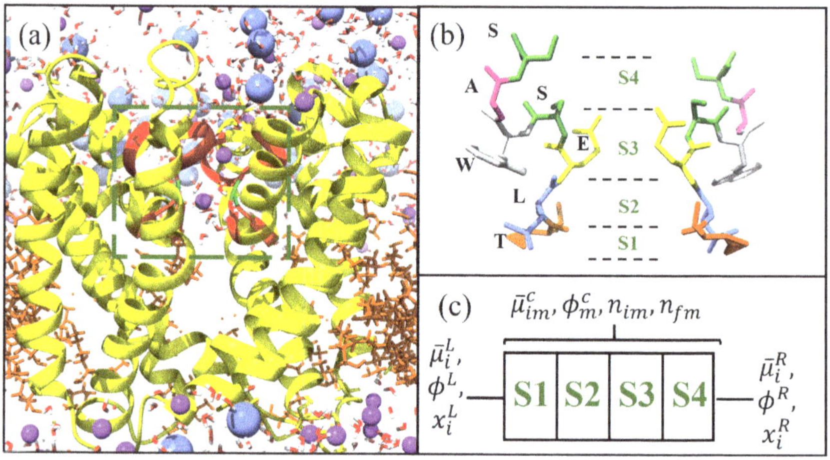

Figure 1. Structure of NaChBac [8] visualised using chimera [11]. (**a**) Yellow ribbons denote the protein spanning a lipid membrane (orange strands) between two aqueous ionic solutions. The selectivity filter (SF) is located within the box and highlighted by the red ribbons. The charged glutamates in the SF are highlighted green, and Na^+ (purple), and Cl^- (blue) ions alongside water molecules are included. (**b**) Structure of the SF for NaChBac with each amino acid highlighted and labelled by colour. The positions of the binding sites are included and labelled S1–S4 from the intra- to the extra-cellular side respectively. In (**c**) we show the lattice model used to define the system.

Table 1. Table of averaged radii and length of each binding site, obtained through the homology based structural model of NaChBac from [8]. The corresponding surface areas and volumes were estimated by assuming that each site was spheroidal in shape. The binding site is identified from a minima in the potential of mean force (PMF), and its length is estimated from the distance between maxima in the PMF. The radius is estimated from the average calculated radius in this region. These lengths and radii are given in the table.

Site	Estimated Average Radius	Estimated Length	Estimated Surface Area	Estimated Volume
S1	3.06 Å	3 Å	116 (Å)^2	117 (Å)^3
S2	2.77 Å	4 Å	126 (Å)^2	129 (Å)^3
S3	2.75 Å	3 Å	90 (Å)^2	80 (Å)^3
S4	2.77 Å	2 Å	78 (Å)^2	63 (Å)^3
Mean	2.8 Å	3 Å	103 (Å)^2	97 (Å)^3

These results are consistent with the results of MD simulations that have been performed on a variety of similar bacterial NaV channels. Chakrabarti et al. [12] conducted a 21.6 µs-long MD simulation of NavAb, observing a variable number of ions in the pore, mainly two or three (rarely four) and spontaneous and reversible ionic diffusion along the pore axis. Ulmschneider et al. [13] simulated the open state of the pore domain of NavMs with a voltage applied, and calculated the conductance which at $\sim 33pS$ was in agreement with experimental results.

The SF has a nominal charge of $-4e$ arising from the fixed gluatamte ring. However, determining the exact charge contribution from these pores is challenging due to the potential partial charges from remaining uncharged amino acids and the protonation that may occur at physiological pH levels. That latter is suspected to be true in voltage-gated Ca^{++} channels which share a ring of glutamates [14,15]. As a result, protonation of the glutamate ring in Navs has been studied fairly extensively [5,16–19]. Corry and Thomas [17] investigated the pore when only a single glutamate residue was protonated. The slightly protonated pore showed little difference in the potential of mean force vs. the normal pore. However, the doubly-protonated state showed a larger barrier for permeation to the pore, and reduced affinity for ion binding. Boiteux et al. [18] found a slight difference in the average number of Na^+ ions in the SF at 2.3 and 2.0 in the fully deprotanated and slightly-protonated states, respectively; however, both states were conducting. In simulations with two protonated residues, the authors observed the existence of a non-conducting state forming as a result of stable hydrogen bonds between the glutamates. As the number of protonated residues increased to three and four, Chloride Cl^- ions started to bind and the pore became non-conductive for Na^+. A similar study with shorter biased simulations suggested that protonation of a single Glu residue would diminish the conductance [16]. Meanwhile, a recent [19] study found that, at physiological pH, the pore may exist in the full deprotonation state but that it could also exist in the single or double-protonation states as well. Furthermore, the calculated pKa value decreases with each additional bound ion, implying that the presence of ions inside the pore leads to protonation of the SF. Thus, in [5] we introduced the notion of an effective charge describing the total charge in the pore as felt by the conducting ion, and its values were estimated by fitting Brownian dynamics simulations to experimental data for wild-type (WT) NaChBac and for a large selection of mutants. In our earlier work we studied NaChBac and its mutants theoretically and by Brownian dynamics simulation [5,20].

In earlier publications [5,21], we reported studies of Na^+ and Ca^{++} permeation in NaChBac, using Brownian dynamics models. The key result of modelling was that ionic conduction is analogous to electron transport in a quantum dot. As a function of the value of fixed charge, we observed a set of resonant conduction peaks separated by regions of blockade where the ions could not enter/leave the pore. This phenomenon is called *ionic Coulomb blockade* (ICB) [22], by analogy with (electronic) Coulomb blockade in quantum dots, for which the physics and the governing equations are essentially the same. Each resonant peak corresponds to an $n \rightarrow n+1$ barrier-less transition, which is of the knock-on kind when $n > 0$ [23], and the regions of blockade are when the charge carrier cannot pass. The occurrence of ICB has also been confirmed experimentally in artificial nanopores [24,25]. Although the ICB model explained immediately the role of the fixed charge, and accounted convincingly for the effect of mutations in which the fixed charge is altered, it is only a good approximation when electrostatic forces are dominant, that is, for divalent and trivalent ions. Furthermore, it does not contain affinities in the pore or excess chemical potentials in the bulk and so it cannot describe selectivity between ions of the same charge. It is also not connected to the results of Molecular simulation (MD) or the structure, and it cannot describe the absolute magnitude of the permeating current.

To provide a more accurate description, we needed a more fundamental model. We therefore developed a kinetic model [20], to investigate Na^+ vs. K^+ selectivity. This model was based on a simplified two site model of NaChBac and it was made self-consistent through the form of its transition rates. These were chosen such that the kinetic model and an earlier statistical and linear response theory had the same form of conductivity at low voltages. However, this did not include the complete structure or any comparison to results from MD simulation. It also did not include the binding site conductivities, or account for the correlations between ions at different binding sites. These two properties are expected to be important for fully describing the permeation properties and making quantitative predictions of the function of biological channels because it is known that small mutations in structure can lead to significant changes in function, for example, [5,26,27]. This was shown in [28], where we introduced a statistical and linear response theory fully accounting for structure and the properties of each binding site, and used it to analyse a point mutation in KcsA exploring the reasoning behind its drop in conductivity and occupancy.

In the present paper, we apply this recently developed statistical and linear response theory [28] to NaChBac with a more accurate model based on the structure introduced in [8]. The theory will include all four binding sites and their estimated volumes and surface areas, and the excess chemical potentials at each site. Furthermore, we extend this theory by deriving the conductivity at linear response in the presence of an electrochemical gradient. The theory is successfully compared to experimental single-channel and whole-cell recordings (some of which published in [5,20]), and results from MD simulations [8]. Finally, the theory allows us to make quantitative predictions of the current-concentration and current-voltage relations, and the effective open probability of the channel; as a function of the energy profile, experimental bulk concentration and structure of the pore.

In what follows, with SI units e is the unit charge, T the temperature, z the ionic valence, and k Boltzmann's constant.

2. Experimental Methods and Data

To apply the theory to NaChBac, and to compare with experimental recordings and make predictions, we consider two experiments. For further details of the experimental methods, including generation of the mutant channels and their expression, as well as details of the electro-physiological experiments, we refer to [5], and here we only present a concise summary. The first of these data sets is single-channel current-voltage recordings originally published in [20]. In these experiments identical bath and pipette solutions containing (in mM: 137 NaCl, 10 HEPES and 10 glucose, pH 7.4 adjusted with 3.6 mM NaOH) were used. Single-channel recordings are possible because Na^+ is the preferred substrate with sufficiently high conductance to provide a single-channel current amplitude which significantly exceeded noise (i.e., a favorable signal-to-noise ratio). In Figure 2a we plot the current-voltage curve, and in (b,c) we provide a current-time trace made at +100 mV. Trace (c) begins at the end of trace (b). There are at least three active channels passing currents with the magnitudes shown by the dashed lines.

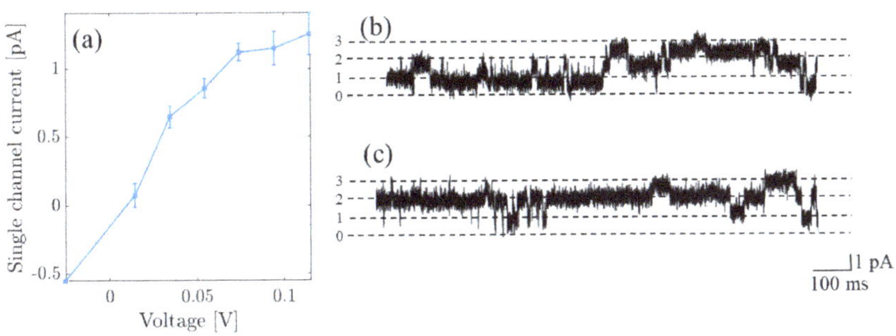

Figure 2. (**a**) Single channel currents recorded from NaChBac (originally published in [20]). (**b**,**c**) The original recording made at +100 mV in the 140 mM NaCl solution; the trace contains contributions from at least three active channels; and (**c**) represents a continuation in time of trace (**b**). The dashed lines show the amplitude level per channel, the numbers on the ordinate denoting the number of open channels.

In the second series of experiments, we performed whole-cell current measurements through NaChBac, in different Na^+/K^+ concentrations (see Figure 3). The black and purple curves in (a) (and the curve in (c)), that is, with 0M and 0.14M of NaCl solutions in the bath solution respectively (or 0.1M and 0M of KCl), were published in [5]. An identical experiment on a mutant was performed and described in [20]. In each case, the pipette solution contained (in mM) 120 Cs-methanesulfonate, 20 Na-gluconate, 5 CsCl, 10 EGTA, and 20 HEPES, pH 7.4 adjusted with 1.8 CsOH, meanwhile the bath solution contained (in mM); 137 NaCl, 10 HEPES and 10 glucose, pH 7.4 (adjusted with 3.6 mM NaOH). Permeability to K^+ was investigated by incrementally replacing the NaCl bath solution with an equivalent KCl solution such that the total ionic concentration was fixed at 140 mM. Total current across the cell was then normalized and, because one can assume that the total number of channels, their type and their open probability is conserved in each cell for the duration of the recording, it can effectively be modeled as a single channel. This normalization was with respect to the absolute value of peak current and is shown in Figure 3a. In (b) we show the current-concentration behaviour at -10 mV, which corresponds to the peak current. The reversal potential is plotted in (c); in cases where inward current was not detected, estimated values were determined from the voltage at which outward current could be detected. Finally, in (d) and (e) we provide the corresponding current-time traces.

Since NaChBac is highly impermeable to K^+ and Cl^- we have neglected the presence of these ions in the pore and in our theory we shall simply consider a single ion species, that is, Na^+ inside the pore.

Comparison of NaChBac Structures

In this subsection we shall compare the structure of NaChBac from the homology model which was used in [8], and the Cryo-EM structures 6vx3.pdb and 6vwx.pdb from [4].

In Figure 4 we provide an overlay of the homology model (yellow ribbons) and the 6vx3.pdb structure (green ribbons), using all of the backbone atoms. (a) provides the overlay of the whole pore and (b) provides a snap-shot of the selectivity filter (SF). From visual inspection there is clearly good agreement between the structures. In the pore the root-mean-square distance between structures (computed using the backbone atoms) is 17.47 Å and 7.14 Å in the SF.

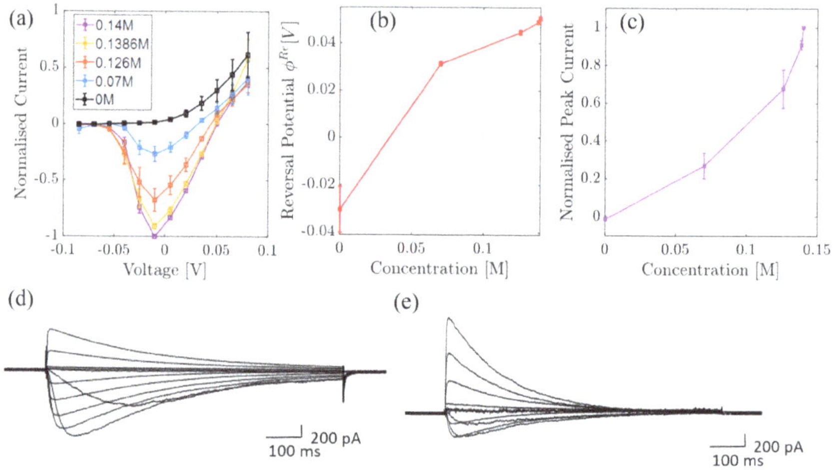

Figure 3. (**a**) Mean peak whole cell voltage-current relationships from cells expressing NaChBac channels, obtained in the bath solution with decreasing Na^+ content ranging from 140 mM to 0 mM (with NaCl being replaced with equimolar KCl). The peak currents were determined from time vs. current traces (examples shown in parts (**d**,**e**). Peak currents are normalized to the peak current recorded from the same cell in 140 mM NaCl-containing solution in the absence of K^+; error bars represent the standard error of the mean (SEM), determined from at least 4 independent cells. In (**b**) we show mean reversal potentials (±SEM) determined from data plotted in part (**a**). In cases where inward current was not detected, the reversal potential was assumed to be the voltage at which outward current could be detected. In (**c**) we plot the mean (±SEM) peak whole cell current (determined from data plotted in part a) as a function of Na concentration. Parts (**d**,**e**) are examples of time-dependent NaChBac currents recorded in 140 mM NaCl (**d**) and 126 mM NaCl and 14 mM KCl (**e**).

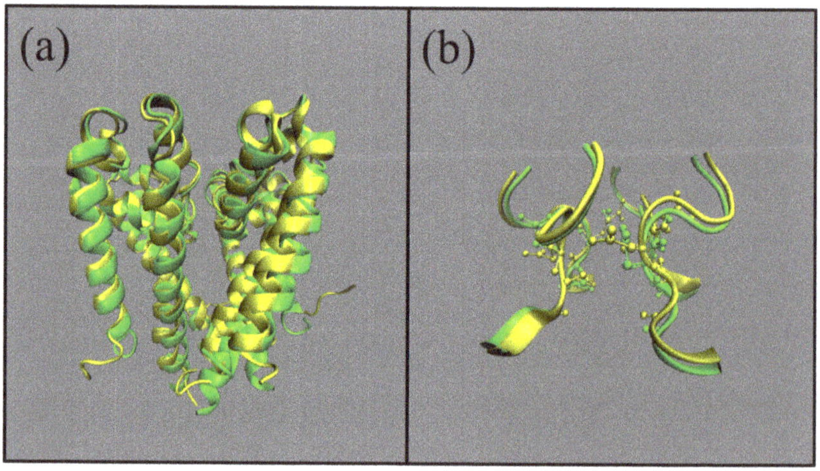

Figure 4. Comparison of NaChBac structures from the homology model (yellow) introduced in [8] and the Cryo-Em structure in green (6vx3.pdb) from [4]. (**a**) represents the whole pore and (**b**) is a snapshot of the (half) selectivity filter.

To further explore these structures we considered the pore radius which can be compared using the HOLE program. In Figure 5 we show a comparison between structures.

The homology model is more open than the Cryo-EM structures (6vx3 and 6vwx) both at the level of the cytosolic mouth (minimum centered on $z = -15$ Å) and in the region of the SF (around z = 0–12 Å). This is confirmed by volume filling representations of the pores which show a bottleneck close to the cytosolic mouth of 6vx3. The SF of 6vwx is narrower because the SF is occupied by two Na^+ ions, and these attract the side chains of the glutamates and the backbone carbonyls of the leucines, moving them towards the centre of the pore. Hence, there are two distinct minima in the pore radius which cannot be spotted in the radius profile of the homology model because this structure was obviously empty. However, the fact that the SF in 6vx3 (whose SF is empty) is also narrower than that of the model suggests that the structural differences might reflect different functional states in the channel cycle. In fact in the paper [4], Gao comments on the narrow radius of the cytosolic mouth, and on the arrangement of the Voltage Sensor Domain, suggesting that these structures might represent an inactivated conformation of the pore. By contrast, our homology model was built using the fully open conformation of NavMs from *Magnetococcus sp.* (PDB ID: 4F4L) as a template. As a result, our homology model probably represents an open conformation of NaChBac. This choice was deliberately taken on the assumption that an open conformation would be more suitable for the computational study of permeation and selectivity. In summary, the good agreement in overlayed structures, along with the choice to use an open conformation of NavMs as a template, makes us confident our model is a reliable system for the study of the selectivity and permeation of NaChBac.

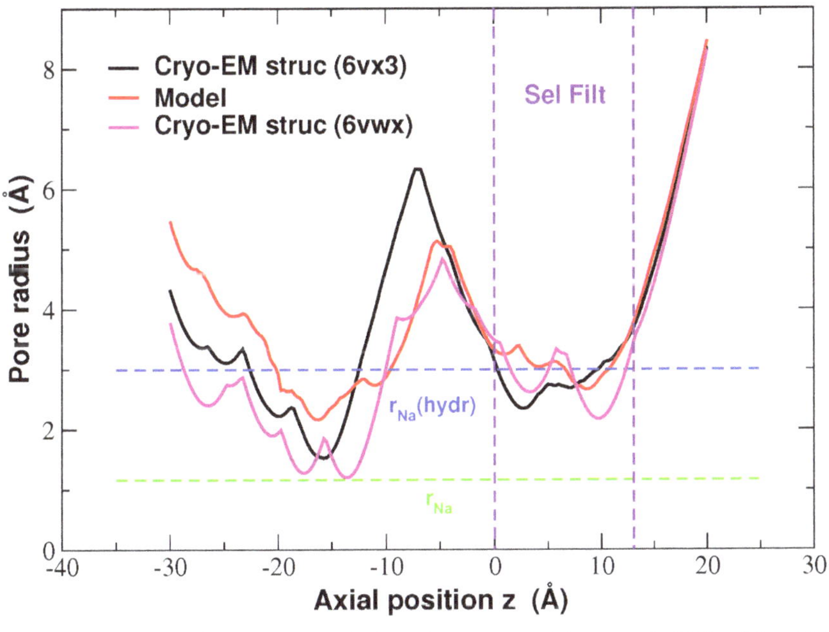

Figure 5. Comparison of average pore radius in the homology model structure (red) [8] and Cryo-EM structures 6vx3.pdb (black) and 6vwx.pdb (pink) [4]. The green and blue dashed lines denote the ionic Na^+ and hydrated Na^+ radii, respectively, and the purple dashed lines at $z = 0, 13$ Å highlight the selectivity filter region.

3. Theory

To model the SF we consider a system comprised of a pore thermally and diffusively coupled at either entrance to bulk reservoirs. This system and the effective grand canonical ensemble was considered and rigorously derived for multi-ion species in [28], and here we only present the necessary details needed to describe a single-species system. This pore is represented as a 1-dimensional lattice with 4 sites that may be occupied by a single ion at most. These are labelled S1–4 starting from the intracellular side in (c) of Figure 1. This figure also provides in (a) an overview of the system and (b) a snapshot of the SF which is highlighted by the red ribbons in (a). Clearly each configuration of Na^+ ions in the pore represent a distinct state of the system with total state space $\{n_j\}$. In this system ions inside the pore interact electrostatically with each other and charges on the surface of the pore via $\mathcal{E}$. Furthermore, they also interact locally at each binding site, m, via short-range contributions $\bar{\mu}^c_m$ and may experience an applied potential ϕ^c_m. Thus, with only Na^+ in the pore we can write the following distribution function, $P(\{n_j\})$,

$$P(\{n_j\}) = \mathcal{Z}^{-1}\frac{(x^b_{Na})^{n_{Na}}}{n_0!n_{Na}!}\exp[-(\mathcal{E}(\{n_j\}) - \sum_m n_{Nam}(\Delta\bar{\mu}_{Nam} + ez\Delta\phi^b_m))/kT]. \quad (1)$$

We have introduced Δ to denote the difference between bulk and site m in the pore such that $\Delta\bar{\mu}^b_m = \bar{\mu}^b - \bar{\mu}^c_m$ and $\Delta\phi^b_m = \phi^b - \phi^c_m$. In these cases $\bar{\mu}$ and ϕ denote the excess chemical potential and applied voltage in the bulk or at site m respectively. The prefactor contains factorial terms due to the indistinguishably of ions n_{Na} and empty sites n_0 in the pore, and x_{Na} denotes the mole fraction. For clarity we will drop the Na subscript. The necessary statistical properties such as site or pore occupancy can be derived from the partition function $\mathcal{Z}$ or Grand potential $\Omega = -kT\log(\mathcal{Z})$.

In [28] we demonstrated that the response to an applied electric field can be calculated following Kubo and Zwanzig [29–31]. We showed that the susceptibility density at each site can easily be derived and related to the conductivity at each site following the Generalised Einstein relation. The total conductivity through the pore is thus calculated by summing the reciprocals of the site-conductivity, in analogy to resistors in series. As a result all sites must be conducting for the total conductivity to be non-negligible. This effect partly explains the reduced conduction of a KcsA mutant [26], although we have to be mindful that the overall pore charge also decreases, increasing the overall energy barrier for conduction, and contributing to the reduced conductivity. We shall extend this derivation here by considering the response to an electrochemical gradient comprised of an electric potential gradient $\delta\phi$ and a concentration gradient δc. We shall assume that both bulk reservoirs are perturbed symmetrically so that the left (+) and right (−) electrochemical potentials, μ^b, can be written,

$$\mu^b = kT\log((c \pm \delta c/2)/c_w) + \bar{\mu}^0 + ez\phi^0 \pm ez\delta\phi/2, \quad (2)$$

where c_w is the concentration of the solvent which is much larger than that of the ions at around ∼55M, and c is the concentration of the solute, $\bar{\mu}^0$ is the equilibrium bulk excess potential which we assume to be unperturbed by the electrochemical gradient and ϕ^0 is the equilibrium electrical potential (which we will consider to be 0). In the following derivation we will write c/c_w as the mole fraction x. Thus following [28] we can write the following free energy, $G(\{n_j\},\delta\phi,\delta c)$, in the presence of this gradient by linearising μ^b about small δc,

$$\begin{aligned} G(\{n_j\},\delta\phi,\delta c) = \mathcal{E}(\{n_j\}) - \sum_{m=1}^{M} n_m(kT\log(x) + \Delta\bar{\mu}^0_m \pm \frac{kT}{2c}\delta c \pm ezv^b_m\delta\phi) \\ + kT\ln(n_0)! + kT\ln n!. \end{aligned} \quad (3)$$

In this expression we have rewritten $\delta\phi^b_m = v^b_m\delta\phi$ where v^b_m is a function representing the fraction of the voltage drop to move from either the left or right bulk to site m in the pore (see [28] for details). In a symmetrically distributed pore (which we assume), the average of v^b_m is equal to $1/2$. In this regime the probability distribution function can be written as

$$P(\{n_j\},\delta\phi,\delta c) = Z^{-1}\frac{x^n}{n_0!n!}\exp[-(\mathcal{E} - \sum_m n_m(\Delta\bar{\mu}^0_m \pm ezv^b_m\delta\phi \pm \frac{kT}{2c}\delta c)/kT]. \quad (4)$$

Here the partition function Z is defined in the standard manner from the conservation of probability and distinguished from the equilibrium partition function $\mathcal{Z}$. Both the free energy and distribution function can also be expressed in terms of the chemical gradient $\eta^L - \eta^R$ because

$$kT\log(x^L/x^R) = \delta\eta = \frac{kT}{c}\delta c. \tag{5}$$

The distribution (4) can be linearised about both small $\delta\phi$ and δc. When calculating the average particle density at each site $\langle n_m\rangle_{\delta c,\delta\phi}/V_m$, where V_m is the site volume, one can obtain relations for the susceptibilities due to the electrical gradient $\chi_m^{\delta\phi}$ and the chemical or concentration gradient $\chi_m^{\delta\eta}$. The former is defined in [28], since we assume a symmetrical pore the latter is defined as,

$$\chi_m^{\delta\eta} = \frac{1}{2kT}\left(\left\langle n_m\left(\sum_m n_m\right)\right\rangle - \left\langle\left(\sum_m n_m\right)\right\rangle\langle n_m\rangle\right)\frac{1}{V_m}. \tag{6}$$

It is worth noting that this expression is similar to $\chi_m^{\delta\phi}$ and is proportional to the variance of particle number at site m plus the covariance between sites m and the remaining sites in the pore. These susceptibilities are also proportional to the electrical conductivity, σ_m, at each binding site, which can be defined from the Einstein relation as: $\sigma_m = ze^2 D_m\chi_m$ where D_m and χ_m correspond to the diffusivity and susceptibility at each site respectively. As a result, the total current across the pore can be calculated as [28]

$$I = \left(\sum_m \frac{1}{\frac{A_m}{L_m}\sigma_m}\right)^{-1}(\delta\phi + \delta\eta/e), \tag{7}$$

where we recall that $\delta\phi$ is the voltage gradient in V, $\delta\eta$ is the chemical gradient in kT, and A_m and L_m are the surface area and length of site m respectively. Finally, the conductivity at each site is calculated from

$$\sigma_m = ze^2 D_m\left(\chi_m^{\delta\phi} + \chi_m^{\delta x}\right), \tag{8}$$

which is a function of the equilibrium bulk chemical potential.

4. Application to NaChBac

In Figure 6a, we consider the free energy spectra for selected (most favoured) pore configurations of NaChBac calculated from Equation (3) (when $\delta\phi = 0$ and $\delta c = 0$). We consider 0.14M NaCl solutions, and 0–3 ions inside the pore. In Equation (3) the total electrostatic energy, $\mathcal{E}$, is calculated by approximating the pore as a capacitor of total charge n_f and capacitance C taking the form $\mathcal{E} = U_c(n_f + n)^2$ where $U_c = \frac{e^2}{2C}$ [21,22]. Since the permitivitiy of water inside the pore is not known (though it must be less than the bulk value of 80) we consider $U_c = 10kT$. This approximation is discussed in detail in [28]. The energy spectra are parabolic vs. n_f, and each n-ion state has multiple configurations (15 in total) and we only highlight the most favoured. These states are determined by the values of $\Delta\bar{\mu}_{Nam}$, and their exact values are determined from fitting to experimental data (see Section 4.1). Differences in this term lead to energy splitting between possible configurations because the site occupied, in addition to the total number of ions inside the pore, determines the energy, *conducting* states correspond to the degeneracies where the lowest energy levels intersect, cf. [23], and this was shown to be the case in KcsA [28]. In NaChBac, the circle highlighting the 2–3 resonant transition occurs at around $n_f \sim -2.7$. Importantly, this differs from $n_f = -2.5$, suggesting that the the 3rd-ion faces an energy barrier to enter each site. If the concentration of the solutions was increased the energy barrier would decrease and the location of the resonant conduction would shift along the abscissa towards $n_f - 2.5$. It is worth reiterating that n_f here represents the total pore charge, and so differences from the fixed glutamate ring charge of $-4e$ can be explained from the additional contribution of all other charges and possible protonation inside the pore. Extended discussions of this point are provided in [5,16–19].

In Figure 6b we plot the energy spectra of the favoured 2 and 3-ion states, vs. n_f but also vs. bulk concentration. From the explanation above it is clear that the latter affects the value of n_f at which the two energy levels intersect. At low concentrations the energy barrier to add an ion to the pore is large. Thus, strong negative pore charge is required to reduce the barrier to attract the ion. Conversely at large concentrations the barrier is small and so less negative charge is needed. Thus one would expect the experimental current to be larger for measurements at higher concentrations, if these could be made.

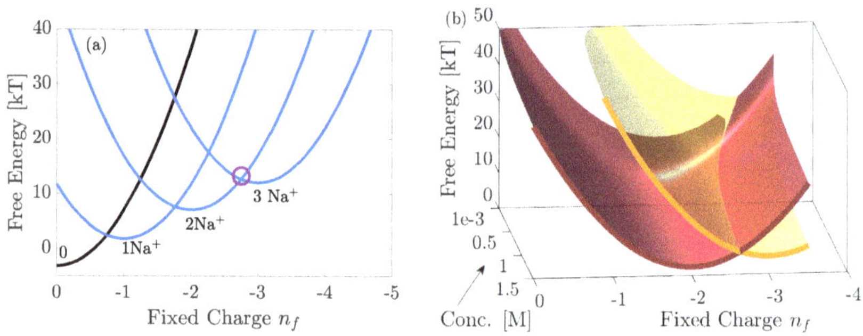

Figure 6. Free energy of the favoured states, plotted with $\Delta\bar{\mu}_{Na,1-4} \sim 2.3, 3.4, 2.8, 2.4kT$. In (**a**) it is plotted vs. n_f with 0.14M NaCl bulk solutions and in (**b**) vs. both n_f and bulk concentration. In (**a**) the blue curves correspond to the occupied $n > 0$ states of the pore, and black denotes the empty state. The purple circle highlights the location at which the two most favoured 2 and 3 ion states coincide, and we see that at $n_f = -2.5$ there is a small energy barrier. As bulk concentration increases this energy barrier reduces and the purple circle would shift towards $n_f = -2.5$. This is further clarified by (**b**) which shows only the 2 and 3 ions states.

To obtain the values of $\Delta\bar{\mu}_{Na,1-4}$ we performed fitting to two data sets, and this will be explained in the following subsection.

4.1. Comparison to Single Channel Data and MD

The values of $\Delta\bar{\mu}_{Na,m}$ used in Figure 6 are obtained by fitting, performed using the LSQCURVEFIT function in Matlab. We fit theory to the equilibrium site occupancies $\langle n_{Na,m} \rangle$ calculated from simulation data [8] (see Figure 2c), and the current at 35 mV. Current is needed here so that we can ensure it is of the correct order of magnitude. We also note that the difference in bulk NaCl concentration between the current and occupancy data is taken into account during fitting. To minimise the number of free parameters we also assumed that the diffusivity in the pore was constant, and equal to a tenth of the bulk value at $\sim 1.33 \times 10^{-10}$ m^{2}s^{-1}, and calculated $\Delta\bar{\mu}_{Na,m}$, relative to $n_f = -2.5$. The diffusivity is expected to be smaller within a confined pore due to the nature of the binding sites [32,33] and, although this value may appear small, it produces a barrier-less conduction rate through the pore of $\sim 0.9 \times 10^8$ ions per second which is of the order of tens of pA. We choose $n_f = -2.5$ because the electrostatic contribution to add a third ion is zero, that is, $\mathcal{E}(3) - \mathcal{E}(2) = 0$.

Both data sets are in excellent agreement with the theory, with currents only starting to differ at relatively large voltages when the experimental data deviate from Ohmic behaviour. Clearly beyond this regime, the system is far from equilibrium and our theory will need to be extended accordingly. After fitting we obtain $\Delta\bar{\mu}_{Na,1-4} \sim 2.3, 3.4, 2.8, 2.4kT$ when $n_f = -2.5$, with the sum of squared residuals being small at 10^{-4}. When the concentration is 0.14M the ions face the following barriers to enter each site: $\sim 4.0, 2.9, 3.6, 4.0kT$. These barriers are fairly similar to each other, although it is clear that S2 is the more favoured site and this is shown by its occupancy. As already discussed and observed in Figure 6, the energy barrier at each site reduces when the bulk concentration increases from 0.14M, resulting in a larger ionic current. This is confirmed by predicted current-voltage dependencies for 0.25 and 0.5M solutions respectively as showing increases in current; and the current-concentration behaviour in Figure 7b. In this latter case the bulk solutions are assumed to be symmetrical, with the driving force originating from a 50 mV voltage drop. This curve clearly demonstrates increasing conduction with concentration and we note that the current is relatively small <10 pA and is continuing to increase even at 2M because the overall energy barrier to enter the pore is large. We expect that these predictions can be further refined if more experimental measurements can be made.

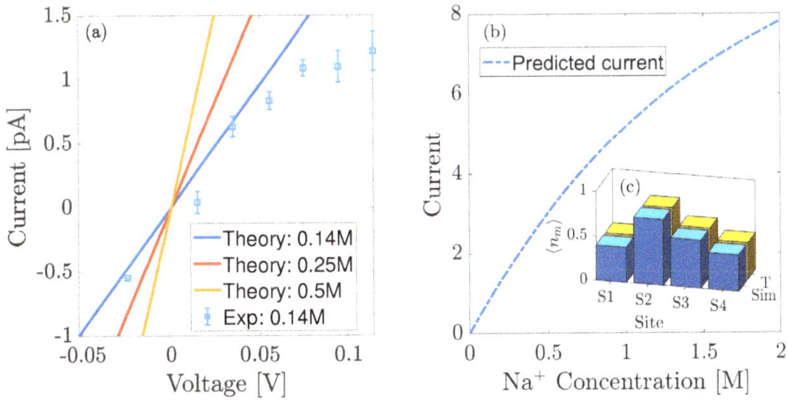

Figure 7. (**a**) Comparison of theoretical current *vs.* experimental data (squares) taken from [20] with symmetrical 0.14M NaCl solutions. (**b**) Predicted current-concentration curve at 50 mV across the pore. (**c**) Comparison of equilibrium occupancy at each site vs. simulation data with 0.5M NaCl solutions [8]. In doing this fitting we find that $\Delta\bar{\mu}_{Na,1-4} \sim 2.3, 3.4, 2.8, 2.4kT$, corresponding to energy barriers of $\sim 4, 2.9, 3.6, 4kT$ at 0.14M and we find the pore diffusivity to be $\sim 1.33 \times 10^{-10}\mathrm{m^2s^{-1}}$.

4.2. Comparison to Whole Cell Data

The theory can now be compared to the experimental whole-cell current-voltage recordings outlined earlier. In this experiment the data are normalised against the maximal current which is calculated when -10 mV is applied across the pore, and the bath solution contains 0.14M of Na^+ ions. We note that in Figure 8a this normalisation is with respect to the absolute value of this maximal value.

Under experimental conditions only the bath solution was varied. As a result, the theoretical equilibrium concentration and (chemical potential) used to calculate the conductivity σ and hence current varies slightly at each experimental point. This is because they are defined from the average concentration (or chemical potential) from both bulk solutions. Since the chemical gradient is calculated from the difference in bulk concentrations, we consider the lower limit of bulk concentration to be 0.1 mM rather than 0, to avoid the gradient diverging at low concentrations. Even at with the lowest concentration being 0.1 mM, the gradient is $\sim 5kT$ and so at the edge of applicability of our theory.

In Figure 8a, we plot the normalised current-voltage curves for the range of bath solutions. Overall we see good agreement between theory and data, but with two exceptions. NaChBac is a voltage-gated channel so that, at negative voltages, the number of open channels is reduced because the open probability decreases resulting in a smaller overall current [5,34]. Thus, at voltages below -10 mV our current diverges from the experimental data, and hence serves as a prediction of the normalised current in a single open channel. This prediction is given by the dashed lines, which we note increase in magnitude as voltage becomes more negative because the gradient increases. Furthermore, when the bath solution contains no Na^+ (black dashed curve) we observe poor agreement between theory and experiment and so highlight the curve with a dashed line. Finally, the inset curve shows the current closest to equilibrium.

The system is in equilibrium when the net current is zero, and this occurs when the applied voltage is equal to the reversal potential ϕ^{Re}. This was measured experimentally and is compared to the theoretical current in (b). In the theory the reversal potential is calculated from,

$$e\phi^{Re} = kT\log(x^L/x^R), \tag{9}$$

where L, R again refer to the left and right pipette/bath solutions respectively. We see good agreement except when the bath solution contains no Na^+. Even, our reduced concentration of 0.1 mM yields a reversal potential smaller than -35 mV. This is echoed by the current at this concentration which is not in good agreement with the experiment (see the black dashed curve in Figure 8a). A possible explanation for these disagreements is that, in the absence of Na^+ in the bath solution, K^+ ions enter the pore but do not conduct, consequently blocking the pore. Furthermore, at this concentration we are at the limits of applicability because the chemical gradient is still relatively large $\sim 5kT$. We plan to discuss this in a future manuscript after further investigations.

In Figure 9a we estimate the effective open probability P_{eff}. This is defined relative to the open probability at peak current P_{max}, from the ratio of theoretical and experimental current for each of the given concentrations. We neglect the estimate in the absence of Na^+ because the theoretical current did not agree with experimental data. We observe that P_{eff} takes values between 0 and 1.5 except for three concentrations all at +50 mV of applied voltage. At 0.126M, 0.1386M and 0.14M bath concentration the theoretical current was below the experimental values and in the latter two concentrations of different sign. This produced estimated effective open probabilities, P_{eff}, taking the values of 2.5, -15 and -0.5 for the three concentrations respectively (only $P_{\text{eff}} \sim -0.5$ is shown). Apart from these points however we observe it to be broadly sigmoidal and being 0 at negative voltages as anticipated. We expect, that the actual open probability, P_{Open}, can be calculated through the following definition,

$$P_{\text{Open}}(V) = P_{\text{eff}} \times P_{\text{max}}, \tag{10}$$

if the open probability of the maximal current is known.

In Figure 9b we highlight the current-concentration ($I - C$) behaviour by plotting the $I - C$ curve at the peak voltage (-10 mV). Note that, unlike Figure 8a, the current is normalised to the maximum current at 0.14M (and not to the absolute value). As expected the theoretical current agrees fairly well with the experimental one except at low concentrations ($\lesssim$5 mM). The curve takes a quasi-linear shape because the current comprises two terms: (1) the conductivity prefactor and (2) the electrochemical gradient. The second term is of the standard form, but our conductivity is a function of the equilibrium bulk chemical potential, which through our derivation must take the averaged concentration between the two bulks and thus slightly varies with bath concentration as well.

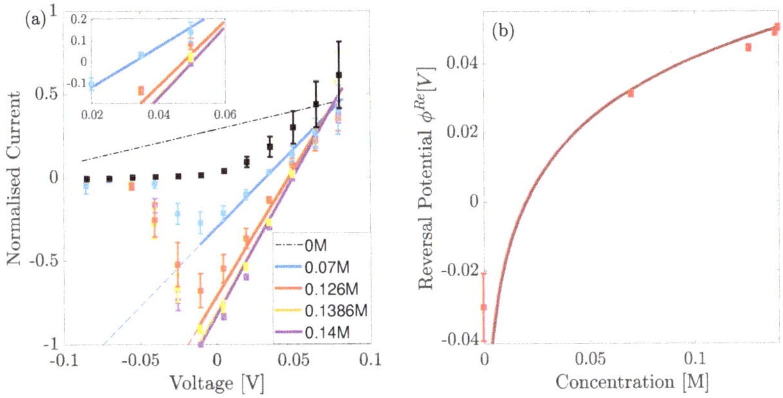

Figure 8. (**a**) Comparison of theoretical (solid line) to experimental (squares) data of normalised (to absolute value) whole-cell current in the presence of an electrochemical gradient, for a range of extra-cellular bulk solutions. The peak occurs at −10 mV, and below this voltage the current reduces due to the reduction in the open probability. Dashed lines predict the normalised currents if the open probability remained unchanged from the value at the peak current. (**b**) Theoretical (solid) and experimental (squares) of the reversal potential (ϕ^{Re}) for a range of concentrations. Theory only differs when the right bulk is absent of Na^+.

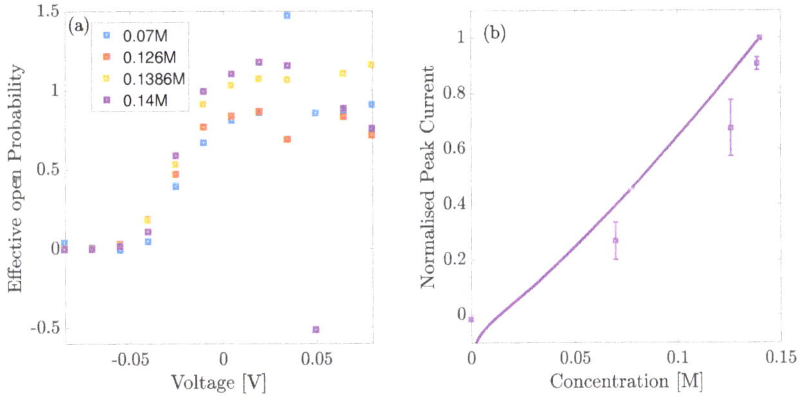

Figure 9. (**a**) Estimated open probability from the ratio of experimental to theoretical current. Below −40 mV the open probability is close to zero indicating that the channels are closed. (**b**) Comparison of normalised theoretical current (solid line) and experimental (squares) data vs. bulk concentration, at −10 mV of applied voltage.

5. Conclusions and Summary

In summary, we have taken the statistical and linear response theory, originally derived in [28] and applied to KcsA and a mutant, and applied it to investigate Na^+ conduction in NaChBac. Importantly, in order to compare with experimental and simulation data see Figures 2 and 3), we needed to extend the theory to take account of a chemical gradient. In doing so, we derived the conductivity at each site and the total through the pore in the presence of an electrochemical gradient. The main result of the paper is the quantitative predictions of pore function that we make as a function of the energy profile, experimental bulk conditions, and the pore structure.

In Figure 7 we compared the theoretical current-voltage and equilibrium site occupancies to experimental and simulation data. This comparison allowed us to extract the following values of $\Delta\bar{\mu}_{Na,1-4} \sim 2.3, 3.4, 2.8, 2.4kT$. At the experimental concentration 0.14M, the 3rd ion faces an energy barrier to enter each site within the pore of ~4, 2.9, 3.6, $4kT$. Although these values are not barrier-less as observed in KcsA [28], they are not expected to be because the experimental current is smaller in NaChBac. Furthermore, these parameters lead to barrier heights consistent with [8,20]. Using these parameters we have predicted the current for higher concentrations, including the current-concentration behaviour with 50 mV of applied voltage and current-voltage dependencies for 0.25 and 0.5M solutions. As expected both show an increase of current as the bulk solution increases. We expect that with more experimental data, we could refined these parameters.

In Figures 8 and 9 we compared the theory to normalised whole-cell data, under the assumption that the normalisation effectively renders it a single-channel for the point of comparison. The theory was found to be in good agreement with experiment except for when the bath solution was devoid of Na^+. A possible explanation is that in the absence of Na^+, K^+ ions enter the pore but do not conduct, subsequently blocking the pore. Furthermore, at this concentration we are at the limits of applicability because the chemical gradient is still relatively large $\sim 5kT$. We plan to investigate this in a future manuscript by introducing a far-from equilibrium kinetic model that accounts for both Na^+ and K^+ ions. Such a model was briefly introduced in [20]. However, it failed to account properly for the correlations between ions at different sites, and only considered a 2 site pore; and so further development is needed.

Finally, we expect our theory to be applicable to the study of mixed-valence, that is, Na^+/Ca^{++} selectivity in NaChBac and related voltage gated Ca^{++} channels, alongside artificial nano-pores.

Author Contributions: Data curation, O.A.F.; Formal analysis, W.A.T.G.; Project administration, D.G.L. and P.V.E.M.; Supervision, D.G.L., P.V.E.M. and S.K.R.; Writing—original draft, W.A.T.G.; Writing—review & editing, W.A.T.G., O.A.F., C.G., M.L.B., T.M., S.K.R., D.G.L. and P.V.E.M. All authors have read and agreed to the published version of the manuscript.

Funding: The work was funded in part by a PhD Scholarship from the Faculty of Science and Technology of Lancaster University, the Engineering and Physical Sciences Research Council (grants EP/M016889/1 and EP/M015831/1), and by a Leverhulme Trust Research Project Grant RPG-2017-134. C.G. is currently supported by a project that has received funding from the European Research Council (ERC) under the European Union's Horizon 2020 research and innovation programme (grant agreement No. 803213).

Data Availability Statement: Data related to this research are openly available from the University of Lancaster Research Directory at (https://doi.org/10.17635/lancaster/researchdata/421 (accessed on 17 February 2021)). Gibby, W.A.T.; Fedorenko, O.A.; Guardiani, C.; Barabash, M.L.; Mumby, T.; Roberts, S.K.; Luchinsky, D.G.; McClintock, P.V.E. (2020) Data for Application of a Statistical and Linear Response Theory to Multi-Ion Na^+ Conduction in NaChBac [Dataset].

Acknowledgments: We acknowledge valuable discussions with Bob Eisenberg, Igor Khovanov and Aneta Stefanovska.

Conflicts of Interest: The authors declare no conflict of interest.

References

1. Zheng, J.; Trudeau, M.C. (Eds.) *Handbook of Ion Channels*; CRC Press Taylor & Francis Group: Boca Raton, FL, USA, 2015.
2. Hille, B. *Ion Channels of Excitable Membranes*, 3rd ed.; Sinauer Associates: Sunderland, MA, USA, 2001.
3. Ashcroft, F.M. *Ion Channels and Disease*; Academic Press: London, UK, 1999.
4. Gao, S.; Valinsky, W.C.; On, N.C.; Houlihan, P.R.; Qu, Q.; Liu, L.; Pan, X.; Clapham, D.E.; Yan, N. Employing NaChBac for cryo-EM analysis of toxin action on voltage-gated Na^+ channels in nanodisc. *Proc. Natl. Acad. Sci. USA* **2020**, *117*, 14187–14193.
5. Fedorenko, O.A.; Kaufman, I.K.; Gibby, W.A.T.; Barabash, M.L.; Luchinsky, D.G.; Roberts, S.K.; McClintock, P.V.E. Ionic Coulomb blockade and the determinants of selectivity in the NaChBac bacterial sodium channel. *Biochim. Biophys. Acta (BBA)-Biomembranes* **2020**, *1862*, 183301.
6. Yue, L.X.; Navarro, B.; Ren, D.J.; Ramos, A.; Clapham, D.E. The cation selectivity filter of the bacterial sodium channel, NaChBac. *J. Gen. Physiol.* **2002**, *120*, 845–853.
7. Coates, L. Ion permeation in potassium ion channels. *Acta Crystallogr. D* **2020**, *76*, 326–331.

8. Guardiani, C.; Rodger, P.M.; Fedorenko, O.A.; Roberts, S.K.; Khovanov, I.A. Sodium binding sites and permeation mechanism in the NaChBac channel: A molecular dynamics study. *J. Chem. Theor. Comp.* **2017**, *13*, 1389–1400.
9. Guardiani, C.; Fedorenko, O.A.; Roberts, S.K.; Khovanov, I.A. On the selectivity of the NaChBac channel: an integrated computational and experimental analysis of sodium and calcium permeation. *Phys. Chem. Chem. Phys.* **2017**, *19*, 29840–29854.
10. Guardiani, C.; Fedorenko, O.A.; Khovanov, I.A.; Roberts, S.K. Different roles for aspartates and glutamates for cation permeation in bacterial sodium channels. *Biochim. Biophys. Acta (BBA)-Biomembranes* **2019**, *1861*, 495–503.
11. Pettersen, E.F.; Goddard, T.D.; Huang, C.C.; Couch, G.S.; Greenblatt, D.M.; Meng, E.C.; Ferrin, T.E. UCSF Chimera—A visualization system for exploratory research and analysis. *J. Comput. Chem.* **2004**, *25*, 1605–1612.
12. Chakrabarti, N.; Ing, C.; Payandeh, J.; Zheng, N.; Catterall, W.A.; Pomès, R. Catalysis of Na^+ permeation in the bacterial sodium channel NaVAb. *Proc. Natl. Acad. Sci. USA* **2013**, *110*, 11331–11336.
13. Ulmschneider, M.B.; Bagnéris, C.; McCusker, E.C.; DeCaen, P.G.; Delling, M.; Clapham, D.E.; Ulmschneider, J.P.; Wallace, B.A. Molecular dynamics of ion transport through the open conformation of a bacterial voltage-gated sodium channel. *Proc. Natl. Acad. Sci. USA* **2013**, *110*, 6364–6369.
14. Root, M.J.; MacKinnon, R. Two identical noninteracting sites in an ion channel revealed by proton transfer. *Science* **1994**, *265*, 1852–1856.
15. Corry, B.; Allen, T.W.; Kuyucak, S.; Chung, S. Mechanisms of permeation and selectivity in calcium channels. *Biophys. J.* **2001**, *80*, 195–214.
16. Furini, S.; Barbini, P.; Domene, C. Effects of the protonation state of the EEEE motif of a bacterial Na^+-channel on conduction and pore structure. *Biophys. J.* **2014**, *106*, 2175–2183.
17. Corry, B.; Thomas, M. Mechanism of ion permeation and selectivity in a voltage gated sodium channel. *J. Am. Chem. Soc.* **2012**, *134*, 1840–1846.
18. Boiteux, C.; Vorobyov, I.; Allen, T.W. Ion conduction and conformational flexibility of a bacterial voltage-gated sodium channel. *Proc. Natl. Acad. Sci. USA* **2014**, *111*, 3454–3459.
19. Damjanovic, A.; Chen, A.Y.; Rosenberg, R.L.; Roe, D.R.; Wu, X.; Brooks, B.R. Protonation state of the selectivity filter of bacterial voltage-gated sodium channels is modulated by ions. *Proteins* **2020**, *88*, 527–539.
20. Gibby, W.A.T.; Barabash, M.L.; Guardiani, C.; Luchinsky, D.G.; Fedorenko, O.A.; Roberts, S.K.; McClintock, P.V.E. Theory and experiments on multi-ion permeation and selectivity in the NaChBac ion channel. *Fluct. Noise Lett.* **2019**, *18*, 1940007.
21. Kaufman, I.K.; McClintock, P.V.E.; Eisenberg, R.S. Coulomb blockade model of permeation and selectivity in biological ion channels. *N. J. Phys.* **2015**, *17*, 083021.
22. Krems, M.; Di Ventra, M. Ionic Coulomb blockade in nanopores. *J. Phys. Condens. Matter* **2013**, *25*, 065101.
23. Yesylevskyy, S.O.; Kharkyanen, V.N. Barrier-less knock-on conduction in ion channels: Peculiarity or general mechanism? *Chem. Phys.* **2005**, *312*, 127–133.
24. Feng, J.; Liu, K.; Graf, M.; Dumcenco, D.; Kis, A.; Di Ventra, M.; Radenovic, A. Observation of ionic Coulomb blockade in nanopores. *Nature Mater.* **2016**, *15*, 850–855.
25. Chernev, A.; Marion, S.; Radenovic, A. Prospects of Observing Ionic Coulomb Blockade in Artificial Ion Confinements. *Entropy* **2020**, *22*, 1430.
26. Zhou, M.; MacKinnon, R. A mutant KcsA K^+ channel with altered conduction properties and selectivity filter ion distribution. *J. Mol. Biol.* **2004**, *338*, 839–846.
27. Tilegenova, C.; Cortes, D.M.; Jahovic, N.; Hardy, E.; Hariharan, P.; Guan, L.; Cuello, L.G. Structure, function, and ion-binding properties of a K^+ channel stabilized in the 2,4-ion bound configuration. *Proc. Natl. Acad. Sci. USA* **2019**, *116*, 16829–16834.
28. Gibby, W.A.T.; Barabash, M.L.; Guardiani, C.; Luchinsky, D.G.; McClintock, P.V.E. Physics of selective conduction and point mutation in biological ion channels. *arXiv* **2020**, arXiv:2010.08450.
29. Kubo, R. Statistical-mechanical theory of irreversible processes. I. General theory and simple applications to magnetic and conduction problems. *J. Phys. Soc. Jpn.* **1957**, *12*, 570–586.
30. Kubo, R. The fluctuation-dissipation theorem. *Rep. Progr. Phys.* **1966**, *29*, 255–284.
31. Zwanzig, R. *Nonequilibrium Statistical Mechanics*; Oxford University Press: Oxford, UK, 2001.
32. Tieleman, D.P.; Biggin, P.C.; Smith, G.R.; Sansom, M.S.P. Simulation approaches to ion channel structure-function relationships. *Quart. Rev. Biophys.* **2001**, *34*, 473–561.
33. Roux, B.; Allen, T.; Berneche, S.; Im, W. Theoretical and computational models of biological ion channels. *Quart. Rev. Biophys.* **2004**, *37*, 15–103.
34. Jo, A.; Hoi, H.; Zhou, H.; Gupta, M.; Montemagno, C.D. Single-molecule study of full-length NaChBac by planar lipid bilayer recording. *PLoS ONE* **2017**, *12*, e0188861.

entropy

MDPI

Review

Dynamics of Ion Channels via Non-Hermitian Quantum Mechanics

Tobias Gulden [1] and Alex Kamenev [2,3,*]

[1] Institute of Science and Technology Austria, 3400 Klosterneuburg, Austria; tgulden@ist.ac.at
[2] School of Physics and Astronomy, University of Minnesota, Minneapolis, MN 55455, USA
[3] William I. Fine Theoretical Physics Institute, University of Minnesota, Minneapolis, MN 55455, USA
* Correspondence: kamenev@physics.umn.edu

Abstract: We study dynamics and thermodynamics of ion transport in narrow, water-filled channels, considered as effective 1D Coulomb systems. The long range nature of the inter-ion interactions comes about due to the dielectric constants mismatch between the water and the surrounding medium, confining the electric filed to stay mostly within the water-filled channel. Statistical mechanics of such Coulomb systems is dominated by entropic effects which may be accurately accounted for by mapping onto an effective quantum mechanics. In presence of multivalent ions the corresponding quantum mechanics appears to be non-Hermitian. In this review we discuss a framework for semiclassical calculations for the effective non-Hermitian Hamiltonians. Non-Hermiticity elevates WKB action integrals from the real line to closed cycles on a complex Riemann surfaces where direct calculations are not attainable. We circumvent this issue by applying tools from algebraic topology, such as the Picard-Fuchs equation. We discuss how its solutions relate to the thermodynamics and correlation functions of multivalent solutions within narrow, water-filled channels.

Keywords: non-Hermitian Hamiltonians; algebraic topology; semiclassical methods; nanopores; ion transport; statistical mechanics

check for updates

Citation: Gulden, T.; Kamenev, A. Dynamics of Ion Channels via Non-Hermitian Quantum Mechanics. *Entropy* **2021**, *23*, 125. https://doi.org/10.3390/e23010125

Received: 2 December 2020
Accepted: 15 January 2021
Published: 19 January 2021

1. Introduction

Transport of ions through narrow channels plays a big role in many biological and technological systems. Many pathogens attack cells by forming nanopores in the cell membrane by using pore-forming toxins (PFTs) [1,2]. This punches holes in the cell membrane through which ions diffuse to the outside, effectively killing the cell. Physically this is similar to artificial nanopores in, e.g., silicon [3,4]. These are heavily used in genetic sequencing techniques where high-throughput of selective transport is the most important factor [5]. Other similar examples include free-standing silicon nanowires [6,7] and water-filled nanotubes [8,9]. These systems play various different roles in biology and technology. However they all follow the same underlying physics of a quasi-1D statistical system formed by ions confined to move in a narrow water-filled tube inside a lipid membrane or solid medium [10–16]. What makes this system special is the large ratio between the dielectric constants of water, $\kappa_1 \simeq 80$, and the surrounding media (e.g., for lipids or silicon oxide $\kappa_2 \simeq 2-4$). Because of this, the electric field created by an ion within is confined to stay mostly inside the water-filled channel and does not leak into the surrounding medium. As several numerical simulations in three dimensions point out the flow field also follows almost entirely the channel direction [17–19]. This simple observation has profound consequences.

First, as was noticed by Parsegian [20], there is a potential barrier for an ion to enter the channel. This barrier is equal to the energy difference between an ion being inside and outside the channel. For a channel of radius a the electric field created by an ion of charge e in the middle of the channel is $E_0 = 2e/(\kappa_1 a^2)$. The corresponding field energy integrated over the channel volume is $U_0 = \frac{\kappa_1}{8\pi} E_0^2 \pi a^2 L = k_B T(\lambda_B L)/(2a^2)$, where L is the length of the channel and $\lambda_B = e^2/(\kappa_1 k_B T) \approx 7$Å is the Bjerrum length at ambient temperature [11]. For a typical channel with $L \approx 40$Å and $a \approx 5$Å the corresponding (self-)energy barrier exceeds ambient temperature $k_B T$ by a factor of 5 or 6. This means that such a channel would block the transport of ions. However, there are at least two mechanisms which can be employed to overcome this issue. One is placing charged radicals along the channel path. The other is entropic screening of the barrier by a collective effect of multiple cations and anions inside the channel. In this review we focus on this latter phenomena, while the former is addressed in References [13,14,21,22].

The second consequence of the mismatch of dielectric constants is that the mutual interactions between the ions within the channel acquire the form of the 1D Coulomb potential

$$\Phi(x_i - x_j) = eE_0|x_i - x_j|, \tag{1}$$

where x_i are 1D coordinates of the ions along the channel axis. As illustrated in Figure 1, the electric field lines emanating from a charge are bent to run along the channel. Only after a characteristic length ξ given by the implicit relation $\xi^2 = a^2\kappa_1/(2\kappa_2)\ln(2\xi/a)$ the field lines start penetrating the lipid membrane and escaping the channel [11]. For a water-filled channel in a lipid membrane this gives $\xi \approx 7a$. Hence, for a sufficiently short channel with $L < \xi$ or (as considered in Section 3) a large concentration of salt ions where the characteristic distance between two ions is smaller than ξ, the interactions effectively follow the 1D Coulomb potential. The linear nature of the potential (1) leads to the curious observation that the energy barrier of transporting a charge through the channel can't be less than U_0, irrespective of how many other ions are present in the channel [20]. Indeed, for the most favorable arrangement of alternating positive and negative ions, the electric field along the channel alternates between $\pm E_0$. This leads back to the value of U_0 for the electrostatic energy of adding a single ion to the channel in the presence of the other ions. This may seem as a predicament that collective screening can't lower the transport barrier. Such conclusion is premature, however. The resolution of this apparent paradox is that in a system of multiple particles at a finite temperature it is the *free energy* (rather than the *energy*) which determines the transport barrier. The difference between the two is given by the entropy, i.e., it is the entropy of the ion gas within the channel which provides the screening mechanism. The nature of entropic suppression of the transport barrier can be traced to the aforementioned independence of the energy U_0 of the positions of individual ions. This observation implies that there is a large number of microscopic configurations which are close in energy. This is the hallmark of a state with large entropy and thus lower free energy.

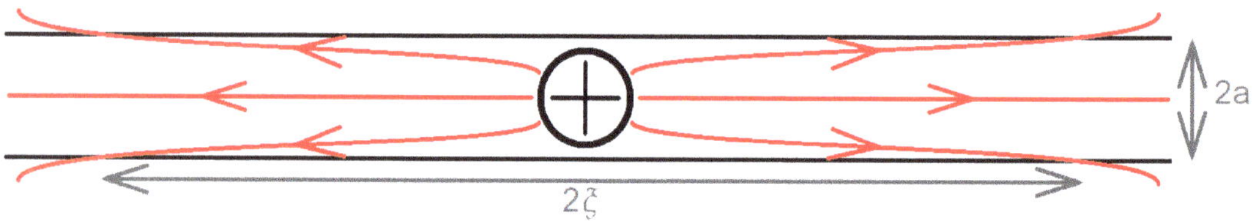

Figure 1. This is an illustration of the electric field lines emanating from an ion inside a water-filled channel of radius a which is surrounded by a medium with lower dielectric constant. Due to the mismatch in dielectric constants the field lines run mostly along the channel which means that another charge would feel an effective 1D Coulomb potential. The ratio is finite however, i.e., a distance ξ away from the ion the field lines start permeating the outside medium. If the channel is shorter than this critical length scale, $L < \xi$, or the typical spacing between charges is smaller than ξ, then all interactions are well-described by the 1D Coulomb potential.

Formalizing these observations is not entirely straightforward. As was first realized by Edwards and Lenard (EL) in 1962 [10] it requires mapping of the 1D statistical system onto an effective quantum mechanics with cosine potential. In fact, this is a particular case of the generic correspondence between D-dimensional statistical mechanics of the

Coulomb gas and (D-1)-dimensional sine-Gordon field theory [23]. The D=2 version of this mapping is well-known in the physics of the Berezinskii-Kosterlitz-Thouless transition. The less appreciated fact is that the Hermitian potential of the form $2\cos\theta = e^{i\theta} + e^{-i\theta}$ is a consequence of having a neutral plasma of monovalent ions with charge $\pm e$. In the EL mapping the $e^{\pm i\theta}$ operators shift the value of the electric field in the channel (the variable canonically conjugated to θ) by a quanta $\pm 2E_0$, which corresponds to the electric field generated by a unit charge $\pm e$.

What happens in the presence of a multivalent dissociated salt, such as, e.g., $CaCl_2$ which produces a plasma with positive charges $+2e$ and twice as many negative charges $-e$? It is not difficult to see that the EL mapping leads to an effective Hamiltonian with the potential $\frac{1}{2}e^{2i\theta} + e^{-i\theta}$. Such a Hamiltonian is non-Hermitian and thus admits a complex-valued spectrum. This may present a problem for the interpretation of the original statistical mechanics of the Coulomb plasma. For example, the free energy density (a manifestly real quantity) is given by the logarithm of the partition function which therefore needs to be real and positive. Fortunately the effective non-Hermitian quantum operator obeys the so-called $\mathcal{PT}$-symmetry [24], which ensures that all eigenvalues are real or appear as complex-conjugate pairs. When calculating the partition function, which includes summing over all eigenvalues, the imaginary parts cancel and we obtain a real, physical result [25]. However, in general there exist complex eigenvalues (spontaneously broken $\mathcal{PT}$-symmetry). This translates to an oscillatory character of certain correlation functions, reflecting short-range charge density wave correlations within the channel.

To model the transport of ions through the channel in this framework we use the concept of boundary charges which was developed in Reference [11]. From now on we assume that the channel is sufficiently short so that all field lines stay inside the channel. If there are no ions inside the channel (or the sum of all charges is zero), then there is no electric field emanating from the channel. If a single ion is added in the center of the channel, then half of its electric field lines are exiting the channel on the left and the other half on the right, cf. Figure 1. This is akin to having two image boundary charges $q, q' = \frac{1}{2}$ at the two ends of the channel (charges are measures in units of e). These charges are provided by polarization effects in the well-conducting reservoirs. There are only integer charges inside the channel. Hence, if the boundary charge at one end is q (the ion emits a fraction q of its field lines at one end), then the other boundary charge is $q' = 1 - q$. Reference [11] shows that moving a unit probe charge through the channel (while allowing the other ions to equilibrate) creates boundary charges which change from zero to one. Once the boundary charges reach an integer value they may either be released from the end points and join the bulk, or enter into the channel. This makes thermodynamic properties periodic functions of q with unit period. In Section 2 we show that the boundary charge q takes the role of the quasi-momentum in the effective quantum mechanics. Hence, the bandwidth of the lowest quantum-mechanical band translates directly to the transport barrier.

This review is devoted to the mathematical apparatus needed to treat the non-Hermitian operators appearing in the physics of multivalent 1D plasmas. However, we want to stress that these methods can be applied more broadly to a wide range of non-Hermitian systems. In particular we focus on semiclassical methods applicable for relatively large concentrations of the dissociated salts. Our central observation is that the corresponding (complex) semiclassical trajectories may be viewed as closed cycles on Riemann surfaces of non-zero genus. The action integrals along such cycles are given by solutions of the Picard-Fuchs differential equation, allowing for their analytic evaluation. As a result one obtains asymptotically exact thermodynamic and correlation functions of the 1D multivalent Coulomb plasmas. Of particular interest is the transport barrier, given by the width of the lowest Bloch band (i.e., energy difference between anti-periodic and periodic ground-states of the Schrödinger equation). We obtain analytic results for the transport barriers for various combinations of ion valencies as functions of salt concentration and temperature.

The structure of this paper is as follows: in Section 2 we discuss the EL mapping of statistical mechanics to an effective quantum mechanics with a cosine potential and its generalizations to the non-Hermitian cases. Section 3 is devoted to the semiclassical treatment of the corresponding non-Hermitian operators using the Picard-Fuchs equation. At the end of that section we go beyond the usual semiclassical formulas and describe how to obtain second- and higher-order corrections with little computational effort. We provide a brief summary and discussions in Section 4.

2. Thermodynamic Description and Equivalent Quantum Mechanics

In this section we discuss the relationship between statistical mechanics of the ion channel and (non-Hermitian) quantum mechanics. We start with a thermodynamic description of the ion channel in terms of the grand-canonical partition function. Then we review how to map the partition function onto a Feynman propagator and derive a Hamilton operator from there. This mapping was pioneered by Edwards and Lenard [10] and subsequently used in several works as starting point [11,25–27]. If the system consists of cations and anions with the same valency and concentration, then the resulting Hamilton operator is Hermitian. However, if the positive and negative charges have different valency, for example solutions of the divalent salts $MgCl_2$ or $CaCl_2$, non-Hermitian terms appear. Hence, the spectrum of the resulting operator also contains complex eigenvalues. We discuss how reality and positivity of the partition function is ensured. In the end we comment on the case if charge neutrality is violated.

2.1. Derivation of the Hamilton Operator

As discussed in Section 1 charged ions inside the channel interact with the effective 1-dimensional Coulomb potential $\Phi(x) = -eE_0|x|$, where $E_0 = 2e/\kappa_1 a^2$ is the electric field strength generated by a single ion with charge e inside a channel of radius a and dielectric constant κ_1 [11]. The total interaction energy of all ions in the channel is given by

$$U = \frac{1}{2}\iint_0^L dx dx' \rho(x)\Phi(x-x')\rho(x'). \tag{2}$$

Here we write the charge density for point charges in terms of δ-functions,

$$\rho(x) = \sum_{j=1}^{N_1+N_2} \sigma_j \delta(x-x_j) + q(\delta(x) - \delta(x-L)), \tag{3}$$

where $\sigma_j = n_1$ for $1 \le j \le N_1$ and $\sigma_j = -n_2$ for $N_1+1 \le j \le N_1+N_2$. This charge density represents N_1 cations with valency n_1 and N_2 anions with valency $-n_2$, and the two fractional boundary charges $\pm q$ at $x = 0, L$. The channel is open and can exchange particles with two 3D bulk reservoirs at the ends. Therefore the thermodynamic properties are given by the grandcanonical partition function,

$$\mathcal{Z} = \sum_{N_1,N_2=0}^{\infty} \frac{f_1^{N_1} f_2^{N_2}}{N_1! N_2!} \prod_{j=1}^{N_1+N_2} \int_0^L dx_j e^{-U/k_B T}, \tag{4}$$

where $f_{1,2}$ are the fugacities of the two charge species. As shown in References [10,11] and in Appendix A, the partition function can be converted into a functional integral by introducing an auxiliary field $\theta(x)$ as conjugate to the charge density $\rho(x)$. Through this process all integrals over the variables x_j decouple, bringing them to the form $\sum_N [f \int dx\, e^{i\sigma\theta(x)}]^N/N! = \exp\{f \int dx\, e^{i\sigma\theta(x)}\}$. The interaction potential (2), being inverse of the 1D Laplace operator, leads to an additional term $\exp\{(k_B T/eE_0)\int dx\, \theta\partial_x^2\theta\}$. As a result the partition function (4) is identically written in terms of the Feynman path integral with an "imaginary time" x, describing quantum mechanics with the Hamiltonian

$$\hat{H} = (i\partial_\theta - q)^2 - \left(\alpha_1 e^{in_1\theta} + \alpha_2 e^{-in_2\theta}\right), \tag{5}$$

where $\alpha_{1,2} = f_{1,2}k_BT/eE_0$ are dimensionless ion concentrations. The Feynman integral is the expectation value of the evolution operator over the imaginary "time" L,

$$\mathcal{Z}_L = \left\langle q \middle| \mathcal{X} e^{-\frac{eE_0}{k_BT}\int_0^L dx\,\hat{H}} \middle| q \right\rangle = \sum_m |\langle q|m\rangle|^2 e^{-\frac{eE_0 L}{k_BT}\varepsilon_m(q)}, \tag{6}$$

where $\mathcal{X}$ is the x-ordering operator. Here $\{\varepsilon_m(q)\}_m$ is the spectrum of the effective Hamiltonian $\hat{H}$, and $|m\rangle = \psi_m(\theta)$ are its eigenvectors in the Hilbert space of periodic functions, $\psi_m(\theta) = \psi_m(\theta + 2\pi)$. The matrix elements are $\langle q|m\rangle = \int_0^{2\pi} d\theta e^{iq\theta}\psi_m(\theta)$. The boundary charge q plays the role of the Bloch quasi-momentum and the spectrum is periodic in q with unit period.

Note that for $\alpha_1 = \alpha_2$ and $n_1 = n_2$ the potential in Equation (5) reduces to the cosine function and the Hamiltonian becomes the well-known Mathieu Hamiltonian [10]. However, if these conditions are violated the potential is non-Hermitian [25]. We discuss implications of this in the following section.

2.2. Physical Observables

The partition function in Equation (6) gives the thermodynamic properties of the ion gas. However, to be physically meaningful the partition function needs to be real and positive, while the spectrum of the non-Hermitian Hamiltonian (5) may contain non-real eigenvalues. This issue is resolved because the Hamiltonian obeys a symmetry akin to $\mathcal{PT}$-symmetry. The combined action of the "parity operator" $\mathcal{P} : \theta \to -\theta$ and "time reversal" $\mathcal{T} : i \to -i$ leaves the Hamiltonian in Equation (5) unchanged. Bender et al. [24] proved that all eigenvalues of a $\mathcal{PT}$-symmetric Hamiltonian are either real or appear in complex conjugated pairs. Hence, summing over all eigenvalues in Equation (6) gives a real result. In [25] is was shown that for positive values of concentrations $\alpha_{1,2} > 0$ the lowest energy band $\varepsilon_0(q)$ is entirely real-valued, ensuring positivity of the partition function. The higher bands $\varepsilon_m(q)$ are in general complex-valued.

Hence we obtain a physically meaningful partition function, and can connect it to thermodynamic observables. The pressure of the Coulomb gas is its free energy per unit length

$$P = k_BT\frac{\partial \ln \mathcal{Z}_L}{\partial L} \overset{L\to\infty}{\longrightarrow} -eE_0\varepsilon_0(q), \tag{7}$$

which for a long channel is determined by the eigenvalue with the smallest real part, $\varepsilon_0(q)$. In equilibrium the system minimizes its free energy by choosing an appropriate boundary charge q. In [25,26] this minimum was found to generally be the non-polarized state of the channel, i.e., $q = 0$. Adiabatic charge transfer through the channel is associated with the boundary charge q sweeping through its full period. As a result, the (free) energy barrier for ion transport is

$$U_0 = eE_0L(\Delta\varepsilon)_0, \tag{8}$$

where $(\Delta\varepsilon)_0$ is the width of the lowest Bloch band. Therefore the ground state energy and the width of the lowest Bloch band of the Hamiltonian (5) give the leading thermodynamic and transport properties of the (n_1, n_2) Coulomb gas. In Section 3 we discuss analytic results for the eigenvalues and the bandwidth.

2.3. Charge Non-Neutrality

In [10] it was shown that for arbitrary values of $\alpha_{1,2}$ the Hamiltonian (5) is always isospectral to a similar charge-neutral Hamiltonian. This can be seen by shifting the coordinate as $\theta \to \theta + \theta_0$. Upon such transformation the dimensionless concentrations

$\alpha_{1,2}$ renormalize as $\alpha_1 \to \alpha_1 e^{in_1\theta_0}$ and $\alpha_2 \to \alpha_2 e^{-in_2\theta_0}$. Notice that the combination $\alpha_1^{n_2}\alpha_2^{n_1}$ remains invariant. Hence, the family of Hamiltonians (5) with

$$\alpha_1^{n_2}\alpha_2^{n_1} = \text{const} \tag{9}$$

is isospectral [10,25]. Therefore one may choose one representative from each isospectral family. A convenient choice is taking the representative with charge neutrality in the bulk reservoirs, i.e., $n_1\alpha_1 = n_2\alpha_2 \equiv \alpha$. The physical reason for this symmetry is that the interior region of the channel always preserves charge neutrality due to the large self-energy of charges. The edge regions screen charge imbalances of the reservoirs. Therefore, irrespective of the relative fugacities of cations and anions in the reservoirs, the thermodynamics of the long channel are equivalent to the one in contact with neutral reservoirs with an appropriate salt concentration α. This brings the Hamiltonian (5) to the form

$$\hat{H} = \alpha\left[\hat{p}^2 - \left(\frac{1}{n_1}e^{in_1\theta} + \frac{1}{n_2}e^{-in_2\theta}\right)\right], \tag{10}$$

where we define the momentum operator as

$$\hat{p} = \alpha^{-1/2}(-i\partial_\theta + q); \qquad [\theta, \hat{p}] = i\alpha^{-1/2}. \tag{11}$$

The commutation relation shows that $\alpha^{-1/2}$ plays the role of the effective Planck constant. Hence, a large concentration of charges corresponds to the semiclassical limit of the Hamiltonian (10). We further rescale the eigenvalues ε as

$$u \equiv \frac{n_1 n_2}{n_1 + n_2}\frac{\varepsilon}{\alpha}. \tag{12}$$

This keeps the classical minimum of the potential at $u = -1$, irrespective of the concentration α and the valencies n_1, n_2. In Section 3 we discuss the spectral properties of the Hamiltonian (10) in the semiclassical limit.

3. Large Charge Concentration

In Section 2 we mapped the grand-canonical partition function of the Coulomb gas onto an equivalent quantum system. The resulting Hamiltonian, Equation (10), contains one free parameter α which is proportional to the concentration of charged ions. In this section we analyze the spectral problem of this Hamiltonian in the limit of large α. As argued after Equation (11), this is the semiclassical limit of the equivalent quantum problem. We use the main semiclassical results, Bohr-Sommerfeld quantization and Gamow's formula, to calculate the eigenvalues and bandwidths of the Hamiltonian for several different cases of valencies (n_1, n_2). In the case of equal valencies, $n_1 = n_2$, the Hamiltonian (10) is the well-known Mathieu Hamiltonian which we discuss in Section 3.1. It's spectral properties were calculated using several different approaches [10,11,25–28]. In this review we focus on an approach based on integration on a complex Riemann surface [25,26,28]. We choose this method because it can also be applied to the cases with different valencies, $n_1 \neq n_2$, see Section 3.2. In that situation the Hamiltonian is non-Hermitian, and the required action integrals are not attainable by straightforward integration. Instead we show how to relate them to integrals along closed cycles on a Riemann surface. Then we use powerful tools from algebraic topology to derive a differential equation for the action integrals. This is known as the Picard-Fuchs equation. The required actions are a combination of the solutions of this differential equation. Through this procedure we bypass the use of direct integration methods. From the actions we obtain the eigenvalues and the bandwidths, which are directly related to the ion pressure and transport barrier for ions in the channel. In Section 3.3 we go one step further. We use the same concepts to calculate the second-order corrections in the WKB series. Most importantly we show that these can be expressed in terms of the already-calculated action and its derivatives, and therefore can be obtained with minimal computational effort. This gives an improved semiclassical approximation of the eigenvalues. Relating this to the pressure in the ion channel we find that beyond the ideal-gas pressure and the Debye-Hueckel correction there is another correction which only depends on the geometry of the channel but not on the concentration of ions. We compare these results to numerical calculations.

3.1. *Equal Valency*

As mentioned in Section 2 the Hamiltonian in Equation (5) is Hermitian if the valencies of the two charges are equal, $n_1 = n_2$. Indeed, in this case it reduces to the well-known Mathieu Hamiltonian,

$$\mathcal{H} = \alpha\left[\hat{p}^2 - 2\cos\theta\right]. \tag{13}$$

In literature there exist several studies of the Coulomb gas with charges of equal valency. In [10] it was first noted that the Coulomb gas is mapped onto the Mathieu equation. In [27] the authors perform a semiclassical calculation on this equation via direct integration. From this they obtain the required actions and analytic approximations of the eigenvalues and bandwidths. [11] provides additional qualitative arguments which lead to the same results. However, as mentioned above, in this section we will follow the Riemann surface methods developed in [25] because in that framework one can also study the case of unequal valencies $n_1 \neq n_2$ in Section 3.2, and these concepts form the basis of our considerations for higher-order corrections in Section 3.3.

3.1.1. Construction of the Riemann Surface

In the semiclassical ansatz we look for wave functions of the form $\psi = e^{i\alpha^{1/2}S}$, where S is the action of the classical problem with the normalized Hamiltonian (13). The semiclassical trajectories satisfy the classical Hamilton equations of motion and thus conserve the (complex) energy u in Equation (12),

$$2u = p^2 - 2\cos\theta\,. \tag{14}$$

In this normalization $u = \mp 1$ corresponds to the bottom (top) of the cosine potential. Our approach to calculate the action integrals $S = \oint_\gamma p(\theta,u)d\theta$ is based on complex algebraic topology. First we set $z = e^{i\theta}$ and consider (z,p) as complex variables. Energy conservation, Equation (14), defines a family of complex algebraic curves parametrized by u and satisfying

$$\mathcal{E}_u: \qquad \mathcal{F}(p,z) = p^2 z - (z^2 + 2uz + 1) = 0. \tag{15}$$

For $u \neq \pm 1$ it can be checked that $(\partial\mathcal{F}/\partial z, \partial\mathcal{F}/\partial p)$ does not vanish on $\mathcal{E}_u$, so each $\mathcal{E}_u$ is nonsingular. Then $\mathcal{F}(p,z)$ implicitly defines a locally holomorphic map $p = p(z)$. The exceptions to this occur at $z = 0, \infty, z_\pm$, where $z_\pm = -u \pm i\sqrt{1-u^2}$ are the roots of $p^2 = 0$ (i.e., classical turning points). In a vicinity of these four branch points $p(z)$ behaves as

$$\begin{aligned} &p \sim z^{-1/2}, && (z \sim 0)\\ &p \sim z^{1/2}, && (z \sim \infty)\\ &p \sim (z - z_\pm)^{1/2}, && (z \sim z_\pm) \end{aligned} \tag{16}$$

respectively, i.e., $p(z)$ is locally double-valued. Note that we added the point at infinity to have an even number of branch points. This compactifies the complex plane and makes it topologically equivalent to a Riemann sphere, cf. Figure 2.

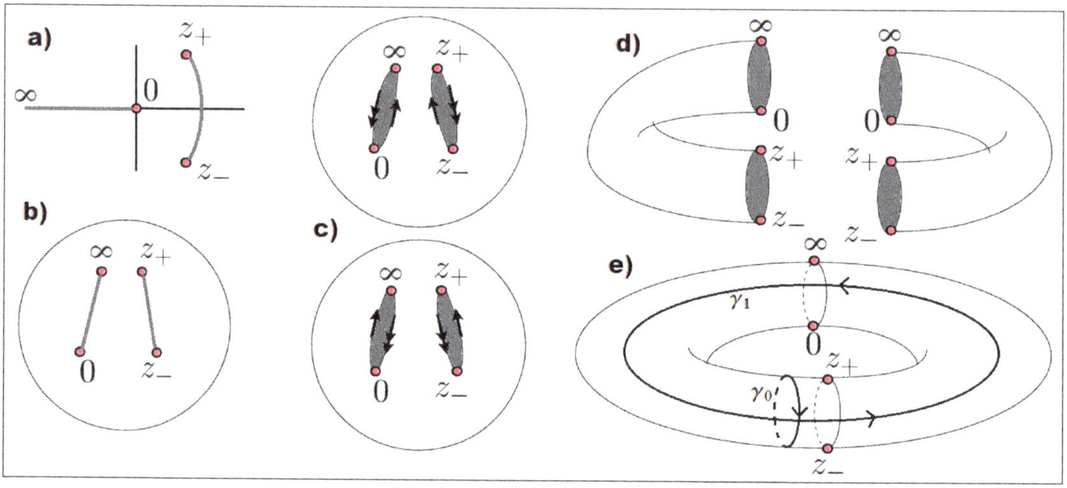

Figure 2. Construction of the Riemann surface of genus 1, as defined by Equation (15). (**a**) In the z-plane there are four branch points at $0, z_\pm, \infty$ which are pairwise connected by two branch cuts (gray). (**b**) Considering $z = \infty$ as a regular point the complex plane compactifies to a Riemann sphere with two cuts on the sphere. (**c**) The double-valued nature of the function $p(z)$ is resolved by defining two copies of the Riemann sphere. The branch cuts are opened and the spheres are deformed into tubes (**d**) and glued together to form a torus (**e**). The arrows are used to signify the edges that are glued together. There are two fundamental cycles γ_0, γ_1 which are topologically different and non-trivial, i.e., they can not be smoothly transformed into each other or a point. Reproduced with permission from Reference [26].

To avoid dealing with $p(z)$ as a double-valued function we introduce a second copy of the complex z-plane and the corresponding Riemann sphere. On both sheets we define two branch cuts connecting the four branch points, between $0, \infty$ and the turning points $z_\pm$ respectively. $p(z)$ is analytically continued across the branch cuts, i.e., when crossing a branch cut we jump from the first sheet to the second and vice versa. Identifying the branch cuts as edges we can deform the two Riemann spheres into tubes and glue them together to form a torus. This construction is visualized in Figure 2. Thus, the complex algebraic curve $\mathcal{E}_u$ in Equation (15) defines a torus which is a compact Riemann surface of genus $g = 1$. (Generically, every compact Riemann surface is topologically equivalent to a sphere with some number of handles g, or a (multi-)torus with g holes, called the genus of the surface).

3.1.2. Integrals on the Riemann Surface and the Picard-Fuchs Equation

The action integrals can be understood as integrals over closed cycles γ, $S(u) = \oint_\gamma \lambda(u)$, where

$$\lambda(u) = p(\theta)\, d\theta = p(z)\frac{dz}{iz} = \frac{(z^2 + 2uz + 1)^{1/2}}{iz^{3/2}}\, dz \tag{17}$$

is the action 1-form which, by construction, is holomorphic on the Riemann surface.

To visualize the relevant trajectories we momentarily return to θ and consider it as complex. In this representation one has square-root branch cuts along the real axis, connecting the classical turning points along the classically allowed region. The integration trajectories run just above or below the real axis and connect the turning points. After combining them to form closed cycles one can push these cycles off the real axis and away from the turning points without altering the integrals (by Cauchy's theorem). We call these the classical cycle γ_0 and the instanton cycle γ_1, as shown in Figure 3. Translating these two cycles to the complex z-plane yields the contours in the right panel of that figure.

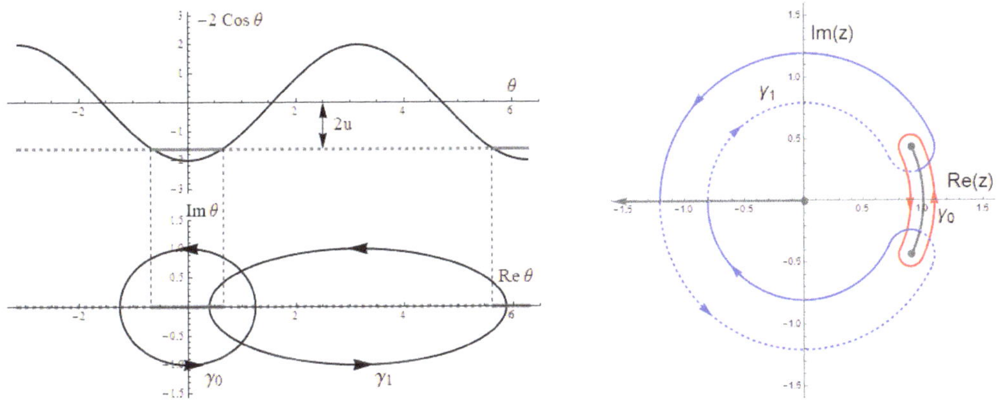

Figure 3. Left: The classically allowed (forbidden) regions along the θ-axis at energy u are shown by the solid (dashed) gray line. Deforming the classical (instanton) orbits into the complex plane leads to the cycles $\gamma_0(\gamma_1)$. **Right**: Cycles γ_0 (red) and γ_1 (blue) in the complex z-plane for $u = -0.9$. Notice that the cycle γ_1 crosses the two cuts from the first sheet (solid line) to the second sheet (dashed line) and back. Reproduced with permission from Reference [25].

Cauchy's theorem is also valid on the Riemann surface since the action form (17) is, by construction, holomorphic on the torus. Therefore all closed cycles can be deformed without changing the integrals, and can be expressed as a combination of an integer number of these two basis cycles. This leads to our key idea how to calculate the action integrals: for this we employ a central theorem of algebraic topology, de Rham's theorem. It states that on a Riemann surface there are exactly as many linearly independent holomorphic 1-forms to integrate upon as there are independent closed cycles to integrate along. This is valid up to exact forms, i.e., 1-forms which integrate to 0 along any closed cycle, and boundaries, i.e., closed curves which can be continuously deformed to a point. Hence, there are exactly two independent holomorphic 1-forms on the Riemann surface. Any set of three 1-forms is linearly dependent modulo an exact form which integrates to 0 upon integration along any closed cycle. (A full explanation of the mathematical concepts is beyond the scope of this review. A detailed discussion of relevant and related concepts is in [29], basic definitions and additional background are in [30,31]. All concepts can also be found online at [32]. A simplified derivation specifically for complex-valued Riemann surfaces is in chapter 2 of [28]).

Equipped with this we look at a set which contains the action 1-form (17) and its first two derivatives with respect to energy u, $\{\lambda(u), \lambda'(u), \lambda''(u)\}$. Taking derivatives does not change the structure of branch points, therefore these are three 1-forms which are all defined on the same Riemann surface. Hence, we know that there must exist a linear combination of these which is an exact form. Reference [25] explains in detail how to find the linear combination and the exact form as

$$\left((u^2-1)\partial_u^2 + \frac{1}{4}\right)\lambda(u) = \frac{d}{dz}\left[\frac{i}{2}\,\frac{1-z^2}{z^{1/2}(z^2+2uz+1)^{1/2}}\right]dz\,. \tag{18}$$

It is evident from Stokes' theorem that the right-hand-side integrates to 0 along any closed cycle on the Riemann surface. Hence, we obtain

$$\oint_\gamma \left((u^2-1)\partial_u^2 + \frac{1}{4}\right)\lambda(u) = (u^2-1)S''(u) + \frac{1}{4}\,S(u) = 0\,. \tag{19}$$

This differential equation for the action $S(u)$ is called the Picard-Fuchs Equation [29]. Integration is performed along a closed cycle γ, which can be the classical or the instanton cycle, $\gamma_{0,1}$ in Figure 3. Therefore both the classical and instanton actions $S_{0,1}(u)$ are solutions of the Picard-Fuchs Equation (19). This equation is a second-order ordinary differential equation, therefore it admits two independent solutions. These can be found in the form $F_0(u^2)$ and $uF_1(u^2)$, where

$$F_0(u^2) = {}_2F_1\left(-\frac{1}{4}, -\frac{1}{4}; \frac{1}{2}; u^2\right),$$
$$F_1(u^2) = {}_2F_1\left(+\frac{1}{4}, +\frac{1}{4}; \frac{3}{2}; u^2\right), \tag{20}$$

are hypergeometric functions [33,34]. These solutions form a basis out of which $S_{0,1}(u)$ must be composed, so we write

$$S_0(u) = C_{00}F_0(u^2) + C_{01}uF_1(u^2),$$
$$S_1(u) = C_{10}F_0(u^2) + C_{11}uF_1(u^2). \tag{21}$$

To find the correct coefficients C_{jk}, $j,k = 0,1$ it is sufficient to evaluate the periods at one specific value of u. Employing the fact that the hypergeometric functions (20) are normalized and analytic at $u = 0$, i.e., $F_k(u^2) = 1 + \mathcal{O}(u^2)$, one notices that $S_j(u) = C_{j0} + uC_{j1} + \mathcal{O}(u^2)$. Thus, to identify C_{jk} we expand the integrand $\lambda(u)$ to first order in u and evaluate the integrals $S_j(u)$ at $u = 0$. Straightforward calculation yields

$$C_{00} = e^{-i\pi/2}C_{10} = 8\pi^{-1/2}\Gamma(3/4)^2,$$
$$C_{01} = e^{+i\pi/2}C_{11} = \pi^{-1/2}\Gamma(1/4)^2. \tag{22}$$

The relations between C_{0k} and C_{1k} are not accidental. They originate from the fact that the cycle γ_1 transforms into γ_0 by substitution $z' = e^{-i\pi}z$ and $u' = e^{i\pi}u$, and vice versa. This gives a global symmetry between the two periods,

$$S_0(u) = e^{-i\pi/2}S_1(e^{i\pi}u)\,. \tag{23}$$

Equations (20)–(23) fully determine the classical and instanton actions $S_{0,1}(u)$. We now proceed to relate them to physical observables.

3.1.3. Semiclassical Results

We seek semiclassical results for the sequence of low-energy bands terminating at $u = -1$. Therefore we quantize the classical action $S_0(u)$ according to the Bohr-Sommerfeld rule to determine the normalized energies u_m as solutions of the equation

$$S_0(u_m) = 2\pi\alpha^{-1/2}(m + 1/2)\,, \qquad m = 0, 1, \ldots \tag{24}$$

We see that the cycle γ_0 contracts to a point when the energy goes to the bottom of the potential, $u \to -1$. This corresponds to vanishing of the classical action, $S_0(u = -1) = 0$. To obtain an approximate analytic expression for the lowest energy levels $\varepsilon_m = 2\alpha u_m$ we expand the classical action to first order near the bottom of the potential,

$$S_0(u) = 2\pi(u + 1). \tag{25}$$

Equations (24) and (25) combined imply $\varepsilon_m = -2\alpha + 2\alpha^{1/2}(m + 1/2)$. As a result the pressure (7) of a monovalent gas is

$$P = -eE_0\varepsilon_0 = 2k_BTf - \sqrt{k_BTeE_0f}. \tag{26}$$

The two terms here are the pressure of the ideal gas with fugacity f and the mean-field Debye-Hueckel interaction correction [22].

The instanton action $S_1(u)$ determines the bandwidth $(\Delta u)_m$ according to Gamow's formula,

$$(\Delta u)_m = \frac{\omega}{\pi\sqrt{\alpha}}\, e^{i\alpha^{1/2}S_1(u_m)/2}\,. \tag{27}$$

Here $\omega = 2$ is the frequency of the harmonic-oscillator approximation of the potential near the classical minimum. We expand the instanton action near the classical minimum and at the quantized energies $u_m = -1 + \alpha^{-1/2}(m + 1/2)$ to obtain

$$S_1(u_m) = 16i + 2i\left(m + \frac{1}{2}\right)\ln\left(\frac{m+1/2}{32e\alpha^{1/2}}\right). \tag{28}$$

Applying this to Gamow's Formula (27) leads to

$$(\Delta\varepsilon)_m = 2\alpha(\Delta u)_m = \frac{4}{\pi}\left(\frac{32e}{m+1/2}\right)^{m+1/2} e^{-8\alpha^{1/2}+(m/2+3/4)\ln\alpha}, \tag{29}$$

This coincides with the known asymptotic results for the Mathieu Equation [27,35,36]. As explained below Equation (3), adiabatic charge transport is associated with a change of the boundary charge q (i.e., quasi-momentum) across the interval $0 < q < 1$ (i.e., the Brillouin zone). Therefore the *free energy* transport barrier is given by the width of the lowest Bloch band, $(\Delta\varepsilon)_0$. One notices that increasing the concentration of salt ions leads to an exponential entropic suppression of the transport barrier, $(\Delta\varepsilon)_0 \propto \alpha^{3/4}e^{-8\sqrt{\alpha}}$.

3.2. Multivalent Ions

So far we worked with the Hermitian example of the Mathieu Hamiltonian, i.e., when both ion species are monovalent, $n_1 = n_2 = 1$. With that we could validate the Riemann surface method by comparing the results to literature. In this section we discuss four different cases with multivalent ions (assuming $n_1 > n_2$ without loss of generality). In such a scenario the Hamiltonian (5) is non-Hermitian. This leads to complex values in the spectrum, which we present in Section 3.2.1. Furthermore, in classical motion the coordinate and momentum acquire complex values. This results in a phase space (θ, p) with two complex dimensions (instead of two real dimensions). The classical (instanton) action is obtained by integrating the momentum $p(\theta)$ along the trajectory which connects two turning points and solves the classical equations of motion with real (imaginary) time. However, solving the equations of motion in complex phase space (θ, p) is non-trivial, if at all attainable. Therefore we go from an integral along the trajectory to an integral along a closed cycle in the plane of complex $z = e^{i\theta}$ which encloses the trajectory, similar to the mapping in Figure 3. With that we connect the non-Hermitian problem to the method that we validated in the previous section. We discuss this calculation for four different combinations of charge valencies in Section 3.2.2. In Section 3.2.3 we connect the results to the classical and instanton actions and physical observables.

3.2.1. Spectrum of the Non-Hermitian Hamiltonian

Non-Hermiticity of the Hamiltonian (10) has a significant effect on its spectrum. Namely, not all eigenvalues are real. In Figure 4 we show numerical results for the eigenvalues at large concentration α, for four different combinations of the integers (n_1, n_2). Most importantly all non-real eigenvalues appear as complex conjugate pairs. This is a consequence of the $\mathcal{PT}$-symmetry of the Hamiltonian and crucial to obtain a physically meaningful partition function, as discussed in Section 2. Furthermore we see sequences of narrow bands which emerge from $u = -\nu$ with $\nu^{n_1+n_2} = 1$. These sequences approximately follow the lines connecting $u = -\nu$ and $u = 1$, but avoid the special point $u = 1$. At some point all of these branches merge. Beyond this the nature of the spectrum changes drastically, instead of narrow bands and large gaps we see wide bands separated by small gaps. This feature is similar to the case of a periodic Hermitian potential: as long as the energy lies below the maximum of the potential there are narrow bands, while for energies exceeding the maximum there are wide bands. Hence, we associate the point where the spectral branches meet with the top of the potential. (It is important to bear in mind that for a complex-valued potential there is no proper definition of a "maximum".) The energy variable u is normalized so that in the Hermitian $(1,1)$ case this point lies at $u=1$. In the non-Hermitian cases we observe $u \approx 0.96$ for $(2,1)$, $u \approx 1.09$ for $(3,1)$, $u \approx 1.20$ for $(4,1)$, and $u \approx 0.84$ for $(3,2)$. These values are independent of α, so this must be a consequence of the underlying classical mechanics.

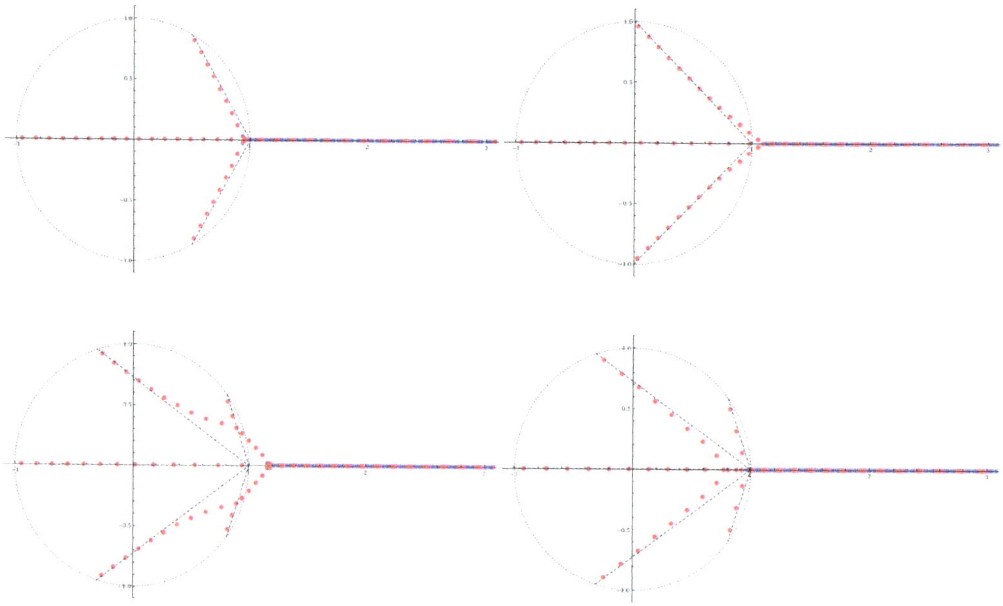

Figure 4. The bands of the non-Hermitian Hamiltonian in space of complex energy u. Blue stands for $q=0$, while red stands for $q=\frac{1}{2}$. The dotted circle marks $|u|=1$. In all cases we see multiple branches of narrow bands with complex values which terminate near the unit circle. The dashed line is a guide to the eye which connects the termination points of the branches, $u=-(1)^{1/(n_1+n_2)}$, to $u=1$. **Top left**: $(n_1,n_2)=(2,1)$, $\alpha=200$; **top right**: $(3,1)$, $\alpha=300$; **bottom left**: $(4,1)$, $\alpha=400$; **bottom right**: $(3,2)$, $\alpha=400$. Reproduced with permission from References [25,26].

To calculate the statistical partition function in Equation (6) the most important eigenvalues are those with small real part. Therefore we will focus on the narrow bands and treat them in semiclassical approximation.

3.2.2. Riemann Surface and Picard-Fuchs Equation

We use the rescaled energy variable u in Equation (12), substitute $z=e^{i\theta}$ in the Hamiltonian (10), and write the classical energy-momentum relation as

$$u\frac{n_1+n_2}{n_1 n_2}=p^2-\left(\frac{1}{n_1}e^{in_1\theta}+\frac{1}{n_2}e^{-in_2\theta}\right). \tag{30}$$

The generalization for the complex algebraic curve in Equation (15) is the family of curves

$$\mathcal{E}_u: \qquad \mathcal{F}(p,z)=n_1 n_2 p^2 z^{n_2}-\left(n_2 z^{n_1+n_2}+(n_1+n_2)u z^{n_2}+n_1\right)=0. \tag{31}$$

This defines implicitly a double-valued function $p(z)$. It is easy to see that $(\partial\mathcal{F}/\partial z, \partial\mathcal{F}/\partial p)$ does not vanish on $\mathcal{E}_u$ unless $u = -e^{\frac{2\pi i m}{n_1+n_2}}$ for an integer m. For the non-singular values of u the function $p(z)$ is locally holomorphic except for the points $z = 0, \infty, z_j$, where $z_j, j = 1, ..., n_1 + n_2$ are the roots of $p^2 = 0$. The z_j are the turning points of classical motion in complex coordinates. Near these special points $p(z)$ behaves as

$$
\begin{aligned}
p &\sim z^{-n_2/2}, && (z \sim 0) \\
p &\sim z^{n_1/2}, && (z \sim \infty) \\
p &\sim (z - z_j)^{1/2}. && (z \sim z_j)
\end{aligned}
\tag{32}
$$

The z_j are $n_1 + n_2$ branch points. If n_2 (n_1) is odd, then 0 (∞) is an additional branch point; for even n_2 (n_1) there is a normal pole at 0 (∞). Hence, there are $n_1 + n_2 + 1$ branch points on the Riemann sphere if one of the integers is odd, and $n_1 + n_2 + 2$ branch points if both are odd. (Here we ignore the case that n_1, n_2 are both even, because if both integers can be divided by the same number n we can define $z' = e^{in\theta}$ to obtain a simpler algebraic curve.) In all cases there is an even number of branch points which can be connected pairwise to form branch cuts. For $(n_1, n_2) = (2, 1)$ we obtain four branch points and two branch cuts and a Riemann surface of genus 1, as in Figure 2. For $(n_1, n_2) = (3, 1), (4, 1), (3, 2)$ the asymptotic expansions (32) give six branch points. Consequently there are three branch cuts in the complex plane. Through a similar construction as in Figure 2 one obtains a Riemann surface which is topologically equivalent to a figure "8", i.e., a figure with two holes and genus 2 [26,28]. In the following we consider these four cases because there are no naturally occurring ions with larger charge. However, mathematically the algebraic curves for higher values of the integers can be constructed in the same way, yielding Riemann surfaces with larger genus.

In Figure 5 we show the structure of branch points in the z-plane for these four cases. On a Riemann surface with genus $g = 1(2)$ there are two (four) independent closed cycles [29]. In Figure 5 we define three cycles for the $(2, 1)$ case, and five cycles for $(4, 1)$ and $(3, 2)$. This is done for convenience and symmetry reasons. The superfluous cycle can be expressed by the other cycles. For $(2, 1)$ the linear combination $\gamma_0 - \gamma_1 - \gamma_2$ does not contain any of the branch points and is contractible to a point. For $(4, 1)$ the trivial cycle is $\gamma_0 - \gamma_1 + \gamma_2 + \gamma_3 - \gamma_4 \cong 0$, and for $(3, 2)$ we see that $\gamma_0 + \gamma_1 - \gamma_2 - \gamma_3 + \gamma_4 \cong 0$. We choose to include the additional cycle because it gives an easy representation for the symmetry relation between the corresponding actions $S_j(u)$, akin to Equation (23). By substituting $z' = e^{-i\phi}z$ and $u' = e^{i\phi}u$ the cycles transform $\gamma_j \to \gamma_{j+1}$. For the $(2, 1)$ case the resulting symmetry relation is

$$
S_0(u) = e^{\pi i/3} S_1(e^{-2\pi i/3}u) = e^{-\pi i/3} S_2(e^{2\pi i/3}u)\,. \tag{33}
$$

The analogous symmetry relations for the genus-2 cases are shown in Reference [26].

To calculate the actions $S(u) = \oint_\gamma \lambda(u)$ we continue in the same manner as in Section 3.1. The 1-form (cf. Equation (17)) with general n_1, n_2 is

$$
\lambda(u) = p(\theta)d\theta = p(z)\frac{dz}{iz} = \frac{\left(n_2 z^{n_1+n_2} + (n_1 + n_2)uz^{n_2} + n_1\right)^{1/2}}{i\sqrt{n_1 n_2}\,z^{1+n_2/2}}dz. \tag{34}
$$

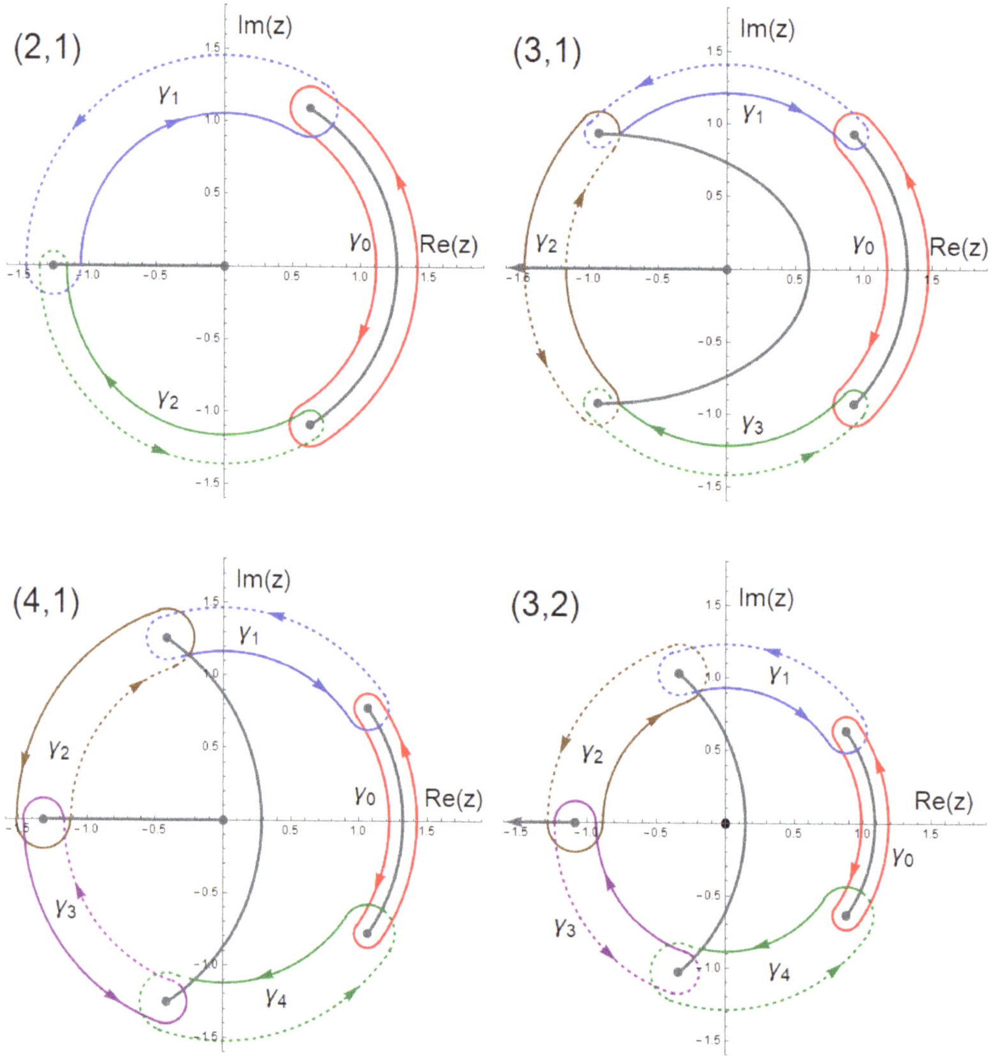

Figure 5. The integration cycles in the complex z-plane for the four non-Hermitian cases that are discussed in Section 3.2. In all images we set $u = 0$. Each color represents one closed cycle of integration. Solid lines denote the sections which lie on the principal sheet, dashed lines the parts on the second sheet. **Top left**: $(2, 1)$; **top right**: $(3, 1)$; **bottom left**: $(4, 1)$; **bottom right**: $(3, 2)$. Note the differences in the structure of the branch cuts: in the $(2, 1)$ case all branch points are finite, while in the $(1, 1)$ case in Figure 3 one branch point lies at ∞. Similar differences exist between the other three figures, whether the branch points are at finite values of z or at ∞, and whether the origin is a branch point or a pole. Reproduced with permission from Reference [26].

On a Riemann surface of genus $g = 1(2)$ there are two (four) independent closed cycles. According to the de Rham theorem, this is equal to the number of linearly independent 1-forms, modulo exact forms. Therefore a set of the 1-form (34) and its first few derivatives, $\{\partial_u^k \lambda(u)\}_{k=0}^{K}$, is linearly dependent if it contains the first $K = 2(4)$ derivatives. We build a linear combination of these which equals an exact form (for details see [26]). The integral of the exact form along a closed cycle gives zero. What is left is a linear combination of the action and its first derivatives, cf. Equation (19). In the $(2,1)$ case we find this Picard-Fuchs equation as

$$(u^3+1)S_j''(u)+\frac{u}{4}S_j(u)=0. \tag{35}$$

This is a second-order differential equation. The Picard-Fuchs equations for the genus-2 cases are fourth-order ODEs which can be found in Reference [26]. Equation (35) admits two solutions $F_0(u^3)$ and $uF_1(u^3)$ which are given in terms of hypergeometric functions [26,34],

$$\begin{aligned} F_0(u^3) &= {}_2F_1\left(-\frac{1}{6},-\frac{1}{6};\frac{2}{3};-u^3\right), \\ F_1(u^3) &= {}_2F_1\left(+\frac{1}{6},+\frac{1}{6};\frac{4}{3};-u^3\right). \end{aligned} \tag{36}$$

The actions are a linear combination of these, $S_j(u)=C_{j0}F_0(u^3)+C_{j1}uF_1(u^3)$. Expanding the hypergeometric functions near the origin, $F_{0,1}(u^3)=1+\mathcal{O}(u^3)$, one notices that $S_j(u)=C_{j0}+uC_{j1}+\mathcal{O}(u^3)$ as $u\to 0$. The constants C_{0k} are therefore given by $C_{00}=S_0(0)$ and $C_{01}=S_0'(0)$. Straightforward integration and the symmetry relation (33) yield

$$\begin{aligned} C_{00}=C_{10}e^{\pi i/3}=C_{20}e^{-\pi i/3} &= \frac{2^{11/6}3\pi^{3/2}}{\Gamma(\frac{1}{6})\Gamma(\frac{1}{3})}, \\ C_{01}=C_{11}e^{-\pi i/3}=C_{21}e^{\pi i/3} &= \frac{3^{1/2}\Gamma(\frac{1}{6})\Gamma(\frac{1}{3})}{2^{11/6}\pi^{1/2}}. \end{aligned} \tag{37}$$

The actions $S_j(u)$ for $(n_1,n_2)=(2,1)$ are fully given by Equations (33), (36), and (37). The analogous expressions for the genus-2 cases with $(n_1,n_2)=(3,1),(4,1),(3,2)$ are given in Reference [26]. In the next section we discuss how to obtain semiclassical results for the physical observables.

3.2.3. Semiclassical Results in the Non-Hermitian Cases

In this section we calculate the eigenenergies and bandwidths of the non-Hermitian Hamiltonian in Equation (10) with the Bohr-Sommerfeld quantization condition and Gamow's formula. To utilize these standard semiclassical results we need to calculate the classical and the instanton actions, $S_{cl,inst}(u)=\oint_{\gamma_{cl,inst}}\lambda(u)$. The crucial part hereby is identifying the correct cycle of integration. In Section 3.1, when discussing the case of a Hermitian Hamiltonian, we identified these with trajectories which connect the classical turning points through the classically allowed or forbidden region respectively, cf. Figure 3. In the non-Hermitian case this is not so clear, because there exist more than two turning points, and in the space with complex coordinate, momentum, and energy the concept of classically allowed or forbidden regions doesn't apply. Instead, to identify the correct actions $S_{cl,inst}(u)$ we look at the analytic behavior of these actions near special values of the energy u.

The Bohr-Sommerfeld condition requires that the classical action goes to zero at the classical minimum of the potential. This happens when two turning points collide which causes the corresponding cycle of integration to collapse to a point. We can easily check that in all four cases in Figure 5 the cycle γ_0 collapses to a point as $u\to -1$. The corresponding action goes to zero, $S_0(-1)=0$. Therefore we identify $S_0(u)$ as the classical action which quantizes into the branch of eigenstates that terminates at $u=-1$. For $(n_1,n_2)=(2,1)$ it follows immediately from the symmetry relation (33) that at the singular point $u=e^{i\pi/3}$ ($e^{-i\pi/3}$) the cycle γ_1 (γ_2) collapses to a point and the action $S_1(u)$ ($S_2(u)$) goes to zero. It should be thus identified with the classical action for the spectral branch terminating at $u=e^{i\pi/3}$ ($e^{-i\pi/3}$). In the same manner the analogous symmetry relations for the genus-2 cases in Reference [26] allow us to identify the classical actions for all the spectral branches in Figure 4. Quantizing these classical actions according to the Bohr-Sommerfeld rule,

$$S_j(u_m^{(j)}) = 2\pi\alpha^{-1/2}(m + 1/2), \qquad m = 0, 1, \ldots, \tag{38}$$

one finds the semiclassical energies $u_m^{(j)}$ determining the $q = 0$ edges of the narrow bands in the complex plane. These results are compared with numerical data in Figure 6. The excellent agreement holds all the way up to the point where all spectral branches coalesce. Beyond this point the semiclassical approximation breaks down, which manifests in e.g., the appearance of wide Bloch bands.

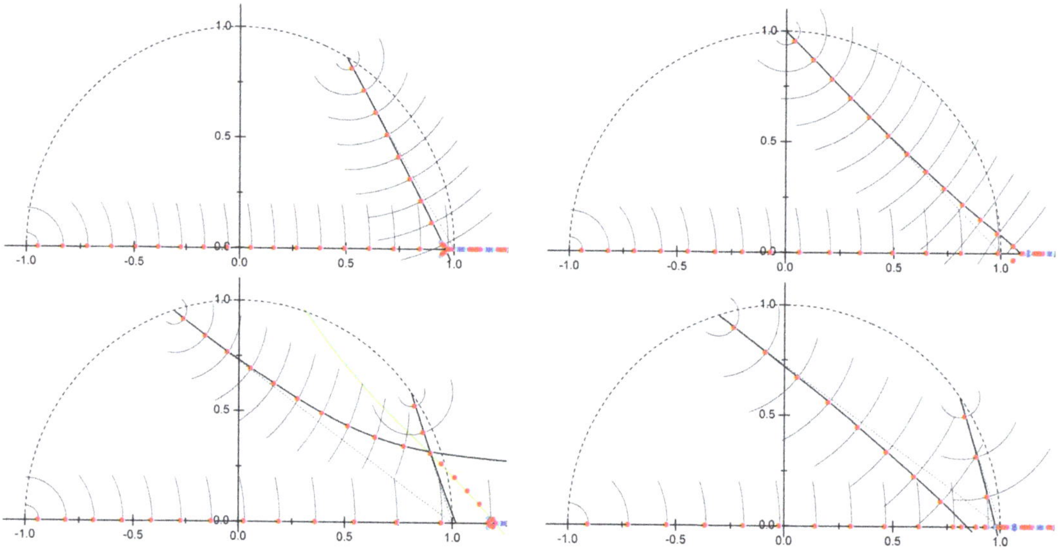

Figure 6. Narrow energy bands (red dots) in the upper half-plane of complex energy u for large α, cf. Figure 4. In all four cases, $Im\, S_0(u) = 0$ along the real axis, where the thin lines mark $|S_0(u)| = 2\pi\alpha^{-1/2}(m + 1/2)$, the quantization condition. The other black lines mark $Im\, S_j(u) = 0$ for the other actions $S_j(u)$, and the thin lines mark $|S_j(u)| = 2\pi\alpha^{-1/2}(m + 1/2)$. In all cases $S_j(u)$ corresponds to an action encircling two branch points. These points coalesce at a singular value of u on the unit circle (dashed) where the spectral branch ends. Near intersections of two lines neither quantization condition holds, cf. $u \approx 0.90 + 0.31i$ in $(4, 1)$ and $u \approx 0.82$ in $(3, 2)$. Beyond this intersection the states are quantized according to the sum of the two corresponding actions, $S_1 + S_2$ in $(4, 1)$ and $S_2 + S_3$ in $(3, 2)$, marked in green. To the right all lines coalesce and beyond this point we observe wide bands with narrow gaps. The lower half-plane shows the mirror image (i.e., complex conjugate) of the upper half plane. **Top left**: $(n_1, n_2) = (2, 1)$, $\alpha = 200$; **top right**: $(3, 1)$, $\alpha = 300$; **bottom left**: $(4, 1)$, $\alpha = 400$; **bottom right**: $(3, 2)$, $\alpha = 400$. Reproduced with permission from Reference [26].

All graphs exhibit spectral branches along the lines where one of the actions $S_j(u)$ is real, while the narrow bands lie at the points determined by the Bohr-Sommerfeld condition (38). For $(2, 1)$ and $(3, 1)$ there exists a total of three spectral sequences, for $(4, 1)$ and $(3, 2)$ five sequences due to a higher number of special energies. In the $(4, 1)$ case the two complex-valued branches intersect at $u \approx 0.90 + 0.32i$. Beyond this point the two sequences merge into one, for which the quantization condition is neither determined by S_1 nor S_2 individually, but instead by the sum $S_1 + S_2$ (shown in green). For $(3, 2)$ the two lines for the complex-conjugate pair S_2 and S_3 collide at $u \approx 0.84$, the other pair collides at $u \approx 0.98$ where the semiclassical approximation breaks down. A closer look at the state at $u \approx 0.89$ reveals that this cannot be explained by the quantization of S_0 along the real axis. However, it meets the Bohr-Sommerfeld condition (38) for $S_2 + S_3$ with $m = 17$. Thus we may conclude that the spectral branches can be derived from the

Bohr-Sommerfeld condition for one of the actions, or upon intersection of two branches by the sum of the two actions of these branches.

To calculate the width of these bands with Gamow's formula,

$$(\Delta u)_m = \frac{\omega}{\pi\sqrt{\alpha}} e^{i\alpha^{1/2} S_{inst}(u_m)/2}, \tag{39}$$

we need to identify the instanton actions. The classical frequency ω is determined from the harmonic oscillator approximation, i.e., by expanding the potential around $\theta = 0$. In Hermitian quantum mechanics the instanton trajectory connects the two classical turning points through the classically forbidden region, cf. Figure 3. Hence, we identify the instanton cycle as the other possible cycle that connects the same two turning points. This is a combination of all other integration cycles γ_i. The instanton actions that correspond to the classical actions $S_0(u)$ are

$$\begin{aligned} S_{inst}(u) &= -S_1(u) + S_2(u), & (2,1); \\ S_{inst}(u) &= -S_1(u) - S_2(u) + S_3(u), & (3,1); \\ S_{inst}(u) &= -S_1(u) - S_2(u) + S_3(u) + S_4(u), & (4,1); \\ S_{inst}(u) &= -S_1(u) + S_2(u) - S_3(u) + S_4(u), & (3,2). \end{aligned} \tag{40}$$

From the symmetry relation (33) between the actions and its analogons for the genus-2 cases it is easy to check that these combinations are purely imaginary, which makes the bandwidth in Equation (39) real, as required.

More can be said when considering the analytic structure of the classical and instanton action in a vicinity of $u = -1$. Therefore we use a concept called monodromy [29,32], which is visualized in Figure 7. We choose some $u \gtrsim -1$ and allow u to wind around -1 (i.e., $(u+1) \to (u+1)e^{2\pi i}$). The two branch points inside the cycle γ_0 in Figure 5 are exchanged by this transformation via a counter-clockwise half-turn; the branch cut in effect rotates by 180°. For γ_0 this has no effect, the cut turns within it. Not so for γ_1: if this cycle is never to intersect the branch points, it is continuously deformed and as a result of this monodromy transformation we obtain $\gamma_1 \to \gamma_1 + \gamma_0$, thus S_1 picks up a contribution of S_0. This effect is visualized in Figure 7. While we have returned to the initial value of u, the period S_1 does not return to its original value and thus can't be analytic. This occurs for every monodromy cycle near $u = -1$. The only function which monotonically increases as the phase of its argument grows is the complex logarithm. Thus, S_1 must have a logarithmic dependence on $1 + u$. One can check that

$$S_1(u) = Q_1(u) - \frac{i}{2\pi} S_0(u) \ln(1+u) \tag{41}$$

yields the correct behavior, where $Q_1(u)$ and $S_0(u)$ are analytic functions of $(1+u)$. The same applies to the other cycle which is connected to the same branch cut. Therefore the instanton action S_{inst} in Equation (40) picks up a contribution of $-2S_0$. Hence, we can derive the Bohr-Sommerfeld quantization condition (38) from the requirement that the monodromy transformation leaves the bandwidth (39) unchanged.

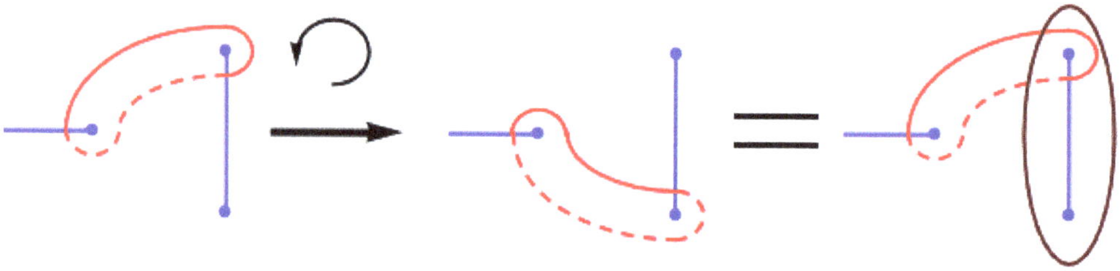

Figure 7. In a monodromy transformation the parameter u is smoothly changed around a critical value in parameter space and returned to its original value, e.g., $(1+u) \to (1+u)e^{2\pi i}$. During the transformation the branch points (blue) move in the complex plane, and the same structure of branch points is recovered. However, if a special value of the parameter u is enclosed by the trajectory in parameter space, e.g., $u = -1$, then the two branch points which collide at $u = -1$ are exchanged. During the transformation the integration cycle (red) is not allowed to cross a branch point, hence they are pulled along with the branch points. To restore the original cycle a closed cycle enclosing the two branch points has to be added.

A comparison of the results for the bandwidth with numerical simulations is shown in Figure 8 for the four non-Hermitian cases and the Hermitian $(1,1)$ case. All cases show good agreement with the numerical data already for moderate values of the parameter α. (Note however, that for the genus-2 cases Gamow's formula had to be multiplied by an overall factor of $3/2$ (in $(3,1)$ case) or 2 (in $(4,1)$ and $(3,2)$ cases), respectively. The origin of this preexponential factor is beyond the scope of this paper.)

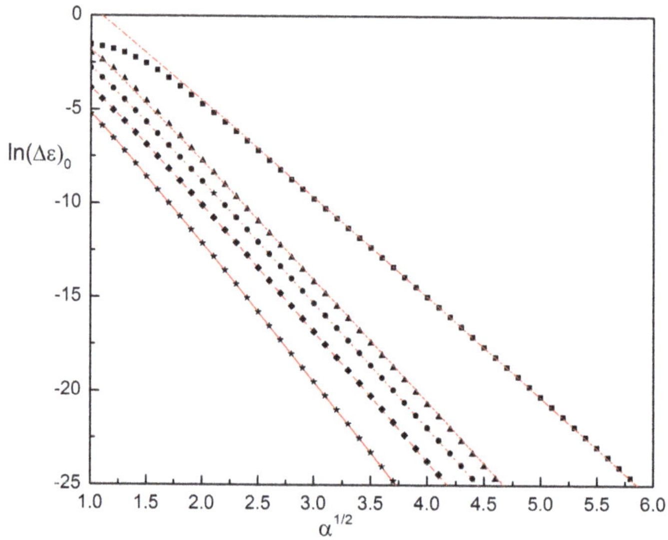

Figure 8. Analytic (numerical) results for the logarithm of the bandwidth of the lowest band, $\ln(\Delta\varepsilon)_0$, as a function of $\alpha^{1/2}$, for all five cases with Riemann surfaces of genus 1 or 2. $(1,1)$: solid line (stars), $(2,1)$: dashed line (diamonds), $(3,1)$: dotted line (circles), $(4,1)$: short-dashed line (triangles), and $(3,2)$: dash-dotted line (squares). Reproduced with permission from Reference [26].

To summarize, we find that in all cases the bandwidth is of the form

$$(\Delta\varepsilon)_m = A \times \left(\frac{k}{m+1/2}\right)^{(m+1/2)} \times \exp(-b\sqrt{\alpha} + (m/2+3/4)\ln\alpha). \tag{42}$$

The pressure, which is calculated from the lowest eigenvalue, contains the ideal gas pressure and the Debye-Hueckel correction,

$$P = Ck_BTf - c\sqrt{k_BTeE_0f}. \tag{43}$$

Here A, k and b, and C and c, are numerical factors that can be calculated directly by expanding S_0 and S_{inst}:

(n_1, n_2)	A	k	b	C	c
$(1,1)$	$4/\pi$	$32e$	8	2	1
$(2,1)$	$2\sqrt{6}/\pi$	$36\sqrt{6}e$	$3\sqrt{6}$	$3/2$	$\sqrt{3/2}$
$(3,1)$	$4\sqrt{2}/\pi$	6.35	7.06	$4/3$	$\sqrt{2}$
$(4,1)$	$5\sqrt{5/2}/2\pi$	1303.46	6.90	$5/4$	$\sqrt{5/2}$
$(3,2)$	$5\sqrt{5/2}/3\pi$	6740.06	5.65	$5/6$	$\sqrt{5/2}$

These values quantify the thermodynamic properties of the ion channels for all five different combinations of charged ions which give a Riemann surface of genus 1 or 2. With a maximum valency of 4 these are also the physically relevant cases. Most importantly we show that the Coulomb gas with unequal valency $n_1 \neq n_2$ has the same qualitative behavior as the standard gas with ions of equal valency, $n_1 = n_2$. In all cases the pressure consists of the ideal gas pressure and the Debye-Hueckel correction, see Equation (43). Crucially for transport through the ion channel, in all cases the bandwidth shows exponential decay with the square-root of the fugacity α and has a universal pre-exponential factor of $\alpha^{3/4}$. However the factor b in the exponent shrinks when the valency is increased, meaning that the transport barrier falls off slower with increased charge concentration when transporting ions with larger valency.

3.3. Higher-Order Corrections from Exact Wkb Method

The approximations for the eigenvalues of the non-Hermitian Hamiltonian can be improved further by considering second- and higher-order terms in the WKB series. The inspiration comes from the exact WKB method which was studied extensively in the context of resurgence theory [37,38]. We use this to get a better approximation for the eigenvalues, and with that the pressure of the Coulomb gas, at moderate values of the charge concentration $\alpha \gtrsim 1$. The key is that the $q = 0$ band edge, which gives the pressure in equilibrium, is determined by an infinite series in α^{-1} (i.e., $\hbar^2$ in usual quantum mechanics),

$$\sum_{n=0}^{\infty} \frac{(-1)^n}{\alpha^n} \oint_{\gamma_{cl}} \rho_{2n}(\theta, u_m) d\theta = \frac{2\pi(m + 1/2)}{\sqrt{\alpha}}. \tag{44}$$

$\rho_0(\theta, u) = p(\theta, u)$ is the classical momentum, and the other terms can be found through a recursive relation [37]. Equation (44) is sometimes also referred to as the generalized Bohr-Sommerfeld quantization condition. Reference [38] shows a calculation of the exact WKB series at all orders for a class of Hermitian genus-1 cases which include the cosine potential, i.e., the $(1,1)$ case in our notation. Here we follow the ideas in [39] and chapter 5 of [28] which give a general procedure to calculate the terms order-by-order for any potential, and can also be applied to non-Hermitian Hamiltonians.

It is evident that truncation of Equation (44) at the $n = 0$ term leads to the usual Bohr-Sommerfeld quantization condition. To improve upon this we include the $n = 1$ term. The integrand is given by

$$\rho_2(\theta, u) d\theta = \left(\frac{\partial_\theta^2(\rho_0(\theta, u)^2)}{48\rho_0(\theta, u)^3} + \frac{5}{24}\partial_\theta \frac{\rho_0'(\theta, u)}{\rho_0(\theta, u)^2} \right) d\theta, \tag{45}$$

where the prime denotes a derivative with respect to θ [37]. The second term is an exact form which integrates to zero. We drop this exact form, use the expression (30) for the classical momentum $p = \rho_0$, and perform the coordinate transformation $z = e^{i\theta}$ to write the second-order 1-form as

$$\tilde{\rho}_2(z,u)dz = \frac{-n_1 z^{n_1} - n_2 z^{-n_2}}{48\left(u\frac{n_1+n_2}{n_1 n_2} + \frac{1}{n_1}z^{n_1} + \frac{1}{n_2}z^{-n_2}\right)^{3/2} iz} dz. \tag{46}$$

A comparison with Equation (34) shows that the second-order 1-form $\tilde{\rho}_2(z,u)dz$ has the same branch points as the action 1-form $\lambda(u)$. Therefore it is defined on the same Riemann surface. As discussed in the preceding sections, on the Riemann surfaces of genus $g = 1(2)$ there exist two (four) linearly independent 1-forms, up to an exact form. We take $\{\partial_u^k \lambda(u)\}_{k=0}^K$ as this maximal independent set with $K = 1(3)$. This forms a basis for the space of all 1-forms. Hence, the second-order correction can be written as a linear combination of these basis 1-forms, modulo an exact form. We find this linear combination in the same way as in the derivation of the Picard-Fuchs Equations (19) and (35) and integrate it along the classical cycle γ_{cl} to get

$$\oint_{\gamma_{cl}} \tilde{\rho}_2(z,u)dz = -a\left(S_0'(u) + 2uS_0''(u)\right), \quad \begin{array}{c|c|c|c|c|c} (n_1,n_2) & (1,1) & (2,1) & (3,1) & (4,1) & (3,1) \\ \hline a & 1/48 & 1/18 & 3/32 & 2/15 & 3/10 \end{array}. \tag{47}$$

These expressions fully define the second-order corrections in terms of the classical action and its derivatives with respect to u. These are easily obtained from the previous results, Equations (20)–(22), (36) and (37) (see Reference [26] for the genus-2 cases). Note that in the genus-1 cases the second derivative $S_0''(u)$ can be replaced with $S_0(u)$ by using the Picard-Fuchs Equations (19) and (35).

Here we want to stress that calculation of the second-order (and any higher) correction is only as computationally demanding as deriving the Picard-Fuchs equation. It does not require solving the differential equation and matching boundary conditions because the correct classical action was already identified. Therefore this can also be used as a simple method to simply calculate the higher-order WKB terms if the classical action was obtained in a different manner. The improvement in the approximation of the lowest eigenvalue is shown in Figure 9.

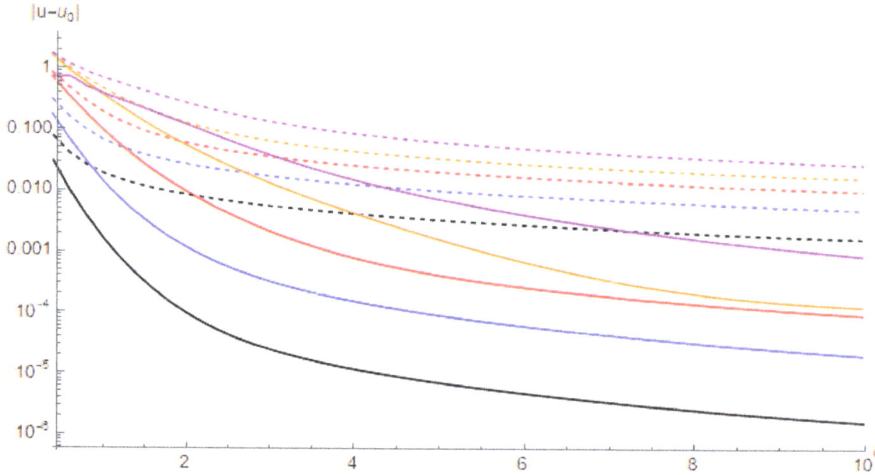

Figure 9. Log-plot of the deviation of the first-order (dashed line) and second-order (solid line) WKB result from the exact numerical result for the lowest eigenvalue as a function of α. We show the five different cases: $(1,1)$ in black, $(2,1)$ in blue, $(3,1)$ in red, $(4,1)$ in orange, $(3,2)$ in purple. The error drops by several orders of magnitude when taking the second-order WKB term into account. The approximations converge to the exact result as $\alpha \to \infty$; however, already at moderate values of $\alpha \gtrsim 1$ the approximations give quite accurate results.

With the second-order result we can calculate the eigenvalues u up to order α^{-1}. Therefore we expand the classical action $S_0(u)$ for $u \gtrsim -1$ to order $(u+1)^2$ and solve for u. Taking the lowest eigenvalue u_0 and applying this to the formula for the pressure (7) gives

$$P = c_0 k_B T f - c_1 \sqrt{e E_0 k_B T f} - c_2 e E_0, \tag{48}$$

with the following constants:

(n_1, n_2)	c_0	c_1	c_2
$(1,1)$	2	1	1/16
$(2,1)$	3/2	$\sqrt{3/2}$	1/9
$(3,1)$	4/3	$\sqrt{2}$	19/144
$(4,1)$	5/4	$\sqrt{5/2}$	1/8
$(3,2)$	5/6	$\sqrt{5/2}$	13/36

This gives the ideal gas pressure and the Debye-Hueckel correction from the usual Bohr-Sommerfeld condition. The second-order WKB term gives an additional correction which is independent of the fugacity but only depends on the geometric properties of the channel which are included in the definition of E_0.

4. Summary of Semiclassical Results

In this review we discussed analytic calculations of the thermodynamic properties of an ion channel at large charge concentrations, with an extension to moderate concentrations. We started with discussing a standard mapping of a statistical system onto an effective quantum system [10,23]. When performing this mapping there is no guarantee that the resulting effective Hamiltonian is Hermitian and has a purely real spectrum. Physically one needs to obtain a real and positive partition function. This is e.g., guaranteed if the Hamiltonian obeys $\mathcal{PT}$-symmetry and its lowest eigenvalue is purely real.

Translation between the quantum results and thermodynamic observables is straightforward. Most importantly, the pressure (i.e., free energy density) is given by the quantum mechanical ground-state energy. The adiabatic transport barrier is the width of the lowest Bloch band. The complex energies of excited states, c.f. Figures 4 and 6, describe higher order correlation functions. Their imaginary part is responsible for spatial oscillations, while the real part yields an overall exponential decay. Such decaying oscillatory correlation functions reflect short-range charge density wave ionic order within the channel. As seen in Figures 4 and 6, the onset of complex eigenvalues happens at lower energies for ions with larger valencies, which implies stronger charge density fluctuations. In all cases we observe that an increase of the charge concentration leads to an exponential reduction of the transport barrier, however this decay is slower if the ion valencies are large. This is visualized in Figure 8.

The approximation with the effective 1D Coulomb potential, Equation (1), works best at large ion concentration. Electric field lines leak out of the channel after a characteristic length ξ which is given by $\xi^2 = a^2\kappa_1/(2\kappa_2)\ln(2\xi/a)$, where a is the radius of the channel and κ_1, κ_2 are the dielectric constants of water and the surrounding medium. Therefore the 1D Coulomb potential best approximates the situation where the characteristic distance between the ions is small. This is the case of large charge concentration, which is also the case when then semiclassical approximation is applicable.

Here we discuss a method how to perform semiclassical calculations without the need to solve the classical equations of motion and without direct integration. This is particularly useful in the non-Hermitian cases when the solutions to the equations of motions are hardly attainable. Instead we derive and solve the Picard-Fuchs differential equation, which is a tool from algebraic topology. The power of the Picard-Fuchs equation is that it is a coordinate-free expression, i.e., one does not need to know the classical trajectories. In the last part we extend our calculations to second- and higher-order terms in the WKB series.

These provide a clearly improved approximation for the eigenvalues especially at moderate charge concentrations, see Figure 9.

The applicability of the Picard-Fuchs method extends far beyond the case of ion channels. It can be a powerful tool for Hermitian and non-Hermitian systems alike, as it can be applied to generic Hamiltonians. Especially the extension to second- and higher-order terms in the WKB series requires very little computational effort once the classical action has been calculated. Mappings of a generic statistical system onto an effective quantum system can lead to a non-Hermitian Hamiltonian for which semiclassical calculations with direct integration are difficult. We believe that the Picard-Fuchs method can be especially useful in these cases, as it allows us to circumvent the complications associated with direct integration like solving equations of motion with complex coordinates.

Author Contributions: Both authors contributed equally to all aspects of the manuscript. Both authors have read and agreed to the published version of the manuscript.

Funding: A.K. was supported by NSF grants DMR-2037654. T.G. acknowledges funding from the Institute of Science and Technology (IST) Austria, and from the European Union's Horizon 2020 research and innovation program under the Marie Skłodowska-Curie Grant Agreement No. 754411.

Institutional Review Board Statement: Not applicable.

Informed Consent Statement: Not applicable.

Acknowledgments: We are indebted to Boris Shklovskii for introducing us to the problem, and Alexander Gorsky and Peter Koroteev for introducing us to the Picard-Fuchs methods. A very special thanks goes to Michael Janas for several years of excellent collaboration on these topics. TG thanks Michael Kreshchuk for introduction to the exact WKB method and great collaboration on related projects. Figures 3 and 4 are reproduced from Reference [25] with friendly permission by the Russian Academy of Sciences. Figures 2, 4, 5, 6, and 8 are reproduced from Reference [26] with friendly permission by IOP Publishing. ©IOP Publishing. All rights reserved.

Conflicts of Interest: The authors declare no conflict of interest.

Appendix A. Equivalence between Partition Function and Quantum Mechanics

Here we present details of the mapping between the statistical system of charged ions and an effective single-particle quantum mechanical problem. Our goal is to evaluate the grandcanonical partition function of the Coulomb gas, Equation (4),

$$\mathcal{Z} = \sum_{N_1,N_2=0}^{\infty} \frac{f_1^{N_1} f_2^{N_2}}{N_1!N_2!} \prod_{j=1}^{N_1+N_2} \int_0^L dx_j \, e^{-U/T}, \tag{A1}$$

where the gas potential energy U is given by Equations (2) and (3). To this end we first consider an auxiliary identity:

$$\begin{aligned} 1 &= \int \mathcal{D}\rho(x) \, \delta\left(\rho(x) - \sum_{j=1}^{N_1+N_2} \sigma_j \delta(x - x_j) - q(\delta(x) - \delta(x - L))\right) \\ &= \int \mathcal{D}\rho(x) \int \mathcal{D}\theta(x) \, e^{-i\int_0^L dx\,\theta(x)\left(\rho(x) - \sum_{j=1}^{N_1+N_2} \sigma_j \delta(x-x_j) - q(\delta(x)-\delta(x-L))\right)} \\ &= \iint \mathcal{D}\rho(x)\mathcal{D}\theta(x) \, e^{-i\left(\int_0^L dx\,\theta(x)\rho(x) - \sum_{j=1}^{N_1+N_2} \sigma_j \theta(x_j) - q(\theta(0)-\theta(L))\right)} . \end{aligned} \tag{A2}$$

Here $\rho(x)$ is a continuous field for the charge density, and $\theta(x)$ its conjugate field. Substituting this identity into the expression for the partition function, one finds:

$$\begin{aligned}\mathcal{Z} = & \iint\limits_{-\infty}^{\infty} \frac{d\theta_0 d\theta_L}{(2\pi)^2}\, e^{iq(\theta_0-\theta_L)} \iint \mathcal{D}\rho(x)\mathcal{D}\theta(x)\, e^{-\frac{1}{2T}\iint_0^L dx dx' \rho(x)\Phi(x-x')\rho(x') - i\int_0^L dx\,\theta(x)\rho(x)} \\ & \times \sum_{N_1=0}^{\infty} \frac{1}{N_1!}\left(f_1 \int_0^L dx\, e^{i\sigma_1\theta(x)}\right)^{N_1} \times \sum_{N_2=0}^{\infty} \frac{1}{N_2!}\left(f_2 \int_0^L dx\, e^{i\sigma_2\theta(x)}\right)^{N_2} \\ = & \iint\limits_{-\infty}^{\infty} \frac{d\theta_0 d\theta_L}{(2\pi)^2}\, e^{iq(\theta_0-\theta_L)} \int \mathcal{D}\theta(x)\, e^{-\frac{T}{2}\iint_0^L dx dx' \theta(x)\Phi^{-1}(x-x')\theta(x') + \int_0^L dx\left(f_1 e^{in_1\theta(x)} + f_2 e^{-in_2\theta(x)}\right)} .\end{aligned} \tag{A3}$$

The integral over $\theta(x)$ runs over all functions with the boundary conditions $\theta(0) = \theta_0$ and $\theta(L) = \theta_L$. We also use that the valencies of the charges are $\sigma_j = n_1$ for $1 \le j \le N_1$ and $\sigma_j = -n_2$ for $N_1 + 1 \le j \le N_1 + N_2$. It is straightforward to verify that the inverse of the interaction operator is given by $\Phi^{-1}(x - x') = -(2eE_0)^{-1}\delta(x - x')\partial_x^2$, because the Coulomb potential in any dimension is a resolvent of the Poisson equation and therefore its inverse is the Laplacian. As a result, the functional integral on the r.h.s. of the last expression takes the form of the Feynman propagator

$$G(\theta_0, \theta_L; L) \equiv \int \mathcal{D}\theta(x)\, e^{-\frac{x_T}{4}\int dx\left[(\partial_x\theta)^2 - \left(\alpha_1 e^{in_1\theta(x)} + \alpha_2 e^{-in_2\theta(x)}\right)\right]} , \tag{A4}$$

where $x_T = T/(eE_0)$ and $\alpha_{1,2} = 4f_{1,2}/x_T$. Expression (A4) represents the "quantum mechanical" probability to propagate from θ_0 to θ_L during the (imaginary) "time" L. The corresponding stationary "Schrödinger equation" for the eigenfunction $\Psi_m(\theta, x) = \Psi_m(\theta)\exp\{-2\varepsilon_m x/x_T\}$ has the form:

$$-\frac{\partial^2\Psi_m(\theta)}{\partial\theta^2} - \left(\alpha_1 e^{in_1\theta(x)} + \alpha_2 e^{-in_2\theta(x)}\right)\Psi_m(\theta) = \varepsilon_m\Psi_m(\theta) . \tag{A5}$$

In terms of the stationary eigenfunctions of this equation the propagator takes the form

$$G(\theta_0, \theta_L; L) = \sum_m \Psi_m(\theta_0)\bar{\Psi}_m(\theta_L)\, e^{-2\varepsilon_m L/x_T} . \tag{A6}$$

Finally the partition function (A3) is nothing but the Fourier transform of the propagator with respect to θ_0 and θ_L and thus may be written as

$$\mathcal{Z} = \sum_m \Psi_m(q)\bar{\Psi}_m(q)\, e^{-2\varepsilon_m L/x_T} , \tag{A7}$$

where $\Psi_m(q) \equiv \int d\theta/(2\pi)\Psi_m(\theta)\exp\{i\theta q\} = \langle q|m\rangle$ is the quasi-momentum representation of the wavefunction in the m-th Bloch band with the energy ε_m. Instead of dealing with Bloch wavefunctions with the boundary condition $\Psi_m(\theta + 2\pi) = e^{i2\pi q}\Psi_m(\theta)$ one may perform a gauge transformation to deal with periodic wavefunctions and having q as the vector potential in the Schrödinger equation. This way we arrive at Equations (5) and (6) in the main text.

References

1. Bischofberger, M.; Gonzalez, M.R.; van der Goot, F.G. Membrane injury by pore-forming proteins. *Curr. Opin. Cell Biol.* **2009**, *21*, 589–595.
2. Los, F.C.O.; Randis, T.M.; Aroian, R.V.; Ratnera, A.J. Role of Pore-Forming Toxins in Bacterial Infectious Diseases. *Microbiol. Mol. Biol. Rev.* **2013**, *77*, 173–207.
3. Li, J.; Stein, D.; McMullan, C.; Branton, D.; Aziz, M.J.; Golovchenko, J.A. Ion-beam sculpting at nanometre length scales. *Nature* **2001**, *412*, 166–169.

4. Storm, A.J.; Chen, J.H.; Ling, X.S.; Zandbergen, H.W.; Dekker, C. Fabrication of solid-state nanopores with single-nanometre precision. *Nat. Mater.* **2003**, *2*, 537.
5. Demming, A. High-throughput DNA Sequencing Etches Forwards. KTH Sweden. 2015. Available online: https://www.aphys.kth.se/photonics/nas/nanopores-in-si-1.583068 (accessed on 18 January 2021).
6. Cui, Y.; Lauhon, L.J.; Gudiksen, M.S.; Wang, J.; Lieber, C.M. Diameter-controlled synthesis of single-crystal silicon nanowires. *Appl. Phys. Lett.* **2001**, *78*, 2214–2216.
7. Zimmerman, J.F.; Murray, G.F.; Wang, Y.; Jumper, J.M.; Austin, J.R.; Tian, B. Free-Standing Kinked Silicon Nanowires for Probing Inter- and Intracellular Force Dynamics. *Nano Lett.* **2015**, *15*, 5492–5498.
8. Takaiwa, D.; Hatano, I.; Koga, K.; Tanaka, H. Phase diagram of water in carbon nanotubes. *Proc. Natl. Acad. Sci. USA* **2008**, *105*, 39–43.
9. Dalla Bernardina, S.; Paineau, E.; Brubach, J.B.; Judeinstein, P.; Rouzière, S.; Launois, P.; Roy, P. Water in Carbon Nanotubes: The Peculiar Hydrogen Bond Network Revealed by Infrared Spectroscopy. *J. Am. Chem. Soc.* **2016**, *138*, 10437–10443.
10. Edwards, S.; Lenard, A. Exact Statistical Mechanics of a One-Dimensional System with Coulomb Forces. II. The Method of Functional Integration. *J. Math. Phys.* **1962**, *3*.
11. Kamenev, A.; Zhang, J.; Larkin, A.; Shklovskii, B. Transport in one-dimensional Coulombgases: From ion channels to nanopores. *Phys. A* **2006**, *359*, 129–161.
12. Démery, V.; Dean, D.S.; Hammant, T.C.; Horgan, R.R.; Podgornik, R. The one-dimensional Coulomb lattice fluid capacitor. *J. Chem. Phys.* **2012**, *137*, 064901.
13. Kaufman, I.K.; Luchinsky, D.G.; Tindjong, R.; McClintock, P.V.E.; Eisenberg, R.S. Energetics of discrete selectivity bands and mutation-induced transitions in the calcium-sodium ion channels family. *Phys. Rev. E* **2013**, *88*.
14. Kaufman, I.K.; McClintock, P.V.E.; Eisenberg, R.S. Coulomb blockade model of permeation and selectivity in biological ion channels. *New J. Phys.* **2015**, *17*.
15. Feng, J.; Liu, K.; Graf, M.; Dumcenco, D.; Kis, A.; DiVentra, M.; Radenovic, A. Observation of ionic Coulomb blockade in nanopores. *Nat. Mater.* **2016**, *15*, 850–855.
16. Schoch, R.B.; Han, J.; Renaud, P. Transport phenomena in nanofluidics. *Rev. Mod. Phys.* **2008**, *80*, 839.
17. Movahed, S.; Li, D. Electrokinetic transport through nanochannels. *Electrophoresis* **2011**, *32*, 1259–1267.
18. Airoldi, P.; Mauri, A.G.; Sacco, R.; Jerome, J.W. Three-dimensional numerical simulation of ion nanochannels. *J. Coup. Syst. Mult. Dyn.* **2015**, *3*, 57–65.
19. Kavokine, N.; Marbach, S.; Siria, A.; Bocquet, L. Ionic Coulomb blockade as a fractional Wien effect. *Nat. Nanotechnol.* **2019**, *14*, 573–578.
20. Parsegian, A. Energy of an Ion crossing a Low Dielectric Membrane: Solutions to Four Relevant Electrostatic Problems. *Nature* **1969**, *221*, 844–846.
21. Zhang, J.; Kamenev, A.; Shklovskii, B. Conductance of Ion Channels and Nanopores with Charged Walls: A Toy Model. *Phys. Rev. Lett.* **2005**, *95*,148101.
22. Zhang, J.; Kamenev, A.; Shklovskii, B.I. Ion exchange phase transitions in water-filled channels with charged walls. *Phys. Rev. E* **2006**, *73*, 051205.
23. Altland, A.; Simons, B. *Condensed Matter Field Theory*; Cambridge University Press: Cambridge UK, 2007.
24. Bender, C.M.; Brody, D.C.; Jones, H.F. Must a hamiltonian be hermitian? *Am. J. Phys.* **2003**, *71*, 1095–1102.
25. Gulden, T.; Janas, M.; Koroteev, P.; Kamenev, A. Statistical Mechanics of Coulomb Gases as Quantum Theory on Riemann Surfaces. *Sov. Phys. JETP* **2013**, *117*, 517–537.
26. Gulden, T.; Janas, M.; Kamenev, A. Riemann surface dynamics of periodic non-Hermitian Hamiltonians. *J. Phys. A Math. Theor.* **2014**, *47*.
27. Connor, J.N.L.; Uzer, T.; Marcus, R.A.; Smith, A.D. Eigenvalues of the Schrödinger equation for a periodic potential with nonperiodic boundary conditions: A uniform semiclassical analysis. *J. Chem. Phys.* **1984**, *80*.
28. Gulden, T. A Semiclassical Theory on Complex Manifolds with Applications in Statistical Physics and Quantum Mechanics. Ph.D. Thesis, University of Minnesota, Minneapolis, MN, USA, 2016.
29. Miranda, R. *Algebraic Curves and Riemann Surfaces*; Birkhauser: Boston, MA, USA, 1992.
30. do Carmo, M. *Riemannian Geometry*; American Mathematical Society: Providence, RI, USA, 1995.
31. Ahlfors, L.V. *Complex Analysis*; McGraw-Hill: New York, NY, USA, 1979.
32. Rehmann, U. Encyclopedia of Mathematics. Available online: http://www.encyclopediaofmath.org (accessed on 18 January 2021).
33. Heckman, G. *Tshinghua Lectures on Hypergeometric Functions*; 2011. Available online: http://www.math.ru.nl/~heckman/tsinghua.pdf (accessed on 18 January 2021).
34. Weisstein, E.W. Wolfram MathWorld: Hypergeometric Function. Available online: https://mathworld.wolfram.com/HypergeometricFunction.html (accessed on 18 January 2021).
35. Meixner, J.; Schäfke, F.W. *Mathieusche Funktionen und Sphaeroidfunktionen mit Anwendungen auf Physikalische und Technische Probleme*; Springer: Berlin, Germany, 1954.
36. Abramowitz, M.; Stegun, I. *Handbook of Mathematical Functions, with Formulas, Graphs, and Mathematical Tables*; Dover Publications: New York, NY, USA, 1972.

37. Bender, C.M.; Orszag, S.A. *Advanced Mathematical Methods for Scientists and Engineers*; McGraw-Hill: New York, NY, USA, 1978.
38. Başar, G.; Dunne, G.V.; Ünsal, M. Quantum geometry of resurgent perturbative/nonperturbative relations. *J. High Energy Phys.* **2017**, *5*, 87.
39. Kreshchuk, M.; Gulden, T. The Picard-Fuchs equation in classical and quantum physics: application to higher-order WKB method. *J. Phys. A Math. Theor.* **2019**, *52*, 155301.

entropy

MDPI

Article

Changes in Ion Selectivity Following the Asymmetrical Addition of Charge to the Selectivity Filter of Bacterial Sodium Channels

Olena A. Fedorenko [1,*,†], Igor A. Khovanov [2], Stephen K. Roberts [1] and Carlo Guardiani [3,‡]

1 Division of Biomedical and Life Sciences, Lancaster University, Lancaster LA1 4YE, UK; s.k.roberts@lancaster.ac.uk
2 School of Engineering, University of Warwick, Coventry CV4 7AL, UK; i.khovanov@warwick.ac.uk
3 Department of Physics, Lancaster University, Lancaster LA1 4YW, UK; carlo.guardiani@uniroma1.it
* Correspondence: olena.fedorenko@nottingham.ac.uk; Tel.: +44-7507-437847
† Current affiliation: School of Life Sciences, University of Nottingham, Nottingham NG7 2UH, UK.
‡ Current affiliation: Department of Mechanical and Aerospace Engineering, Sapienza University of Rome, 00184 Roma, Italy.

Received: 28 October 2020; Accepted: 7 December 2020; Published: 9 December 2020

Abstract: Voltage-gated sodium channels (NaVs) play fundamental roles in eukaryotes, but their exceptional size hinders their structural resolution. Bacterial NaVs are simplified homologues of their eukaryotic counterparts, but their use as models of eukaryotic Na^+ channels is limited by their homotetrameric structure at odds with the asymmetric Selectivity Filter (SF) of eukaryotic NaVs. This work aims at mimicking the SF of eukaryotic NaVs by engineering radial asymmetry into the SF of bacterial channels. This goal was pursued with two approaches: the co-expression of different monomers of the NaChBac bacterial channel to induce the random assembly of heterotetramers, and the concatenation of four bacterial monomers to form a concatemer that can be targeted by site-specific mutagenesis. Patch-clamp measurements and Molecular Dynamics simulations showed that an additional gating charge in the SF leads to a significant increase in Na^+ and a modest increase in the Ca^{2+} conductance in the NavMs concatemer in agreement with the behavior of the population of random heterotetramers with the highest proportion of channels with charge $-5e$. We thus showed that charge, despite being important, is not the only determinant of conduction and selectivity, and we created new tools extending the use of bacterial channels as models of eukaryotic counterparts.

Keywords: ion channel; selectivity; permeability; patch-clamp; computer simulations

1. Introduction

Voltage-gated sodium and calcium channels (NaVs and CaVs, respectively) are involved in a multitude of processes, including electrical signaling, secretion, and synaptic transmission [1]. The malfunction or dysregulation of NaVs and CaVs leads to a wide range of neurological, cardiovascular, and muscular disorders, including periodic paralysis [2], arrhythmia [3], and epilepsy [4], which highlights the importance of these molecules.

Eukaryotic NaVs and CaVs have similar structure and comprise a pore-forming $\alpha 1$ subunit of approximately 190–250 kDa, which co-assembles with a number of auxiliary subunits. The $\alpha 1$ subunit is organized in four domains, each comprising a voltage sensor (encompassing helices S1–S4) and a pore domain (including helices S5–S6) [5–9]. The four domains arranged around the pore are not identical, resulting in a channel structure that is asymmetric and pseudo-tetrameric [10]. The atomic level resolution of the structure of these molecules is essential to understand their structure-function relationships, but this task is particularly challenging for eukaryotic NaV channels, due to them being membrane integral proteins [11] and is exacerbated by their particularly large size. Consequently, to date only the structure of a single eukaryotic NaV has been resolved at the atomic level (3.8 Å), but the channel was not electrophysiologically characterized [12].

Bacterial NaVs and CaVs are simplified homologues of eukaryotic channels. They are homotetrameric channels formed by four identical monomers corresponding to the four domains of the $\alpha1$ subunit of eukaryotic channels [13–16]. Their minimalist structure has enabled the determination of high-resolution atomistic structures, which has allowed extensive structure-functional characterization with respect to their cation selectivity, gating, and binding of anesthetics e.g., [17–19]. Although a complete understanding of selectivity has not been arrived at, the availability of high-resolution structures has provided a detailed understanding of the atomistic-level interaction of ions in the Selectivity Filter (SF). Despite this, there is still no widely accepted predictive model for cation selectivity, representing gaps in our knowledge of the molecular mechanism of ion permeation and selectivity. Mutation studies suggest the fixed charge (Q_f) of the SF to be one of the major determinants of selectivity and permeation [15,20–32]. The Q_f charge is at the core of many theoretical models attempting to explain the physical origins of cation selectivity in ion channels. For example, cation conduction has recently been modelled within the framework of Ionic Coulomb Blockade (ICB) [24,25], an electrostatic model with the aim of predicting Na^+ and Ca^{2+} permeability based on knowing the actual Q_f value of the SF.

Although prokaryotic and eukaryotic channels show the same general architecture along the axis of the pore (an outer vestibule and an inner water filled cavity separated by a narrow SF), the use of prokaryotic channels as models of their eukaryotic counterparts is limited by their lack of radial asymmetry. In the case of the homotetrameric bacterial NaVs, four identical monomers form the channel pore; in contrast, their eukaryotic counterparts are composed of four non-identical domains which introduce significant radial asymmetry [14,26–31]. This difference becomes evident at the level of the SF, where conduction and selectivity are controlled by a DEKA ring ($Q_f = -1e$) in eukaryotic NaVs and an EEEE ring ($Q_f = -4e$) in prokaryotic NaVs. Another puzzling fact is that the EEEE locus is typical of bacterial sodium-selective channels but also characterizes calcium-selective eukaryotic channels [16,32–34], thus leading to NaChBac being initially predicted to be Ca^{2+}-selective. The existence of disparate sequences indicates that bacterial and eukaryotic channels enforce their ion preferences through different molecular strategies [15,34,35]. As a result, the selectivity and conduction mechanisms discovered in prokaryotes are not readily transferable to eukaryotes.

The puzzling functional similarity between bacterial NaVs and eukaryotic CaVs has been termed the "EEEE paradox" [36]. The paradox arises as a result of the violation of the assumption that Q_f is the main driving force of cation selectivity. A possible resolution of the paradox is related to the existence of a conserved D residue in domain 2 of CaVs in the neighborhood of the EEEE ring. Monte Carlo simulations predicted this D residue (termed D2p51 in [37]) to occupy a position in close proximity to the EEEE locus. This observation led to the hypothesis that the locus imparting Ca^{2+} permeability is actually EEEED, with a Q_f value of $-5e$ [25,37]. Moreover, when this conserved D residue in domain 2 of Cav1.2 (referred to as D707 in [29]) was replaced with neutral residues, a striking reduction in Ca^{2+} binding to the SF was measured [29]. These results suggest D707 to be an important cation binding determinant of eukaryotic channels.

The different behavior of prokaryotic and eukaryotic voltage-gated sodium and calcium channels highlights the importance of incorporating radial asymmetry in the SF of prokaryotic channels. In our previous work [38] we reported the creation of a concatenated bacterial NaV, in which four NavMs monomers were covalently linked to form a stable single polypeptide chain, resembling the general structure of a eukaryotic NaV. This allowed the targeted mutagenesis of individual domains introducing radial asymmetry in the bacterial channel with the aim to gain further insight on the role of Q_f as a determinant of ion selectivity. In the present study, we report the first attempt to mutate the concatemer and generate a bacterial sodium channel with radial asymmetry in the SF. In order to obtain atomistic-level detail of selectivity and permeation, the electrophysiological characterization was integrated with Molecular Dynamics (MD) simulations of wild-type NavMs ($Q_f = -4e$) and a mutant with an additional negative charge in the SF ($Q_f = -5e$).

In the present study, we have also employed an independent yet complementary approach to introduce radial asymmetry into the SF of a bacterial sodium channel. Namely, a number of combinations of NaChBac monomers (differing in their amino acid composition and Q_f value of the SF) were transfected into Chinese Hamster Ovary (CHO) cells to generate a random population of heterotetrameric channels with radial asymmetry in the SF. The mixed monomer approach using NaChBac monomers showed that Ca^{2+} conduction is increased in channels with a $Q_f > -4e$ (consistent with the proposed explanation for the EEEE paradox). Our data confirm the key role of the SF charge as the major determinant of conduction and selectivity. However, the failure to completely overturn the sodium selectivity of the NavMs concatemer to Ca^{2+} selectivity (with much smaller relative Ca^{2+} permeability exhibited by the $-5e$ mutant NavMs concatemers compared to that for eukaryotic CaVs) suggests the existence of fine-tuning mechanisms of structural origin.

2. Materials and Methods

2.1. Materials Generation of Mutant Bacterial Channels

cDNA constructs encoding NaChBac (GenBank accession number BAB05220) and NavMs (GenBank accession number WP_011712479) bacterial sodium channels were synthesized by EPOCH Life Science (www.epochlifescience.com). NavMs concatemer was subcloned into pTRACER-CMV2 (Invitrogen) downstream of CMV promoter, as described previously [38].

Site-directed mutagenesis was performed using specific primers containing the sequence for the desired amino acid substitutions (according to Q5® Site-Directed Mutagenesis Kit; New England BioLabs Inc., Hitchin, UK). For the generation of LEDWAS mutant from wild-type NaChBac, we used the forward primer CACGCTAGAGgatTGGGCGAGCG and the reversed primer ACCACTTGGAACAATGTTAAC; for LASWAS mutant, we used the forward primer GGTCACGCTAgccTCATGGGCGAGcggc and the reversed primer ACTTGGAACAATGTTAACAAACtaagc.

$Q_f = -5e$ NavMs mutants were generated from NavMs concatemer, which was designed with restriction sites delimiting each domain (Supplementary Figure S3A). Domain I (KpnI/EcoRI) and Domain II (EcoRI/EcoRV) were excised by restriction digest. The domain fragments were re-amplified by PCR using primer pairs to regenerate the restriction site prior to subcloning into vector pCR Blunt II-TOPO (Invitrogen): primers are Kpn1_NavMs_F (CCCGGTACCAGCCGCCA CCATGTCACGCAAAATAAG)/EcoRI_ NavMs_R (CCCGAATTCGGGCTCGTCCTCCCAGATG) for Domain I and EcoRI_ NavMs_F (CCCGAATTCATGTCTAGGAAGATCC)/EcoRV_ NavMs_F (CCCGATATCGGGCTCGTCCTCCCAGATG) for Domain II. Site-directed mutagenesis (for S179D, according to NavMs monomer residue nomenclature) was performed on each domain using primers LEDWSM_ NavMs_F (GACCTTAGAGgatTGGTCTATGGGC) and LEDWSM_ NavMs_R (ATCACCTGAAATAGTGTG) prior to the restriction enzyme-mediated excision and ligation (T4 DNA ligase; NEB) of the Domain DNA fragment in the NavMs concatemer at sites KpnI/EcoRI (for Domain I) and EcoRI/EcoRV (Domain II).

All the clones were sequenced to check for correct construction and to ensure that no unwanted PCR induced mutations had been introduced. DNA for the transfection of cells was prepared using Midi Plasmid Kit (Qiagen, Manchester, UK).

2.2. Cell Culture and Transfection

Chinese hamster ovary (CHO) and human embryonic kidney (HEK293T) cell lines were obtained from Dr. Stephen K. Roberts. Cells were cultured in DMEM high glucose with L-glutamine (Lonza, Slough, UK) supplemented with 10% Fetal Bovine Serum (Thermo Scientific, Loughborough, UK) with the addition of 50 U/mL of penicillin and 50 μg/mL of streptomycin (Sigma, Irvine, UK). Cells were maintained in a T25 flask (Thermo Scientific) at 37 °C in a 5% CO_2 incubator and passaged twice a week. Then, 24 h before transfection, the cells were seeded in 6-well plates (Corning, Deeside, UK) containing No.1 coverslips (Scientific Laboratory Supplies, Nottingham, UK). A total of 10 µL of transfection reagent (Mirus, Cardiff, UK) and 5 µg of plasmid DNA or a mixture of DNAs in defined proportions were equilibrated separately in 250 µL of UltraMEM™ Reduced Serum Medium (Lonza, Slough, UK) at room temperature for 5 min and then mixed and incubated at room temperature for 20 min to form the DNA-reagent complex. Treated cells (at 80% confluency) were supplemented with DNA-reagent complex and incubated at 37 °C and 5% CO_2 for 24–48 h before experiments.

2.3. Electrophysiology

Whole-cell voltage clamp recordings were performed at room temperature (20 °C) using an Axopatch 200A (Molecular Devices, Inc., Wokingham, UK) amplifier. Patch-clamp pipettes were pulled from borosilicate glass (Kimax, Kimble Company, Dover, USA) to resistances of 2–3 MOhm. Shanks of the pipette's tip were coated with bee's wax to reduce the pipette capacitance. The pipette solution contained (in mM) 15 Na-gluconate, 5 NaCl, 90 NMDG, 10 EGTA, and 20 HEPES, pH 7.4 adjustedwith 3 mM HCl). To record Na^+ influx currents the bath solution was (in mM) 140 Na-methanesulfonate, 5 CsCl, 10 HEPES and 10 glucose (pH 7.4 adjusted with 4.8 mM CsOH); for the measurement of the Ca^{2+} influx currents, 140 mM of Na-methanesulfonate was replaced with 100 mM of Ca-methanesulfonate.

Data collection was initiated 3 min after obtaining the whole cell configuration to ensure the complete equilibration of the pipette solution and cytosol. The bath solution was grounded using a 3 M KCl agar bridge; the liquid junction potential determined experimentally (as described by [39]) agreed with that calculated (using JPCalc program, Clampex, Axon Instruments, Inc., Wokingham, UK) and was negligible. To ensure the complete exchange of the bath solution, electrophysiological recordings were initiated after >4 min of solution change. The rate of the gravity-fed perfusion system for the bath solution exchange was approximately 0.7 mL/min in a chamber volume of approximately 200 µL.

The results were analyzed using the Clampfit 10.1 software (Molecular Devices, Wokingham, UK) and OriginPro8 (OriginLab Corporation, Wellesley, MA, USA). Pooled data are presented as means ± SEM (n), where n is the number of independent experiments.

2.4. Equilibrium Simulations of NavMs Channel

The initial structure of wild-type NavMs was taken from the Protein Data Bank (PDB ID: 3ZJZ). Mutation S179D on chain A, and embedding in a membrane of 248 POPC molecules was performed using the CHARMM membrane builder [40,41]. The membrane was bathed on both sides by a 0.14 M NaCl solution or a 0.1 M $CaCl_2$ solution. The size of the simulation box was 102 × 102 × 86 Å and the total number of atoms in the four simulated systems was a little short of 90,000. All the acidic residues have been assigned a charge $-1e$, while basic residues have been assigned a charge $+1e$ based on an analysis of the pKa values with the PROPKA program (server.poissonboltzmann.org/pdb2pqr). All the simulations were performed with the NAMD 2.11b2 [42] suite of programs using the ff14SB [43] force field for the protein and the Lipid14 force-field [44] for the phospholipids. As already observed in [45], in the absence of harmonic restraints the pore rapidly closes at the cytoplasmic gate. In order to avoid this behavior that likely results from the absence of the Voltage Sensor Domain in the simulated system, harmonic restraints (50 kcal/mol/Å^2) were applied to the backbone atoms of the transmembrane helices (residues 131–154 and 194–222) throughout the simulation. The four systems first underwent 10,000 steps of conjugate gradient minimization.

During equilibration, harmonic restraints were applied to non-hydrogen atoms of the protein backbone and side-chains (outside the transmembrane helices; residues 155–193), as well as to the phospholipid heads. A harmonic restraint was also applied to the dihedral angle formed by carbons 8, 9, 10, 11 of oleoyl acid and to the improper dihedral C1–C3–C2–O2 involving the three carbons of the glycerol unit and the hydroxyl oxygen linked to its central carbon. The equilibration was organized in six stages, whereby the constraints were gradually released. The values of the force constants used in the six stages can be found in Supplementary Table S1. The production run was carried out in the isothermal isobaric (NPT) ensemble for 100 ns (in NaCl solution) or 150 ns (in $CaCl_2$ solution). The pressure was kept at 1 atm by the Nose–Hoover Langevin piston method, while the temperature was kept at 300 K by coupling to a Langevin thermostat with a damping coefficient of 1 ps^{-1}. Long-range electrostatic interactions were evaluated with the smooth particle mesh Ewald algorithm. For the short-range non-bonded interactions, we used a cutoff of 12 Å with a switching function at 10.0 Å. The integration time step was 2 fs, and the bonds between hydrogen and heavy atoms were fixed to eliminate the most rapid oscillatory motions. The Potential of Mean Force (PMF) was computed using equation $F(z) = -k_B T \log(\rho(z)/\rho_b)$, where k_B is the Boltzmann constant, T is the absolute temperature, $\rho(z)$ is the density profile of sodium or calcium ions, and ρ_b is the density of these ions in the bulk. Since the ion density in the channel is typically higher than in the bulk, the PMF normally has negative values. To avoid a divergence in the logarithmic expression of the PMF, we assigned $F(z) = 0$ when $\rho(z) = 0$—that is, in the regions of the channel that are never visited by ions.

2.5. Current-Voltage Curves Calculation

Current–voltage (IV) curves in NavMs were attained using the collective diffusion model introduced in [46], where the time-course $Q(t)$ of the net charge transported across the channel at equilibrium is thought of as an unbiased random walk. The net charge transported in the time interval Δt between two consecutive frames of the trajectory is $\Delta Q = \sum_{z_1 \le z \le z_2} \frac{e_i \Delta z_i}{L_z}$, where the sum runs over all ions i, such that $z_1 \le z \le z_2$, $z_1 = -4.5$ Å and $z_2 = 16.5$ Å are the axial limits of the filter region somewhat extended in the vestibule and central cavity. The use of this extended SF gives us the opportunity to exploit the fluctuations due to ions exploring the vestibule region without entering into the SF as well as the aborted permeation events where the ion crosses the mouth of the SF but is immediately pulled back in due to the attraction of the acidic residues. In the expression, Δz_i is the axial displacement of the ion in the time interval Δt and $L_z = z_2 - z_1$ is the length of the SF. The time course of the charge, $Q(t)$, can then be attained as $Q(t) = \sum_{t_i < t} \Delta Q(t_i)$.

Diffusion theory predicts that, for sufficiently long times, the mean square displacement of the charge $\langle Q^2(t) \rangle$ grows linearly with a slope proportional to the diffusion coefficient, $\langle Q^2(t) \rangle \sim 2D_Q t + Const$. Applying linear response theory, the steady current induced by a small constant voltage V can be computed as $I_{steady} = D_Q V / k_B T$. Using such an approach, the linear region of an IV curve can be computed based on the spontaneous ion fluctuations at equilibrium in the absence of any applied electric field.

3. Results

3.1. Experimental Results

To introduce radial asymmetry in the SF of NaChBac, two approaches were adopted. First, mixed populations of NaChBac monomers (differing in their amino acid composition and the Q_f value of the SF) were co-transfected into CHO cells to generate hetero-tetrameric channels exhibiting radial asymmetry in the SFs. Second, we used a concatenated NavMs tetramer [38] to generate radial asymmetry in the SF by the targeted mutation of one of the four repeats.

3.1.1. Na^+/Ca^{2+} Selectivity for Randomly Mixed Populations of NaChBac Monomers

The random assembly of channel tetramers can be demonstrated taking advantage of the different electrophysiological properties of WT NaChBac and the L226P mutant illustrated by the recordings in Supplementary Figure S1. The L226P mutation causes conspicuous alterations in channel gating of NaChBac from depolarization-activated whole-cell currents to non-inactivating hyperpolarization-activated whole-cell currents [47], (Supplementary Figure S1A,B). The mutation shifts the voltage of the maximal current from −10 mV in WT to −180 mV in the mutant (Supplementary Figure S1D), thus currents at −10 mV originating from separate channel populations of WT and L226P homotetramers can be easily separated. The current recordings from CHO cells co-transfected with NaChBac-encoding WT:L226P cDNAs in a ratio 3:1 (Supplementary Figure S1C,E) exhibited unique currents at −10 mV, which cannot be explained by the simple addition of whole current traces from homotetramer channels formed from either L226P or wild-type NaChBac (note that there is no current at −10 mV from L226P channels), indicating that unique heterotetramers are being formed. Assuming that the assembly of heterotetramers is formed without bias, the proportions of channel types can be determined by binomial distribution. It is noteworthy that this assumption is in agreement with previous findings [47,48], showing no bias for heterotetramer formation in CHO cells expressing a mixture of WT and G219P mutant NaChBac monomers and dimers.

Using this approach, CHO cells were co-transfected with cDNAs of NaChBac-encoding WT and mutants, with varied Q_f in the SF, in different ratios. Note that the open probabilities and single channel conductances for the WT NaChBac (LESWAS) and LEDWAS homotetramers were equivalent (Supplementary Figure S2), and that the whole-cell Na^+ currents from cells expressing homotetramer WT and LEDWAS were similar in magnitude (Figure 1a,c), consistent with the expression of the channel (i.e., number of channels) being independent of the single amino acid mutations introduced into the SF. Figure 1 shows the currents recorded from cells transfected with defined mixtures of NaChBac monomers; see Table 1 for the probabilities of different charged species, assuming that the assembly follows a binomial distribution.

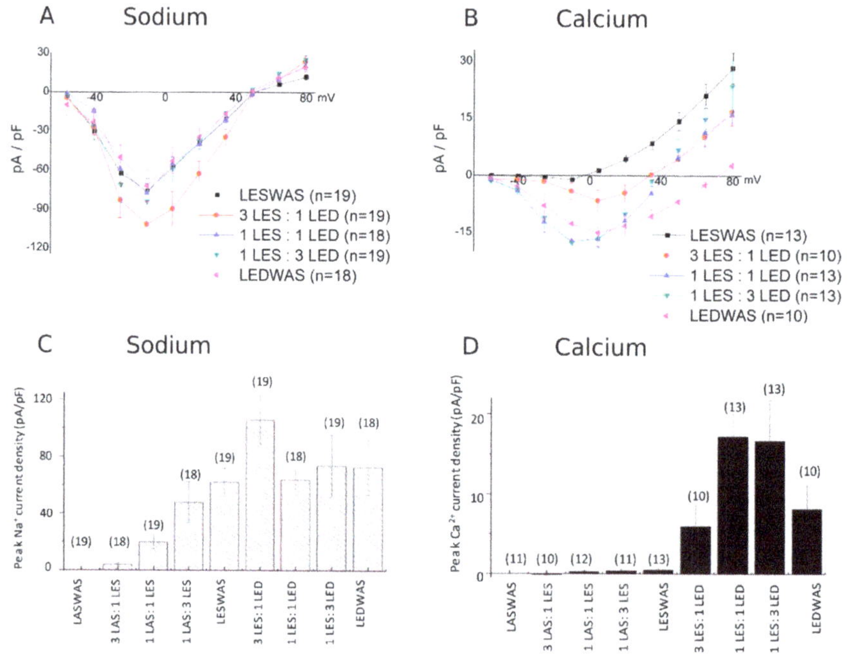

Figure 1. Na^+/Ca^{2+} selectivity for NaChBac monomer mixtures. The voltage–current relations (**A**,**B**) and the mean (+/− SEM) whole-cell peak current density at −10 mV (**C**,**D**) for Na^+ (**A**,**C**) and Ca^{2+} (**B**,**D**) in CHO cells transfected with cDNAs encoding for NaChBac channels possessing either a wild-type selectivity filter (LESWAS/LES) or a mutated selectivity filter (LASWAS/LAS or LEDWAS/LED); 5 µg of total DNA was used per transfection and was composed of either a mixture of types of cDNA at defined ratios, as indicated in Table 1 and on the *X*-axis, or a single cDNA type. Numbers in parentheses indicate the number of replicates.

Table 1. Probabilities of homo- and hetero-tetramer NaChBac channel formation in CHO cells co-transfected with 5 µg (total) of cDNAs encoding for wild-type (LESWAS) NaChBac and mutated NaChBac, in which the selectivity filter amino acid sequence was LASWAS or LEDWAS. Note that the Q_f values for LASWAS, LESWAS, and LEDWAS monomers are 0, −1, and −2, respectively. Probabilities for channel formations were determined by Binomial distribution $P(n, N-n) = C(n, N-n)p^n(1-p)^{N-n}$.

cDNA Population Transfected into CHO Cells	LASWAS	LASWAS LESWAS (3:1)	LASWAS LESWAS (1:1)	LASWAS LESWAS (1:3)	LESWAS	LESWAS LEDWAS (3:1)	LESWAS LEDWAS (1:1)	LESWAS LEDWAS (1:3)	LEDWAS
Probabilities and Q_f values for tetramer formation	$Q_f = 0$: 100%	$Q_f = 0$; 32% $Q_f = -1$; 42% $Q_f = -2$; 21% $Q_f = -3$; 4% $Q_f = -4$; 0.3%	$Q_f = 0$; 6.2% $Q_f = -1$; 25% $Q_f = -2$; 37% $Q_f = -3$; 25% $Q_f = -4$; 6.2%	$Q_f = 0$; 0.3% $Q_f = -1$; 4% $Q_f = -2$; 21% $Q_f = -3$; 42% $Q_f = -4$; 32%	$Q_f = -4$; 100%	$Q_f = -4$; 32% $Q_f = -5$; 42% $Q_f = -6$; 21% $Q_f = -7$; 4% $Q_f = -8$; 0.3%	$Q_f = -4$; 6.2% $Q_f = -5$; 25% $Q_f = -6$; 37% $Q_f = -7$; 25% $Q_f = -8$; 6.2%	$Q_f = -4$; 0.3% $Q_f = -5$; 4% $Q_f = -6$; 21% $Q_f = -7$; 42% $Q_f = -8$; 32%	$Q_f = -8$; 100%

Whole-cell currents were initially recorded in bath solution containing 140 mM of Na-methanesulfonate, followed by recordings after the complete replacement of bath Na^+ with 100 mM of Ca-methanesulfonate. Na^+ permeation appears relatively insensitive to changes in Q_f values between $-4e$ and $-8e$ and equivalent in cells expressing only LESWAS and/or LEDWAS monomers. Focusing on channels exhibiting Q_f values less than $-4e$, it is interesting to note that, despite co-transfection with a 1:1 ratio of LASWAS:LESWAS resulting in an expected only 6.2% of the channel population being homotetramers of LESWAS ($Q_f = -4e$), the Na^+ current density was approximately 30% of that recorded from cells expressing only LESWAS homotetramer channels (Figure 1a,c). An equivalent interpretation can be made for measurements of current density from cells transfected with a 1:3 ratio of cDNAs encoding LASWAS:LESWAS: the sodium current density was equivalent to that recorded from cells expressing only LESWAS homotetramers, despite only 32% of the channel population being predicted to be homotetrameric LESWAS. It is also interesting to compare the current density of the 3:1 LASWAS:LESWAS-expressing cells. Note that these cells show about 25% current density compared to the LESWAS-only cells (5 and 20 pA/pF, respectively). If one looks at the binominal predictions, 25% of channels are predicted to have $Q_f = -2e$ and greater and this is consistent with a $Q_f = -1e$ and 0 being non-conducting (Figure 1c). The simplest explanation for the disproportionately large Na^+ current in cells expressing mixtures of LESWAS and LASWAS monomers is that functional NaChBac channels possessing a SF with a Q_f value less than $-4e$ are functional and able to mediate the Na^+ influx.

Extending this type of analysis to the Ca^{2+} currents, cells transfected with a 1:3 ratio of LESWAS:LEDWAS encoding cDNAs (in which 0.3% of expressed functional channels were predicted to be LEDWAS homotetramers, with a $Q_f = -8e$) exhibited a similar current density for Ca^{2+} influx as that from cells expressing only LEDWAS channels (Figure 1d). Thus, functional NaChBac channels possessing SFs with a Q_f value of less than $-8e$ appear to be able to mediate Ca^{2+} influx, with the possibility that a Q_f value of $-5e$ is sufficient to permit Ca^{2+} permeation. This explanation is also consistent with the observation that the Ca^{2+} current density is greatest in cells transfected with equal and 1:3 ratios of LESWAS:LEDWAS (Figure 1b,d). Note that the Na^+ influx current density remains relatively constant in cells transfected with both LESWAS and LEDWAS encoding cDNAs, indicating that the effect of varying Q_f between $-4e$ and $-8e$ was specific to the Ca^{2+} current density.

3.1.2. Na^+/Ca^{2+} Selectivity for Concatenated NavMS Channels

Although the use of a mixed population of cDNAs encoding for NaChBac and its mutants suggested the value of Q_f to be a major determining factor for Na^+/Ca^{2+} selectivity, the results are subject to the caveat that the whole-cell currents result from the cumulative current from an unknown but predictable range of different channel types. To address this complication, we attempted to generate a stable concatenation of NaChBac to enable the expression of a homogeneous population of NaChBac mutants; however, we have previously shown [38] the NaChBac oligomer to be unstable and not to remain intact in the plasma membrane. In contrast, an equivalent intact NavMs oligomer could be stably expressed in HEK293T cells [38] and thus enable the generation of a homogeneous population of bacterial channels, in which the Q_f value of the SF can be altered in steps of $1e$. The SF of eukaryotic CaVs is formed by a ring of glutamates (the EEEE locus) and a conserved aspartate residue in domain II (D2p51 [37]). The D2p51 residue is suggested to form a binding site for a third incoming Ca^{2+} from the extracellular side of the pore and thus bring an additional positive charge to the SF region necessary for the release of a bound Ca^{2+} to the cytosolic side (i.e., a knock-on mechanism [49]). Although direct evidence for the role of the D2p51 in Ca^{2+} permeation remains elusive, replacing the D2p51 residue in Cav1.2 (aka D707) with a neutral amino acid residue significantly reduces the Ca^{2+} binding of the SF [29]. Thus, to gain further insight into the role of the D2p51 in Ca^{2+} permeation, we used site-directed mutagenesis targeted to repeat I or II in the NavMs oligomer to generate a bacterial NaV with an "EEEED" locus ($Q_f = -5e$) in the SF (Supplementary Figure S3). NavMs has a high homology (45% sequence identity) to NaChBac [28,45], which should enable comparison with the results from NaChBac.

The WT NavMs SF is defined by ^{177}LESWSM^{182}, and we generated NavMs tetramers (Supplementary Figure S3) with the S179D mutation in either repeat I (mutant DI) or repeat II (mutant DII). Both mutants are therefore expected to carry a charge of $-5e$ in the SF. Figure 2 shows typical whole-cell currents from WT and mutant NavMs in bath solution containing either 140 mM of Na^+ or 100 mM of Ca^{2+}.

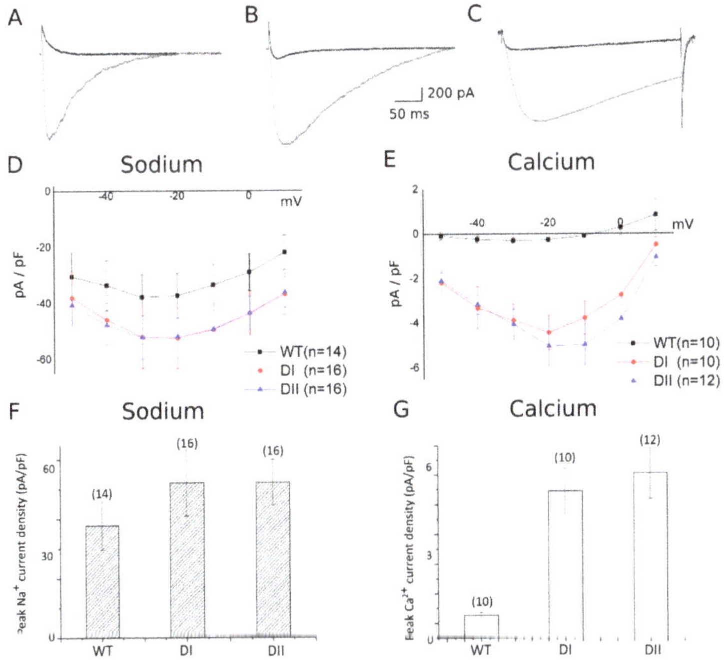

Figure 2. Na^+/Ca^{2+} selectivity for NavMs concatemer possessing varied Q_f values in their SF. The original recordings representatives of wild-type NavMS (**A**) and its DI (**B**) and DII (**C**) mutants in 140 mM Na^+ solution (grey traces) and in 100 mM Ca^{2+} solution (black traces). The voltage–current relations (**D**,**E**) and the mean (+/− SEM) whole-cell peak current density at −10 mV (**F**,**G**) for Na^+ (**D**,**F**) and Ca^{2+} (**E**,**G**) in HEK 293T cells transfected with cDNAs encoding for either wild-type or mutated NavMS. Numbers in parentheses indicate the number of replicates; and in HEK293T cells transfected with wild-type NavMS concatemer (WT) and mutant NavMS concatemer (DI and DII).

In order to make quantitative comparisons between the electrophysiological behavior of NaChBac and NavMs mutants, the peak calcium and sodium currents as well as their ratio are tabulated in Supplementary Table S3 for the NaChBac heterotetramer populations and in Supplementary Table S2 for the NavMs mutants. The comparison of the data of the two tables shows that the ratio of the peak current densities for Na^+ and Ca^{2+} in wild-type NavMs (0.018) is comparable to that for wild-type NaChBac (0.010). The tables also show that the ratio of peak current densities for Ca^{2+} and Na^+ in the two NavMs mutants with an SF charge of $-5e$ (0.080 for the DI mutant and 0.097 for the DII mutant) is similar to that for mutant channels formed from the expression of the 3LES:1LED mixture of NaChBac (0.054) in CHO cells, which yields the highest probability of occurrence (42%) of heterotetramers with an SF charge equal to $-5e$. Although both data sets support the increased Ca^{2+} permeability in $-5e$ mutant bacterial sodium channels, the difference in the Ca^{2+} current magnitude that is evident upon comparing NavMs and NaChBac channels clearly indicates that factors other than the value of Q_f are important in determining the Ca^{2+} permeability.

3.2. Computational Results

In an attempt to gain a molecular-level understanding of the different behavior of WT NavMs ($Q_f = -4e$) and its mutant with charge $Q_f = -5e$, we ran equilibrium MD simulations in a 100 mM solution of $CaCl_2$ or 140 mM NaCl (for 150 and 100 ns, respectively). The initial structure of WT NavMs was taken from the Protein Data Bank (ID: 3ZJZ). Mutation S179D on chain A and embedding in a membrane of 248 POPC molecules was performed using the CHARMM membrane builder [40,41].

In 140 mM of NaCl solution, the WT NavMs SF is stably occupied by a single Na^+ even if transient events of occupation by a second ion can also be spotted (Supplementary Figure S4a). $Q_f = -5e$ mutants SF is almost immediately occupied by two Na^+ ions and, after 40 ns, the filter becomes stably occupied by three sodium ions (Supplementary Figure S4b). The different behavior of the two species is also reflected in the PMF profile (Figure 3a,b), which is characterized by a single deep minimum centered at z = 4–5 Å for WT NavMs, and a minimum split into three sub-basins at z = 2.0 Å, z = 5.0 Å, and z = 8–9 Å, corresponding to three different binding regions, for the $Q_f = -5e$ mutant. The barriers between the sub-basins are in the order of 1–2 kcal/mol and can be easily overcome at the simulation temperature, yet the sodium ions linger in each binding site for longer than they would in case of a uniform probability distribution of occupancy.

The nature of these binding sites can be better characterized by analyzing the conformation of the SF in the last frame of the 100 ns simulations (Figure 3e–h). A notable feature of wild-type NavMs is that the side chains of E178 residues do not point toward the center of the channel, but they are aligned along the channel wall pointing towards the extracellular side. As a result, the distance between the resident sodium ion and the ε-oxygen of E178 always exceeds 4.0 Å. This means that there are no direct sodium-protein interactions; Na^+ interacts with the protein via water molecules in its hydration shell. Indeed, the withdrawn placement of E178 side chains leaves sufficient space in the SF for Na^+ to fully keep its first hydration shell of six water molecules. In contrast, the conformation of the SF of $Q_f = -5e$ mutant revealed three sodium ions that directly interact with the residues of the SF; specifically, the extracellular one interacting with D179 and E178 both located on chain A, the central one with E178, and the intracellular sodium with the backbone carbonyl group of one of the L177 residues. The additional negative charge thus determines an enhanced ability of the NavMs mutant to capture sodium ions from the bulk. This, combined with the possibility of a knock-on mechanism deriving from the simultaneous presence of three Na^+ ions in the SF, possibly explains the larger sodium current density for NavMs channels with $Q_f = -5e$. As a result of this structural arrangement (and in contrast to that for the WT; Figure 3c), Na^+ ions accessing the SF of the mutant lose on average 3.5 water molecules. However, this loss is compensated by the interactions with the oxygens provided by the acidic residues (2 oxygens) and by other protein residues (1 oxygen), such that the total number of coordinating oxygens is maintained equivalent to that for sodium in bulk solution (Figure 3c,d). Note that the interactions of the resident ions with all other residues of the SF are water-mediated.

In 100 mM of $CaCl_2$ within the timescale of our simulations, no Ca^{2+} gains access to the SF of the WT channel, while a single Ca^{2+} enters into the SF of the $Q_f = -5e$ mutant during the early stages of the simulation and thereafter remains locked inside, while also repelling other potentially incoming calcium ions (Supplementary Figure S5a,b). This pattern is in keeping not only with the behavior of WT and mutant NavMs concatemer, but also with the results of the experiments on mixed populations of NaChBac heterotetramers. In fact, while the calcium peak current of the LESWAS homotetramers ($Q_f = -4e$) is just 0.66 pA/pF, that of the 3LES:1LED population, where we expect the highest proportion of channels with $Q_f = -5e$, is tenfold higher (6.0 pA/pF). The seeming mismatch between the currents recorded in experiments and the total block of the Ca^{2+} ion revealed by the simulations is (at least in part) due to the fact that, in the latter, no electric field was applied. Moreover, experimental recordings are performed on a timescale of hundreds of milliseconds, one million times longer than that covered by simulations, allowing time for slow, activated events of ion permeation. A comparison between our computational results and those by other groups is discussed in Supplementary Figure S6.

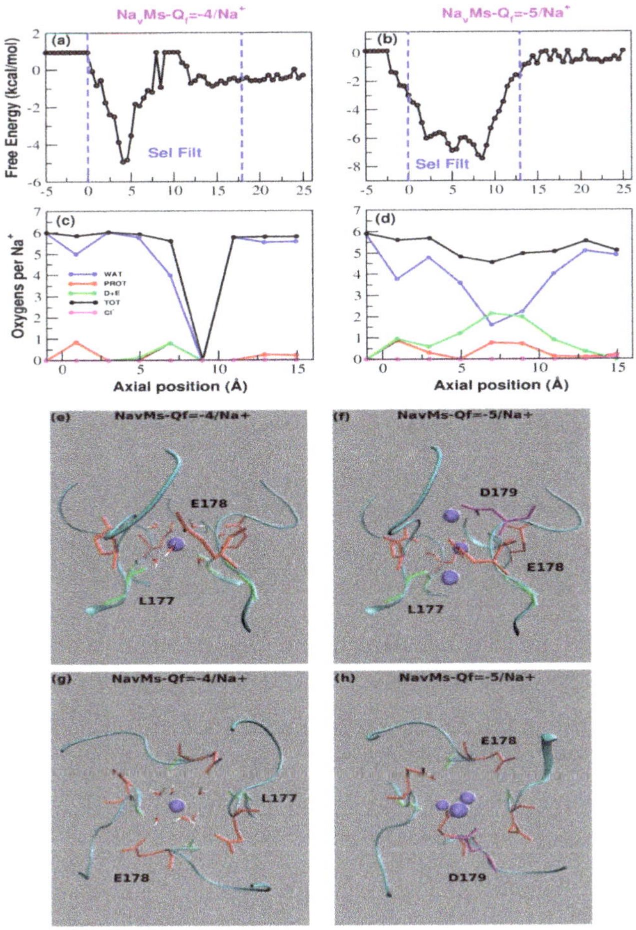

Figure 3. MD simulations for the WT and mutant NavMS in NaCl 140 mM. (**a**,**b**) Potential of Mean Force of Na^+ as a function of the axial position in WT NavMS (**a**) and the mutant with charge $Q_f = -5e$ (**b**). (**c**,**d**) Average number of coordinating oxygens per sodium ion in axial bins with a thickness of 2.0 Å. (**c**) Wild-type NavMs; (**d**) NavMs mutant with $Q_f = -5e$. The distance cutoff to identify sodium-chloride interactions was set to 3.5 Å, and for sodium-oxygen to 3.2 Å. Color code is as follows. Blue line: number of coordinating water-provided oxygens; green line: number of coordinating oxygens provided by aspartate and glutamates; red line; number of coordinating oxygens provided by other protein residues; black line: total number of coordinating oxygens; magenta line: number of coordinating chlorides. (**e**,**h**) Configuration of the selectivity filter of wild-type NavMS (**e**,**g**) and the mutant with charge $Q_f = -5e$ (**f**,**h**). All the structures correspond to the last frame of a 100 ns simulation in 0.14 M NaCl. Panels (**e**,**f**) show a side view of the SF; panels (**g**,**h**) show the top view. Glu178 is shown in red, while Asp179 is shown in purple. The backbone of Leu177 is shown in green. Sodium ions are portrayed as blue beads. Panels (**e**,**g**) also show the water molecules that mediate the interactions between the resident sodium ion and the protein in wild-type NavMS.

The position of the ion in the SF revealed by the Potential of Mean Force (PMF) shows that Ca^{2+} ions do visit the vestibule region of the WT channel, but they never enter into the SF (Figure 4a). The presence of an additional negative charge in the SF (S179D) is sufficient to pull in a Ca^{2+} ion that occupies a binding site centered at z = 6.5 Å (Figure 4b). The PMF minimum corresponding to this binding site has a depth of approximately 7.0 kcal/mol, which, at the simulation temperature of 300 K, corresponds to 11.5 $k_B T$. The energy well is thus so deep that a single Ca^{2+} cannot leave the SF. Thus, to be consistent with the experimental recording of Ca^{2+} current (Figure 2e,g), calcium permeation must involve some sort of knock-on mechanism. The role of the aspartate residue in the SF is immediately highlighted by Figure 4e–h, which shows the configuration of the SF in the last frame of the simulation. The resident calcium ion appears to be directly bonded to D179 and to E178, both located on chain A (Figure 4). The interactions with the other glutamates of the SF are all water-mediated. In order to better characterize calcium hydration, in Figure 4c,d we plot the average number of coordinating oxygen atoms per calcium ion in axial bins with a thickness of 2.0 Å. Figure 4 shows that when a calcium ion enters into the SF, the number of hydrating water molecules drops from approximately 8 to 4.5. This dehydration is compensated by an increase in the number of coordinating oxygens provided by aspartate and glutamate residues (approximately 3). Thus, when Ca^{2+} enters the SF, the total number of coordinating oxygens remains roughly unchanged (Figure 4c,d).

A collective diffusion model approach was adopted to approximate the Ca^{2+} currents [46]. The algorithm relates the spontaneous permeation events at equilibrium with steady currents induced by small voltages. This approach thus enables the estimation of currents from equilibrium simulations; however, as it is based on linear response theory, its predictions are reliable only in a small voltage range. The results of the calculation are summarized in Table 2.

Table 2. Current estimates through linear response theory. The first column shows the NavMs species analyzed, either the wild-type form EEEE with an SF charge $Q_f = -4e$ or the mutant EEEED with an additional negative charge in the SF ($Q_f = -5e$). The second column shows the ion carrying the current, the third column reports the estimated conductance in pS, and the fourth column lists the estimated current at V = −20 mV. This voltage corresponds to the peak current in the current–voltage plots determined from whole-cell patch-clamp experiments.

Species	Ion	Conductance (pS)	Currents (pA; −20 mV)
EEEE	Ca^{2+}	1.69	−0.033
EEEED	Ca^{2+}	4.87	−0.097
EEEE	Na^+	23.06	−0.46
EEEED	Na^+	35.37	−0.70

Notwithstanding the limitations of our calculations, the collective diffusion modelling predictions are in reasonable agreement with the experimental observations (Figure 2). For example, (1) whole-cell recordings showed that peak sodium currents increased by approximately 1.5-fold in the $Q_f = -5e$ NavMs channel (−35 to −55 pA/pF); this is mirrored by a 1.5-fold increase in sodium conductance (from 23.06 to 35.37 pS) predicted by linear response theory calculations. (2) Experimental measurements of peak calcium currents in the $Q_f = -5e$ mutant are approximately 10 times smaller than those for sodium. This is consistent with the modelling in the $Q_f = -5e$ mutant, in which a seven-fold greater sodium (35.37 pS) conductance is predicted compared to that for calcium (4.87 pS). (3) The small finite Ca^{2+} influx predicted in the WT NavMs (Table 2; 1.69 pS) can be observed in the electrophysiological recordings (Figure 2e).

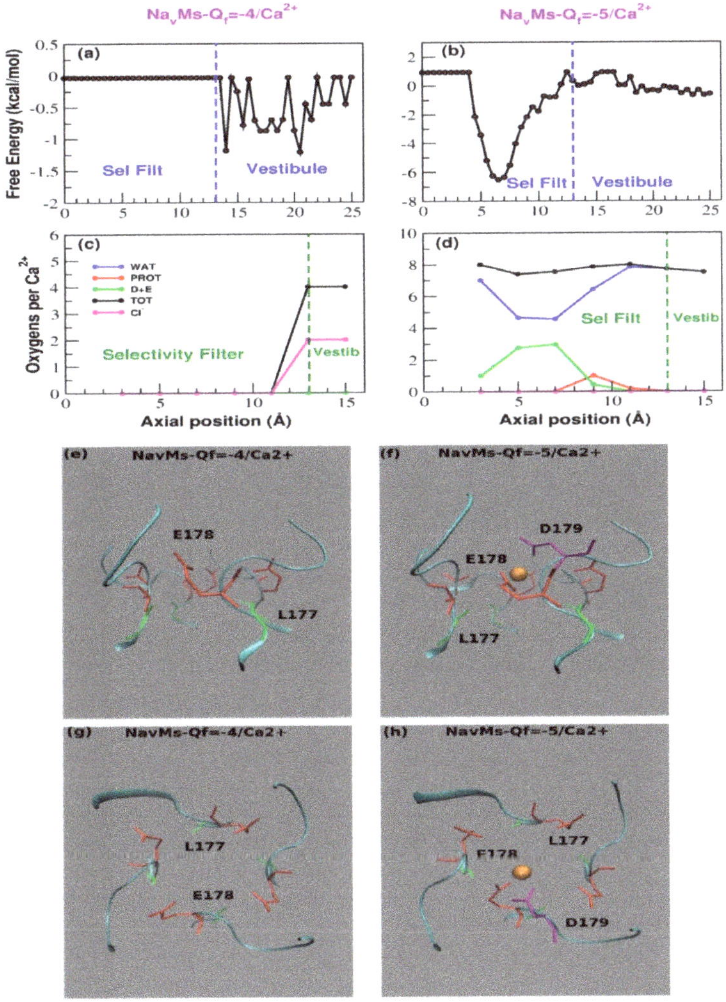

Figure 4. MD simulations for WT and mutant NavMs in $CaCl_2$ 100 mM. (**a**,**b**) Potential of Mean force of Ca^{2+} as a function of the axial position in WT NavMS (**a**) and the mutant with charge $Q_f = -5e$ (**b**). (**c**,**d**) Average number of coordinating oxygens per calcium ion in axial bins with a thickness of 2.0 Å. (**c**) Wild-type NavMs; (**d**) NavMs mutant with $Q_f = -5e$. The distance cutoff to identify both calcium-chloride and calcium-oxygen interactions was set to 3.5 Å. Color code is as follows. Blue line: number of coordinating water oxygens; green line: number of coordinating oxygens provided by aspartate and glutamates; red line; number of coordinating oxygens provided by other protein residues; black line: total number of coordinating oxygens; magenta line: number of coordinating chlorides. (**e**–**h**) Configuration of the selectivity filter of wild-type NavMS (**e**,**g**) and the mutant with charge $Q_f = -5e$ (**f**,**h**). All the structures correspond to the last frame of a 150 ns simulation in 0.10 M of $CaCl_2$. Panels (**e**,**f**) show a side view of the SF; panels (**g**,**h**) show the top view. Glu178 is shown in red, while Asp179 is shown in purple. The backbone of Leu177 is shown in green. Calcium ions are portrayed as orange beads.

4. Discussion and Conclusions

In this work, we engineered radial asymmetry in the bacterial NaChBac and NavMs channels as a first attempt to mimic the features of eukaryotic voltage-gated sodium and calcium channels. It is well known that prokaryotic sodium channels are characterized by a glutamate ring that imparts a charge $-4e$ to the SF and endows the channel with Na^+ selectivity. It is also well established that an increase in the negative charge of the SF makes the channel progressively more calcium-selective. Pioneering studies by the Clapham group, for instance, showed that mutating into aspartate either serine of the SF sequence TLESWAS of NaChBac decreases the P_{Na}/P_{Ca} ratio, while a mutation of both serines makes the channel completely calcium-selective [20]. Using the same strategy, more recently Tang et al. replaced the TLESWSM sequence in the SF of NavAb with TLDDWSD, causing a complete shift from sodium to calcium selectivity [21]. It is noteworthy that, due to the tetrameric symmetry of prokaryotic NaVs, in all these studies the charge of the SF was always varied in $-4e$ steps and radial symmetry was maintained. It is thus known that a charge $Q_f = -4e$ is typical of a Na+-selective channel, while a charge $-8e$ or $-12e$ leads to calcium selectivity. This change in the Q_f value is rather coarse and does not address the fact that the SF of eukaryotic channels is asymmetric. Therefore, it is important to investigate the influence on the selectivity of charge changes by $-1e$ steps.

The study of random heteroteramers in our work indicated that channels with an SF charge smaller than $-4e$ mediate Na^+ currents and channels with an SF charge in the $-4e < Q_f < -8e$ range conduct Ca^{2+}. Furthermore, the study of the NavMs concatemer showed that the presence of an additional negative charge in the SF leads to a significant increase in the Na^+ and Ca^{2+} current.

The electrophysiological behavior of the NaChBac populations of randomly assembled heterotetramers appears to be in reasonable agreement with the predictions of the Ionic Coulomb Blockade (ICB) model [23–25]. According to this model, ion permeation and selectivity through channels mainly depend on the Q_f of the SF. If calcium permeation is plotted as a function of Q_f, a pattern of alternating conductance and stop bands can be observed. In contrast, the same plot for sodium predicts a steep increase in current magnitude up to values in Q_f of $<-2e$, followed by a plateau and the absence of stop bands (Figures 2 and 3, in [25]), consistent with sodium permeation being relatively insensitive to changes in Q_f. Thus, the predictions of the ICB model appear to be compatible with the plot of peak sodium currents in Figure 1c. Furthermore, it is tempting to envisage the pattern of Ca^{2+} current density shown in Figure 1d as an oscillation in the calcium conductance (i.e., conductance and stop bands), which would repeat over a wider range of Q_f values. The ICB model, however, appears to be less successful in explaining the so-called "EEEE paradox"—that is, the apparently shared "EEEE" motif in both the sodium-selective bacterial NaVs and the calcium-selective eukaryotic CaVs. Kaufman et al. tentatively reconciled this inconsistency, noting the presence of a conserved aspartate close to the EEEE ring of eukaryotic CaVs, thus redefining the motif as EEED, which raises the SF charge to $-5e$ [36]. Our experiments on the NavMs concatemer go some way towards confirming this prediction, but also highlight that other factors in addition to the value of Q_f are important. The ICB model predicts that a charge $-5e$ allows the access of a third Ca^{2+} ion when the SF is already occupied by two resident calcium ions. Our MD simulations, however, show that while no calcium ion gains access to the SF of WT NavMs, only a single Ca^{2+} ion stably occupies the filter of the mutant with charge $-5e$. This calcium ion is strongly bound to D179 and E178 located on the same subunit, and sits in a free energy well so deep that it cannot leave the SF. At the same time, the resident ion probably exerts an electrostatic repulsion on other potentially incoming Ca^{2+} ions, preventing a knock-on mechanism in a similar fashion as that described for NaChBac [49].

Our experiments and simulations thus suggest that the extra negative charge is effective in the capture of cations from the bulk, but it does not promote permeation. Contrary to that postulated by simplified physical models (in which the channel atomic structure is not considered), the charge of the SF is not the only determinant of conduction and selectivity. It is possible that calcium flow in eukaryotic Cavs requires some sort of fine modulation of charge effects. Flood et al., for instance, performed an interesting computational study grafting the SF and external vestibular region of the human NaV1.2 channel into the scaffold of the NavRh bacterial channel [35]. Their multi-microsecond MD simulations revealed that permeation and selectivity depend on the close interplay of the DEKA and EEDD rings, so that the charge of the extended filter region is $-5e$ as in our NavMs mutant. In its protonated state, the lysine residue of the DEKA ring acts like a built-in sodium ion involved in the formation of multi-carboxylates/multi-ion complexes. When the charged ammonium group of lysine is in the HFS site, where the electrostatic potential is most negative, it creates a smooth electrostatic environment leading into the cavity, whereas, when it is bent toward the central cavity, it creates a zone of high electrostatic potential that cuts the cavity off from the SF. Our recent work [38], showing the possibility to create stable concatemers of the bacterial NavMs channel, offers the opportunity to experimentally test these computational predictions by creating a bacterial channel chimera where the SF and vestibule of the human Nav1.2 channel are grafted onto the NavMs concatemer.

Since no positively charged residue appears to be located close to the SF of eukaryotic CaVs, the fine modulation of the charge might rely on the differential protonation of the acidic residues of the EEEED locus. The effect of protonation has been extensively studied through MD simulations. Furini et al., for instance, showed that the glutamate side chains in NavAb can adopt two different orientations pointing either towards the extracellular environment or towards the central cavity [34]. Interestingly, they found that the likelihood of the inwardly directed arrangement increases when E177 residues are protonated. Moreover, the presence of a glutamate residue with the side chain directed to the central cavity increases the energy barrier for the translocation of sodium ions. Since E177 was observed to adopt an alternative conformation in MD simulations with Ca^{2+} ions [50], it is possible that these protonation-induced configurations also affect selectivity. While the control of the protonation state of the filter is a trivial task in MD simulations, it is a challenging endeavor in biophysical experiments.

This leads us to the methodological aspect of our work. Our study not only tested the importance of SF charge in controlling ion selectivity and permeation, but created new tools extending the use of bacterial channels as models of eukaryotic ones. Indeed, the current work is the first one to report experiments on a NaV channel in which the pore region has been mutated to have radial asymmetry, and thus it represents an important first step in bridging the major limitation in using bacterial sodium channels to investigate their eukaryotic counterparts. Our methodology will enable us to design physical experiments to investigate the mechanisms of the fine modulation of charge effects that are likely to occur in asymmetric eukaryotic channels, such as that predicted by Flood et al. [35].

A further methodological merit of our approach is its relevance in understanding the effect of pH on channel permeation and selectivity. In fact, when the pH is varied, the four glutamates of the SF are unlikely to be protonated or deprotonated simultaneously. A more probable scenario is that they are protonated or deprotonated one at a time, resulting in $+1e$ or $-1e$ changes in the SF charge [23]. Finally, our combination of molecular dynamics and electrophysiological approaches provided fresh insight into the molecular mechanisms of cation permeation in bacterial sodium channels, and gave insight into understanding the molecular mechanisms that underlie the function of NaVs and CaVs.

Supplementary Materials: The following are available online at http://www.mdpi.com/1099-4300/22/12/1390/s1: Supplementary Table S1 Parameters of NavMs equilibration. Supplementary Table S2. Peak currents ratios of WT and mutant NavMs channels. Supplementary Table S3. Peak currents ratios of NaChBac heterotetramers. Supplementary Figure S1. The tetramer formation via co-transection is proved to be a random process without bias for homo- or hetero-tetramer formation. Supplementary Figure S2. The single channel currents recorded from WT NaChBac. Supplementary Figure S3. Schematic representation of NavMs concatemer. Supplementary Figure S4. Ion occupancy of Selectivity Filter in MD simulations of WT NavMs and its mutant with $Q_f = -5e$ in NaCl 140 mM. Supplementary Figure S5. Ion occupancy in Selectivity Filter in MD simulations of WT NavMs and its mutant with $Q_f = -5e$ in $CaCl_2$ 100 mM. Supplementary Figure S6. Current-voltage plot calculations: comparison of constant electric field simulations and equilibrium simulations in conjunction with linear response theory.

Data Availability: Data related to this research are openly available from the University of Warwick archive at (https://wrap.warwick.ac.uk/143573). Fedorenko, Olena A., Khovanov, Igor A., Roberts, Stephen K., and Guardiani, Carlo (2020) Data for *Changes in ion selectivity following asymmetrical addition of charge to the selectivity filter of bacterial sodium channels* [Dataset].

Author Contributions: Conceptualization, O.A.F., C.G., S.K.R., and I.A.K.; methodology, O.A.F. and C.G.; formal analysis, O.A.F. and C.G.; investigation, O.A.F. and C.G.; writing—original draft preparation, O.A.F. and C.G.; writing—review and editing, S.K.R. and I.A.K. All authors have read and agreed to the published version of the manuscript.

Funding: This research was funded by EPSRC; project: Ionic Coulomb blockade oscillations and the physical origins of permeation, selectivity, and their mutation transformations in biological ion channels (grant numbers: EP/M015831/1 and EP/M016889/1). CG is currently supported by a project that has received funding from the European Research Council (ERC) under the European Union's Horizon 2020 research and innovation programme (grant agreement No. 803213).

Acknowledgments: We are grateful to Huaping Sun for the generation of the $-5e$ NavMs constructs.

Conflicts of Interest: The authors declare no conflict of interest.

References

1. Catterall, W.A. Forty years of sodium channels: Structure, function, pharmacology, and epilepsy. *Neurochem. Res.* **2017**, *42*, 2495–2504.
2. Flucher, B.E. Skeletal muscle CaV1.1 channelopathies. *Pflug. Arch. Eur. J. Physiool.* **2020**, *472*, 739–754.
3. Zhang, Q.; Chen, J.; Qin, Y.; Wang, J.; Zhou, L. Mutations in voltage-gated L-type calcium channel: Implications in cardiac arrhythmia. *Channels* **2018**, *12*, 201–218.
4. Oyrer, J.; Maljevic, S.; Scheffer, I.E.; Berkovic, S.F.; Petrou, S.; Reid, C.A. Ion channels in genetic epilepsy: From genes and mechanisms to disease-targeted therapies. *Pharmacol. Rev.* **2018**, *70*, 142–173.
5. Catterall, W.A. Structure and regulation of voltage-gated Ca^{2+} channels. *Annu. Rev. Cell. Dev. Biol.* **2000**, *16*, 521–555.
6. Tyson, J.R.; Snutch, T.P. Molecular nature of voltage-gated calcium channels: Structure and species comparison. *Wiley Interdiscrip. Rev. Membr. Transp. Signal.* **2013**, *2*, 181–206.
7. Catterall, W.A.; Swanson, T.M. Structural basis for pharmacology of voltage-gated sodium and calcium channels. *Mol. Pharmacol.* **2015**, *88*, 141–150.
8. Zamponi, G.W.; Striessnig, J.; Koschak, A.; Dolphin, A.C. The physiology, pathology, and pharmacology of voltage-gated calcium channels and their future therapeutic potential. *Pharmacol. Rev.* **2015**, *67*, 821–870.
9. Xu, L.; Ding, X.; Wang, T.; Mou, S.; Sun, H.; Hou, T. Voltage-gated sodium channels: Structures, functions, and molecular modeling. *Drug Discov. Today* **2019**, *24*, 1389–1397.
10. Wu, J.; Yan, Z.; Li, Z.; Qian, X.; Lu, S.; Dong, M.; Zhou, Q.; Yan, M. Structure of the voltage-gated calcium channel Cav1.1 at 3.6 Å resolution. *Nature* **2016**, *537*, 191–196.
11. Carpenter, E.P.; Beis, K.; Cameron, A.D.; Iwata, S. Overcoming the challenges of membrane protein crystallography. *Curr. Opin. Struct. Biol.* **2008**, *18*, 581–586.
12. Shen, H.; Zhou, Q.; Pan, X.; Li, Z.; Wu, J.; Yan, N. Structure of a eukaryotic voltage-gated sodium channel at near-atomic resolution. *Science* **2017**, *355*, 4326.
13. Ren, D.; Navarro, B.; Xu, H.; Yue, L.; Shi, Q.; Clapham, D.E. A prokaryotic voltage-gated sodium channel. *Science* **2001**, *294*, 2372–2375.
14. Zhang, X.; Ren, W.; DeCaen, P.; Yan, C.; Tao, X.; Tang, L.; Wang, J.; Hasegawa, K.; Kumasaka, T.; He, J.; et al. Crystal structure of an orthologue of the NaChBac voltage-gated sodium channel. *Nature* **2012**, *486*, 130–134.
15. Finol-Urdaneta, R.K.; Wang, Y.; Al-Sabi, A.; Zhao, C.; Noskov, S.Y.; French, R.J. Sodium channel selectivity and conduction: Prokaryotes have devised their own molecular strategy. *J. Gen. Physiol.* **2014**, *143*, 157–171.
16. Catterall, W.A.; Zheng, N. Deciphering voltage-gated Na(+) and Ca(2+) channels by studying prokaryotic ancestors. *Trends Biochem. Sci.* **2015**, *40*, 526–534.
17. Naylor, C.E.; Bagnéris, C.; DeCaen, P.G.; Sula, A.; Scaglione, A.; Clapham, D.E.; Wallace, B.A. Molecular basis of ion permeability in a voltage-gated sodium channel. *EMBO J.* **2016**, *35*, 820–830.
18. Bagneris, C.; DeCaen, P.G.; Naylor, C.E.; Pryde, D.C.; Nobeli, I.; Clapham, D.E.; Wallace, B.A. Prokaryotic NavMs channel as a structural and functional model for eukaryotic sodium channel antagonism. *Proc. Natl. Acad. Sci. USA* **2014**, *111*, 8428–8433.
19. Guardiani, C.; Rodger, P.M.; Fedorenko, O.A.; Roberts, S.K.; Khovanov, I.A. Sodium binding sites and permeation mechanism in the NaChBac channel: A molecular dynamics study. *J. Chem. Theory Comput.* **2017**, *13*, 1389–1400.

20. Yue, L.; Navarro, B.; Ren, D.; Ramos, A.; Clapham, D.E. The cation selectivity filter of the bacterial sodium channel, NaChBac. *J. Gen. Physiol.* **2002**, *120*, 845–853.
21. Tang, L.; Gamal El-Din, T.M.; Payandeh, J.; Martinez, G.Q.; Heard, T.M.; Scheuer, T.; Zheng, N.; Catterall, W.A. Structural basis for Ca^{2+} selectivity of a voltage-gated calcium channel. *Nature* **2014**, *505*, 56–61.
22. Guardiani, C.; Fedorenko, O.A.; Khovanov, I.A.; Roberts, S.K. Different roles for aspartates and glutamates for cation permeation in bacterial sodium channels. *Biochim. Biophys. Acta (BBA)-Biomembr.* **2019**, *1861*, 495–503.
23. Fedorenko, O.A.; Kaufman, I.K.; Gibby, W.A.T.; Barabash, M.L.; Luchinsky, D.G.; Roberts, S.K.; McClintock, P.V.E. Ionic coulomb blockade and the determinants of selectivity in the NaChBac bacterial sodium channel. *Biochim. Biophys. Acta (BBA)-Biomembr.* **2020**, *1862*, 183301.
24. Kaufman, I.K.; McClintock, P.V.E.; Eisenberg, R.S. Coulomb blockade model of permeation and selectivity in biological ion channels. *New J. Phys.* **2015**, *17*, 083021.
25. Kaufman, I.K.; Fedorenko, O.A.; Luchinsky, D.G.; Gibby, W.A.T.; Roberts, S.K.; McClintock, P.V.E.; Eisenberg, R.S. Ionic Coulomb blockade and anomalous mole fraction effect in the NaChBac bacterial ion channel and its charge-varied mutants. *Nonlinear Biomed. Phys.* **2017**, *5*, 4–12.
26. Payandeh, J.; Scheuer, T.; Zheng, N.; Catterall, W.A. The crystal structure of a voltage-gated sodium channel. *Nature* **2011**, *475*, 353–358.
27. Payandeh, J.; Gamal El-Din, T.M.; Scheuer, T.; Zheng, N.; Catterall, W.A. Crystal structure of a voltage-gated sodium channel in two potentially inactivated states. *Nature* **2012**, *486*, 135–139.
28. Bagnéris, C.; Decaen, P.G.; Hall, B.A.; Naylor, C.E.; Clapham, D.E.; Kay, C.W.; Wallace, B.A. Role of the C-terminal domain in the structure and function of tetrameric sodium channels. *Nat. Commun.* **2013**, *4*, 2465.
29. Shaya, D.; Findeisen, F.; Abderemane-Ali, F.; Arrigoni, C.; Wong, S.; Nurva, S.R.; Loussouarn, G.; Minor, D.L., Jr. Structure of a prokaryotic sodium channel pore reveals essential gating elements and an outer ion binding site common to eukaryotic channels. *J. Mol. Biol.* **2014**, *426*, 467–483.
30. Bagnéris, C.; Naylor, C.E.; McCusker, E.C.; Wallace, B.A. Structural model of the open-closed-inactivated cycle of prokaryotic voltage-gated sodium channels. *J. Gen. Physiol.* **2015**, *145*, 5–16.
31. McCusker, E.C.; Bagnéris, C.; Naylor, C.E.; Cole, A.R.; D'Avanzo, N.; Nichols, C.G.; Wallace, B.A. Structure of a bacterial voltage-gated sodium channel pore reveals mechanisms of opening and closing. *Nat. Commun.* **2012**, *3*, 1102.
32. Dudev, T.; Lim, C. Why voltage-gated Ca^{2+} and bacterial Na^{+} channels with the same EEEE motif in their selectivity filters confer opposite metal selectivity. *Phys. Chem. Chem. Phys.* **2012**, *14*, 12451–12456.
33. Furini, S.; Barbini, P.; Domene, C. Effects of the protonation state of the EEEE motif of a bacterial Na+ channel on conduction and pore structure. *Biophys. J.* **2014**, *106*, 2175–2183.
34. Boiteux, C.; Flood, E.; Allen, T.W. Comparison of permeation mechanisms in sodium-selective ion channels. *Neurosci. Lett.* **2019**, *700*, 3–8.
35. Flood, E.; Boiteux, C.; Allen, T.W. Selective ion permeation involves complexation with carboxylates and lysine in a model human sodium channel. *PLoS Comput. Biol.* **2018**, *14*, e1006398.
36. Kaufman, I.K.; Luchinsky, D.G.; Gibby, W.A.T.; McClintock, P.V.E.; Eisenberg, R.S. Putative resolution of the EEEE selectivity paradox in L-type Ca^{2+} and bacterial Na^{+} biological ion channels. *J. Stat. Mech.* **2016**, 054027.
37. Cheng, R.C.; Tikhonov, D.B.; Zhorov, B.S. Structural modeling of calcium binding in the selectivity filter of the L-type calcium channel. *Eur. Biophys. J.* **2010**, *39*, 839853.
38. Sun, H.; Zheng, Z.; Fedorenko, O.A.; Roberts, S.K. Covalent linkage of bacterial voltage-gated sodium channels. *BMC Biophysics.* **2019**, *12*, 1.
39. Neher, E. Correction for liquid junction potentials in patch clamp experiments. *Methods Enzymol.* **1992**, *207*, 123–131.
40. Jo, S.; Kim, T.; Iyer, V.G.; Im, W. CHARMM-GUI: A web-based graphical user interface for CHARMM. *J. Comput. Chem.* **2008**, *29*, 1859–1865.
41. Wu, E.L.; Cheng, X.; Jo, S.; Rui, H.; Song, K.C.; Dávila-Contreras, E.M.; Qi, Y.; Lee, J.; Monje-Galvan, V.; Venable, R.M.; et al. CHARMM-GUI membrane builder toward realistic biological membrane simulations. *J. Comput. Chem.* **2014**, *35*, 1997–2004.
42. Phillips, J.C.; Braun, R.; Wang, W.; Gumbart, J.; Tajkhorshid, E.; Villa, E.; Chipot, C.; Skeel, R.D.; Kale, L.; Schulten, K. Scalable molecular dynamics with NAMD. *J. Comput. Chem.* **2005**, *26*, 1781–1802.

43. Maier, J.A.; Martinez, C.; Kasavajhala, K.; Wickstrom, L.; Hauser, K.E.; Simmerling, C. ff14SB: Improving the accuracy of protein side chain and backbone parameters from ff99SB. *J. Chem. Theory Comput.* **2015**, *11*, 3696–3713.
44. Dickson, C.J.; Madej, B.D.; Skjevik, A.A.; Betz, R.M.; Teigen, K.; Gould, I.R.; Walker, R.C. Lipid14: The amber lipid force field. *J. Chem. Theory Comput.* **2014**, *10*, 865–879.
45. Ulmschneider, M.B.; Bagneris, C.; McCusker, E.C.; DeCaen, P.G.; Delling, M.; Clapham, D.E.; Ulmschneider, J.P.; Wallace, B.A. Molecular dynamics of ion transport through the open conformation of a bacterial voltage-gated sodium channel. *Proc. Natl. Acad. Sci. USA* **2013**, *110*, 6364–6369.
46. Liu, Y.; Zhu, F. Collective diffusion model for ion conduction through microscopic channels. *Biophys. J.* **2013**, *104*, 368–376.
47. Zhao, Y.; Scheuer, T.; Catterall, W.A. Reversed voltage-dependent gating of a bacterial sodium channel with proline substitutions in the S6 transmembrane segment. *Proc. Natl. Acad. Sci. USA* **2004**, *101*, 17873–17878.
48. Zhao, Y.; Yarov-Yarovoy, V.; Scheuer, T.; Catterall, W.A. A gating hinge in sodium channels; a molecular switch for electrical signaling. *Neuron* **2004**, *41*, 859–865.
49. Guardiani, C.; Fedorenko, O.A.; Roberts, S.K.; Khovanov, I.A. On the selectivity of the NaChBac channel: An integrated computational and experimental analysis of Na^+ and Ca^{2+} permeation. *Phys. Chem. Chem. Phys.* **2017**, *19*, 29840–29854.
50. Ke, S.; Zangerl, E.M.; Stary-Weinzinger, A. Distinct interactions of Na^+ and Ca^{2+} ions with the selectivity filter of the bacterial sodium channel NavAb. *Biochem. Biophys. Res. Commun.* **2013**, *430*, 1272–1276.

entropy

MDPI

Perspective

Prospects of Observing Ionic Coulomb Blockade in Artificial Ion Confinements

Andrey Chernev, Sanjin Marion and Aleksandra Radenovic *

Laboratory of Nanoscale Biology, Institute of Bioengineering, School of Engineering, EPFL, 1015 Lausanne, Switzerland; andrey.chernev@epfl.ch (A.C.); sanjin.marion@epfl.ch (S.M.)

* Correspondence: aleksandra.radenovic@epfl.ch

Received: 9 November 2020; Accepted: 11 December 2020; Published: 18 December 2020

Abstract: Nanofluidics encompasses a wide range of advanced approaches to study charge and mass transport at the nanoscale. Modern technologies allow us to develop and improve artificial nanofluidic platforms that confine ions in a way similar to single-ion channels in living cells. Therefore, nanofluidic platforms show great potential to act as a test field for theoretical models. This review aims to highlight ionic Coulomb blockade (ICB)—an effect that is proposed to be the key player of ion channel selectivity, which is based upon electrostatic exclusion limiting ion transport. Thus, in this perspective, we focus on the most promising approaches that have been reported on the subject. We consider ion confinements of various dimensionalities and highlight the most recent advancements in the field. Furthermore, we concentrate on the most critical obstacles associated with these studies and suggest possible solutions to advance the field further.

Keywords: nanofluidics; ionic Coulomb blockade; 2D materials; nanopores; nanotubes; angstrom slits

---

1. Introduction

In the past fifteen years, various artificial nanofluidic platforms have become highly compelling for fundamental studies of physical phenomena and numerous practical applications where the transport of the confined ions plays a crucial role [1,2]. Among the most exciting practical applications are power generation [3–7], filtration, and molecular separation [8–10]. The last five years have witnessed remarkable progress in the fabrication of nanofluidic devices, enabling researchers to develop artificial nanofluidic systems with the confinement of one to a few water molecules (below 1 nm) [11]. Such nanofluidic platforms have been realized in zero-dimensional (0D), 1D, or 2D geometry [7,12–14] (Figure 1). These platforms exhibit giant permeability and ion selectivity comparable to biological ion channels, excluding anions and macromolecules, and closely mimic functionalities previously observed only in biological channels. Superb selectivity of the sodium/calcium family of channels has fascinated scientists due to its physiological relevance and underlying physical mechanism [15–17]. A simplified electrostatic and Brownian dynamics model of the prototypical model Ca^{2+} or Na^{+} channel has been used in describing its conduction and selectivity [18], echoing the phenomenology of Coulomb blockade [17]. The term ionic Coulomb blockade (ICB) was suggested first as a counterpart of the electronic Coulomb blockade (ECB) by Krems and Di Ventra [19]. Confinement below 1 nm dictates a departure from the mean-field assumptions as the correlations between the ions and finite-size effects cannot be neglected and complicates the insight into the ionic charge transfer at the nanoscale [17,20,21].

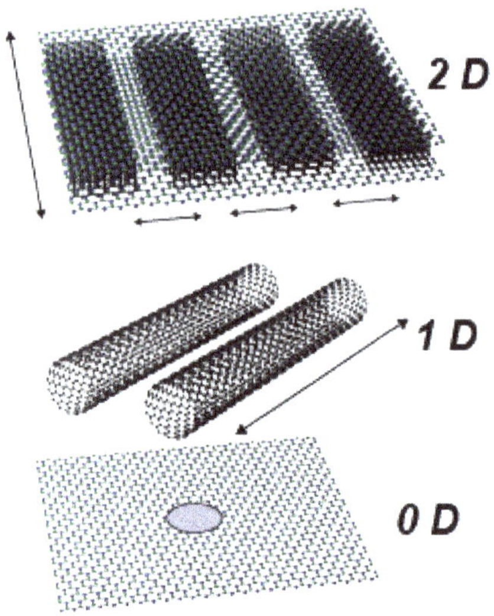

Figure 1. Different geometries are considered for confined nanofluidic systems from two-dimensional (2D) nanoslits to 1D nanowires and quantum dot-like nanopores in the atomically thin material matrix, i.e., 2D nanopores.

Artificial nanofluidic devices developed for single-ion conductivity measurements provide a unique opportunity to test the proposed models and reveal the many-body effects in ionic systems. This perspective article aims to discuss practical challenges in verifying the models that evoke ionic Coulomb blockade in a variety of settings that are now available due to recent progress in nanofabrication [12,22,23]. Furthermore, it also provides suggestions for the integration of additional approaches such as modulation of the charges, pressure, pH, ionic strength, temperature, and potential in the range not attainable on lipid bilayers.

As suggested initially, ionic Coulomb blockade is based on the relation of ion kinetic and barrier energies and manifests as the nonlinear transport of ions. This behavior was later experimentally observed in atomically thin, sub-nanometer-sized molybdenum disulfide (MoS_2) nanopores [13]. Although theoretical models predict ICB and suggest particular conditions to reveal the effect [17,20,24], irrefutable experimental observation of this phenomenon is still challenging. The reason for that is a whole set of issues that have to be taken into account before considering this complex process. The most critical issues that could mask ICB observation are related to the fact that ionic transport is measured at room temperatures, causing large charge fluctuations, which leads to the increased noise, and instability of the nanofluidic devices. Together with regular wetting and contamination challenges in nanofluidics [25], we can argue that the ionic systems' nonlinear current-voltage characteristics are insufficient to prove the ICB effect. Unlike in the case of ECB [26], there are still no convincing data showing conductance oscillations or single-ion devices.

Therefore, it is essential to critically assess the suitability of different nanofluidic platforms for the observation of ICB and propose novel solutions that will allow us to untangle the role of the issues mentioned above. New geometries like nanopores in 2D materials, single-digit nanotubes and angstrom slits play a key role and are essential in verifying the different parameters required to observe ICB reproducibly. The complex nature of ionic transport can be considered with different settings and thus allow direct observation of experimental parameter variation such as pore length, diameter, access resistance, surface charge, gating voltage temperature, and many others [27].

2. Prerequisites for Ionic Coulomb Blockade

The basic prerequisites for ionic Coulomb blockade are shared with its electronic counterpart [28,29]. The geometrical requirement for ICB to occur is represented by two reservoirs with charge carriers, in our case ions, with a channel in between with a large resistance to ion transport at least in one direction [17,19,20,24]. More specifically, the channel has to provide the type of confinement that allows ions to dwell inside of it as it was suggested originally [15]. Ion transport between the two reservoirs is inhibited due to strong Coulomb interactions and an ion dwelling at or near the channel, so that there is a limit on the possible charge, which can be transferred between the two sides of the chamber [17,19,20,24]. When the energy barrier for ions to enter the channel, and subsequently traverse it, is larger than the system's thermal energy, ionic Coulomb blockade can be observed. We identify two main mechanisms that are expected to result in ionic Coulomb blockade signatures in nanopore or nanochannel systems. One is linked to capacitive charging [19,30] and analogous to the electronic case of a tunnel junction. The second mechanism is linked to the existence of an "island" corresponding to a quantum dot in ECB, which can be gated [15,16,20] and can be further developed towards the analogy of the single-electron transistor [20].

The first case, which we call the capacitive ICB, is relatively similar to the ECB case with a single tunneling junction. It occurs if an ion transits between the two reservoirs and charges it analogously to a capacitor, thus producing a barrier for further ions to transit [19,30].

This case requires that the channel/pore be ion-selective so that a capacitive barrier can form for an ion of valence z, that the thermal energy is lower than the capacitive self-energy of the channel $U = Q^2/2C_s$, and that the transferred charge $Q = nze$ dwells inside the pore or next to the pore entrance for a sufficiently long time for it to interact with the transport of further ions. We note that the ICB effect is expected to be stronger for ions carrying more than one unit of charge, as the self-energy has a quadratic dependence on the valence z. The capacity of the channel/pore of length L and radius R to store charge is the self-capacitance C_s and can, neglecting any fine effects, be approximated by $C_s = 4\pi\epsilon_0\epsilon_r R^2/L$, with ϵ_r the relative dielectric constant of water. In general, the capacitive self-energy is a good measure if ICB is detectable as it gives information on the resulting energy barriers for ion transport.

Coarse-grained molecular dynamics simulations [19,30] indicate capacitive ICB could occur in sub-nm pores in quasi-2D or 2D materials but also predict a weak nonlinear current-to-voltage dependence uncharacteristic for the classical electronic Coulomb blockade. In contrast, Brownian dynamics simulations show that the role of screening and ion-pair formation via Bjerrum pairs are paramount in these systems [20]. Here, both parameters are strongly modulated by the dielectric constant of water, which has been shown to be reduced in confinement in respect to its bulk value [31,32], and which would decrease the capacitive self-energy and require lower temperatures or smaller and thinner pores for ICB to be detectable. Moreover, finite-size effects would need to be considered, including the peculiarities of pore-wall interactions, which can strongly influence ion transport in quasi-2D or 2D membranes [33–36].

In the second case, ionic Coulomb blockade happens in the systems that are reminiscent of a quantum dot, confined with two energy barriers—schemes that are used for ECB observations and applications. As for the ICB counterpart, it can be observed in the setting where the ion occupancy of the channel limits ion transport [17,20] and where it has a role analogous to an isolated quantum dot in the electronic case [29]. Here, ions need to have an energetically favorable position inside the channel, which binds them there in spite of thermal motion so that their presence would block other ions traversing the channel. This stable position can be formed due to ion-channel interactions, either by electrostatic gating voltage modulation or by the presence of surface charges [24,37]. In the latter case, the surface charge will then attract oppositely charged ions and thus block ion transport through the channel.

When a fixed voltage bias is applied between the reservoirs, the ion current oscillates with peaks at certain quantized values of these gating surface charges. A typical signature of ICB is when this neutralizing charge has a stepwise dependence on the gating voltage or surface charge, and when the conductance of the channel has peaks at certain values with conductance almost completely suppressed otherwise [17].

Microscopically, there are several distinctions between the ionic and the electronic Coulomb blockades. While electrons are negatively charged by default, ions can be both negative and positive, with different valences available. These ions are also present in a solution so that the electrostatic interaction between them is screened, losing its long-range components. Thus, such an effect will be local and will require extreme confinements for size exclusion and Coulomb interactions to dominate transport properties. In the case of strong 1D confinement, ion transport through such channels is expected to go through the dissociation of pairs of ions (second Wien effect) [38,39].

Nevertheless, in the case of gating via voltage or wall charge inhibiting the movement of ions due to tight binding, some of the ions are tightly bound, and only under certain conditions are they able to dissociate, a situation recently termed as a "fractional Wien effect" [20]. Under this model, it is also predicted that when electrostatic screening becomes too large, the ionic Coulomb blockade effect becomes suppressed. This also provides an upper limit on the size of a confining channel to about 2.5 nm. This means that, in certain geometries, ICB could be detected even in pores larger than one nanometer, depending on the magnitude of electrostatic screening.

Providing unquestionable proof for the ionic Coulomb blockade is not an easy task. Most experimental "wet" implementations cannot easily change the interaction between the ions and the channel interior, i.e., by changing its surface charge or applying a gating voltage to it. One of the major indicators of ICB is nonlinear current-to-voltage curves [17,20,24,35]. As the bias voltage is increased until a certain threshold is crossed, there is no ion transport through the system. After that, a nonlinear increase in current is expected until linear (ohmic) behavior is obtained. In the classical electronic CB, several steps called the Coulomb staircase in the current versus voltage curves are expected in this nonlinear regime [29]. Recent work indicates that, in the case of ICB, there could be no such steps, with a direct transition to an ohmic regime [20]. However, it is problematic to identify ICB only with current-voltage characteristics (IV curves), especially in the absence of a staircase-like pattern. Realistic devices are known to exhibit leakage currents that will exceed the current in the blocked ICB conditions [13], possibly overshadowing any conduction steps.

Furthermore, similar nonlinear IV curves are known to occur during the electrowetting of channels and are caused by remnant gas inside the pore system [25,40–42]. Partial wetting of these systems presents with nonlinear activation-like IV curves, similar to the ones that have been associated with ionic current blockade [13]. There are many open questions regarding our understanding of ICB, which could help to design better experiments. Variations of the ambient temperature could be used to probe the activation energy barriers and possibly promote ICB. Varying the solution's pH could be used to change the gating charges, thus enabling the variation of the gating charge in larger channels. However, the main requirement is still to manufacture a device with such properties to optimize the conditions for the observation of ICB, a topic we tackle in the following sections.

3. ICB in 2D Nanopores

Over the last decade, 2D materials have become a rich area of research and are showing tentative signs of impacting our everyday life [43]. In bulk, these materials have the form of layered crystals, with van der Waals interaction holding together 2D layers with a thickness starting from 0.3 nm. These 2D sheets are under intense study because of their fascinating electronic properties, spanning the range from isolating and semiconducting to superconducting. Nanopores formed in 2D membranes from graphene [44–46], hexagonal boron nitride (h-BN) [47], transition metal dichalcogenides (TMDCs) [4,48], and MXenes [49] have been used to investigate nanofluidic phenomena or can act as a single molecular sensor.

2D nanopores have the peculiarity that due to their thickness, or lack of it, the pore sizes need to be small for the interactions between a single ion and the pore to dominate ion transport. For capacitive ICB to occur, we need ion selectivity, a sufficient capacitive barrier, and ions to remain next to the pore entrance for a sufficient time. 2D nanopores are known for their selectivity to ions, which is expected to come from the surface charges and the resulting ion enhancement zones near the pore [5]. Simple classical estimates of the self-capacitance would support 2D material nanopores having a sufficient energy barrier [13], but there is no evidence of ion retention for sufficiently long times to cause this blockade effect unless the ion is retained inside the pore itself. Classical bulk models for nanopore conductance break down as we approach the nanometer scale [21,33,50], requiring us to include the specifics of the pore atomic structure and its interactions with the ions and water.

Ion transport in 2D systems is a rich topic of research [33,51], but little is known experimentally about the peculiarities of transport in sub-nm pores. When the degree of confinement approaches the ion's size, i.e., sub-nm size ranges, then an effect called hydration layer shredding occurs [19,37,52]. If an ion needs to lose a part of its hydration layer to traverse the pore, it experiences an energy barrier that can also result in a nonlinear IV curve, one of the expected signatures for ICB. Unless there is some sort of weakly bound state for the ion in the middle of the channel, this hydration layer shredding is not enough by itself to induce a Coulomb blockade effect. This can be achieved by tailoring pore interior bond edges, which could temporarily bind an ion inside the pore, causing it to block further ion transport [50]. Pore edge termination is known to influence ion transport [53–55] and could be tailored by binding other chemical species and modifying the pore interior [54–58]. Ion transport through pores is determined by bulk solution and solvent-mediated interactions between the ions and the pore interior edges. This interaction has been shown to cause ions to dwell inside the pore [24,35]; however, it is unclear if this influences ion transport via the Coulomb blockade mechanism.

2D nanopore systems have several unresolved issues that still require further study. Control of the pore shapes and sizes is not easy to achieve as pores are known to be highly sensitive to the fabrication method [47,59] (Figure 2). However, this can be addressed by the predefined pore shapes. Together with electrostatic gating, it is expected to be a highly effective approach for ion selectivity and controlled transport [24]. In this case, the trapped cations themselves not only act as natural clogs of the pores, but also "protect" the clogged pore states by creating the repulsive potential around each pore that essentially suppresses the knock-on mechanism by mobile cations dancing around.

Solvation of pore systems is expected to be able to change the pore structure, and etching of pores and their enlargement due to chemical species or dissolved reactive oxygen. In addition, 2D systems are prone to surface contaminants (e.g., hydrocarbons) that originate from the 2D material transfer process [60]. These contaminants can clog pores or change their properties, making controlled experiments difficult. Additionally, wetting or clogging issues can plague these systems [25] and possibly even produce nonlinear IV curves due to the presence of nanobubbles [40,41,61]. Moreover, the standard method of wetting all nanoscale devices uses prewetting with water-ethanol mixtures, a procedure known to produce nanobubbles on hydrophobic surfaces and which has been linked to the same effect on mildly hydrophobic 2D material surfaces [25]. In addition, remnants of bound ethanol, or other molecules, have been shown to modify surface properties and can persist after subsequent flushing [62]. Differentiating these effects from ICB nonlinear IV curves requires additional probes and separate gating control.

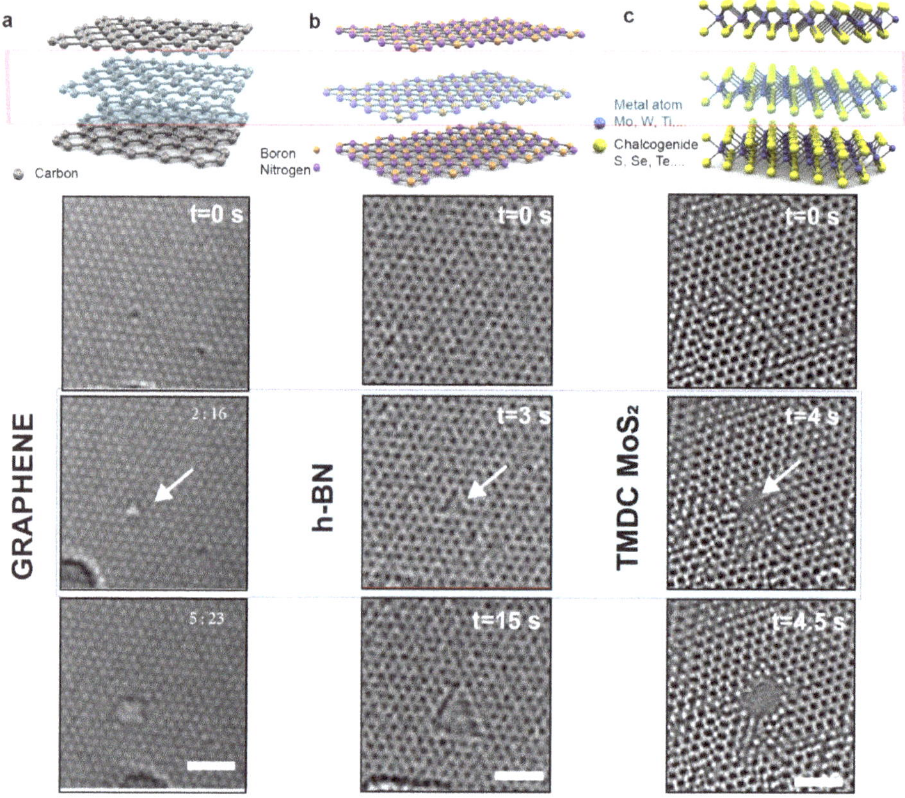

Figure 2. (**a–c**) Selection of the most popular 2D materials, i.e., graphene, hexagonal boron nitride, and molybdenum disulfide structures in bulk and monolayer form and generation and evolution of defects in 2D materials under e-beam. Defects and pores are created under 80 keV e-beam irradiation while imaging with a transmission electron microscope. Images adapted from (**a**) [59] and (**b**,**c**) [48].

Gating can be applied to nanofluidic platforms for modulating ICB in 2D pores (Figure 3). Notably, it can be applied in various ways, and not only limited to electrostatic gating. For example, one approach would be applying mechanical stress directly to nanopores [63] as the strain has been predicted to modulate ion transport through sub-nm 2D nanopores by promoting or removing a stable dwelling spot for ions in the center of the pore [31–33]. Strain moves the pore edge atoms out of their equilibrium positions, changing ion–atom and ion-water interactions, thus modifying the free energy landscape for the ions, possibly promoting or destroying a bound state for the ion inside the pore. Chemical gating of these systems is limited, and the surface charge can be modulated via pH modulation [11] and photo-gating [6], but it is not expected to make significant changes to the interaction of ions with the pore edges and could promote pore growth due to etching. Another useful probe would be studying the effect of temperature variation to provide information about the energy barriers for transport and possibly promote or suppress ICB effects. Although promising, this strategy has a reduced temperature range, in essence, from the melting to the boiling point of the solution.

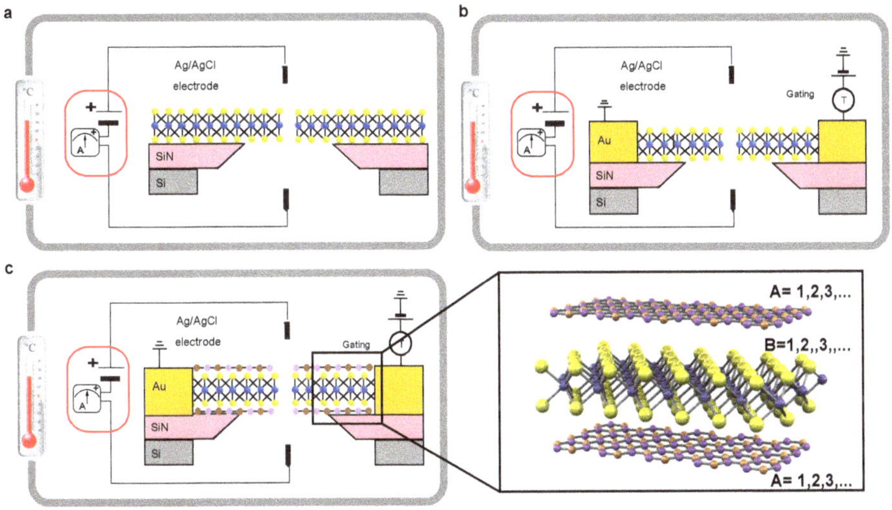

Figure 3. Common 2D nanopore measurement system with temperature control (**a**) equipped with transverse contacts for additional measurement channel or gating (**b**) and 2D material encapsulation (hBN-MoS_2-hBN) (**c**) for the improved signal-to-noise ratio and decoupling of the measurement circuits (cis-trans and transverse).

Heterostructures

A promising approach may be to use layered heterostructures of 2D materials [64]. Here, a stable position for an ion inside the pore could be designed through a combination of different materials in A-B or A-B-A scheme where the monolayers are held together by van der Waals forces (Figure 3c). The major challenge for this is the existence of hydrocarbon contaminants complicating the reproducible production of such heterostructures using material transfer, where one 2D material (B) is transferred over another (A). This disadvantage would not be an issue in the case of in situ growth of heterostructures. Similar heterostructures have already been opened in several different research avenues, such as graphene interacting with h-BN allowed several groups to study the Hofstadter butterfly effect while numerous optoelectronic devices were based on the heterostructures from semiconducting monolayers [65–67].

Van der Waals heterostructures could be used to achieve direct gating by sandwiching a semiconductor such as MoS_2 in between two layers of a wider bandgap energy 2D material such as h-BN. This suggested geometry could be considered either on a stand-alone basis or in combination with the in-plane transverse gating similar to the previous nanopore field-effect transistor (FET) approaches [68,69].

Another promising setting could be achieved by using stacked membranes (i.e., graphene oxide), where the percolating path of ions through the porous membranes and between the layers could be modified due to ions getting stuck and blocking ion transport [70].

However, to progress in the aforementioned systems, one has to have solid fabrication approaches for each of them. It is important to advance in nanopore fabrication techniques to reliably produce sub-nm pores through layered 2D membranes using existing techniques, namely an electron beam in transmission electron microscope [50,60,61,71] or electron beam lithography [72,73]. This approach is known to be the most common practice in nanopore research. However, complex multiple-step processes often result in relatively high contamination and low device yield; thus, alternative solutions have to be applied.

One possible solution is a 2D material growth over the aperture in a suspended SiN membrane [73]. However, the first attempts show that the SiN pore ends up inducing growth, and thus, the island of growth is located exactly above it. Another fine point is that the resulting 2D material might be polycrystalline, which makes its structure controversial for pore stability. Following that, this approach requires a significant improvement in order to benefit from the potential cleanliness of such a process.

On the contrary, the electrochemical reaction allows the precise control of the pore size in the flow cell [74]. However, this technique does not allow the pore size confirmation other than using ionic current-voltage characteristics. The reason for that is a poor quality of nanopore samples after the nanofluidic experiments, as they appear to be irreversibly contaminated after the measurements in a flow cell.

4. ICB in 1D Nanowires

The 1D channels represent the next promising nanofluidic platform (Figure 4). Such systems, reminiscent of narrow nanotubes, have been proposed as a very effective ion-confining solution capable of showing the single-ion transport and consequently able to reveal an ICB [20]. Furthermore, based on the geometrical confinement properties, Fermi-Dirac distribution for ions can be achieved, and thus, single-ion transport was predicted [17] and explained in detail for biological channels [17] and infinite 1D channels [20].

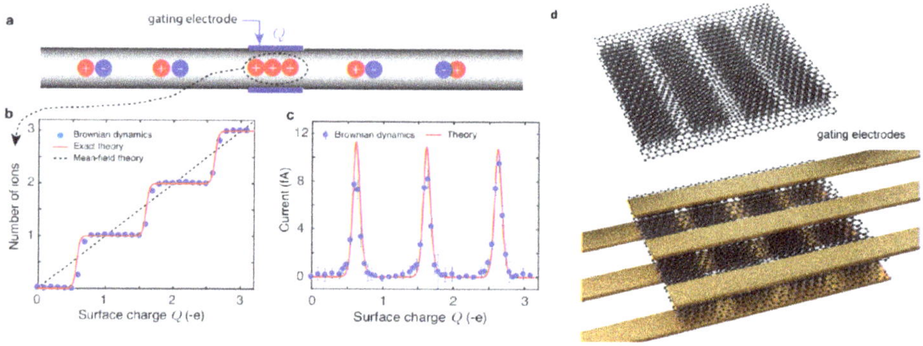

Figure 4. 1D experimental geometry suggested for the ionic Coulomb blockade observation (**a**–**c**) in 1D nanochannels and (**d**) 2D nanoslits by the application of an external electric field and a consequent electrostatic gating of the ion-confining channel. Pictures (**a**–**c**) reprinted from Kavokine et al. [20].

State-of-the-art techniques that allow single nanotube manipulation [3,75], if applied to the ultra-narrow nanotubes, could allow fabrication of a single conductive channel for ions, 1D ionic channels that show great potential for the observation of ICB. Inspired by ECB applications [76], 1D ionic channels were also suggested as the charge-carrier pumping systems, where the gating voltage on multiple gating contacts oscillates out of phase and thus allows a turnstile mechanism and a single-ion passage through the channel [20]. The unifying point of the aforementioned theoretical works [17,20] is the fixed wall charge that has to be controlled in order to manipulate the charge carrier transport at a single-ion level. However, despite the great success in nanofabrication [3,75], theoretically predicted effects are still to be confirmed experimentally, partly due to an even higher complexity of nanofabrication required (point contact, low leakage current, room temperature of measurements, high device yield). However, from the technical point of view, nanofabrication approaches developed for carbon nanotubes and nanowire field-effect transistors could be of help when applied to one-dimensional ion channels [75,77].

5. ICB in 2D Nanoslits

First of all, 2D nanoslits inherited the flexibility of material stacking that was suggested for the first time in graphene research [66]. Following that, as in the case with 2D nanopores, this type of ion confinement in 2D nanoslits could benefit from a wide variety of available materials [59]. Furthermore, 2D nanoslits attract a particular interest as a first-ever system to experimentally show the value of the dielectric constant of water in confinement [32], frictionless water transport [78], and unmatched selectivity, allowing only water and protons to pass through the narrow 2D nanoslits [7]. However, despite the significant flexibility given by the fabrication process of 2D nanoslits, one particularly important property has to be considered—these systems are represented by multiple nanofluidic channels, and thus, read-out of a single-channel ion transport and control over the transport at a single-ion level is still complicated.

One possible solution to that would be an extensive application of the most advanced fabrication techniques allowing the electrostatic control over each channel (Figure 4d) and thus, limiting the ionic current to a single channel of interest, a similar level of control could also be achieved mechanically similarly as in microfluidics [79]. Moreover, the 2D nanoslit fabrication approach [12] lacks control over channel side wall quality, and despite the atomically smooth surface of top and bottom of the channel, this might significantly affect the transport in narrow nanoslits with the comparable aspect ratio in both horizontal and vertical directions. Nonlinear transport of ions in such systems has been shown [14]; however, in that case, it is not associated with the Coulomb blockade of the channels due to a multiple input from a selection of nanoscale effects that might affect the conductance [80]. One possible solution to that would be an improved electrostatic control over the electric field in the channel as mentioned earlier (Figure 4d), similarly as proposed for other geometries.

6. Summary

The complex nature of ion transport in nano-confined systems is usually challenged with natural artifacts, such as improper wetting of the pore, pore instability (especially in the case of a 2D nanopore), inability to confirm the pore size after the measurement, and many others. Development of a well-defined approach that would be able to address these issues one by one would propel the research field. However, this conception sounds much easier than it is; one would think of the experiments where the local temperature is controlled, together with the further gating of the confining geometry (either electrical or optical).

Temperature control is important in such systems since it has a significant impact on the ions' mobility, the potential barrier size, and charge fluctuations, which might increase noise and the instability of nanofluidic devices. Therefore, in both ICB and ECB cases, the temperature criterion has to be fulfilled, i.e., $kT << Ec$.

Electrostatic or optical gating of the confinement is expected to provide a crucial transition from a closed to an opened state within the system. The current status of nanofabrication allows us to create wafer-scale batches of devices showing a corresponding quality [22], and to combat one of the most important bottlenecks in nanopore research field—the small amount of nanopore devices and consequently low statistics of the measurements. Together with the demonstrated pioneer approaches [68,69,81] the nanopore research field is reaching the limitations and heading towards the achievement of the controlled confinement that would allow us to probe charged states within the pore on a single-ion level.

However, we assume that it is important to show the most controversial points that still have to be resolved in the field. For example, nanoscale contamination is starting to play an even more crucial role in nanoelectronic devices. Leakage currents and electrochemical reactions happening at the nanoscale in 2D materials and gating contacts of nanopore field-effect transistors come into play [68,69]. With irreproducible point-like defects and contaminants, ICB observation is even more challenging since, unlike in ECB, one cannot simply "freeze" the unwanted states and free charges. This problem keeps growing as long as we need more fabrication steps to produce a better-controlled ion-confining system.

Furthermore, recent studies [25] have shown that nanoscale contamination may lead to partial rewetting, which adds a high level of complexity for these studies. In particular, we have to admit that the most common practice of pore wetting with ETOH solution with subsequent electrolyte flushing and exchange may lead directly to the formation of nanoscale gas bubbles [61]. Apart from that, the typical ETOH wetting process has been shown to change the hydrogen distribution over the surface, and thus, change the ion-solvation part of the interactions of the ions with the surface and consequently mask the ICB.

Nevertheless, we would like to draw attention to the most promising future prospects and applications of this field to probe each contribution to ion transport phenomena and to improve the control over these platforms. That would allow us to achieve a synthetic model of a cell's ion channels on a discrete channel level and benefit from step-by-step additions to a solid-state platform's controls. As a result, we may expect to obtain crucial insights on gating, strain, temperature control, etc., to use as inputs in the bottom-up computation design of biopores [82], which have already proven themselves as a unique platform for sensing and single-molecule experiments [83]. So far, no further ICB signatures have been experimentally demonstrated as conductance oscillations and Coulomb staircase, even though these have been predicted for biological nanopores [17].

Author Contributions: Writing-Original Draft Preparation, A.C. and S.M.; Writing-Review & Editing, A.C., S.M., A.R.; Visualization, A.R.; Supervision, A.R.; Project Administration, A.R.; Funding Acquisition, A.R. All authors have read and agreed to the published version of the manuscript.

Funding: This research was funded by Swiss National Science Foundation (SNSF) support through grant number [200021_192037]" and "The A.C. was funded by Eurotech Postdoc Programme (European Commission Horizon 2020. Grant Agreement number 754462).

Conflicts of Interest: The authors declare no conflict of interest.

References

1. Bocquet, L.; Charlaix, E. Nanofluidics, from bulk to interfaces. *Chem. Soc. Rev.* **2010**, *39*, 1073–1095.
2. Bocquet, L. Nanofluidics coming of age. *Nat. Mater.* **2020**, *19*, 254–256.
3. Siria, A.; Poncharal, P.; Biance, A.-L.; Fulcrand, R.; Blase, X.; Purcell, S.T.; Bocquet, L. Giant osmotic energy conversion measured in a single transmembrane boron nitride nanotube. *Nat. Cell Biol.* **2013**, *494*, 455–458.
4. Feng, J.; Graf, M.; Liu, K.; Ovchinnikov, D.; Dumcenco, D.; Heiranian, M.; Nandigana, V.; Aluru, N.R.; Kis, A.; Radenovic, A. Single-layer MoS2 nanopores as nanopower generators. *Nat. Cell Biol.* **2016**, *536*, 197–200.
5. Macha, M.; Marion, S.; Nandigana, V.V.R.; Radenovic, A. 2D materials as an emerging platform for nanopore-based power generation. *Nat. Rev. Mater.* **2019**, *4*, 588–605.
6. Graf, M.; Lihter, M.; Unuchek, D.; Sarathy, A.; Leburton, J.-P.; Kis, A.; Radenovic, A. Light-Enhanced Blue Energy Generation Using MoS2 Nanopores. *Joule* **2019**, *3*, 1549–1564.
7. Gopinadhan, K.; Hu, S.; Esfandiar, A.; Lozada-Hidalgo, M.; Wang, F.; Yang, Q.; Tyurnina, A.V.; Keerthi, A.; Radha, B.; Geim, A.K. Complete steric exclusion of ions and proton transport through confined monolayer water. *Science* **2019**, *363*, 145–148.
8. Venkatesan, B.M.; Bashir, R. Nanopore sensors for nucleic acid analysis. *Nat. Nanotechnol.* **2011**, *6*, 615–624.
9. Dekker, C. Solid-state nanopores. *Nat. Nanotechnol.* **2007**, *2*, 209–215.
10. Sofos, F.; Karakasidis, T.E.; Spetsiotis, D. Molecular dynamics simulations of ion separation in nano-channel water flows using an electric field. *Mol. Simul.* **2019**, *45*, 1395–1402.
11. Wang, M.; Hou, Y.; Yu, L.; Hou, X. Anomalies of Ionic/Molecular Transport in Nano and Sub-Nano Confinement. *Nano Lett.* **2020**, *20*, 6937–6946.
12. Radha, B.; Esfandiar, A.; Wang, F.C.; Rooney, A.P.; Gopinadhan, K.; Keerthi, A.; Mishchenko, A.; Janardanan, A.; Blake, P.; Fumagalli, P.B.L.; et al. Molecular transport through capillaries made with atomic-scale precision. *Nat. Cell Biol.* **2016**, *538*, 222–225.

13. Feng, J.; Liu, K.; Graf, M.; Dumcenco, D.; Kis, A.; Di Ventra, M.; Radenovic, A. Observation of ionic Coulomb blockade in nanopores. *Nat. Mater.* **2016**, *15*, 850–855.
14. Esfandiar, A.; Radha, B.; Wang, F.; Yang, Q.; Hu, S.; Slaven, G.; Nair, R.R.; Geim, A.K.; Gopinadhan, K. Size effect in ion transport through angstrom-scale slits. *Science* **2017**, *358*, 511–513.
15. Doyle, D.A.; Cabral, J.M.; Pfuetzner, R.A.; Kuo, A.; Gulbis, J.M.; Cohen, S.L.; Chait, B.T.; MacKinnon, R. The Structure of the Potassium Channel: Molecular Basis of K+ Conduction and Selectivity. *Science* **1998**, *280*, 69–77.
16. Payandeh, J.; Scheuer, T.; Zheng, N.; Catterall, W.A. The crystal structure of a voltage-gated sodium channel. *Nat. Cell Biol.* **2011**, *475*, 353–358.
17. Kaufman, I.K.; McClintock, P.; Eisenberg, R.S. Coulomb blockade model of permeation and selectivity in biological ion channels. *New J. Phys.* **2015**, *17*, 83021.
18. Fedorenko, O.; Kaufman, I.; Gibby, W.; Barabash, M.; Luchinsky, D.; Roberts, S.K.; McClintock, P. Ionic Coulomb blockade and the determinants of selectivity in the NaChBac bacterial sodium channel. *Biochim. Biophys. Acta* **2020**, *1862*, 183301.
19. Krems, M.; Di Ventra, M. Ionic Coulomb blockade in nanopores. *J. Phys. Condens. Matter* **2013**, *25*, 065101.
20. Kavokine, N.; Marbach, S.; Siria, A.; Bocquet, L. Ionic Coulomb blockade as a fractional Wien effect. *Nat. Nanotechnol.* **2019**, *14*, 573–578.
21. Kavokine, N.; Netz, R.R.; Bocquet, L. Fluids at the Nanoscale: From Continuum to Subcontinuum Transport. *Annu. Rev. Fluid Mech.* **2020**, *53*.
22. Thakur, M.; Macha, M.; Chernev, A.; Graf, M.; Lihter, M.; Deen, J.; Tripathi, M.; Kis, A.; Radenovic, A. Wafer-Scale Fabrication of Nanopore Devices for Single-Molecule DNA Biosensing Using MoS_2. *Small Methods* **2000**.
23. Thiruraman, J.P.; Das, P.M.; Drndic, M. Stochastic Ionic Transport in Single Atomic Zero-D Pores. *ACS Nano* **2020**, *14*, 11831–11845.
24. Smolyanitsky, A.; Paulechka, E.; Kroenlein, K. Aqueous Ion Trapping and Transport in Graphene-Embedded 18-Crown-6 Ether Pores. *ACS Nano* **2018**, *12*, 6677–6684.
25. Marion, S.; Macha, M.; Davis, S.J.; Chernev, A.; Radenovic, A. Wetting of Nanopores Probed with Pressure. *arXiv* **2019**, arXiv:1911.05229.
26. Vivitasari, P.U.; Azuma, Y.; Sakamoto, M.; Teranishi, T.; Majima, Y. Coulomb blockade and Coulomb staircase behavior observed at room temperature. *Mater. Res. Express* **2017**, *4*, 024004.
27. Xue, L.; Yamazaki, H.; Ren, R.; Wanunu, M.; Ivanov, A.P.; Edel, J.B. Solid-state nanopore sensors. *Nat. Rev. Mater.* **2020**, *5*, 931–951.
28. Beenakker, C.W.J. Theory of Coulomb-blockade oscillations in the conductance of a quantum dot. *Phys. Rev. B* **1991**, *44*, 1646–1656.
29. Nazarov, Y.V.; Nazarov, Y.; Blanter, Y.M. *Quantum Transport: Introduction to Nanoscience Quantum Transport: Introduction to Nanoscience*; Cambridge University Press: Cambridge, UK, 2009.
30. Tanaka, H.; Iizuka, H.; Pershin, Y.V.; Di Ventra, M. Surface effects on ionic Coulomb blockade in nanometer-size pores. *Nanotechnology* **2017**, *29*, 025703.
31. Schlaich, A.; Knapp, E.W.; Netz, R.R. Water Dielectric Effects in Planar Confinement. *Phys. Rev. Lett.* **2016**, *117*, 048001.
32. Fumagalli, L.; Esfandiar, A.; Fabregas, R.; Hu, S.; Ares, P.; Janardanan, A.; Yang, Q.; Radha, B.; Taniguchi, T.; Watanabe, K.; et al. Anomalously low dielectric constant of confined water. *Science* **2018**, *360*, 1339–1342.
33. Sahu, S.; Zwolak, M. Colloquium: Ionic phenomena in nanoscale pores through 2D materials. *Rev. Mod. Phys.* **2019**, *91*, 021004.
34. Fang, A.; Kroenlein, K.; Smolyanitsky, A. Mechanosensitive Ion Permeation across Subnanoporous MoS2 Monolayers. *J. Phys. Chem. C* **2019**, *123*, 3588–3593.
35. Fang, A.; Kroenlein, K.; Riccardi, D.; Smolyanitsky, A. Highly mechanosensitive ion channels from graphene-embedded crown ethers. *Nat. Mater.* **2019**, *18*, 76–81.
36. Sahu, S.; Elenewski, J.; Rohmann, C.; Zwolak, M. Optimal transport and colossal ionic mechano-conductance in graphene crown ethers. *Sci. Adv.* **2019**, *5*, eaaw5478.
37. Zwolak, M.; Lagerqvist, J.; Di Ventra, M. Quantized Ionic Conductance in Nanopores. *Phys. Rev. Lett.* **2009**, *103*, 128102.
38. Onsager, L. Deviations from Ohm's Law in Weak Electrolytes. *J. Chem. Phys.* **1934**, *2*, 599–615.

39. Kaiser, V.; Bramwell, S.T.; Holdsworth, P.C.; Moessner, R. Onsager's Wien Effect on a Lattice. *Nat. Mater.* **2013**, *12*, 1033–1037.
40. Smirnov, S.N.; Vlassiouk, I.V.; Lavrik, N.V. Voltage-Gated Hydrophobic Nanopores. *ACS Nano* **2011**, *5*, 7453–7461.
41. Powell, M.R.; Cleary, L.; Davenport, M.; Shea, K.J.; Siwy, Z.S. Electric-field-induced wetting and dewetting in single hydrophobic nanopores. *Nat. Nanotechnol.* **2011**, *6*, 798–802.
42. Radenovic, A.; Trepagnier, E.; Csencsits, R.; Downing, K.H.; Liphardt, J. Fabrication of 10 nm diameter hydrocarbon nanopores. *Appl. Phys. Lett.* **2008**, *93*, 183101.
43. Wang, Q.H.; Kalantar-Zadeh, K.; Kis, A.; Coleman, J.N.; Strano, M.S. Electronics and optoelectronics of two-dimensional transition metal dichalcogenides. *Nat. Nanotechnol.* **2012**, *7*, 699–712.
44. Slaven, G.; Hubbard, W.J.; Reina, A.; Kong, J.; Branton, D.; Golovchenko, J.A. Graphene as a subnanometre trans-electrode membrane. *Nat. Cell Biol.* **2010**, *467*, 190–193.
45. Schneider, G.F.; Kowalczyk, S.W.; Calado, V.E.; Pandraud, G.; Zandbergen, H.W.; Vandersypen, L.M.K.; Dekker, C. DNA Translocation through Graphene Nanopores. *Nano Lett.* **2010**, *10*, 3163–3167.
46. Merchant, C.A.; Healy, K.; Wanunu, M.; Ray, V.; Peterman, N.; Bartel, J.; Fischbein, M.D.; Venta, K.; Luo, Z.; Johnson, A.T.C.; et al. DNA Translocation through Graphene Nanopores. *Nano Lett.* **2010**, *10*, 2915–2921.
47. Liu, K.; Lihter, M.; Sarathy, A.; Caneva, S.; Qiu, H.; Deiana, D.; Tileli, V.; Alexander, D.T.L.; Hofmann, S.; Dumcenco, D.; et al. Geometrical Effect in 2D Nanopores. *Nano Lett.* **2017**, *17*, 4223–4230.
48. Liu, K.; Feng, J.; Kis, A.; Radenovic, A. Atomically Thin Molybdenum Disulfide Nanopores with High Sensitivity for DNA Translocation. *ACS Nano* **2014**, *8*, 2504–2511.
49. Mojtabavi, M.; VahidMohammadi, A.; Hejazi, D.; Kar, S.; Shahbazmohamadi, S.; Wanunu, M. Wafer-Scale Lateral Self-Assembly of Mosaic Ti3C2Tx (MXene) Monolayer Films. *arXiv* **2020**, arXiv:2006.12740.
50. Thiruraman, J.P.; Fujisawa, K.; Danda, G.; Das, P.M.; Zhang, T.; Bolotsky, A.; Perea-López, N.; Nicolaï, A.; Senet, P.; Terrones, M.; et al. Angstrom-Size Defect Creation and Ionic Transport through Pores in Single-Layer MoS2. *Nano Lett.* **2018**, *18*, 1651–1659.
51. Pérez, M.D.B.; Nicolaï, A.; Delarue, P.; Meunier, V.; Drndić, M.; Senet, P. Improved model of ionic transport in 2-D MoS2 membranes with sub-5 nm pores. *Appl. Phys. Lett.* **2019**, *114*, 023107.
52. Jain, T.; Rasera, B.C.; Guerrero, R.J.S.; Boutilier, M.S.; O'hern, S.C.; Idrobo, J.-C.; Karnik, R. Heterogeneous Sub-Continuum Ionic Transport in Statistically Isolated Graphene Nanopores. *Nat. Nanotechnol.* **2015**, *10*, 1053–1057.
53. Rollings, R.C.; Kuan, A.T.; Golovchenko, J.A. Ion selectivity of graphene nanopores. *Nat. Commun.* **2016**, *7*, 11408.
54. Lee, J.; Yang, Z.; Zhou, W.; Pennycook, S.J.; Pantelides, S.T.; Chisholm, M.F. Stabilization of graphene nanopore. *Proc. Natl. Acad. Sci. USA* **2014**, *111*, 7522–7526.
55. Yang, T.; Lin, H.; Zheng, X.; Loh, K.P.; Jia, B. Tailoring pores in graphene-based materials: From generation to applications. *J. Mater. Chem. A* **2017**, *5*, 16537–16558.
56. Surwade, S.P.; Smirnov, S.N.; Vlassiouk, I.V.; Unocic, R.R.; Veith, G.M.; Dai, S.; Mahurin, S.M. Water desalination using nanoporous single-layer graphene. *Nat. Nanotechnol.* **2015**, *10*, 459–464.
57. Walker, M.I.; Ubych, K.; Saraswat, V.; Chalklen, E.A.; Braeuninger-Weimer, P.; Caneva, S.; Weatherup, R.S.; Hofmann, S.; Keyser, U.F. Extrinsic Cation Selectivity of 2D Membranes. *ACS Nano* **2017**, *11*, 1340–1346.
58. Liu, C.; Jin, Y.; Li, Z. Water transport through graphene and MoS2 nanopores. *J. Appl. Phys.* **2019**, *126*, 024901.
59. Jiang, J.; Xu, T.; Lu, J.; Sun, L.; Ni, Z.-H. Defect Engineering in 2D Materials: Precise Manipulation and Improved Functionalities. *Research* **2019**, *2019*, 1–14.
60. Graf, M.; Lihter, M.; Thakur, M.; Georgiou, V.; Topolancik, J.; Ilic, B.R.; Liu, K.; Feng, J.; Astier, Y.; Radenovic, A. Fabrication and practical applications of molybdenum disulfide nanopores. *Nat. Protoc.* **2019**, *14*, 1130–1168.
61. Cantley, L.; Swett, J.L.; Lloyd, D.; Cullen, D.A.; Zhou, K.; Bedworth, P.V.; Heise, S.; Rondinone, A.J.; Xu, Z.; Sinton, S.; et al. Voltage gated inter-cation selective ion channels from graphene nanopores. *Nanoscale* **2019**, *11*, 9856–9861.
62. Daukiya, L.; Seibel, J.; De Feyter, S. Chemical modification of 2D materials using molecules and assemblies of molecules. *Adv. Phys. X* **2019**, *4*, 1625723.
63. Davis, S.J.; Macha, M.; Chernev, A.; Huang, D.M.; Radenovic, A.; Marion, S. Pressure-Induced Enlargement and Ionic Current Rectification in Symmetric Nanopores. *Nano Lett.* **2020**, *20*, 8089–8095.

64. Zhang, F.; Wang, Y.; Erb, C.; Wang, K.; Moradifar, P.; Crespi, V.H.; Alem, N. Full orientation control of epitaxial MoS2 on hBN assisted by substrate defects. *Phys. Rev. B* **2019**, *99*, 155430.
65. Withers, F.; Del Pozo-Zamudio, O.; Mishchenko, A.; Rooney, A.P.; Gholinia, A.; Watanabe, K.; Taniguchi, T.; Haigh, S.J.; Geim, A.K.; Tartakovskii, A.I.; et al. Light-emitting diodes by band-structure engineering in van der Waals heterostructures. *Nat. Mater.* **2015**, *14*, 301–306.
66. Novoselov, K.S.; Mishchenko, A.; Carvalho, A.; Neto, A.H.C. 2D materials and van der Waals heterostructures. *Science* **2016**, *353*, aac9439.
67. Jariwala, D.; Marks, T.J.; Hersam, M.C. Mixed-dimensional van der Waals heterostructures. *Nat. Mater.* **2017**, *16*, 170–181.
68. Heerema, S.J.; Vicarelli, L.; Pud, S.; Schouten, R.N.; Zandbergen, H.W.; Dekker, C. Probing DNA Translocations with Inplane Current Signals in a Graphene Nanoribbon with a Nanopore. *ACS Nano* **2018**, *12*, 2623–2633.
69. Graf, M.; Lihter, M.; Altus, D.; Marion, S.; Radenovic, A. Transverse Detection of DNA Using a MoS2 Nanopore. *Nano Lett.* **2019**, *19*, 9075–9083.
70. Zhan, H.; Xiong, Z.; Cheng, C.; Liang, Q.; Liu, J.Z.; Li, D. Solvation-Involved Nanoionics: New Opportunities from 2D Nanomaterial Laminar Membranes. *Adv. Mater.* **2019**, *32*, e1904562.
71. Danda, G.; Drndić, M. Two-dimensional nanopores and nanoporous membranes for ion and molecule transport. *Curr. Opin. Biotechnol.* **2019**, *55*, 124–133.
72. Verschueren, D.V.; Yang, W.; Dekker, C. Lithography-based fabrication of nanopore arrays in freestanding SiN and graphene membranes. *Nanotechnology* **2018**, *29*, 145302.
73. Waduge, P.; Bilgin, I.; Larkin, J.; Henley, R.Y.; Goodfellow, K.; Graham, A.C.; Bell, D.C.; Vamivakas, N.; Kar, S.; Wanunu, M. Direct and Scalable Deposition of Atomically Thin Low-Noise MoS2 Membranes on Apertures. *ACS Nano* **2015**, *9*, 7352–7359.
74. Feng, J.; Liu, K.; Graf, M.; Lihter, M.; Bulushev, R.D.; Dumcenco, D.; Alexander, D.T.L.; Krasnozhon, D.; Vuletic, T.; Kis, A.; et al. Electrochemical Reaction in Single Layer MoS2: Nanopores Opened Atom by Atom. *Nano Lett.* **2015**, *15*, 3431–3438.
75. Marcotte, A.; Mouterde, T.; Niguès, A.; Siria, A.; Bocquet, L. Mechanically Activated Ionic Transport across Single-Digit Carbon Nanotubes. *Nat. Mater.* **2020**, *19*, 1057–1061.
76. Fujiwara, A.; Nishiguchi, K.; Ono, Y. Nanoampere charge pump by single-electron ratchet using silicon nanowire metal-oxide-semiconductor field-effect transistor. *Appl. Phys. Lett.* **2008**, *92*, 042102.
77. Chen, Z.; Appenzeller, J.; Knoch, J.; Lin, Y.-M.; Avouris, P. The Role of Metal−Nanotube Contact in the Performance of Carbon Nanotube Field-Effect Transistors. *Nano Lett.* **2005**, *5*, 1497–1502.
78. Mouterde, T.; Keerthi, A.; Poggioli, A.; Dar, S.A.; Siria, A.; Geim, A.K.; Bocquet, L.; Radha, B. Molecular Streaming and Its Voltage Control in Ångström-Scale Channels. *Nature* **2019**, *567*, 87–90.
79. Thorsen, T.; Maerkl, S.J.; Quake, S.R. Microfluidic Large-Scale Integration. *Science* **2002**, *298*, 580–584.
80. Muthukumar, M. *Polymer Translocation*; CRC Press: Boca Raton, FL, USA, 2016.
81. Xie, P.; Xiong, Q.; Fang, Y.; Qing, Q.; Lieber, C.M. Local electrical potential detection of DNA by nanowire–nanopore sensors. *Nat. Nanotechnol.* **2012**, *7*, 119–125.
82. Xu, C.; Lu, P.; El-Din, T.M.G.; Pei, X.Y.; Johnson, M.C.; Uyeda, A.; Bick, M.J.; Xu, Q.; Jiang, D.; Bai, H.; et al. Computational design of transmembrane pores. *Nat. Cell Biol.* **2020**, *585*, 129–134.
83. Derrington, I.M.; Craig, J.M.; Stava, E.; Laszlo, A.H.; Ross, B.C.; Brinkerhoff, H.; Nova, I.C.; Doering, K.; I Tickman, B.; Ronaghi, M.; et al. Subangstrom single-molecule measurements of motor proteins using a nanopore. *Nat. Biotechnol.* **2015**, *33*, 1073–1075.

MDPI
St. Alban-Anlage 66
4052 Basel
Switzerland
Tel. +41 61 683 77 34
Fax +41 61 302 89 18
www.mdpi.com

Entropy Editorial Office
E-mail: entropy@mdpi.com
www.mdpi.com/journal/entropy

www.ingramcontent.com/pod-product-compliance
Lightning Source LLC
Chambersburg PA
CBHW061807210926
51850CB00006B/639
9783036516462